D0154469

Handbook of Groundwater Remediation Using Permeable Reactive Barriers

Applications to Radionuclides, Trace Metals, and Nutrients

Handbook of Groundwater Remediation Using Permeable Reactive Barriers

Applications to Radionuclides, Trace Metals, and Nutrients

Edited by

David L. Naftz
U.S. Geological Survey
Salt Lake City, Utah

Stan J. Morrison
Environmental Sciences Laboratory
Grand Junction, Colorado

Christopher C. Fuller
U.S. Geological Survey
Menlo Park, California

James A. Davis
U.S. Geological Survey
Menlo Park, California

Amsterdam Boston London New York Oxford Paris
San Diego San Francisco Singapore Sydney Tokyo

This book is printed on acid-free paper. 

Academic Press
An imprint of Elsevier Science.
525 B Street, Suite 1900, San Diego, California 92101-4495, USA
http://www.academicpress.com

Academic Press
84 Theobalds Road, London W1CX 8RR , UK
http://www.academicpress.com

Library of Congress Catalog Card Number: 2002101651

International Standard Book Number: 0-12-513563-7

PRINTED IN THE UNITED STATES OF AMERICA
02 03 04 05 06 07 MM 9 8 7 6 5 4 3 2 1

Contents

Chapter 2
Design and Performance of Limestone Drains to Increase pH and Remove Metals from Acidic Mine Drainage

Charles A. Cravotta III. and George R. Watzlaf

Chapter 3
Preliminary Investigation into the Suitability of Permeable Reactive Barriers for the Treatment of Acid Sulfate Soils Discharge

T. David Waite, Rosalind Desmier, Michael Melville, and Bennett Macdonald

Chapter 4
Permeable Reactive Barrier/ GeoSiphon Treatment for Metals-Contaminated Groundwater

W. E. Jones, M. E. Denham, M. A. Phifer, F. C. Sappington, and F. A. Washburn

Chapter 7

Development and Performance of an Iron Oxide/Phosphate Reactive Barrier for the Remediation of Uranium-Contaminated Groundwater

Jennifer L. Joye, David L. Naftz, James A. Davis, Geoff W. Freethey, and Ryan C. Rowland

Chapter 8

Treatability Study of Reactive Materials to Remediate Groundwater Contaminated with Radionuclides, Metals, and Nitrates in a Four-Component Permeable Reactive Barrier

James Conca, Elizabeth Strietelmeier, Ningping Lu, Stuart D. Ware, Tammy P. Taylor, John Kaszuba, and Judith Wright

Part III
Evaluations of Chemical and Biological Processes

Chapter 9

Evaluation of Apatite Materials for Use in Permeable Reactive Barriers for the Remediation of Uranium-Contaminated Groundwater

Christopher C. Fuller, Michael J. Piana, John R. Bargar, James A. Davis, and Matthias Kohler

Chapter 10

Sulfate-Reducing Bacteria in the Zero Valent Iron Permeable Reactive Barrier at Fry Canyon, Utah

Ryan C. Rowland

Chapter 11

Biogeochemical, Mineralogical, and Hydrological Characteristics of an Iron Reactive Barrier Used for Treatment of Uranium and Nitrate

Baohua Gu, David B. Watson, Debra H. Phillips, and Liyuan Liang

Chapter 12
Analysis of Uranium-Contaminated Zero Valent Iron Media Sampled From Permeable Reactive Barriers Installed at U.S. Department of Energy Sites in Oak Ridge, Tennessee, and Durango, Colorado

Leah J. Matheson, Will C. Goldberg, W. D. Bostick, and Larry Harris

Part IV
Case Studies of Permeable Reactive Barrier Installations

Chapter 13
Design and Performance of a Permeable Reactive Barrier for Containment of Uranium, Arsenic, Selenium, Vanadium, Molybdenum, and Nitrate at Monticello, Utah

Stan J. Morrison, Clay E. Carpenter, Donald R. Metzler, Timothy R. Bartlett, and Sarah A. Morris

Chapter 14

Field Demonstration of Three Permeable Reactive Barriers to Control Uranium Contamination in Groundwater, Fry Canyon, Utah

David L. Naftz, Christopher C. Fuller, James A. Davis, Stan J. Morrison, Edward M. Feltcorn, Ryan C. Rowland, Geoff W. Freethey, Christopher Wilkowske, and Michael Piana

Chapter 15

Collection Drain and Permeable Reactive Barrier for Treating Uranium and Metals from Mill Tailings Near Durango, Colorado

Stan J. Morrison, Donald R. Metzler, and Brain P. Dwyer

Chapter 16

In Situ Reduction of Chromium-Contaminated Groundwater, Soils, and Sediments by Sodium Dithionite

Cynthia J. Paul, Faruque A. Khan, and Robert W. Puls

Contributors

Numbers in parenthesis indicate page numbers on which authors contributions begin.

J. E. Amonette (163), Battelle, Pacific Northwest National Laboratory, Richland, Washington.

John R. Bargar (255), Stanford Synchrotron Radiation Laboratory, Stanford, California.

Timothy R. Bartlett (371), Environmental Sciences Laboratory, Grand Junction, Colorado.

Shawn Benner (495), Department of Geological and Environmental Sciences, Stanford University, Stanford, California.

David Blowes (495), Department of Earth Sciences, University of Waterloo, Waterloo, Ontario, Canada.

W. D. Bostick (343), Materials and Chemistry Laboratory, Inc., Oak Ridge, Tennessee.

Clay E. Carpenter (371), Environmental Sciences Laboratory, Grand Junction, Colorado.

C. R. Cole (163), Battelle, Pacific Northwest National Laboratory, Richland, Washington.

James Conca (221), Los Alamos National Laboratory, Carlsbad, New Mexico.

Charles A. Cravotta III (19), U.S. Geological Survey, New Cumberland, Pennsylvania.

James A. Davis (1, 133, 195, 255, 401), U.S. Geological Survey, Menlo Park, California.

M. E. Denham (105), Savannah River Technology Center, Westinghouse Savannah River Company, Aiken, South Carolina.

Rosalind Desmier (67), Centre for Water and Waste Technology, School of Civil and Environmental Engineering, School of Geography, The University of New South Wales, Sydney, Australia.

Brian P. Dwyer (435), Sandia National Laboratories, Albuquerque, New Mexico.

Edward M. Feltcorn (401), U.S. Environmental Protection Agency, Radiation and Indoor Air, Washington, District of Columbia.

Geoff W. Freethey (133, 195, 401), U.S. Geological Survey, Salt Lake City, Utah.

J. S. Fruchter (163), Battelle, Pacific Northwest National Laboratory, Richland, Washington.

Christopher C. Fuller (1, 255, 401), U.S. Geological Survey, Menlo Park, California.

Will C. Goldberg (343), MSE Technology Applications Inc., Butte, Montana.

Baohua Gu (305), Environmental Sciences Division, Oak Ridge National Laboratory, Oak Ridge, Tennessee.

Larry Harris (343), Materials and Chemistry Laboratory, Inc., Oak Ridge, Tennessee.

W. E. Jones (105), Savannah River Technology Center, Westinghouse Savannah River Company, Aiken, South Carolina.

Jennifer L. Joye (195), U.S. Geological Survey, Menlo Park, California.

Faruque A. Khan (465), U.S. Environmental Protection Agency, Office of Research and Development, National Risk Management Research Laboratory, Ada, Oklahoma.

John Kaszuba (221) Los Alamos National Laboratory, Los Alamos, New Mexico.

Matthias Kohler (255), Colorado School of Mines, Environmental Science and Engineering Department/Division, Golden, Colorado.

Liyuan Liang (305), Environmental Sciences Division, Oak Ridge National Laboratory, Oak Ridge, Tennessee.

Ningping Lu (221) Los Alamos National Laboratory, Los Alamos, New Mexico.

Ralph Ludwig (495), U.S. Department of Environmental Protection, Robert S. Kerr Environmental Research Center, Ada, Oklahoma.

Bennett Macdonald (67), School of Geography, The University of New South Wales, Sydney, Australia.

Leah J. Matheson (343), MSE Technology Applications Inc., Butte, Montana.

Rick McGregor (495), XCG Consultants Ltd., Waterloo, Ontario, Canada

Michael Melville (67), School of Geography, The University of New South Wales, Sydney, Australia.

Donald R. Metzler (371, 435), U.S. Department of Energy, Grand Junction, Colorado.

Sarah A. Morris (371), Environmental Sciences Laboratory, Grand Junction, Colorado.

Stan J. Morrison (1, 371, 401, 435), Environmental Sciences Laboratory, Grand Junction, Colorado.

David L. Naftz (1, 133, 195, 401), U.S. Geological Survey, Salt Lake City, Utah.

Cynthia J. Paul (465), U.S. Environmental Protection Agency, Office of Research and Development, National Risk Management Research Laboratory, Ada, Oklahoma.

Debra H. Phillips (305), Environmental Sciences Division, Oak Ridge National Laboratory, Oak Ridge, Tennessee.

Michael Piana (255, 401), NEC Electronics, Process Engineering Division, Roseville, California.

M. A. Phifer (105), Savannah River Technology Center, Westinghouse Savannah River Company, Aiken, South Carolina.

Carol Ptacek (495), Environment Canada, NHRI, Burlington, Ontario, Canada.

Robert W. Puls (465), U.S. Environmental Protection Agency, Office of Research and Development, National Risk Management Research Laboratory, Ada, Oklahoma.

Ryan C. Rowland (133, 195, 281, 401), U.S. Geological Survey, Salt Lake City, Utah.

F. C. Sappington (105), Savannah River Technology Center, Westinghouse Savannah River Company, Aiken, South Carolina.

Elizabeth Strietelmeier (221) Los Alamos National Laboratory, Los Alamos, New Mexico.

J. E. Szecsody (163), Battelle, Pacific Northwest National Laboratory, Richland, Washington.

Tammy P. Taylor (221) Los Alamos National Laboratory, Los Alamos, New Mexico.

V. R. Vermeul (163), Battelle, Pacific Northwest National Laboratory, Richland, Washington.

T. David Waite (67), Centre for Water and Waste Technology, School of Civil and Environmental Engineering, The University of New South Wales, Sydney, Australia.

Stuart D. Ware (221) Los Alamos National Laboratory, Los Alamos, New Mexico.

F. A. Washburn (105), Savannah River Technology Center, Westinghouse Savannah River Company, Aiken, South Carolina.

David B. Watson (305), Environmental Sciences Division, Oak Ridge National Laboratory, Oak Ridge, Tennessee.

George R. Watzlaf (19), U.S. Department of Energy, National Energy Technology Laboratory, Pittsburgh, Pennsylvania.

Christopher Wilkowske (401), U.S. Geological Survey, Salt Lake City, Utah.

M. D. Williams (163), Battelle, Pacific Northwest National Laboratory, Richland, Washington.

Judith Wright (221), UFA Ventures Inc., Carlsbad, New Mexico.

Foreword

The mining, milling, refining, and industrial uses of metals, coal, petroleum products, and nuclear fuels have left a legacy of groundwater contamination. With increasing population and the continuing allocation of groundwater resources, groundwater remediation is increasingly recognized as being paramount to the health and productivity of our industrial society. The extraction and treatment of contaminated groundwater is effective in some situations; however, the costs are often prohibitive. Research and field demonstrations of groundwater remediation are beginning to show promise for more economic removal of contaminants. Permeable reactive barriers is an emerging technology that offers the potential of passive, low-cost approaches to reaching groundwater cleanup goals.

The publication of this book is timely because sufficient hydrologic and geochemical data are now becoming available to evaluate permeable reactive barrier performance more thoroughly. The information presented in this book encompasses a wide variety of reactive materials, design alternatives, flow systems, and evaluation approaches representing the diversity of intellectual endeavors devoted to this topic. Results of a number of bench- and field-scale projects are presented that provide real-world examples on how various hydrologic and geochemical techniques can be used to assess design and performance issues associated with permeable reactive barriers. Information contained in this book will be of use to private sector consultants, university and government researchers, and policy makers dealing with all phases of permeable reactive barriers, including fundamental science, design, construction, monitoring, and performance evaluation.

Gerald Boyd
Deputy Assistant Secretary for
Science and Technology
U.S. Department of Energy

Robert M. Hirsch
Associate Director for Water
U.S. Geological Survey

treatment of acid discharge. Chapter 4 provides a discussion and application of the GeoSiphon passive extraction technology in combination with a PRB for the treatment of acidic groundwater with trace metal and volatile organic carbon contaminants. The GeoSiphon system uses natural hydraulic head to induce passive groundwater flow to a series of *ex situ* treatment cells for cost-effective groundwater remediation. Chapter 5 presents the theory, design, and performance modeling of deep aquifer remediation tools (DARTs). DARTs are used in combination with non-pumping wells to provide an *in situ* treatment wall. Detailed modeling of groundwater flow through DARTs is presented to determine optimal installation and operational parameters. Chapter 6 discusses the design and creation of a subsurface PRB by injecting a chemical reducing agent (potassium dithionite). Placement of reactive material by injection does not require excavation, thereby extending the applicability of PRB technology to deeper contaminant plumes. Additional advantages of injected PRBs include diminished human exposure during barrier placement and the ability to renew the treatment media without excavation.

Section 2 (chapters 7 and 8) describes the design, development, and testing of new reactive materials for use in PRBs. Chapter 7 focuses on the development and testing of engineered solid mixtures of pelletized bone charcoal and hematite pellets coated with amorphous iron oxide for use in PRBs to contain uranium. Chapter 8 reports on the result of a laboratory study utilizing a PRB with four sequential layers to treat a mixed waste plume containing plutonium-239, plutonium-240, americium-241, strontium-90, nitrate, and perchlorate. The layers include a polyelectrolyte-impregnated porous gravel for flocculating colloids, an apatite layer for plutonium, americium, and strontium immobilization, a layer of pecan shells as a biobarrier to nitrate and perchlorate, and a limestone gravel layer for removal of anionic species.

Section 3 (chapters 9 through 12) addresses the chemical and biological processes that occur in PRBs—including data on zero valent iron, one of the most popular reactive media currently (2002) used in PRBs. Chapter 9 presents a detailed evaluation of the geochemical processes that occur during uranium removal by phosphate material, including phosphate rock, bone meal, bone charcoal, and pelletized bone charcoal apatites. X-ray adsorption spectroscopy and X-ray diffraction analysis of apatite materials after exposure to uranium in laboratory and field demonstration tests were used to document uranium removal mechanisms and the potential for remobilization during PRB aging. Chapter 10 investigates the presence and impacts of sulfate-reducing bacteria in a zero valent iron PRB. Zero valent iron PRBs create a chemically reducing and hydrogen gas enriched environment that is favorable for sulfate-reducing bacteria. Isotopic, microbiological, and geochemical modeling methods were used to document the occurrence of sulfate-reducing bacteria and how the bacteria impact the long-term performance of a PRB by reducing the porosity. Chapter 11 evaluates the biogeochemical interactions that occur during contaminant treatment in a zero valent iron PRB. Results confirm the efficient removal of uranium, technetium, and nitrate by zero valent iron and describe the effects of associated chemical and biological reactions. Corrosion of iron in the PRB resulted in the formation of minerals determined to be responsible for a reduction in long-term PRB performance. Increased anaerobic microbial populations also were observed in the vicinity of the PRB and may be partly responsible for sulfate and nitrate removal. Chapter 12 evaluates uranium removal processes by using core samples collected from PRBs in Tennessee and Colorado. X-ray photoelectron spectroscopy was used to identify the oxidation state of uranium on the surface of the zero valent iron. This information was combined with additional solid-phase data to investigate geochemical processes responsible for uranium removal.

Section 4 (chapters 13 through 17) provides data to evaluate PRB performance through a series of case studies at field demonstration sites in the United States (Utah, Colorado, North Carolina, and Tennessee) and Canada (British Columbia and Ontario). Chapters 13–15 focus on the removal of uranium from contaminated groundwater at facilities associated with mining and milling operations. Chapter 13 presents detailed hydrologic and geochemical results during the design and first year of operation of a zero valent iron PRB constructed near Monticello, Utah. Information on the laboratory testing, design, and performance monitoring at this facility are presented. Material presented in this chapter will be of use to readers considering the construction of full-scale PRB installations. Detailed analysis of hydrologic and geochemical data collected from three PRBs installed near Fry Canyon, Utah is presented in chapter 14. The three PRBs contain different reactive materials that were installed adjacent to each other for a direct comparison of contaminant removal efficiencies. Substantial differences in PRB performance were observed during this 3-year field demonstration. Recommendations from this field demonstration can be used to improve the design, operation, and monitoring of future PRB installations. Chapter 15 presents information on designs, construction methods, and aqueous chemistry results during four years of PRB operation at a site near Durango, Colorado. In addition to uranium, contaminants removed by the zero valent iron PRB included arsenic, cadmium, copper, molybdenum, nitrate, radium-226, selenium, vanadium, and zinc. Performance of three forms of zero valent iron (foam plates, granular, and steel wool) and two reactor designs (leach field and baffled tank) were compared during this field project

The final 2 chapters (16 and 17) in section 4 contain detailed field demonstration results on PRBs containing a chemical reductant for the *in situ* removal of chromium (chapter 16) and one containing municipal compost to reduce acidity and remove cadmium, copper, nickel, and zinc from groundwater affected by acid mine drainage (chapter 17). Chapter 16 describes the results from a field project using injection of sodium dithionite into iron-bearing sediments, to create a PRB that chemically reduces the mobile chromium (VI) in groundwater to the less mobile chromium (III). Chapter 17 reports on the use of organic compost PRBs to stimulate bacteria mediated sulfate reduction, resulting in a decrease in net acidity and the attenuation of heavy metals by metal-sulfide precipitation. The effects of fluctuating water tables and groundwater temperature on PRB performance were also evaluated at the two field sites in Canada. A cost analysis for both PRBs is included.

The future is bright for utilizing PRBs during groundwater clean-up projects. This handbook contains the information necessary to assist in the advancement of PRB technology to more routine and cost-effective remediation. The U.S. Geological Survey and U.S. Department of Energy sponsored the production of the handbook.

David L. Naftz
Salt Lake City, Utah

Stan J. Morrison
Grand Junction, Colorado

Christopher C. Fuller
James A. Davis
Menlo Park, California

About the Editors

David L. Naftz is a hydrologist with the U.S. Geological Survey in Salt Lake City, Utah. He received his Ph.D. from the Colorado School of Mines in geochemistry. He began his professional career with the Wyoming Department of Environmental Quality in 1983. Since joining the U.S. Geological Survey in 1984, Dr. Naftz has worked on a variety of water-quality research projects throughout the Rocky Mountain States. His applied research topics during the past 17 years have included water-quality impacts from coal and uranium mining, reservoir construction, irrigation drainage, oil and gas recovery, atmospheric deposition, and explosives. Dr. Naftz began his field-oriented research on permeable reactive barriers in 1996.

Stan J. Morrison is a geochemist and manager of the Environmental Sciences Laboratory (ESL) in Grand Junction, Colorado. MACTEC Environmental Restoration Services manages the ESL for the U.S. Department of Energy. He worked as a uranium exploration geologist from 1975 through 1982 and received his Ph.D. in geochemistry from the University of Utah in 1986. His post-doctoral work on contaminant transport in streambeds was completed in 1987. Since 1987, Dr. Morrison has been at the ESL where he investigates the migration and remediation of groundwater contamination focusing on metals and radionuclides. He has conducted research on permeable reactive barriers since 1991 and was the technical lead on several field applications of this technology.

Christopher C. Fuller is a hydrologist with the National Research Program of the U.S. Geological Survey in Menlo Park, California. He received his M.S. from the University of Southern California in geochemistry. He began his professional career as a research technician in marine geochemistry at the University of Southern California. After joining the U.S. Geological Survey in 1982, Mr. Fuller has worked on a variety of field and laboratory aqueous geochemistry research studies. His research topics during the past 20 years have included geochemical processes affecting metal transport in mine-contaminated streams, laboratory and spectroscopic characterization of metal-ion sorption on mineral surfaces, and use of environmental radioisotopes for sediment chronology. Fuller began his research on permeable reactive barriers in 1996.

James A. Davis is a hydrologist with the U.S. Geological Survey in Menlo Park, California. He received his Ph.D. from Stanford University in environmental engineering and sciences and did post-doctoral study at the Swiss Federal Institute for Water Resources. He began his professional career with the U.S. Geological Survey in 1980 and has worked on research projects involving the fate and transport of metal contaminants throughout the United States and in Canada, Costa Rica, and Australia. His research interests include mineral/water interface geochemistry, the coupling of hydrologic and geochemical models, fate and transport of radionuclides, and the spectroscopic characterization of amorphous mineral phases and contaminants at mineral surfaces.

Acknowledgments

We would like to thank the authors for their dedication and hard work during the preparation of this book. In addition, we would like to thank the technical reviewers for their time and expertise in reviewing individual book chapters: *Shawn Benner*, Stanford University; *Breton Bruce*, U.S. Geological Survey; *Clay Carpenter*, MACTEC-Environmental Restoration Services; *Steven Day*, Geo-Solutions, Inc; *James Farrell*, University of Arizona; *John Fitzpatrick*, U.S. Geological Survey; *Baohua Gu*, Oak Ridge National Laboratory; *Walt Holmes*, U.S. Geological Survey; *Nic Korte*, Private Consultant; *Edward Landa*, U.S. Geological Survey; *Lars Lovgren*, Umeå University, Sweden; *James Mason*, U.S. Geological Survey; *Leah Matheson*, MSE; *Peter McMahon*, U.S. Geological Survey; *George Moridis*, Lawrence Berkeley National Laboratory; *David Nimick*, U.S. Geological Survey; *Robert Smith*, Idaho National Engineering and Environmental Laboratory; and *Bruce Thomson*, University of New Mexico.

Chapter 1

Introduction to Groundwater Remediation of Metals, Radionuclides, and Nutrients with Permeable Reactive Barriers

Stan J. Morrison,* David L. Naftz,† James A. Davis,‡ and Christopher C. Fuller‡

*Environmental Sciences Laboratory,[1] Grand Junction, Colorado 81503
†U.S. Geological Survey, Salt Lake City, Utah 84119
‡U.S. Geological Survey, Menlo Park, California 94025

I. INTRODUCTION

The remediation of metal mining and milling sites and contaminated groundwater at those sites is a worldwide goal to eliminate unacceptable risks to human health and the environment. In addition to metal mining and milling wastes, groundwater contamination can result from *in situ* mining, industrial waste generation, nuclear fuels processing, and coal pile drainage. The most widely used method to remediate groundwater is to extract it from the ground and process it through a water treatment plant, so-called pump-and-treat remediation. To meet mandated maximum contaminant levels in the groundwater usually requires removal of many pore volumes of groundwater for long time periods, resulting in high treatment costs. Few sites have been remediated to regulated levels using pump-and-treat systems; therefore, the cost effective-

[1]Operated by MACTEC Environmental Restoration Services for the U.S. Department of Energy Grand Junction Office.

Handbook of Groundwater Remediation Using Permeable Reactive Barriers

ness of pump-and-treat technology is uncertain (Mackay and Cherry, 1989; Travis and Doty, 1990; National Academy of Sciences, 1994). The most difficult aspect of pump-and-treat remediation is the efficient extraction of contaminants that are highly associated with the aquifer matrix.

A permeable reactive barrier (PRB) is an engineered zone of reactive material placed in an aquifer that removes contamination from groundwater flowing through it (Fig. 1) (Rumar and Mitchell, 1995). PRBs have been used with varying degrees of success to help meet groundwater standards. The ability of reactive materials in PRBs to remove both organic and inorganic contaminants from groundwater is well established. The technology relies on natural gradients to move groundwater through the treatment media and, therefore, has cost benefits compared with active pumping scenarios. Cost effectiveness, however, has not been reliably determined because of a lack of long-term performance data. Groundwater remediation using PRBs reduces exposures of workers to contaminants because the treatment process is conducted underground. Another advantage over pump-and-treat methods is

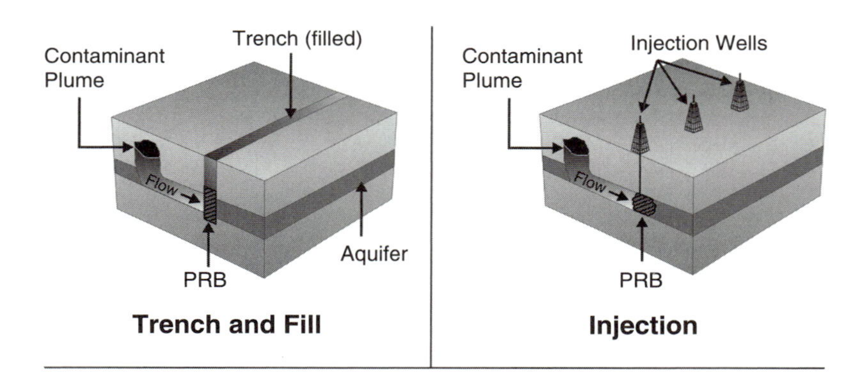

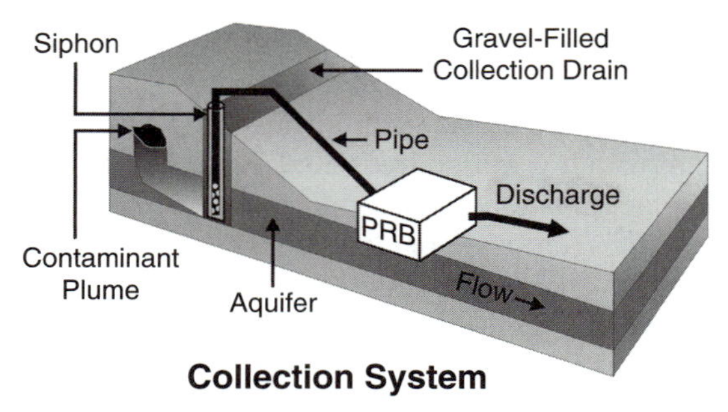

Figure 1 Schematic of permeable reactive barriers.

that the land surface can be used for beneficial purposes during remediation because there are no aboveground structures.

The impetus for the development of innovative treatment technologies in the United States is based on federal law and policy. The United States Environmental Protection Agency (EPA) is required to select "remedial actions that permanently and significantly reduce the volume, toxicity, or mobility of the hazardous substances, pollutants and contaminants" (U.S.C., 2001). "EPA expects to consider using innovative technology when such technology offers the potential for comparable or superior treatment performance or implementability, fewer or lesser adverse impacts than other available approaches, or lower costs for similar levels of performance than demonstrated technologies" (CFR, 2001).

PRB technologies are known by various names and acronyms that have evolved rapidly since 1990. PRBs containing limestone used to remediate groundwater contaminated with acid mine drainage are often referred to as limestone drains (Pearson and McDonnell, 1975; Hedin *et al.*, 1994). Longmire *et al.* (1991) referred to PRBs as geochemical barriers and investigated their use for groundwater remediation at uranium (U) mill sites. EPA formed a multiagency working group within its Remediation Technology Development Forum in 1993 to help promote the rapid development of PRB technology. This group uses "permeable reactive barrier" as the preferred terminology. Other names that have been used include chemical barriers (Spangler and Morrison, 1991; Kaplan *et al.*, 1996), permeable reactive walls (Blowes and Ptacek, 1992; O'Hannesin, 1993), *in situ* treatment curtains (Blowes *et al.*, 1995), and porous reactive walls (Blowes *et al.*, 1997). More specific names depicting the type of reactive material have also been used, including reactive iron walls (Focht *et al.*, 1995), permeable biological reaction barriers (Thombre *et al.*, 1997), in situ iron walls (O'Hannesin and Gillham, 1998), and denitrification barriers (Robertson and Cherry, 1995). The term "funnel and gate" is sometimes used interchangeably with PRB, but funnel and gate refers to the configuration of impermeable walls (in a funnel configuration) that are used to direct groundwater to the PRB or gate (Starr and Cherry, 1994).

II. HISTORY OF PRBS FOR METALS, RADIONUCLIDES, AND NUTRIENTS

PRBs containing limestone have been used since the 1970s to remediate acid mine drainage (Pearson and McDonnell, 1975; Hedin *et al.*, 1994; see Chapter 2 and 3). Most of the limestone installations have been for passive treatment of surface water or shallow groundwater. In 1999, a limestone PRB was installed to treat contaminated groundwater issuing from a coal pile at the U.S. Department of Energy (DOE) Savannah River Site in South Carolina (see Chapter 4).

Researchers at the University of Waterloo, Ontario, Canada, determined that zero-valent iron (ZVI) could be used in PRBs to remediate groundwater contaminated by halogenated organic solvents (Gillham and O'Hannesin, 1992). The first ZVI-based PRB was installed in 1991 at the Canadian Forces Base Borden site at Ontario, Canada (Gillham *et al.*, 1993). The success of the Borden PRB in remediating groundwater contaminated with chlorinated organic solvents initiated a period of activity resulting in the construction of more than 40 ZVI-based PRBs between 1993 and 2001.

Shortly after the Borden demonstration, researchers began conducting laboratory studies to determine the potential of ZVI for precipitating redox-sensitive metals such as chromium (Cr) (Blowes *et al.*, 1995) and radionuclides such as U (Cantrell *et al.*, 1995). The use of ZVI-based PRBs to treat inorganic contaminants has lagged behind development of the technology for organic solvents partly because organic contamination is more prevalent at sites on the EPA's National Priorities List and other highly publicized sites. Another reason is that inorganic contaminants are not degraded or transformed into nontoxic compounds as occurs with chlorinated solvents and, therefore, final disposal of the concentrated waste product is an issue with PRB treatment.

ZVI-based PRBs that treat inorganic contaminants are currently in operation at five sites in the United States: Elizabeth City, North Carolina; Oak Ridge, Tennessee; Monticello, Utah; Fry Canyon, Utah; and Durango, Colorado. At the DOE site near Durango, PRBs using several forms of ZVI were installed in October 1995 to treat contaminated groundwater issuing from a U mill tailings disposal cell (see Chapter 15). In 1996, a 46-m-long PRB was installed at the U.S. Coast Guard station at Elizabeth City to treat groundwater contaminated with Cr (Blowes *et al.*, 2000; see Chapter 16). Two PRBs were installed at the DOE Oak Ridge site in December 1997 to treat groundwater contaminated with U (see Chapter 11). A demonstration-scale PRB containing ZVI was installed at Fry Canyon in September 1997 to treat U (see Chapter 14). In September 1999, a 30-m-long PRB containing ZVI was installed at the Monticello site to treat groundwater contaminated with U and associated contaminants (see Chapter 13).

Some PRB installations in the 1990s contained reactive materials other than limestone and ZVI to treat inorganic contaminants. In 1994, zeolite was used to treat strontium-90 in two PRBs, one installed at the DOE facility at Oak Ridge (Taylor, 1998) and the other at a site in West Valley, New York (Patel *et al.*, 1999). Microbial-based PRBs containing compost to treat acid mine drainage were installed at the Nickel Rim site, Ontario, Canada, in 1995

Reductive precipitation can also b
dithionite into an aquifer (see Chap
causes reduction of solid-phase ferri
et al., 1994; Scott *et al.*, 1998). Dithi
reduced to Fe^{2+}. Ferrous iron is cap
precipitates as solid solution of Cr(I
and some halogenated organic co
Chapter 6).

$$S_2O_4^{2-} + 2\,Fe(III)_{[solid]} + 2\,H_2O$$

$$CrO_4^{2-} + 3\,Fe(II)_{[solid]} + 5\,H^+ =$$

B. Biotic Reduction

Biotic reduction in PRBs is ini
nutrient materials that are used by
wheat straw, and alfalfa hay have b
pal waste or compost has been usec
see Chapters 8 and 17). Dissolved su
organic material by sulfate, consum
reduction are shown in Eq. (5) and E
conditions are produced in the PR
and other redox-reactive inorganic
reduction have successfully treated
ation. Bacteria are living organisms
conditions to exist and multiply.
population of microorganisms and

$$2\,CH_2O_{[solid\ organic]} + SO_4^{2-}$$

$$Me^{2+} + HS^- = MeS_{[so}$$

C. Chemical Precipitation

Limestone is the most common
drainage. Contaminants are remo
associated with an increase in the p
acid mine drainage has pH values
trations of iron and aluminum.
increases the alkalinity, and raises t

e accomplished by introducing dissolved
er 6 and 16). The dithionite ion ($S_2O_4^{2-}$)
iron as illustrated by Eq. (3) (Amonette
onite oxidizes to sulfite (SO_3^{2-}) as Fe^{3+} is
ble of reducing Cr(VI) to Cr(III), which
I) and Fe(III) hydroxide [Eq. (4)]. U(VI)
pounds are also reduced by Fe^{2+} (see

$$= 2\,SO_3{}^{2-} + 2\,Fe(II)_{[solid]} + 4\,H^+ \quad (3)$$

$$r(OH)_{3[solid]} + 3\,Fe(III)_{[solid]} + H_2O \quad (4)$$

ated by supplying electron donor and
microorganisms. Leaf mulch, sawdust,
en used as electron donors, and munici-
as a nutrient source (Benner *et al.* 1997;
fate is an electron acceptor. Oxidation of
tion of acidity, and the coupling to metal
q. (6) (Christensen *et al.*, 1996). Reducing
, resulting in the precipitation of metals
contaminants. PRBs that employ biotic
etal, sulfate, nitrate, and acid contamin-
requiring suitable physical and chemical
roundwater temperature will affect the
he rate of sulfate reduction.

$$+ H^+ = 2\,CO_2 + 2\,H_2O + HS^- \quad (5)$$

$$_{]} + H^+,\ \text{where Me} = \text{metal} \quad (6)$$

naterial used in PRBs to treat acid mine
ed in response to mineral precipitation
value. Groundwater contaminated with
s low as 2 and usually has high concen-
mestone dissolves in the acidic water,
e pH value [Eq. (7)]. Metal contaminants

can precipitate as hydroxides [Eq. (8)] or carbonates [Eq. (9)] if pH has increased sufficiently.

$$CaCO_{3\,[\text{limestone}]} + H^+ = HCO_3^- + Ca^{2+} \tag{7}$$

$$Me^{2+} + 2\,(OH)^- = Me(OH)_2, \text{ where Me} = \text{metal} \tag{8}$$

$$Me^{2+} + HCO_3^- = MeCO_3 + H^+, \text{ where Me} = \text{metal} \tag{9}$$

Apatite has also been used in a PRB to promote chemical precipitation and to accumulate U (see Chapter 14). Commercial sources of apatite include mined phosphatic rock deposits and animal bones. Phosphate release from apatite stabilizes lead by forming hydroxypyromorphite [$Pb_{10}(PO_4)_6(OH)_2$], a low-solubility mineral (Ma *et al.*, 1995). Apatite accumulates cadmium and zinc but not as effectively as it accumulates lead (Chen *et al.*, 1997). Bostick *et al.* (1999) identified U mineralization on the surface of phosphate particles as meta-autunite [$Ca(UO_2)_2(PO_4)_2 \cdot 6.5H_2O$] when apatite was reacted with uranyl nitrate solutions, suggesting that the mechanism of U uptake is mineral precipitation. Examination of phosphatic bone-charcoal pellets by X-ray adsorption spectroscopy indicates that a U–phosphate mineral formed at high U concentrations but that adsorption predominates at U concentrations characteristic of the Fry Canyon site (see Chapter 9).

D. Adsorption and Ion Exchange

Adsorption is a process whereby an aqueous chemical species attaches to a solid surface. Most adsorption reactions are reversible and occur at relatively rapid rates. Some adsorption reactions are specific with the attachment occurring preferentially at particular sites. Other adsorption reactions are less specific and have more of a tendency to compete with other ions for attachment sites. Surface complexation models have been used effectively to predict the reaction chemistry of adsorption processes (Davis and Kent, 1990; Dzombak and Morel, 1990).

Adsorbents can be used in PRBs to remove U and other inorganic contaminants from groundwater. Because of its high surface area per gram, AFO has a high affinity for adsorption of U and metal contaminants, but is dependent on solution concentrations of carbonate and hydrogen ions (Waite *et al.*, 1994; Hsi and Langmuir, 1985). Zeolites have also been used in PRBs to treat inorganic contaminants. Zeolites are hydrated aluminosilicates with a large internal surface area and high sorption capacity that treat inorganic contaminants by adsorption and cation exchange. In cation exchange, a contaminant molecule replaces another molecule at the surface of a zeolite particle.

Chemical mechanisms other than reductive precipitation, as described in Section III, have been suggested for the uptake of U by ZVI. Fiedor *et al.* (1998) suggests that the predominant U removal mechanism is by adsorption to ferric oxides and oxyhydroxides that form from the oxidation of ZVI. Both oxidation states of U have been observed in the ZVI used in the Oak Ridge and Durango projects (see Chapter 12).

IV. LONGEVITY

A. Zero-Valent Iron

The length of time that ZVI-based PRBs will perform efficiently is relatively unknown and is the focus of several current investigations. Cost effectiveness is directly linked to longevity. The ZVI-based PRB at the Borden site operated for 4 years. Examination of PRB reactive media excavated at the Borden site revealed only small amounts of mineral precipitation and no indication of diminished performance (O'Hannesin and Gillham, 1998). A commercial, ZVI-based PRB at Sunnyvale, California, has been operating since early 1995 (Warner *et al.*, 1998). The Borden and Sunnyvale PRBs were installed to treat organic contaminants, but mineral precipitation reactions involving major ions should be similar to ZVI-based PRBs used to treat inorganic contaminants.

The longevity of ZVI is potentially reduced by three phenomena: (1) dissolution of the iron, (2) mineral precipitation leading to permeability reduction, and (3) passivation of the ZVI from alteration of ZVI grain surfaces. ZVI dissolves in the presence of oxygenated groundwater [Eq. (10)]. If sufficient O_2 is available, ZVI oxidizes further and forms a mixture of ferric oxyhydroxides or rust [Eq. (11)]. The reaction of dissolved O_2 with ZVI is relatively rapid, and O_2 is usually depleted from the groundwater within a few centimeters of migrating through the ZVI. If the groundwater is oxygenated, ferric oxyhydroxide minerals will accumulate at the upgradient interface of the PRB with the groundwater. This mineralized zone has been shown to cause low permeability in laboratory column experiments by Mackenzie *et al.* (1997), who suggested that placing high-porosity zones of ZVI mixed with gravel upgradient from the PRB can extend the life of the PRB. Upgradient gravel zones are commonly used in PRBs for this purpose (e.g., see Chapter 13 and 14).

$$2\,Fe^0_{[ZVI]} + 4\,H^+ + O_2 = 2\,H_2O + 2\,Fe^{2+} \tag{10}$$

$$4\,Fe^0_{[ZVI]} + 6\,H_2O + 3\,O_2 = 4\,Fe(OH)_{3[solid]} \tag{11}$$

ZVI will dissolve even in the absence of O_2 [Eq. (12)]. In this reaction, water is an oxidant and gaseous H_2 is produced by the reduction of aqueous protons. The reaction leads to an increase in pH. Hydrogen gas and an increase in pH values are often observed in ZVI-based PRBs, confirming the anaerobic dissolution of ZVI.

$$Fe^0_{[ZVI]} + 2\,H^+ = Fe^{2+} + H_2 \qquad (12)$$

An increase in pH values can cause precipitation of carbonate minerals in alkaline ground waters [Eq. (9)]. Production of iron hydroxide [$Fe(OH)_3$, $Fe(OH)_2$] and carbonate phases can lead to permeability reduction. Sulfide minerals may precipitate in ZVI-based PRBs and further occlude permeability; the influence of sulfate-reducing bacteria on sulfide generation is investigated in Chapter 10.

Little is known about the phenomena of surface passivation. Presumably, minerals will precipitate on the ZVI surface, eventually leading to a surface deposit that is sufficiently thick and continuous to prevent the electron transfer processes to continue, similar to processes that limit the reduction of metal species on magnetite (White and Peterson, 1998).

B. Other Reactive Materials

Few data exist to evaluate the longevity of a dithionite-based PRB. The trial demonstration at the Hanford site is still reducing Cr concentrations after 2 years of operation. Similar to ZVI, an increase in pH values should occur as groundwater is reduced by Fe^{2+} [Eq. (13)]. The increase in pH is likely to cause mineral precipitation, which could decrease the permeability of the PRB. Few data are currently available to evaluate the severity of this effect. As contaminants and other oxidized groundwater species are reduced by the PRB, Fe^{2+} is oxidized to Fe^{3+} and the PRB will eventually become ineffective. Because the emplacement is by injection, it may be cost effective to rejuvenate the PRB with subsequent injections of dithionite.

$$2\ Fe(II)_{[solid]} + 2\ H^+ = 2\ Fe(III)_{[solid]} + H_2 \qquad (13)$$

The longevity of biotic-reduction PRBs depends on the loss of organic material and the reduction in permeability. Organic material degrades during sulfate reduction and other oxidation reactions. Materials such as leaf mulch have relatively low density and compact easily. These materials must be mixed with gravel to provide sufficient permeability and compaction strength. Therefore, only a small proportion (by weight) of organic material can be mixed with gravel, and large-capacity reactor cells may be required to achieve the desired longevity.

Limestone-based PRBs lose efficiency as limestone is armored by Fe or Al precipitates, which can lead to clogging and failure. High concentrations of Fe^{3+}, Al^{3+}, and dissolved oxygen increase the risk of clogging. Techniques are being investigated, such as flushing precipitates out of the PRBs, which may lead to greater longevity (see Chapter 2 and 3).

Few data are available to evaluate the longevity of apatite-based PRBs. A PRB at the Fry Canyon site has been in operation since September 1997 and is still effectively treating U-contaminated groundwater. A buildup of uranyl phosphate on the phosphate minerals could passivate the reactive surfaces and cause the uptake reaction to slow or cease.

The longevity of a zeolite-based PRB will largely depend on its sorptive capacity and the specificity for the contaminant of concern. The sorptive capacity can be accurately calculated from laboratory test results. Reactions that cause clogging and surface passivation are likely to be less significant because the chemical mechanism is primarily cation exchange.

Contaminants [e.g., U(VI)] adsorbed to the outer surfaces of AFO particles and other adsorption-based reactive materials can eventually desorb and reenter the groundwater. Even if the upgradient groundwater is free of contaminants (such as could occur if the contaminant source is removed), the contaminant concentration front will continue to migrate through the reactive medium until it reaches the downgradient edge. Once the front is at that position, the contaminant will be released into the environment at concentrations controlled by the chemistry of the incoming groundwater. This situation is commonly referred to as "contaminant breakthrough." In column experiments with bone charcoal, more than 60% of the sorbed U was released after breakthrough upon flushing with U-free groundwater (see Chapter 9).

Some laboratory results indicate that contaminant concentrations at breakthrough may be lower than those predicted by a simple adsorption model because of mineral aging processes. As AFO ages, some contaminants associate with the AFO at concentrations well above the concentrations that can be achieved by adsorption to the outer surfaces (Ainsworth *et al.* 1994). This effect may be caused by diffusion into inner pores and subsequent adsorption to interior surfaces, as was reported for arsenic uptake on AFO by Fuller *et al.* (1993). Release of contaminants bound in this manner would occur more slowly than release of contaminants adsorbed to the outer surfaces. Up to 40% of sorbed U remained attached to bone charcoal in column experiments conducted by Fuller and colleagues (Chapter 9). Aging reactions may result in more effective retention of contaminants, which would lower future maintenance costs.

V. DESIGNS

Various designs have been used to control the flow of contaminated water through PRBs. In most cases, the reactive material was placed in a shallow trench so that groundwater flows through the buried PRB.

Some systems involve moving the groundwater to reactive media through pipes. The limestone PRB at the Savannah River site uses a limestone trench to remove aluminum and raise the pH value (see Chapter 4). The effluent from the limestone trench flows through a siphon to a system for treatment with conventional water-treatment methods. The siphon does not require active pumping. At the Durango site (see Chapter 15), contaminated groundwater is collected by a collection system and piped to the PRBs. Piped systems allow easy monitoring of the groundwater treatment efficiency because the flow into the PRB can be accurately measured and only two samples are required (influent and effluent) to evaluate treatment efficiency.

Varied designs have been used for limestone-based PRBs. Limestone ponds have been constructed on upwelling seeps or connected to inlet pipes from collection galleries. In diversion wells, acidic water enters tanks filled with fine-grained limestone through an inlet at the bottom of the tank and precipitates flush out over the top of the tank. Perforated pipes placed as subdrains can be used to periodically flush accumulated precipitates from limestone systems (see Chapter 2).

A novel approach to constructing a PRB by injecting dissolved dithionite into the subsurface was developed at the DOE Pacific Northwest National Laboratory, and a PRB was constructed at the site in 1997 (see Chapter 6). The dithionite ion reduces naturally occurring Fe^{3+} in the sediments to Fe^{2+} that is capable of reductive precipitation of aqueous Cr(VI). The injection method permits PRB emplacement at greater depths than conventional trenching.

Cost savings may be realized by installing reactive materials in nonpumping wells (see Chapters 5 and 7). Groundwater streamlines directed into open boreholes increase the capture area. Wilson *et al.* (1997) modeled this effect and suggested that arrays of unpumped wells could be used to form a PRB. The method has increased cost savings for deep plumes where conventional PRB emplacement is difficult.

Monitoring the performance of subsurface PRBs is a challenge. The goal of a monitoring design is to accurately assess the mass flux of contamination passing through the PRB and bypassing the PRB. Many PRBs have closely spaced monitoring wells to help accomplish this goal. Flow velocities are estimated from measured values of hydraulic conductivity and water table gradient; concentration gradients are based on chemical analyses of water samples. Tracer tests, borehole-deployed velocity tools, and *in situ* velocity tools have been used to help estimate groundwater flow velocities. Mineral and chemical analyses of solid samples collected by coring or excavation have also been used to examine performance. The costs of extensive monitoring can be daunting, and efficient methods to accurately determine PRB performance are badly needed. Details of monitoring systems are described for several case studies in Section 4, "Case Studies of Permeable Reactive Barrier Installations."

VI. GOALS OF THIS BOOK

The chapters in this book present information about many of the existing field installations of PRBs for metals, radionuclides, and nutrients. Although the focus is on field applications, results of several laboratory investigations are presented that promote better understanding of reaction mechanisms and treatment efficiencies. Readers will gain a broad understanding of the breadth of research and applications that have been conducted. Construction details and performance data for some PRBs are also provided. Project managers can use the background information to help them make decisions on the applicability of PRBs at their sites, and technical experts will benefit from case study and laboratory results.

REFERENCES

Ainsworth, C. C., Pilon, J. L., Gassman, P. L., and VanDerSluys, W. G., (1994). Cobalt, cadmium, and lead sorption to hydrous iron oxide: Residence time effect. *Soil Sci. Soc. Am. J.* **58**, 1615–1623.

Ammonette, J. E., Szecsody, J. E., Schaef, H. T., Templeton, J. C., Gorby, Y. A., and Fruchter, J. S., (1994). Abiotic reduction of aquifer materials by dithionite: A promising in-situ remediation technology, *In* "Thirty-Third Hanford Symposium on Health and the Environment, in-Situ Remediation: Scientific Basis for Current and Future Technologies" (G. W. Gee and N. R. Wing, eds.), pp. 851–881. Battelle Press, Columbus, OH.

Benner, S. G., Blowes, D. W., and Ptacek, C. J., (1997). A full-scale porous reactive wall for prevention of acid mine drainage. *Ground Water Monit. Remed.* **17**, 99–107.

Blowes, D. W., Gillham, R. W., Ptacek, C. J., and Puls, R. W., (2000). "An in Situ Permeable Reactive Barrier for the Treatment of Hexavalent Chromium and Trichloroethylene in Groundwater, Vol. 1 design and installation, EPA/600/R-99/095a. University of Waterloo, Waterloo, Ontario, Canada.

Blowes, D. W., and Ptacek, C. J., (1992). "Geochemical Remediation of Groundwater by Permeable Reactive Walls: Removal of Chromate by Reaction with Iron-Bearing Solids", pp. 214–216. Subsurface Restoration Conference, U.S. Environmental Protection Agency, Kerr Laboratory, Dallas, TX.

Blowes, D. W., Ptacek, C. J., Hanton-Fong, C. J., and Jambor, J. L., (1995). In situ remediation of chromium contaminated groundwater using zero-valent iron. American Chemical Society, 209th National Meeting, Anaheim, California, April 2–7, 1995, **35**, 780.

Blowes, D. W., Puls, R. W., Bennett, T. A., Gillham, R. W., Hanton-Fong, C. J., and Ptacek, C. J., (1997). In situ porous reactive wall for treatment of Cr(VI) and trichloroethylene in groundwater, *In* "International Containment Conference Proceedings," pp. 851–857. February 9–12, St. Petersburg, FL.

Bostick, W. D., Stevenson, R. J., Jarabek, R. J., and Conca, J. L., (1999). Use of apatite and bone char for the removal of soluble radionuclides in authentic and simulated DOE groundwater. *Adv Environ. Res.* **3**, 488–498.

Cantrell, K. J., Kaplan, D. I., and Wietsma, T. W., (1995). Zero-valent iron for the in situ remediation of selected metals in groundwater. *J. Hazard. Mater.* **42**, 201–212.

Chen, X., Wright, J. V., Conca, J. L., and Peurrung, L. M., (1997). Evaluation of heavy metal remediation using mineral apatite. *Water Air Soil Pollut.* **98**, 57–78.

Christensen, B., Laake, M., and Lien, T., (1996). Treatment of acid mine water by sulfate-reducing bacteria; results from a bench scale experiment. *Water Res.* **30**, 1617–1624.

Code of Federal Regulations (CFR), (2000). National oil and Hazardous Substances Pollution Contingency Plan. Title 40 CFR Part 300.430(a)(1)(iii)(E).

Davis, J. A., and Kent, D. B., (1990). Surface complexation modeling in aqueous geochemistry. *Rev. Minerol.* **23**, 177–260.

Dzombak, D. A., and Morel, F. M. M. (1990). "Surface Complexation Modeling Hydrous Ferric Oxide." Wiley, New York.

Fiedor, J. N., Bostick, W. D., Jarabek, R. J., and Farrell, J., (1998). Understanding the mechanism of uranium removal from groundwater by zero-valent iron using x-ray photoelectron spectroscopy. *Environ. Sci. Technol.* **32**, 1466–1473.

Focht, R. M., Vogan, J. L., and O'Hannesin, S. F. (1995). Field application of reactive iron walls for in situ degradation of volatile organic compounds in groundwater. *In* "Proceedings of the 209th American Chemical Society, Division of Environmental Chemistry, National Meeting," pp. 81–94. Anaheim, CA.

Fuller, C. C., Davis, J. A., and Waychunas, G. A. (1993). Surface chemistry of ferrihydrite. 2. Kinetics of arsenate adsorption and coprecipitation. *Geochim. Cosmochim. Acta* **57**, 2271–2282.

Gillham, R. W., and O'Hannesin, S. F. (1992). Metal-catalyzed abiotic degradation of halogenated organic compounds. *In* "Modern Trends in Hydrogeology," 1992 IAH Conference, Hamilton, Ontario, Canada.

Gillham, R. W., O'Hannesin, S. F., and Orth, W. S. (1993). Metal enhanced abiotic degradation of halogenated aliphatics: Laboratory tests and field trials, HazMat Central Conference, Chicago, IL.

Hedin, R. S., Watzlaf, G. R., and Nairn, R. W. (1994). Passive treatment of acid mine drainage with limestone. *J. Environ. Qual.* **23**, 1338–1345.

Hsi, C. D., and Langmuir, D. (1985). Adsorption of uranyl onto ferric oxyhydroxide: application of the surface complexation site-binding model. *Geochim. Cosmochim. Acta* **49**, 1931–1941.

Kaplan, D. I., Cantrell, K. J., Wietsma, T. W., and Potter, M. A., (1996). Retention of zero-valent iron colloids by sand columns: Application to chemical barrier formation. *J. Environ. Qual.* **25**, 195–204.

Longmire, P. A., Brookins, D. G., Thomson, B. M., and Eller, P. G. (1991). Application of sphagnum peat, calcium carbonate, and hydrated lime for immobilizing U tailings leachate, *In* "Scientific Basis for Nuclear Waste Management XIV" (T. Abrajano, Jr., and L. H. Johnson, eds.), Vol. **212**, pp. 623–631.

Ma, Q. Y., Logan, T. J., and Traina, S. J. (1995). Lead immobilization from aqueous solutions and contaminated soils using phosphate rocks. *Environ. Sci. Technol.* **29**, 1118–1126.

Mackay, D. M., and Cherry, J. A. (1989). Groundwater contamination: Pump-and-treat remediation. *Environ. Sci. Technol.* **23**, 630–636.

Mackenzie, P. D., Sivavec, T. M., and Horney, D. P. (1997). Extending hydraulic lifetime of iron walls. *In* "International Containment Conference Proceedings," pp. 781–787. February 9–12, St. Petersburg, FL.

National Academy of Sciences (1994). Alternatives for ground water cleanup. Report of the National Academy of Sciences Committee on Ground Water Cleanup Alternatives. National Academy Press. Washington, DC.

O'Hannesin, S. F. (1993). "A Field Demonstration of a Permeable Reaction Wall for the in Situ Abiotic Degradation of Halogenated Aliphatic Organic Compounds." M.Sc. dissertation. Department of Earth Sciences, University of Waterloo, Waterloo, Ontario, Canada.

O'Hannesin, S. F., and Gillham, R. W. (1998). Long-term performance of an in situ 'iron wall' for remediation of VOCs. *Ground Water* **36**, 164–170.

Patel, A., Huang, C., Van Benschoten, J. E., Rabideau, A. J., and Hemann, M. (1999). Removal of Sr-90 from groundwater by clinoptilolite barrier, EOS Transactions, American Geophysical Union, 1999 fall meeting, **80**, F325.

Pearson, F. H., and McDonnell, A. J. (1975). Limestone barriers to neutralize acidic streams. *J. Environ. Engin. Div. EE3* 425–441.

Robertson, W. D. and Cherry, J. A. (1995). In situ denitrification of septic-system nitrate using reactive porous media barriers: Field trials. *Ground Water* **33**, 99–111.

Rumar, R. R., and Mitchell, J. K. (eds.) (1995). "Assessment of Barrier Containment Technologies," U.S. Department of Energy. Available from National Technical Information Service, No. PB96–180583, 437.

Scott, M. J., Metting, F. B., Fruchter, J. S., and Wildung, R. E. (1998). Research investment pays off. *Soil Groundwater Cleanup* October 6–13.

Spangler, R. R., and Morrison, S. J. (1991). Laboratory-scale tests of a chemical barrier for use at uranium mill tailings disposal sites. *In* "U.S. Department of Energy Environmental Remediation 91 Conference Proceedings," pp. 739–744. Pasco, Washington.

Starr, R. C., and Cherry, J. A. (1994). In situ remediation of contaminated ground water: The funnel-and-gate system. *Ground Water* **32**, 465–476.

Taylor, P. (1998). Development and operation of a passive-flow treatment system for Sr-90 contaminated ground water, Permeable Reactive Barriers Action Team presentation materials, U.S. Environmental Protection Agency Remediation Technology Development Forum conference at Oak Ridge, Tennessee, November 17–19, 167–172.

Thombre, M. S., Thomson, B. M., and Barton, L. L. (1997). Use of a permeable biological reaction barrier for groundwater remediation at a uranium mill tailings remedial action (UMTRA) site. *In* "International Containment Technology Conference Proceedings," pp. 744–750. February 9–12, St. Petersburg, FL.

Travis, C. C., and Doty, C. B. (1990). Can contaminated aquifers at superfund sites be remediated? *Environ. Sci. Technol.* **24**, 1464–1466.

United States Code (U.S.C) (2001). The Public Health and Welfare, Comprehensive Environmental Response, Compensation, and Liability Act, Title 42 U.S.C. Chapter 103, Section 9621(b).

Waite, T. D., Davis, J. A., Payne, T. E., Waychunas, G. A., and Xu, N. (1994). Uranium (VI) adsorption to ferrihydrite: Application of a surface complexation model. *Geochim. Cosmochim Acta* **58**, 5465–5478.

Warner, S. D., Sorel, D., and Yamane, C. L. (1998). Performance monitoring of the first commercial subsurface permeable reactive treatment zone composed of zero-valent iron,

Permeable Reactive Barriers Action Team presentation materials, U.S. Environmental Protection Agency Remediation Technology Development Forum conference at Oak Ridge, Tennessee, November 17–19, 323–324.

White, A. F., and Peterson, M. L. (1998). The reduction of aqueous metal species on the surfaces of Fe(II)-containing oxides: The role of surface passivation. *Ame. Chem. Soc. Symp.* **715**, 323–341.

Wilson, R. D., MacKay, D. M., and Cherry, J. A. (1997). Arrays of unpumped wells for plume migration control by semi-passive in situ remediation. *Ground Water Monitor. Remediat.* **17**, 185–193.

Part I

Innovations in Design, Construction, and Evaluation of Permeable Reactive Barriers

Chapter 2

Design and Performance of Limestone Drains to Increase pH and Remove Metals from Acidic Mine Drainage

Charles A. Cravotta III* and George R. Watzlaf†
*U.S. Geological Survey, New Cumberland, Pennsylvania 17070
†U.S. Department of Energy, National Energy Technology Laboratory, Pittsburgh, Pennsylvania 15236

Data on the construction characteristics and the composition of influent and effluent at 13 underground, limestone-filled drains in Pennsylvania and Maryland are reported to evaluate the design and performance of limestone drains for the attenuation of acidity and dissolved metals in acidic mine drainage. On the basis of the initial mass of limestone, dimensions of the drains, and average flow rates, the initial porosity and average detention time for each drain were computed. Calculated porosity ranged from 0.12 to 0.50 with corresponding detention times at average flow from 1.3 to 33 h. The effectiveness of treatment was dependent on influent chemistry, detention time, and limestone purity. At two sites where influent contained elevated dissolved Al (>5 mg/liter), drain performance declined rapidly; elsewhere the drains consistently produced near-neutral effluent, even when influent contained small concentrations of dissolved Fe^{3+} (<5 mg/liter). Rates of limestone dissolution computed on the basis of average long-term Ca ion flux normalized by initial mass and purity of limestone at each of the drains ranged from 0.008 to 0.079 $year^{-1}$. Data for alkalinity concentration and flux during 11-day closed-container tests using an initial mass of 4 kg crushed limestone and a solution volume of 2.3 liter yielded dissolution rate constants that were comparable to these long-term field rates. An analytical method is proposed using closed-container test data to evaluate long-term performance (longevity) or to estimate the mass of limestone needed for a limestone treatment. This method considers flow rate, influent alkalinity, steady-state maximum alkalinity of effluent, and desired effluent alkalinity or detention time at a future time(s) and applies first-order rate laws for limestone dissolution (continuous) and production of alkalinity (bounded).

I. INTRODUCTION

Acidic or abandoned mine drainage (AMD) degrades aquatic ecosystems and water supplies in coal- and metal-mining districts worldwide. The AMD typically contains elevated concentrations of dissolved and particulate iron (Fe) and dissolved sulfate (SO_4^{2-}) produced by the oxidation of pyrite (FeS_2) in coal and overburden exposed to atmospheric oxygen (O_2) (Rose and Cravotta, 1998; Nordstrom and Alpers, 1999). Half the acid produced by the stoichiometric oxidation of pyrite results from the oxidation of pyritic sulfur to SO_4^{2-} and the other half results from the oxidation of ferrous (Fe^{2+}) to ferric (Fe^{3+}) iron and its consequent precipitation as $Fe(OH)_3$ and related solids[1] (Bigham *et al.*, 1996; Cravotta *et al.*, 1999). Concentrations of manganese (Mn^{2+}), aluminum (Al^{3+}), and other solutes in AMD commonly are elevated due to aggressive dissolution of carbonate, oxide, and aluminosilicate minerals by acidic water along paths downflow from oxidizing pyrite (Cravotta, 1994; Blowes and Ptacek, 1994).

AMD commonly develops where the carbonate minerals, calcite ($CaCO_3$) and dolomite [$CaMg(CO_3)_2$], are absent or deficient relative to pyrite in coal overburden (Brady *et al.*, 1994, 1998). Dissolution of calcite, which is the principal component of limestone, can neutralize acidity and increase pH and concentrations of alkalinity ($CO_3^{2-} + HCO_3^- + OH^-$) and calcium ($Ca^{2+}$) in mine water (Cravotta *et al.*, 1999). The overall rate of calcite dissolution generally decreases with increased pH, decreased partial pressure of carbon dioxide (P_{CO_2}), and increased activities of Ca^{2+}, bicarbonate (HCO_3^-), and carbonate (CO_3^{2-}) near the calcite surface (Plummer *et al.*, 1979; Morse, 1983; Arakaki and Mucci, 1995).

Acidity and metals can be removed from AMD through various passive treatment systems that increase pH and alkalinity (Hedin *et al.*, 1994a;

[1]Hereafter, $Fe(OH)_3$ indicates the hydrous Fe-oxide and -sulfate compounds that together form ochres in AMD environments, including $Fe(OH)_3$ or ferrihydrite (nominally $Fe_5HO_8 \cdot 4H_2O$), goethite ($\alpha-FeOOH$), schwertmannite [$Fe_8O_8(OH)_6SO_4$], and jarosite [$(H,K,Na)Fe_3(SO_4)_2(OH)_6$] (Taylor and Schwertmann, 1978; Ferris *et al.*, 1989; Murad *et al.*, 1994; Bigham *et al.*, 1996).

Skousen *et al.*, 1998). Many of these systems utilize crushed limestone in packed beds in series with settling ponds or wetlands. For example, anoxic limestone drains (ALDs) are particularly effective for the generation of alkalinity (Turner and McCoy, 1990; Brodie *et al.*, 1991; Hedin and Watzlaf, 1994; Hedin *et al.*, 1994b; Watzlaf *et al.*, 2000a,b). Typically, for an ALD, crushed limestone of uniform size is placed in a buried bed(s) that intercepts net acidic (acidity > alkalinity) AMD before its exposure to atmospheric O_2. Excluding O_2 from contact with the mine water in an ALD minimizes the potential for oxidation of Fe^{2+} and precipitation of $Fe(OH)_3$.

The precipitation of $Fe(OH)_3$ and various other hydroxide and/or sulfate compounds of Fe^{3+}, Al^{3+}, and, possibly, Ca^{2+} and SO_4^{2-} within a bed of crushed limestone can "armor" the limestone surface (strong adhesion and complete pacification by encrustation), decreasing the rate and extent of limestone dissolution and alkalinity production (Watzlaf *et al.*, 1994; Hedin and Watzlaf, 1994; Aschenbach, 1995; Robbins *et al.*, 1999). Furthermore, the accumulation of precipitated compounds can decrease the porosity and permeability of the packed bed (Watzlaf *et al.*, 1994, 2000a,b; Robbins *et al.*, 1996). Hence, design criteria for ALDs as proposed by Hedin *et al.* (1994a) and Hedin and Watzlaf (1994) generally are conservative with respect to influent chemistry (requirement of < 1 mg/liter of dissolved O_2, Fe^{3+}, or Al^{3+}) and sizing (prolonged detention time) to ensure "maximum" alkalinity production over the life of an ALD.

Although most ALDs have been constructed as horizontal flooded systems, where a greater head at the inflow than outflow maintains the hydraulic gradient, some systems have been designed for vertical flow (upward or downward). Continuous inundation and detention of CO_2 within an ALD can enhance calcite dissolution and alkalinity production. By this mechanism, a greater quantity of alkalinity can be generated in an ALD compared to systems such as limestone channels (Ziemkiewicz *et al.*, 1997) or diversion wells (Arnold, 1991; Cram, 1996) that are open to the atmosphere. After treatment by an ALD, effluent is typically routed through ponds and/or wetlands where exposure to the atmosphere facilitates Fe^{2+} oxidation and the precipitation and settling of solid $Fe(OH)_3$.

Stringent requirements for low concentrations of O_2, Fe^{3+}, and Al^{3+} in the influent AMD make ALDs inappropriate for the treatment of oxic or highly mineralized water, which commonly occurs in mined areas. For example, of 140 AMD samples collected in 1999 from bituminous and anthracite coal mines in Pennsylvania (Cravotta *et al.*, 2001), only 17% were net acidic *and* had <1 mg/liter of dissolved O_2, Fe^{3+}, and Al^{3+}. Thus, ALDs could be appropriate for AMD treatment at a minority of the 140 sites, provided the dissolved O_2, Fe^{3+}, and Al^{3+} concentrations remain at low levels *and* resources and space are available for construction of the treatment system. However, for the majority of these discharges that do not meet criteria for an ALD, variations of the basic ALD design could be appropriate. One alternative uses pretreatment through a compost bed to decrease concentrations of dissolved O_2, Fe^{3+}, and Al^{3+} in the mine water to acceptable levels before routing the water through a limestone bed (Kepler and McCleary, 1994; Watzlaf *et al.*, 2000b). Nevertheless, short-term laboratory studies (<2 years) indicate that limestone alone can be as effective as this layered system for the neutralization of mine water containing dissolved O_2 and low-to-moderate concentrations of Fe^{3+} and Al^{3+} (<1–20 mg/liter) (Watzlaf, 1997; Sterner *et al.*, 1998). This variation on the ALD design is essentially an oxic limestone drain (OLD). In an enclosed OLD, oxidation and hydrolysis reactions will not be prevented but must be managed (Cravotta and Trahan, 1999). Despite the potential for armoring and clogging, hydrous oxides can be effective for the sorption of dissolved Mn^{2+} and trace metals (e.g., Kooner, 1993; Coston *et al.*, 1995; Webster *et al.*, 1998; Cravotta and Trahan, 1999). Precipitation of Mn oxides is possible after most dissolved Fe has been precipitated (Watzlaf, 1997; Cravotta and Trahan, 1999). If sufficiently rapid flow rates can be attained, the solid hydrolysis products may be transported through the OLD; however, consensus on design criteria for OLDs generally has not been reached.

This chapter evaluates the performance of limestone drains for the treatment of acidic, metal-contaminated water. Data on the basic design and chemical compositions of influent and effluent of 13 limestone drains that were constructed to treat discharges from abandoned coal mines in Pennsylvania and Maryland are evaluated with respect to flow rate, detention time, limestone dissolution rate, and long-term prospects for effectiveness. Criteria and methods for sizing and evaluation of limestone drains and general considerations for treatment of AMD containing dissolved O_2 and metals are presented.

II. SITE DESCRIPTIONS AND METHODS OF DATA COLLECTION

A. Description of Limestone Drains

The 13 limestone drains in Pennsylvania and Maryland considered for this study had varying designs reflecting site-specific treatment requirements and space availability. Detailed engineering documentation was not available; however, general descriptions of the drains were reported previously (Watzlaf *et al.*, 2000a,b; Cravotta and Trahan, 1999; Cravotta and Weitzel,

2001). Most drains were constructed as horizontal, buried limestone beds with an elongated form. Untreated inflow was intercepted within the trench or was piped into the drain. If possible, trenches were excavated in clay or liners were installed to prevent leakage. Generally, the outflow pipe from each of the flooded drains was extended to a level above the top of the limestone to ensure continuous inundation of the limestone and to minimize airflow into the drains. Additional information is provided here and in Table I.

Howe Bridge 1. Discharge from an abandoned well is captured and piped to the limestone drain. Influent water is sampled via a well prior to contact with limestone. Four sampling wells are spaced evenly along the length of the drain.

Howe Bridge 2. Discharge from another abandoned well is treated in an S-shaped limestone drain. Influent water is sampled via a well as the water flows into the limestone drain. Two sampling wells are along the length of the drain.

Elklick. Water from an abandoned borehole is collected in a bed ($7.0 \times 1.8 \times 0.9$ m) of crushed, low-pyrite sandstone at the head of the limestone drain. Influent water is sampled at a well in this sandstone. Three sampling wells are spaced equally along the length of the drain.

Jennings. The limestone drain treats an abandoned underground mine discharge that is collected in an bed of inert river gravel and piped to the system. Influent water is sampled via a sampling well prior to contact with limestone. The limestone drain consists of a series of six buried limestone-filled cells.

Morrison. Seepage is intercepted at the toe of the spoil of a reclaimed surface mine. An adjacent seep, similar in quality to the preconstruction water, is used to represent influent water quality. Three sampling wells are along the length of the limestone drain.

Filson-R and Filson-L. Seepage is intercepted at the toe of the spoil. A seep, between limestone drains, is similar in quality to the preconstruction raw water and is used to represent influent water quality.

Schnepp, REM-R, and REM-L. At each site, a limestone drain was constructed downgradient from collapsed underground mine entrances. Influent water quality is based on historical data.

Orchard. Three parallel limestone drains were constructed below a collapsed drift. Seepage from the drift was collected behind a wooden dam and then piped into the drains. Influent is accessible before contacting limestone. Valving at the inflow to each drain was used to control inflow rates. Access wells were installed at five locations along the length of each drain.

Table I

Initial Mass, Purity, Bulk Volume, Bulk Density, and Porosity of Crushed Limestone Used for Construction of 13 Limestone Drains in Pennsylvania and Maryland

Limestone drain site[a]	Year built[a] (t=0)	Mass of limestone, M_0 (tonne)[a]	Purity of limestone, X_{CaCO_3} (%)[a,b]	Size range limestone fragments (cm)[a]	Drain dimensions[a] Length (m)	Width (m)	Depth (m)	Bulk volume, V_B, (m^3)[c]	Bulk density, ρ_B, (kg/m^3)[d]	Porosity, ϕ (fractional units)[d]	Void ratio, $e = \phi/(1 - \phi)$ (unitless)
1. Howe Bridge 1	1991	455.0	82	5.1–7.6	36.6	6.1	1.2	267.9	1698	0.359	0.560
2. Howe Bridge 2	1993	132.0	82	5.1–7.6	13.7	4.6	0.9	56.7	2327	0.122	0.139
3. Elklick	1994	165.0	85	5.1–20.3	36.6	3.1	0.9	102.1	1616	0.390	0.640
4. Jennings	1993	365.0	90	15.2	228.0	1.0	1.0	228.0	1601	0.396	0.655
5. Morrison	1990	65.0	92	5.1–7.6	45.7	0.9	0.9	37.0	1756	0.337	0.509
6. Filson-R	1994	590.0	88	5.1–7.6	54.9	6.1	0.9	301.4	1958	0.261	0.354
7. Filson-L	1994	635.0	88	5.1–7.6	54.9	6.1	0.9	301.4	2107	0.205	0.258
8. Schnepp	1993	130.0	90	1.9–2.5	12.2	6.1	0.9	67.0	1941	0.268	0.365
9. REM-R	1992	125.0	82	7.6	16.8	6.1	0.9	95.1	1314	0.504	1.017
10. REM-L	1992	125.0	82	7.6	13.7	7.6	0.9	93.0	1344	0.493	0.972
11. Orchard	1995	38.1	97	6.0–10	73.2	2.4	0.8	16.7	2283	0.139	0.161

(continues)

Table I *(continued)*

Limestone drain site[a]	Year built[a] (t=0)	Mass of limestone, M_0 (tonne)[a]	Purity of limestone, X_{CaCO_3} (%)[a,b]	Size range limestone fragments (cm)[a]	Drain dimensions[a]			Bulk volume, V_B, (m^3)[c]	Bulk density, ρ_B, (kg/m^3)[d]	Porosity, ϕ (fractional units)[d]	Void ratio, $e = \phi/(1-\phi)$ (unitless)
					Length (m)	Width (m)	Depth (m)				
12. Buck Mountain	1997	320.0	92	6.0–10	48.8	1.9	1.9	176.2	1816	0.315	0.459
13. Hegins	2000	800.0	90	24–36	41.6	6.7	1.7	473.8	1688	0.363	0.570
Median									1756	0.337	0.509

[a] Data for year built, limestone mass, purity, size range, and drain dimensions at sites 1–10 from Watzlaf *et al.* (2000a), site 11 from Cravotta and Trahan (1999), site 12 from Cravotta and Weitzel (2002), and site 13 this report (C. A. Cravotta, unpublished data).
[b] Purity expressed as weight percent $CaCO_3$.
[c] Bulk volume computed as product of length·width·depth assuming tabular shape, except for Orchard. Orchard consisted of three parallel, semicircular troughs, each with a cross-sectional area of 0.244 m^2 and a length of 24.4 m (Cravotta and Trahan, 1999).
[d] Bulk density computed as initial limestone mass divided by bulk volume, $\rho_B = M_0/V_B$ [Eq. (3)]. Porosity and void ratio computed on basis of Eq. (4), $\phi = (\rho_S - \rho_B)/\rho_S$, assuming constant particle density, $\rho_S = 2650\,kg/m^3$ (Cravotta and Trahan, 1999).

Buck Mountain. Seepage from a collapsed drainage tunnel is collected at various points where water upwells along the length of the drain. Influent water quality is based on historical data and an adjacent seep. Access wells were installed at seven locations along the length of the drain. Perforated piping was installed along the length of the drain for flushing of accumulated solids.

Hegins. Discharge from a collapsed mine tunnel is piped into the drain, which consists of four cells in series. Influent is accessible before contacting limestone. Outflow spills from the top of each cell into the next. Access wells were installed at four locations in each of the cells. Perforated piping with valves for each cell was installed along the length of the drain for flushing of accumulated solids. At the time of this report, the Hegins drain had not been buried nor completely flooded.

The Orchard and Hegins drains were designed as OLDs for treatment of "oxic" influent that nonetheless contained relatively low concentrations of total Fe (<3 mg/liter). Drains at the other sites were previously classified as ALDs (Hedin and Watzlaf, 1994; Watzlaf *et al.*, 2000a). Data on dissolved O_2 and Fe^{3+} concentrations in *influent* to many of these drains were not available to confirm their ALD classification. However, previously reported data on the difference in concentration of Fe^{3+} between influent and effluent for the Howe Bridge 1, Howe Bridge 2, Morrison, Jennings, and REM-R drains (Hedin and Watzlaf, 1994) indicate that only the first three drains, which treated influent containing <1 mg/liter of Fe^{3+}, were unambiguously ALDs. The Jennings and REM-R drains treated influent that contained ≥ 10 mg/liter of Fe^{3+}, indicating that these should be classified as OLDs (>1 mg/liter of Fe^{3+}, Al^{3+}, or O_2). Hence, in this report, ALDs and OLDs are considered together to characterize limestone drain treatment systems generally, and then differences in treatment effectiveness are examined that could result from different factors, such as detention time and influent compositions, including pH and redox state.

B. Methods of Sampling and Analysis

Details of the field and laboratory methods for water-quality sampling and analysis were reported previously (Watzlaf and Hedin, 1993; Hedin and Watzlaf, 1994; Hedin *et al.*, 1994b; Cravotta and Trahan, 1999; Watzlaf *et al.*, 2000a). Standard methods were used for field analysis of pH, alkalinity (pH 4.5 end point), temperature, and specific conductance on unfiltered samples (e.g., Wood, 1976; Wilde *et al.*, 1998) and for laboratory analysis of acidity (pH 8.3 end point), major ions, and various metals (e.g., Greenberg *et al.*, 1992; Fishman and Friedman, 1989). Samples for analysis of dissolved metals were filtered through 0.45-μm pore-sized filters. Outflow from each of the drains was accessible for volumetric flow measurement and water-quality monitoring; however, only the Howe Bridge 1, Howe Bridge 2, Elklick, Morrison, Orchard, Buck Mountain, and Hegins drains had sampling access wells at intermediate points between inflow and outflow. Samples were retrieved from access wells by use of pumps or bailers. The actual inflow was not accessible at a majority of the drains, thus adjacent untreated seeps that had similar character as the untreated AMD were sampled or historical data for the untreated AMD were used to represent the influent. However, use of historical data to represent influent water quality may overestimate contaminant levels because water quality of untreated AMD commonly improves with time (e.g., Wood, 1996). Generally, flow rate and water quality at each drain were monitored quarterly or more frequently during the first year after the drains were constructed and less frequently during subsequent years.

The long-term averages for flow rate, pH, net acidity (acidity–alkalinity), alkalinity, and concentrations of solutes in influent and effluent of the 13 limestone drains were compiled or computed for subsequent evaluation of effects of limestone drain treatment on the neutralization of acidity and attenuation of metals transport (Table II). On the basis of water-quality data for many of the sites considered, as reported by Hedin and Watzlaf (1994) and Cravotta and Trahan (1999), activities of aqueous species, P_{CO_2}, and mineral saturation indices were calculated using the WATEQ4F computer program (Ball and Nordstrom, 1991). The saturation index (SI) provides a basis for evaluating the potential for dissolution or precipitation of a solid phase (Stumm and Morgan, 1996). Values of SI that are negative (< -0.1), approximately zero (± 0.1), or positive (> 0.1) indicate the water is undersaturated, saturated, or supersaturated, respectively, with the solid phase. If undersaturated, the water can dissolve the solid phase. If supersaturated, the water cannot dissolve the solid phase but can potentially precipitate it.

Additionally, to evaluate dissolution and precipitation effects at the Orchard drain, limestone samples that had been measured for weight, density, porosity, and geometric surface dimensions (thickness, width, length) were suspended with braided nylon chord and then retrieved and remeasured after 5 months to 5 years elapsed time of immersion. The retrieved samples were used to determine the rate of limestone dissolution at a particular location and the chemistry and mineralogy of encrustation. The dissolution rate was normalized with respect to the geometric surface area of the sample so that the field results could be compared with rate estimates from published laboratory experiments (Cravotta and Trahan, 1999).

Finally, empirical test results for the Howe Bridge 1 and Morrison sites that had previously been reported by Watzlaf and Hedin (1993) are used in

Table II

Average[a] Quality of Influent (In) and Effluent (Eff) at 13 Limestone Drains in Pennsylvania and Maryland

Limestone drain site[b]	pH (units)		Net acidity[c] (mg/liter as $CaCO_3$)		Alkalinity (mg/liter as $CaCO_3$)		Sulfate (mg/liter)		Calcium[d] (mg/liter)		Iron[e] (mg/liter)		Manganese (mg/liter)		Aluminum[e] (mg/liter)		Cobalt (mg/liter)		Nickel (mg/liter)		Zinc (mg/liter)	
	In	Eff	In	Eff	In	Eff	In	Eff	In	Eff	In	Eff	In	Eff	In	Eff	In	Eff	In	Eff	In	Eff
1. Howe Bridge 1	5.7	6.3	472	352	33	155	1300	1300	157	209	276	275	41	41	< 0.2	< 0.2	0.48	0.48	0.51	0.50	0.62	0.55
2. Howe Bridge 2	5.4	6.5	411	274	35	163	1200	1200	154	206	250	248	37	36	< 0.2	< .2	0.39	0.39	0.40	0.40	0.42	0.39
3. Elklick	6.1	6.7	52	−63	34	159	330	330	*77*	129	59	53	4.8	4.9	< 0.2	< .2	0.07	0.07	0.10	0.09	0.13	0.08
4. Jennings	3.2	6.2	280	−34	0	139	630	620	*83*	208	76	59	8.4	8.3	20.9	1.1	0.13	0.15	0.40	0.40	0.66	0.54
5. Morrison	5.2	6.4	387	51	29	278	1300	1000	115	223	*207*	156	49	41	0.5	< 0.2	0.86	0.75	0.79	0.65	0.95	0.72
6. Filson-R	5.7	6.5	100	−139	48	299	410	440	69	189	59	55	20	20	0.4	< 10.2	0.23	0.23	0.18	0.18	0.27	0.18
7. Filson-L	5.7	6.6	104	−175	48	317	410	400	69	180	59	55	20	16	0.4	< 10.2	0.23	0.13	0.18	0.10	0.27	0.17
8. Schnepp	3.3	6.2	225	−43	0	168	980	750	*90*	198	*92*	61	28	26	6.7	< 0.2	NA[f]	0.27	NA	0.33	NA	0.34
9. REM-R	4.3	5.5	NA	835	0	54	2800	2400	*210*	232	*589*	447	136	126	4.5	3.2	NA	1.49	NA	1.54	NA	2.46
10. REM-L	NA	6.0	NA	259	NA	113	NA	1300	*150*	201	NA	185	NA	51	NA	< .2	NA	0.60	NA	0.66	NA	0.76
11. Orchard	3.6	6.1	57	−43	0	64	190	170	24	50	1.7	0.2	2.6	1.6	0.9	0.2	0.08	0.04	0.12	0.08	0.11	0.15

(*continues*)

Table II *(continued)*

Limestone drain site[b]	pH (units)		Net acidity[c] (mg/liter as $CaCO_3$)		Alkalinity (mg/liter as $CaCO_3$)		Sulfate (mg/liter)		Calcium[d] (mg/liter)		Iron[e] (mg/liter)		Manganese (mg/liter)		Aluminum[e] (mg/liter)		Cobalt (mg/liter)		Nickel (mg/liter)		Zinc (mg/liter)	
	In	Eff	In	Eff	In	Eff	In	Eff	In	Eff	In	Eff	In	Eff	In	Eff	In	Eff	In	Eff	In	Eff
12. Buck Mountain	4.5	6.5	30	−85	1	86	48	48	3	38	8.5	9.6	.6	0.8	0.5	0.1	0.04	0.03	0.05	0.03	0.06	0.03
13. Hegins	3.8	4.4	40	30	0	1	230	230	10	17	0.3	0.2	1.5	1.7	4.1	3.4	0.08	NA	0.12	NA	0.12	NA

[a] Average for period of record, not adjusted for frequency or time of sample collection.
[b] Data for influent and effluent quality at sites 1–10 from Watzlaf *et al.* (2000a) and at sites 11–13 this report (C. A. Cravotta, unpublished data from USGS NWIS data base). Influent data for sites 5–7 based on quality of nearby seep and for sites 8–10 and 12 based on quality of untreated discharge prior to drain construction.
[c] Net acidity = acidity – alkalinity; negative values indicate net alkaline conditions.
[d] Values in italics for influent calcium estimated on the basis of net alkalinity or net acidity.
[e] Values in italics for influent iron or aluminum indicate influent chemistry (Fe^{3+} or Al^{3+} > 1 mg/liter) that is inconsistent with design criteria for ALD (Hedin *et al.*, 1994a,b; Skousen *et al.*, 1998).
[f] Not available.

this chapter to indicate qualitative and quantitative effects of variable influent compositions, detention times, and limestone purity on limestone drain performance. Collapsible, 3.8-liter polyethylene containers, hereafter called "cubitainers," were loaded with 4 kg of 1.3- by-3.5-cm limestone fragments (two-thirds total volume), filled with the untreated mine water to exclude any air, and then maintained at field water temperature to evaluate the generation of alkalinity (Fig. 1). The untreated mine water at the sites initially contained some alkalinity but was net acidic: Howe Bridge (pH 5.7; $P_{CO_2} = 10^{-1.24}$; alkalinity = 39 mg/liter as $CaCO_3$; acidity = 516 mg/liter as $CaCO_3$) and Morrison (pH 5.4; $P_{CO_2} = 10^{-1.05}$; alkalinity = 29 mg/liter as $CaCO_3$; acidity = 460 mg/liter as $CaCO_3$). The cubitainer tests were conducted in duplicate using several local varieties of limestone with reported purity ranging from 82 to 99% by weight $CaCO_3$. The impurities consisted of $MgCO_3$, as well as noncarbonate components, such as silica, clay, and organic matter. Periodically over 11 days, samples were withdrawn through a valve to fill a 60-ml

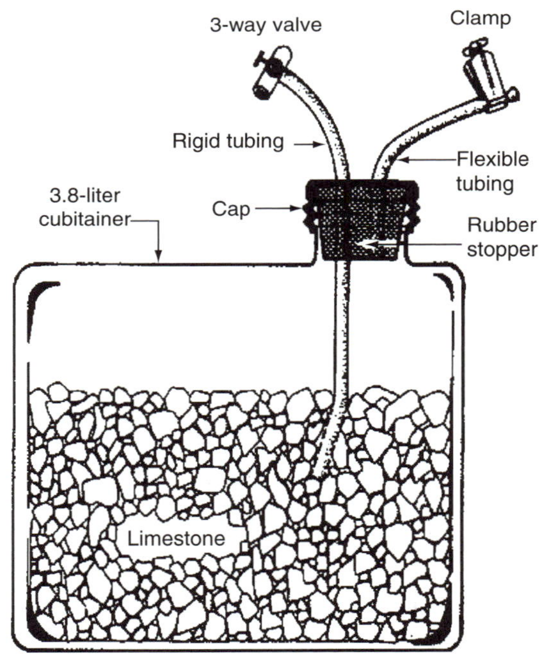

Figure 1 Schematic of a 3.8-liter "cubitainer" containing 4 kg limestone and filled with mine discharge water to evaluate alkalinity production rates (after Watzlaf and Hedin, 1993). Limestone sized to 1.3 × 3.5 cm. Cubitainer and tubing are impermeable to gas.

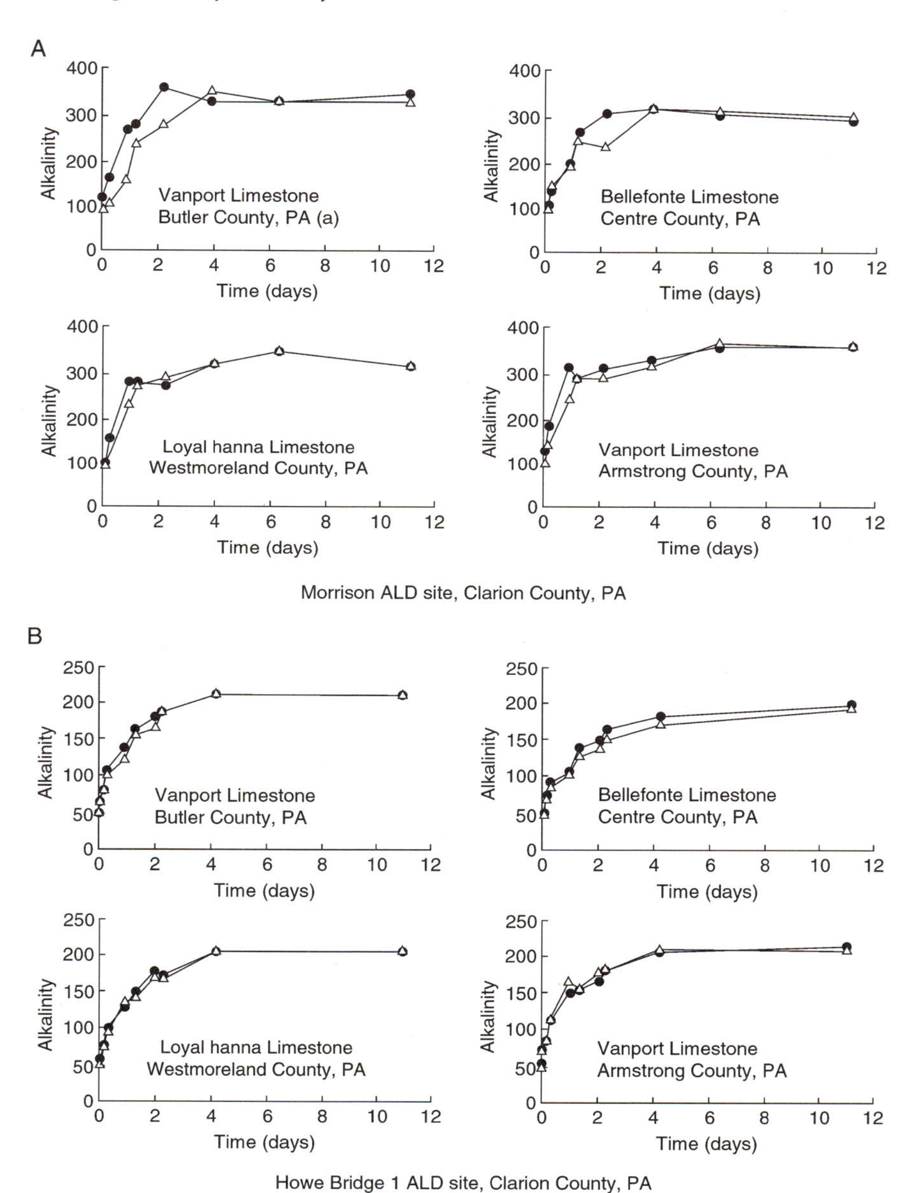

Figure 2 Alkalinity concentration (mg/liter as $CaCO_3$) generated during immersion of four limestone types from various locations by mine drainage from Morrison and Howe Bridge sites (after Watzlaf and Hedin, 1993). Each graph plots the alkalinity concentration in duplicate cubitainers (Δ, ○).

syringe after purging approximately 10 ml of fluid from the sample tubing. Each sample was forced through a 0.45-μm pore-sized filter and then analyzed for alkalinity (pH 4.5 end point) (Watzlaf and Hedin, 1993). Alkalinity data for the cubitainer tests were reported as time-series plots, some of which are shown in Fig. 2.

III. PERFORMANCE AND DESIGN OF LIMESTONE DRAINS

According to laboratory and field studies, such as those by Watzlaf and Hedin (1993), Hedin and Watzlaf (1994), and Cravotta and Trahan (1999), the concentration of alkalinity that is produced in a limestone bed and its effectiveness for increasing pH and removing dissolved metals from contaminated mine drainage depend on the composition of the water; the quantity, purity, and particle size of limestone; and the detention time and rates of chemical reactions. Typically, the pH and concentrations of alkalinity and Ca^{2+} increased asymptotically with increased detention time, or downflow distance, within a limestone bed (Figs. 2, 3A, 4, and 5). These trends generally result from a decrease in the rate of limestone dissolution as calcite equilibrium is approached (Figs. 5 and 6). With increased pH, concentrations of acidity and various dissolved metals may decrease due to precipitation and adsorption reactions within the limestone bed (Figs. 5 and 6). However, the occurrence and efficiency of these processes vary among different treatment systems.

Downgradient trends for pH, alkalinity, acidity, Ca, and other solute concentrations within the Orchard OLDs provide a useful basis for interpreting important chemical variations within limestone drains, generally. Downgradient trends for the first 6 months of treatment (Figs. 5A and 5B) are consistent with those expected for an ALD, except that Fe^{2+} transport generally would be conservative in an ALD (indicated by dashed line in Fig. 5B). In contrast, the downgradient trends after the first 6 months, when the drains began to retain Mn and trace metals (Fig. 5D), are consistent with those expected for OLDs (e.g., Cravotta and Trahan, 1999).

Hedin and Watzlaf (1994), Robbins *et al.* (1999), and Cravotta and Trahan (1999) showed that limestone theoretically could dissolve throughout the limestone drains they investigated because the water was consistently undersaturated with respect to calcite, attaining calcite saturation index ($SI_{CALCITE}$) values from −2.4 to −0.3 under the conditions evaluated. For example, despite increased pH, alkalinity, and Ca through the Orchard OLDs, the $SI_{CALCITE}$ remained negative (Fig. 6A). Solute concentrations, pH, and

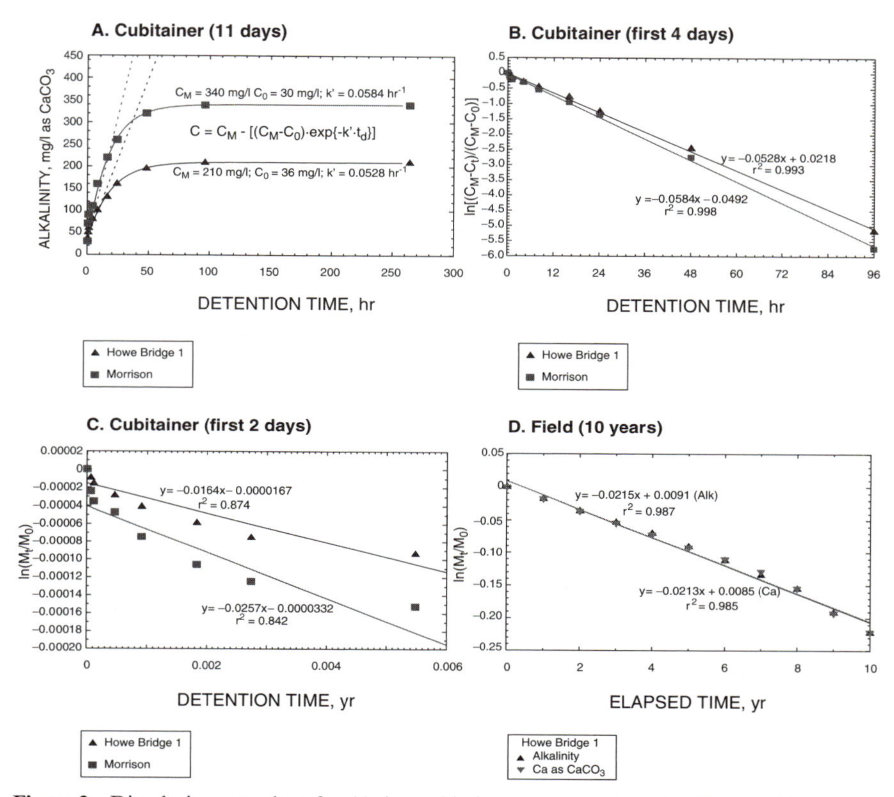

Figure 3 Dissolution rate data for 11-day cubitainer experiments and a 10-year field study. (A) Generalized alkalinity data points for cubitainer tests from Fig. 2. (B) Natural logarithm of the difference between steady-state maximum alkalinity (C_M) and measured alkalinity (C_T) divided by the difference between C_M and initial alkalinity (C_0) versus time. Slope indicates rate constant, k', for computation of alkalinity. (C and D) Natural logarithm of remaining limestone mass (M_t) divided by initial mass (M_0) versus time. Slope indicates rate constant, k, for computation of limestone mass. Field data for the Howe Bridge limestone drain are based on reported initial limestone mass and limestone purity (Table I) and annual averages for flow rate and concentrations of alkalinity and Ca in effluent and influent.

$SI_{CALCITE}$ increased most rapidly near the inflow (Figs. 5 and 6A). Cravotta and Trahan (1999) explained that for a computed detention time of 1 h, the water in the Orchard OLDs was net alkaline and had pH ~ 6 and $SI_{CALCITE}$ -2; however, by tripling the detention time from 1 to 3 h, pH increased only to ~ 6.8, whereas $SI_{CALCITE}$ increased to -1 and alkalinity and Ca concentrations doubled from about 60 to 120 and 90 to 180 mg/liter, respectively. These observations are consistent with the asymptotic trends for

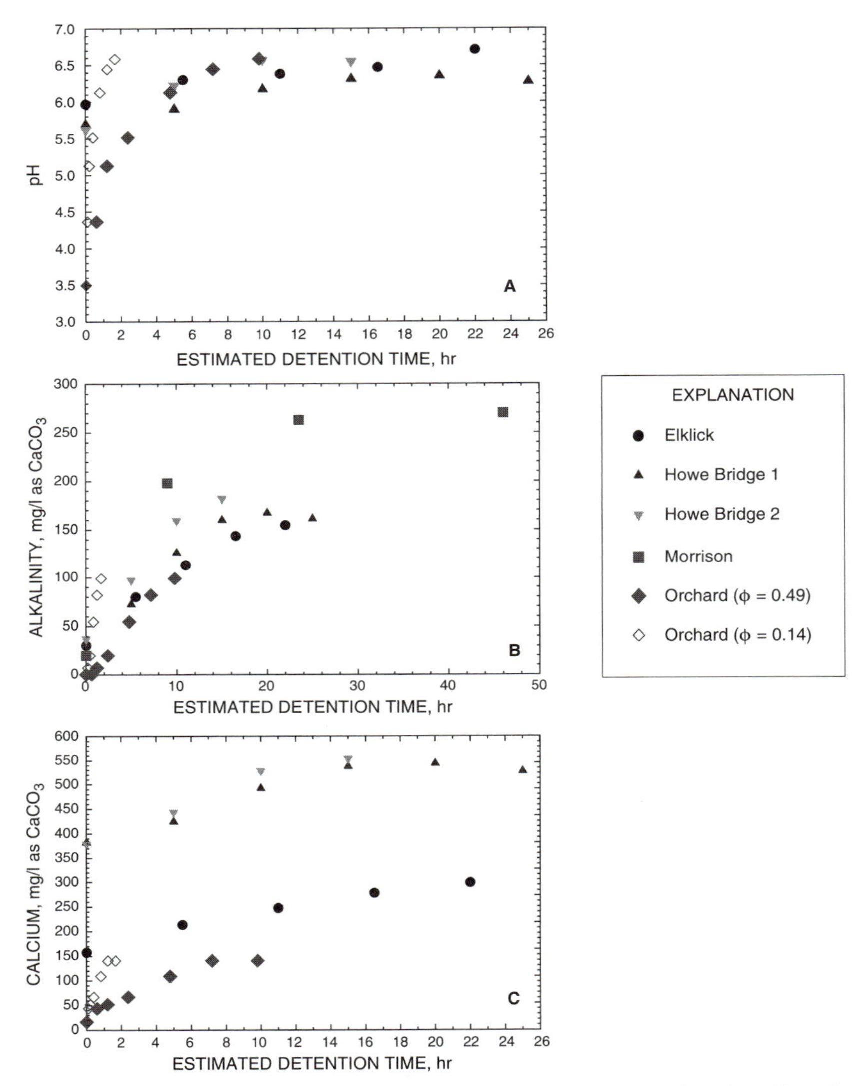

Figure 4 Changes in pH, alkalinity, and Ca concentration with detention time (downflow distance) of contaminated mine drainage within limestone drains at Elklick, Howe Bridge 1 and 2, Morrison, and Orchard sites described in Table I. Detention time computed as a product of porosity (ϕ), downflow distance (L), and cross-sectional area (A) divided by flow rate (Q): $t_d = \phi \cdot L \cdot A/Q$, assuming $\phi = 0.49$ (solid symbols). Data for Orchard also are displayed for $\phi = 0.14$ (open symbols) computed for initial limestone mass and drain volume (Table I).

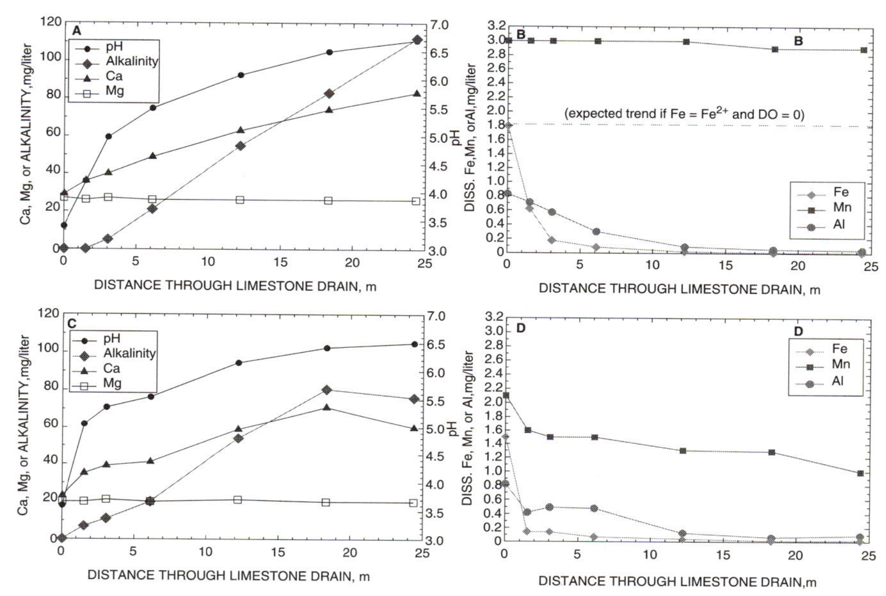

Figure 5 Sequential changes in water chemistry within Orchard limestone drain. Averages for March–August, 1995: (A) pH and concentrations of alkalinity, Ca, and Mg and (B) concentrations of dissolved Fe, Mn, and Al. Averages for September 1995–May 2000: (C) pH and concentrations of alkalinity, Ca, and Mg (D) concentrations of dissolved Fe, Mn, and Al.

a decreased alkalinity production rate with an increased detention time for cubitainer and field tests at other sites (Figs. 2 and 4). Hence, the detention time is a critical variable determining the performance of limestone drains. Furthermore, although the asymptotic trends for alkalinity production with detention time (Figs. 2 and 4) could be interpreted to indicate decreasing rates of limestone dissolution as equilibrium with calcite is approached, they could also indicate decreased limestone dissolution and/or alkalinity production rates because of the accumulation of secondary minerals.

The precipitation of secondary minerals may result in limestone drains due to increases in pH, alkalinity, and Ca concentrations. For example, gypsum ($CaSO_4 \cdot 2H_2O$) and hydrous oxides of Fe(III) and Al could precipitate within limestone beds if concentrations of SO_4^{2-}, Fe^{3+}, and Al^{3+} in influent are elevated (Figs. 6B, 6C, and 6D). Hedin and Watzlaf (1994) and Robbins *et al.* (1999) showed that except where concentrations of SO_4 exceeded about 2000 mg/liter, the water within limestone drains remained undersaturated with respect to gypsum. As the concentration of Ca increased within the Orchard OLD, gypsum saturation index (SI_{GYPSUM}) values increased but

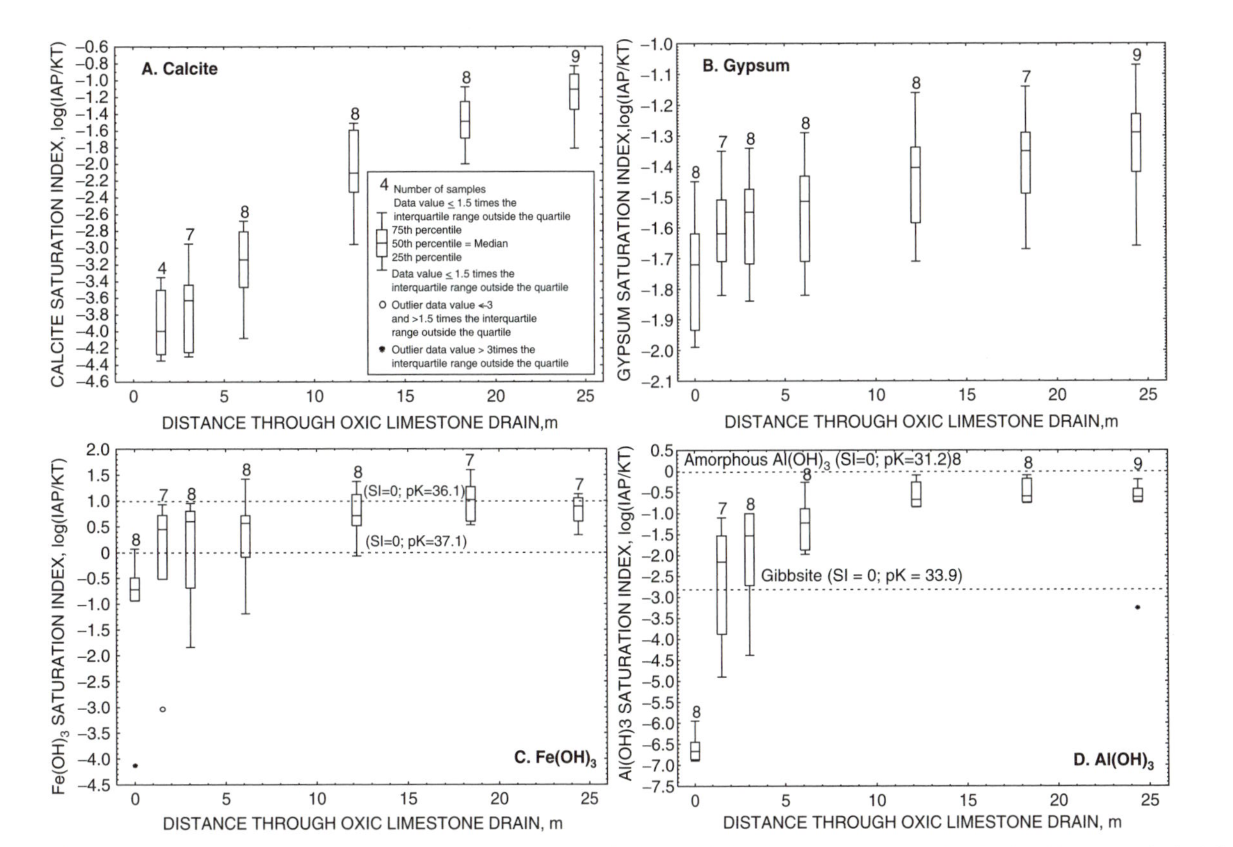

Figure 6 Sequential changes in saturation with respect to various minerals that could dissolve or precipitate within Orchard limestone drain (after Cravotta and Trahan, 1999). Saturation index calculated using the WATEQ4F computer program (Ball and Nordstrom, 1991). Data for samples collected March 1995–March 1996. From Cravotta and Trahan (1999).

remained well below equilibrium (Fig. 6B), hence precipitation of this mineral on limestone surfaces was unlikely. Nevertheless, even at sites where the SI_{GYPSUM} was positive, indicating a potential for gypsum precipitation, SO_4^{2-} transport generally was conservative (Hedin and Watzlaf, 1994), indicating that the precipitation of gypsum was not an effective attenuation mechanism for this constituent. In contrast, the precipitation of Al^{3+} and Fe^{3+} is likely within limestone drains because of the low solubilities of these metals (e.g., Stumm and Morgan, 1996). The precipitation of hydrous oxides of Al and Fe(III) accounted for quantitative removal of the dissolved Al and Fe within the Orchard OLDs. Although dissolution features were visible, calcite and limestone samples immersed in the drain were coated, at least locally, by these secondary minerals (Robbins *et al.*, 1999). The formation and accumulation of any of these minerals within a limestone bed could pose a problem if they were to armor the limestone surfaces or clog pore spaces, impeding flow through the bed.

A. Porosity and Detention Time

Detention time (t_d) and, hence, rates of alkalinity production or other effects of limestone dissolution within limestone beds can be estimated on the basis of volumetric flow rate (Q) and estimated void volume (V_V) within the bed,

$$t_d = V_V/Q \tag{1}$$

or

$$t_d = \phi \cdot V_B/Q \tag{2}$$

where V_B is the bulk volume and ϕ is the porosity ($\phi = V_V/V_B$). If the total mass of limestone (M) and bulk volume (V_B) of the drain are known, porosity can be computed. According to Freeze and Cherry (1979), bulk density (ρ_B) is defined as

$$\rho_B = M/V_B \tag{3}$$

and is proportionally related to porosity (ϕ) by the stone density (ρ_S) where

$$\rho_B = \rho_S \cdot (1 - \phi). \tag{4}$$

Hence, by rearranging Eq. (4) and assuming $\rho_S = 2.65\,g/cm^3$ considered typical for limestone (Freeze and Cherry, 1979; Cravotta and Trahan, 1999), porosity can be determined for various bulk densities and vice versa:

$$\phi = 1 - (\rho_B/\rho_S). \tag{5}$$

Knowing ρ_S, ϕ, and Q, Eq. (2) can be rewritten by substituting $V_B = M/[\rho_S \cdot (1 - \phi)]$ in accordance with Eqs. (3) and (5) to determine the detention time for water flowing through a limestone drain with a given mass

$$t_d = M/[Q \cdot \rho_S \cdot (1 - \phi)/\phi] \quad (6)$$

or the mass of limestone required to achieve a given detention time

$$M = t_d \cdot [Q \cdot \rho_S \cdot (1 - \phi)/\phi]. \quad (7)$$

For the 13 limestone drains evaluated in this chapter, computed porosities ranged from 0.12 to 0.50, with a median of 0.34; the median bulk density was 1756 kg/m^3 (Table I). The corresponding detention times for the given porosities and average flow at these sites ranged from 1.3 to 33 h (Table III). Many of the computed porosity values are less than a reference value of 0.49 based on precise, laboratory measurements; laboratory porosity ranged from 0.38 to 0.50 for well-sorted limestone fragments, with higher values associated with larger (5.1 to 7.6 cm) particles (Rice *et al.*, 1970; Hedin and Watzlaf, 1994). Lower porosities, e.g., 0.14 for the Orchard limestone drain (Table I), were explained by Cravotta and Trahan (1999) as possibly resulting from compaction of mixed-size, tabular limestone fragments; rough or flexible outside walls; and/or the accumulation of secondary minerals. Nevertheless, the computed porosity of 0.12 for the Howe Bridge 2 limestone drain probably is lower than its actual porosity. The authors suspect that reported dimensions for this drain are inaccurate; however, construction data could not be verified. Furthermore, although mass estimates are based on the delivered weight of limestone, accidental spillage and intentional spreading outside the drain could be significant. Some reported weights account for this wastage, whereas others do not. Tracer tests reported by Watzlaf *et al.* (2000a) and Cravotta and Trahan (1999), which directly indicate detention time(s), were interpreted to be consistent with the previously reported porosities at these drains. However, tracer tests and longitudinal sampling at the respective study sites indicated complex, preferential flow paths caused by a range of porosities and other hydrological controls. Hence, different estimates of detention time are possible considering the uncertainty and range of porosities.

Table III and Fig. 4 show that porosity is a critical variable for characterizing a limestone drain. For example, estimated detention times using the computed porosity of 0.36 for the drain at the Howe Bridge 1 site are approximately one-half of those computed using an assumed porosity of 0.49, and detention times for Orchard drain vary by almost a factor of 6 considering the computed porosity of 0.14 and reference value of 0.49. Furthermore, the same mass of limestone would occupy a two-thirds larger volume for $\phi = 0.49$ than for $\phi = 0.14$. Hence, consistent and accurate

Table III

Limestone Dissolution Rate and Associated Changes in Limestone Mass and Detention Time for 13 Limestone Drains in Pennsylvania and Maryland

			Mass limestone[a,b]			Average Ca concentration[a,c]					Detention time[e]			
Limestone drain site[a]	Year built[a] ($t = 0$)	Current age, $t = 2001$ (years)	Initial, M_0 (tonne)	Current, M_{2001} (tonne)	Average flow rate,[a] Q (liters/min)	Influent (mg/liter)	Effluent (mg/liter)	Net as $CaCO_3$ (mg/liter)	Average flux of Ca as $CaCO_3$,[c] J_{CaCO_3} (tonne/year)	Limestone dissolution rate,[d] k ($year^{-1}$)	Initial, t_{d0} (h)		Current, t_{d2001} (h)	
											$\phi =$ est.	$\phi =$ 0.49	$\phi =$ est.	$\phi =$ 0.49
1. Howe Bridge 1	1991	10	455.0	314.8	96.8	157	209	130.0	6.62	0.018	16.6	28.4	13.9	23.8
2. Howe Bridge 2	1993	8	132.0	103.4	48.4	154	206	130.0	3.31	0.031	2.4	16.5	1.9	12.9
3. Elklick	1994	7	165.0	147.2	37.1	77	129	130.0	2.54	0.018	17.9	26.9	15.8	23.7
4. Jennings	1993	8	365.0	308.9	73.4	83	208	312.1	12.06	0.037	20.5	30.0	15.3	22.4
5. Morrison[c]	1990	11	65.0	51.0	8.0	115	223	270.0	1.14	0.019	26.0	49.1	21.1	39.8
6. Filson-R[c]	1994	7	590.0	541.1	44.0	69	189	299.5	6.93	0.013	29.8	81.0	27.2	73.8
7. Filson-L[c]	1994	7	635.0	618.3	31.2	69	180	277.0	4.55	0.008	33.0	123	31.2	116
8. Schnepp	1993	8	130.0	105.0	19.8	*90*	198	270.0	2.81	0.024	15.1	39.7	12.4	32.7
9. REM-R	1992	9	125.0	89.3	112.0	*210*	232	55.0	3.24	0.032	7.1	6.7	5.4	5.1
10. REM-L	1992	9	125.0	73.3	96.2	*150*	201	127.5	6.45	0.063	7.9	7.9	4.5	4.5
11. Orchard	1995	6	38.1	32.3	30.0	24	50	64.3	1.01	0.027	1.3	7.7	1.1	6.5

(*continues*)

Table III *(continued)*

Limestone drain site[a]	Year built[a] ($t = 0$)	Current age, $t = 2001$ (years)	Mass limestone[a,b]: Initial, M_0 (tonne)	Mass limestone[a,b]: Current, M_{2001} (tonne)	Average flow rate,[a] Q (liters/min)	Average Ca concentration[a,c]: Influent (mg/liter)	Average Ca concentration[a,c]: Effluent (mg/liter)	Average Ca concentration[a,c]: Net as $CaCO_3$ (mg/liter)	Average flux of Ca as $CaCO_3$,[c] J_{CaCO_3} (tonne/year)	Limestone dissolution rate,[d] k (year^{-1})	Detention time[e]: Initial, t_{d0} (h), $\phi =$ est.	Detention time[e]: Initial, t_{d0} (h), $\phi = 0.49$	Detention time[e]: Current, t_{d2001} (h), $\phi =$ est.	Detention time[e]: Current, t_{d2001} (h), $\phi = 0.49$
12. Buck Mountain	1997	4	320.0	233.5	504.0	3	38	87.5	23.19	0.079	1.8	3.8	1.3	2.8
13. Hegins	2000	1	800.0	789.7	1000.0	10	17	17.8	9.34	0.013	2.9	4.8	2.8	4.8
Median		8	165.0		48.4			130.0	4.55	0.024	15.1	26.9	12.4	22.4

[a] Influent concentrations for sites 5–7 based on water quality of nearby seep and for sites 8–10 and 12 based on water quality of untreated discharge prior to drain construction. Average for monitoring period not adjusted for frequence or time of sample collection. Values in italics for influent Ca estimated on basis of net alkalinity or net acidity (Table III) for computations of $CaCO_3$ mass flux and limestone dissolution rate.

[b] Current mass of limestone computed on basis of Eq. (14), $M_{2001} = M_0 \cdot \exp\{-k \cdot t\}$, where t is current age and k is limestone dissolution rate.

[c] Net concentration of Ca as $CaCO_3(\Delta Ca_{CaCO_3})$ computed as difference between concentration in effluent and influent, multiplied by 2.5. Average flux of Ca as $CaCO_3$ computed as the product of average flow rate and net Ca concentration, $J_{CaCO_3} = Q \cdot \Delta Ca_{CaCO_3}$ [Eq. (11)].

[d] Limestone dissolution rate, $k = J_{CaCO_3}/(M_0 \cdot X_{CaCO_3})$, computed on the basis of Eq. (12) as average flux of Ca as $CaCO_3$ divided by the product of initial mass of limestone and limestone purity (Table I).

[e] Detention time computed on basis of Eq. (6), $t_d = M_t/[Q \cdot \rho_S \cdot (1 - \phi)/\phi)]$, for average flow rate, using porosity estimates in Table I or assuming porosity of 0.49, and assuming constant particle density, $\rho_S = 2650\,kg/m^3$ (Cravotta and Trahan, 1999). Initial and current detention time computed for initial mass (M_0) and current mass (M_{2001}) of limestone, respectively.

estimates of porosity, bulk volume, and mass of limestone are needed to determine the necessary space required for construction and to obtain the desired detention time for effluent through a limestone drain.

B. Alkalinity Production Rate

Effluent from each of the 13 drains had higher pH, alkalinity, and Ca concentrations and lower acidity than the influent (Table II) because of limestone dissolution within the drains. The extent of change in these parameters varied among the drains due to differences in influent chemistry, detention time, and limestone dissolution rate (Table III). Although concentrations of alkalinity and Ca in effluent were not correlated with detention time among the different drains (Table II), changes in pH and other chemical variables generally were expected to be greatest near the inflow and diminished with increased detention time or distance from the inflow in each of the drains. For example, longitudinal monitoring data for Elklick, Howe Bridge 1 and 2, Morrison, and Orchard limestone drains (Fig. 4) and cubitainer test data for the Howe Bridge 1 and Morrison drains (Figs. 2 and 3A) all indicated that the rate of alkalinity production decreased with increased detention time. The cubitainer tests indicated, for the same influent, that the generation of alkalinity within the containers did not vary with limestone purity. Furthermore, the cubitainer tests showed that alkalinity approached a steady-state, maximum concentration after about 48 h that was equivalent to that for effluent from the respective limestone drains (Figs. 2 and 4). The maximum alkalinity for the Morrison cubitainer tests was approximately 350 mg/liter, whereas that for the Howe Bridge tests was approximately 210 mg/liter (Fig. 3A). Watzlaf and Hedin (1993) attributed differences in alkalinity production between the Morrison and the Howe Bridge tests to differences in influent P_{CO_2} and pH.

According to Lasaga (1981), the first-order kinetics relation for a steady-state condition as observed for the cubitainer tests can be written as

$$dC/dt = k' \cdot (C_M - C), \tag{8}$$

where C is the alkalinity (or Ca) concentration as $CaCO_3$, C_M is the steady-state or maximum concentration, and k' is the rate constant that has units of inverse time (1/time). Integration of Eq. (8) yields:

$$\ln[(C_M - C_t)/(C_M - C_0)] = -k' \cdot t_d, \tag{9}$$

where C_0 is the initial concentration and C_t is the concentration at any detention time. A logarithmic plot of $[(C_M - C)/(C_M - C_0)]$ versus detention time will yield a straight line with slope of $-k'$ for a first-order reaction

(Lasaga, 1981). Figure 3B shows these plots on the basis of generalized time-series data for the Howe Bridge 1 and Morrison cubitainer tests. The linear slopes indicate values for k' of 0.053 and 0.058 h^{-1} for Howe Bridge 1 and Morrison, respectively.

Taking the logarithm and rearranging Eq. (9), alkalinity concentration at any detention time can be computed as a function of these rate constants *and* the initial and maximum alkalinities:

$$C_t = C_M - [(C_M - C_0) \cdot \exp\{-k' \cdot t_d\}]. \tag{10}$$

The solid curves drawn through the generalized cubitainer data points in Fig. 3A were computed on the basis of Eq. (10) and the corresponding generalized data from the cubitainer tests for the Howe Bridge 1 and Morrison limestone drains.

C. Limestone Dissolution Rate

For a limestone-based treatment system, the mass of limestone dissolved over time can be determined by measuring the $CaCO_3$ mass flux (J_{CaCO_3})

$$J_{CaCO_3} = Q \cdot \Delta C_{CaCO_3}, \tag{11}$$

where Q is the average flow rate and ΔC_{CaCO_3} is the difference between effluent and influent concentrations of acidity, alkalinity, or Ca expressed as $CaCO_3$. Generally, data for Ca concentrations are preferred over alkalinity or net acidity for computation of the $CaCO_3$ flux. Alkalinity cannot be measured at pH < 4.5, and acidity data generally are less precise and less accurate than those for Ca. For most sites, only the long-term averages for flow rate and ΔC_{CaCO_3} were available (Tables II and III). Hence, the limestone dissolution rate or decay constant, k, with units of inverse time at each of the drains was estimated as the long-term average Ca ion flux rate, J_{CaCO_3}, normalized by the initial mass and weight fraction $CaCO_3$ (X_{CaCO_3}) of the limestone:

$$k = J_{CaCO_3}/(M_0 \cdot X_{CaCO_3}). \tag{12}$$

For the 13 drains, estimates of k computed using Eq. (12) ranged from 0.008 to 0.079 $years^{-1}$ (Table III). Figure 7 illustrates differences in the projected remaining mass of limestone for this range of k values and assuming linear or exponential decay of the same initial mass. Large differences among the decay trends with increasing time for different decay constants indicate the importance of this parameter.

For a decreasing quantity over time, a first-order decay equation can be written as

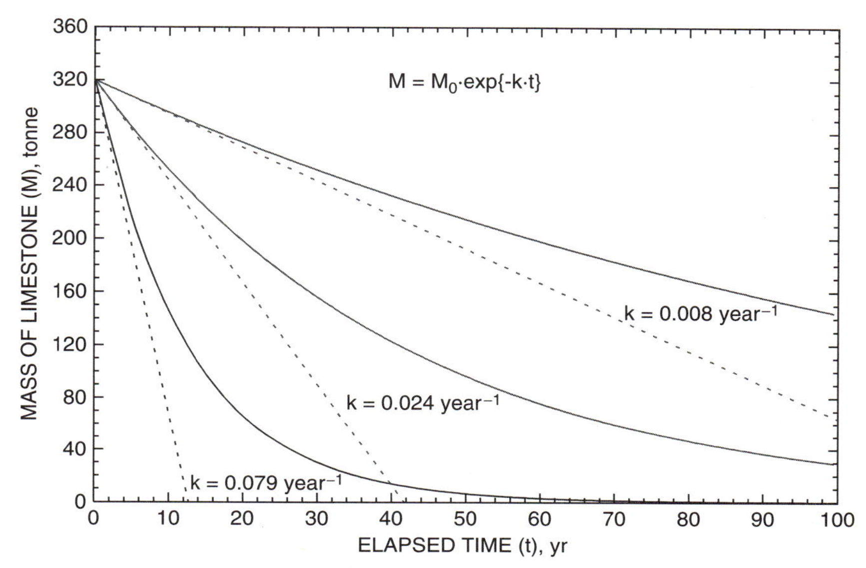

Figure 7 Change in mass of limestone with age of limestone drain considering initial mass of 320 tonne, various rate constants (k), and exponential decay (solid lines) or constant mass flux (dashed lines; x intercept is the inverse of the rate constant).

$$-dM/dt = k \cdot M, \tag{13}$$

where the quantity M typically has units of mass or concentration and k is the rate constant (Lasaga, 1981). Equation (13) applies to a variety of important geochemical processes, including radioactive decay, the oxidation of ferrous iron, and, as demonstrated in this chapter, the dissolution of limestone in an ALD or OLD. Integration of Eq. (13) yields the exponential decay expression

$$M_{\mathrm{t}} = M_0 \cdot \exp\{-k \cdot t\}, \tag{14}$$

where M_0 is the initial quantity and M_{t} is the remaining quantity after continuous dissolution over elapsed time, t.

Although the assumption of constant (linear) $CaCO_3$ flux is reasonable for a relatively large mass of limestone, a small flux of $CaCO_3$, and a short initial time period, the linear and exponential decay trends diverge as time increases (Fig. 7). Because of this divergence, values of k derived using the time-averaged $CaCO_3$ flux [Eq. (12)] could decrease with elapsed time. For example, at the Orchard OLD, the average $CaCO_3$ flux decreased from 1.6 tonne/year for the first year of operation (Cravotta and Trahan, 1999) to 1.0 tonne/year for the 5-year average (Table III). The corresponding dissolution rate constant decreased from an initial estimate of 0.042 $year^{-1}$ (Cravotta and Trahan, 1999) to 0.027 $year^{-1}$ for the 5-year average (Table III). Decreases in alkalinity concentration and corresponding decreases in the $CaCO_3$ flux and estimated limestone dissolution rates with age were also observed for the Howe Bridge 1 ALD. At this site, the average $CaCO_3$ flux for the first 3 years was 6.5 tonne/year (Hedin and Watzlaf, 1994) and that for the 10-year average was 6.1 tonne/year (Table III). The decreased $CaCO_3$ flux is expected as the remaining mass of limestone decreases through time.

The limestone dissolution rate, or exponential decay constant, can be determined by taking the logarithm of Eq. (14) and rearranging to

$$\ln (M_t/M_0) = -k \cdot t. \tag{15}$$

For a first-order process or reaction, a logarithmic plot of M_t/M_0 versus time will yield a straight line with slope $= -k$. For example, plots of $\ln(M_t/M_0)$ versus time were generated using annual averages for 10 years of monitoring at the Howe Bridge 1 drain (Fig. 3D). For each year, estimates of J_{CaCO_3} were computed using data for Ca and alkalinity concentrations [Eq. (11)] and were then divided by the reported limestone purity to estimate the incremental dissolved mass of limestone and hence the remaining mass. The resultant slopes from the plots of $\ln(M_t/M_0)$ versus time indicated k of 0.021 $year^{-1}$. This value is in good agreement with an estimate of 0.018 $year^{-1}$ derived using the long-term average Ca ion flux for Howe Bridge 1 (Table III).

Data for cubitainer experiments were also used to estimate limestone dissolution rate constants for the Howe Bridge 1 and Morrison limestone drains (Fig. 3C). An initial limestone mass of 4.0 kg and an initial solution volume of 2.3 liter were assumed. The initial solution volume was computed as the difference between the initial 3.8-liter total volume and the 1.5-liter limestone volume ($V_S = M_S/\rho_S = 4.0\,kg/2.65\,kg/liter$). Due to the periodic removal of solution, the solution volume was reduced incrementally by 60 ml at each time step when alkalinity was determined. The remaining mass of limestone at each time step was determined by the difference between initial mass and the cumulative $CaCO_3$ mass added to the solution, which was computed as the product of overall change in alkalinity concentration ($\Delta C_{CaCO_3} = C_t - C_0$) and the remaining volume of solution (V_L), corrected for limestone purity:

$$M_t = M_0 - (\Delta C_{CaCO_3} \cdot V_L/X_{CaCO_3}). \tag{16}$$

Because detention times under field conditions were less than 50 h (Table III), data for only the first 2 days of cubitainer tests were used for the logarithmic plot of M_t/M_0 versus time, in years. The resultant slopes of the regression lines indicate that values for k would be 0.016 and 0.026 $year^{-1}$ for the Howe Bridge 1 and Morrison limestone drains, respectively (Fig. 3C). Considering that k values ranged by a factor of 10 among the 13 drains evaluated, these estimates are in good agreement with values of 0.018 and 0.019 $year^{-1}$ derived on the basis of a 10-year average Ca ion flux for the respective limestone drains (Table III).

For the Orchard drain, limestone slabs immersed for 1 to 5 years near the middle of the drain dissolved at rates of 0.044 to 0.057 $year^{-1}$ (Table IV). These rates were comparable to the average k value of 0.027 $year^{-1}$ derived on the basis of long-term average Ca ion flux for the first 5 years of operation (Table III). Nevertheless, because of variations in chemistry along the length of the drain, the directly measured limestone dissolution rates varied by more than a factor of 5 between inflow and midflow points, with greatest rates near the inflow where the pH was least (Table IV). Hence, direct estimates of the limestone dissolution rate must consider the location of the limestone along the flow path through the drain and the possibility for variations in water chemistry at the location. In contrast, the dissolution rate estimated using average Ca or alkalinity flux integrates along the length of the drain and through time, making this estimate more useful as a measure of overall drain performance.

The rate of limestone dissolution within a bed is expected to vary as the chemical composition of the water changes and/or the flow rate changes, affecting detention time and chemistry along the flow path. Directly measured $CaCO_3$ dissolution rates within the Orchard OLD were greatest near the inflow, where $SI_{CALCITE}$ was minimal, and decreased with increased pH and activities of Ca^{2+} and HCO_3^- and decreased P_{CO_2} at downflow points (Table IV, Fig. 6A). These trends as a function of solution composition are consistent with rate laws established by laboratory experiments (Plummer *et al.*, 1979; Morse, 1983; Arakaki and Mucci, 1995). During the first 5 months, the limestone dissolution rate was 0.439 $year^{-1}$ at the inflow and 0.081 $year^{-1}$ at the midflow. However, lower dissolution rates of 0.044 to 0.057 $year^{-1}$ were determined for samples immersed at the midflow for periods of 1 to 5 years (Table IV). This decline in the dissolution rate, directly determined by reduction in mass over elapsed time of immersion, is consistent with the decline in the estimated value of k computed using 1- and 5-year averages for $CaCO_3$ flux.

At the Orchard OLD, the field rates of limestone dissolution were intermediate to those for the other limestone drains evaluated (Table III). However, the field rates for limestone reaction with AMD were about an order of magnitude less than reported laboratory rates of calcite dissolution at comparable pH in hydrochloric acid (HCl) solutions (e.g., Plummer *et al.*, 1979; Morse, 1983; Arakaki and Mucci, 1995). Aschenbach (1995) demonstrated that calcite dissolved more rapidly in HCl solutions than in synthetic AMD (sulfuric acid solutions) for a given pH and that calcite remained indefinitely

Table IV

Dissolution Rate for Limestone Samples Immersed in Limestone Drain at Orchard Drift, Pennsylvania

Elapsed time of immersion[a]	Longitudinal distance downflow,[b] L (m)	Limestone dissolution rate		Mass of Fe–Al–Mn encrustation (mg)	Average hydrochemical values for time interval			
		k (g/g/year)	Log $(mmol/cm^2/s)$[c]		pH	Partial pressure CO_2, log (atm)	Alkalinity (mg/liter as $CaCO_3$)	Net Ca concentration[d] (mg/liter as $CaCO_3$)
5 months	0	0.439	−6.86	800	3.4	> −0.6	0	0
	6.1	0.130	−7.39	415	5.5	−1.2	21	43
	12.2	0.081	−7.60	79	6.1	−1.4	51	78
12 months	0	NA[e]	NA	361[f]	3.4	> −0.6	0	0
	6.1	0.061	−7.74	108[f]	5.5	−1.2	21	47
	12.2	0.044	−7.89	126[f]	6.1	−1.4	54	86
60 months	12.2	0.057[g]	− 7.78[g]	NA	6.1[g]	−1.4	51	82

[a] Data for 5-month (March 23, 1995–August 23, 1995) and 12-month (March 23, 1995–March 26, 1996) elapsed time intervals adapted from Cravotta and Trahan (1999). Data for 60-month (March 23, 1995–March 29, 2000) interval this report (C. A. Cravotta, unpublished data).

[b] Longitudinal distance downflow in meters: 0 is at inflow, 6.1 m is one-fourth distance from inflow to outflow, and 12.2 m is one-half distance from inflow to outflow.

[c] Dissolution rate in units of (g/g/year) normalized by sample surface area to units of $(mmol/cm^2/s)$. Generally, for the Orchard Drift limestone drain, rates normalized by surface area can be computed by dividing k by $10^{6.501}$, which is the product of the average specific surface area of 75 immersed limestone samples ($A_{SP} = 1.004\ cm^2/g$), the millimolar mass of $CaCO_3$ ($m_{CaCO_3} = 0.1004$ g/mmol), and $10^{7.50}$ s/year.

[d] Net concentration of Ca computed as difference between averages for effluent and influent, multiplied by 2.5.

[e] NA, not available. Limestone slabs suspended from tether (noose) at inflow (0-m distance) for a 12-month period dissolved extensively and fell from tether.

[f] Encrustation quantities reported for the 12-month interval should be interpreted with caution because of substantial losses of loosely bound coatings from samples during sample retrieval. Composition of crust that could be salvaged was reported by Cravotta and Trahan (1999).

[g] Due to larger-diameter pipes installed at the inflow to each drain in 1998 (to reduce clogging), flow rates through OLDs actually were increased and, hence, pH near midflow actually was lower during the period 1998 to 2000 than in the preceding years. However, the grand average for pH at the midflow does notreflect this change; the average pH is weighted by more frequent sampling during the first year of operation than during later periods.

undersaturated in the synthetic AMD at near-neutral pH. A similar trend of calcite undersaturation in the Orchard OLD is indicated by the asymptotic, marginal increase in $SI_{CALCITE}$ with increased distance or detention time (Fig. 6A). Slow field dissolution rates and calcite undersaturation could be caused by the precipitation or adsorption of inhibiting agents, which effectively reduce the exposed surface area. The occurrence and possible roles of microbial films and precipitated minerals as inhibiting agents for limestone dissolution have been documented for ALDs and OLDs (Robbins *et al.*, 1996, 1997, 1999). Furthermore, Mg^{2+} and SO_4^{2-}, which are common ions in mine drainage, have been reported to be significant inhibitors of $CaCO_3$ dissolution (Morse, 1983).

D. Limestone Drain Longevity and Size

The longevity of a limestone drain is defined loosely in this chapter as the elapsed time when the mass of remaining limestone will no longer produce sufficient alkalinity to meet treatment goals. Given the empirically derived constants, k and k', the initial alkalinity (C_0), and the maximum alkalinity (C_M), which can be determined with cubitainer tests, the decline in limestone mass through time (age) and any associated decline in alkalinity concentration with decreased mass (detention time) of a limestone drain can be computed. Figure 8 shows the results of computations of mass decay and associated alkalinity for the Howe Bridge 1 and Morrison limestone drains using k and k' derived from cubitainer data for the Howe Bridge and Morrison mine discharges (Figs. 3C and 3B). By overlaying observed data for the actual drains and comparing different results obtained using various values for rate constants, k and k', and porosity, ϕ, the utility and sensitivity of the aforementioned equations to simulate the long-term performance of the limestone drains at these sites can be evaluated.

The projected change in mass of limestone with age of the Howe Bridge 1 and Morrison limestone drains is shown as Fig. 8A. This projection assumes continuous, exponential decay in accordance with Eq. (14) and utilizes the initial mass when constructed (Table I) and the mass-flux decay constant, k, derived from generalized cubitainer data (Fig. 3C). The decay trends are remarkably similar if decay constants derived on the basis of long-term averages for field data are used (dashed curves) because these estimates for k were equivalent to those derived from cubitainer test data. Figure 8B shows the corresponding change in detention time with age as the mass of limestone declines in accordance with Eq. (6). The computed detention time assumes constant flow rate and porosity; detention time decreases if porosity is decreased from 0.49 (solid curves) to lower values (dashed curves).

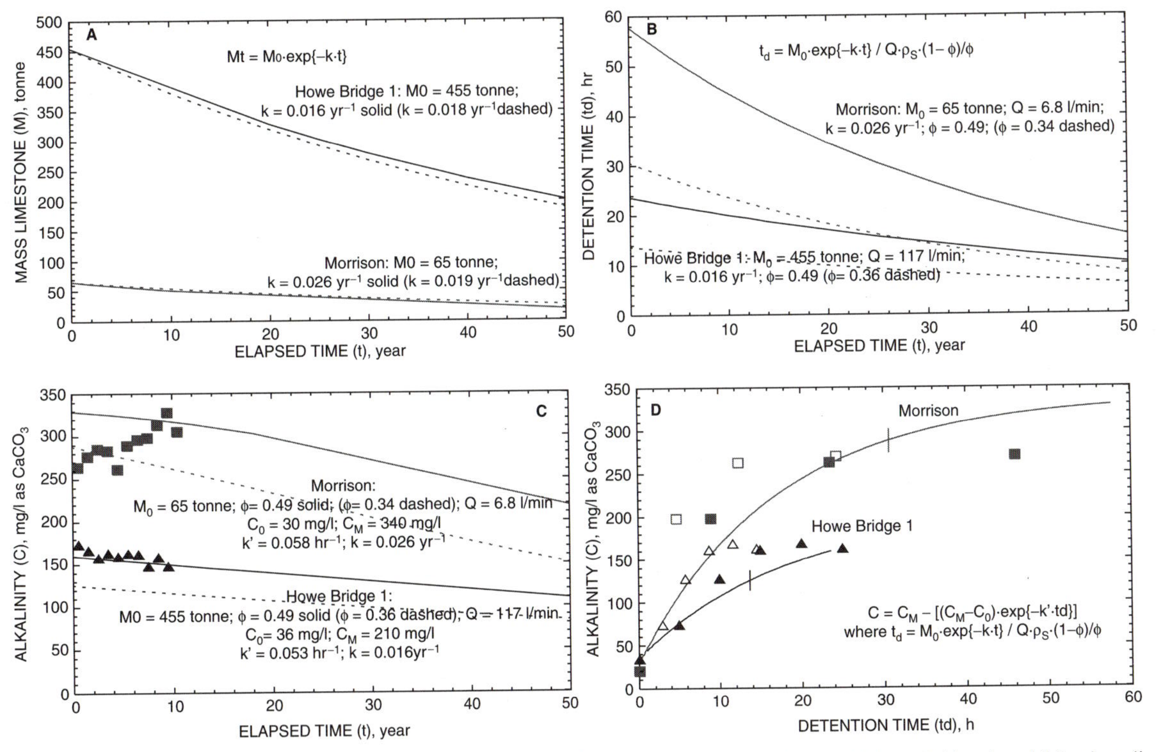

Figure 8 Simulated change in limestone mass, detention time, and alkalinity concentration with age of Howe Bridge 1 and Morrison limestone drains. (A) Mass versus age considering exponential decay and rate constants derived from cubitainer tests (solid) and field flux estimates (dashed). (B) Detention time versus age assuming constant particle density ($\rho_S = 2650\,\mathrm{kg/m^3}$), flow rate ($Q$), porosity = 0.49 (solid), and computed porosity (dashed). (C and D) Simulated (lines) and sampled (points) alkalinity versus age or detention time. Data in C are annual averages for effluent and in D are typical values along longitudinal profile. Solid lines in C and D for porosity = 0.49. Dashed lines in C and open symbols in D are for computed porosity at the respective site.

Figure 8C shows long-term trends for computed and observed alkalinity of effluent from the Howe Bridge 1 and Morrison limestone drains. The simulated alkalinity was computed in accordance with Eq. (10) for progressively declining detention times. The alkalinity simulations (solid curves) used generalized cubitainer data for C_0, C_M, and k' for the respective sites (Figs. 3A, 3B, and 3C). The observed annual average alkalinity of effluent from each of the drains is overlain on the curves. To provide the same baseline influent alkalinity to compare simulated and observed data, the observed values were normalized as the difference between the annual averages for effluent and influent added to the grand average influent concentration. A reasonable match between simulated and observed values for alkalinity is obtained for a porosity of 0.49 at the Howe Bridge site. The lower estimates of porosity based on reported dimensions and mass (Table I) increase the deviation between simulated and observed data trends (dashed lines). However, the simulated and observed trends for the Morrison limestone drain are not closely matched. The Howe Bridge functions as a piston or plug-flow system, with untreated water piped into the limestone drain and detention time of treated water increasing along the length of the drain. In contrast, the Morrison drain intercepts several seeps along its length and hence the effluent is a mixture of water having various detention times. Furthermore, the influent sample for the Morrison drain is collected from an adjacent seep. The sampled seep may not be representative of all the various seeps into the drain.

Figure 8D shows simulated and observed trends for alkalinity with detention time. For the simulations, the greatest detention time for each of the limestone drains is associated with the intial condition (age = 0); detention time and corresponding alkalinity values decrease with increased age. To extend the simulated curves to small detention times at the outflow, the remaining mass and corresponding values for detention time and alkalinity were computed over an elapsed time of 200 years. The resultant estimates for effluent alkalinity after 200 years of continous dissolution correspond with current conditions near the inflow to the drains. Field data for longitudinal samples from monitoring wells within the drains are plotted as individual points in Fig. 8D for comparison with the simulated curves. Assuming a porosity of 0.49, the simulated trend on the basis of the cubitainer tests for the Howe Bridge 1 site matches the observed data for this site. The simulated and observed trends for the Morrison site are not closely matched.

Hedin and Watzlaf (1994) evaluated construction characteristics, detention times, and chemistry of influent and effluent of more than 20 limestone drains to determine the optimum size for maximum alkalinity production. They derived a limestone drain-sizing equation that included a term for longevity and a term for detention time:

$$M = Q \cdot [(t_L \cdot C_M/X_{CaCO_3}) + (t_d \cdot \rho_B \cdot /\phi)], \quad (17)$$

where t_L is the desired longevity of treatment and $\rho_B = \rho_S \cdot (1 - \phi)$ per Eq. (4). With Eq. (17) Hedin and Watzlaf (1994) and Hedin *et al.* (1994a,b) wanted to ensure that the limestone mass computed was sufficient to produce a constant, maximum alkalinity until the specified longevity had elapsed. They solved Eq. (17) for longevity of 20 years and a detention time of 15 h. However, Eq. (17) has several limitations. The first term assumes linear decay (constant mass flux), which is inconsistent with expected exponential decay (Fig. 7), and the second term assumes alkalinity concentration would be constant (maximum) over the effective lifetime of the drain (longevity), after which the remaining limestone mass would be less than that required to produce the specified detention time and alkalinity. The assumption of constant, maximum alkalinity for long detention times (>48 h) is supported by data obtained using cubitainer tests and, to some extent, field data (e.g., Figs. 2 and 4). However, at shorter detention times, alkalinity can vary as a function of influent chemistry, detention time, and/or mass of limestone remaining (Figs. 2, 4, and 8). For large flows that have relatively low net acidity, but still require treatment, the aforementioned approach would indicate excessively large quantities of limestone and large space required for installation.

Alternatively, the size of limestone drains could be estimated using the previously described equations for the exponential decay of limestone and the corresponding alkalinity as a function of detention time (Figs. 3, 7, and 8). The application of these equations requires knowledge of the same variables as Eq. (17) plus the rate constants, k and k', and the initial concentration of alkalinity or Ca of the influent. For example, one may define longevity as the time when the remaining mass of limestone equals that required to achieve a minimum detention time in a drain, such as 15 h that is typically recommended (Hedin and Watzlaf, 1994; Hedin *et al.*, 1994a,b; Watzlaf *et al.*, 2000a). For this computation, Eq. (14) can be rearranged to estimate the elapsed time (t), or age, when the future mass of limestone (M_t) will provide a minimum detention time:

$$t = \ln\,(M_0/M_t)/k \quad (18)$$

and substituting $M_t = Q \cdot [t_d \cdot \rho_S \cdot (1 - \phi)/\phi]$ per Eq. (7)

$$t = \ln\,[M_0/(Q \cdot [t_d \cdot \rho_S \cdot (1 - \phi)/\phi])]/k. \quad (19)$$

Equation (19) can be rearranged and solved for the initial mass of limestone that will be necessary to achieve a desired future detention time:

$$M_0 = \exp\{k \cdot t\} \cdot [Q \cdot t_d \cdot \rho_S \cdot (1 - \phi)/\phi] \quad (20)$$

For example, Fig. 9 shows the computed initial mass of limestone to achieve a range of future desired detention times and, hence the alkalinity of effluent per Eq. (10), after an elapsed time of 20 years. Results are shown for constant particle density, porosity, and flow rate and a range of dissolution rates as determined for the 13 limestone drains evaluated. To compute alkalinity as a function of detention time, $k' = 0.053\,h^{-1}$ was assumed on the basis of cubitainer tests for Howe Bridge (Figs. 3B and 6D). Generally, for the application of Eqs. (20) and (10), site-specific data should be used to determine the rate constants, k and k', and the influent and maximum alkalinity or Ca concentrations, C_0 and C_M. Use of larger values for k and/or smaller values for k' will indicate a larger mass requirement and vice versa.

Several factors could account for the observed decreases in alkalinity production rates and computed limestone dissolution rates with age of the limestone drain(s). Review of long-term data indicated no significant trend in the pH or alkalinity of the untreated AMD at the Orchard or Howe Bridge 1 sites, hence the untreated AMD did not become less aggressive with respect to calcite. The decreased rate of dissolution must have been caused by decreased contact between the water and the rock and/or changes in the limestone surface properties. As the limestone mass declines, the exposed surface area will also decline due to (1) rapid dissolution of erodible edges and exposed defects, (2) a smaller total surface area associated with smaller particles or shrunken cores of unweathered limestone, and (3) the accumulation of various minerals and biofilms on the surface. For example, limestone and calcite samples, which were immersed at the inflow and at downflow points within the Orchard OLD, showed effects of both precipitation and dissolution, particularly near the inflow (Cravotta and Trahan, 1999; Robbins *et al.*, 1999). Microscopic examination of calcite mounts on glass slides showed newly formed Fe and Mn oxide and dissolution etch pits (Robbins *et al.*, 1999). The appearance of these secondary minerals in the etch pits implies that sites of dissolution locally could be rendered inactive. Furthermore, the accumulation of large quantities of hydrous oxides or other secondary minerals between limestone fragments, particularly at the bottom of the drain, could reduce the flow of water and contact between the water and limestone through that zone of the limestone drain. This scenario could explain the greater decline in the limestone dissolution rate at the Orchard OLD compared to the Howe Bridge ALD.

E. Attenuation of Acidity and Dissolved Metals

The goal of AMD treatment is to attenuate acidity and metals transport, generally by promoting the formation of solids that can be safely retained on site or disposed. However, metals that accumulate and cannot be flushed

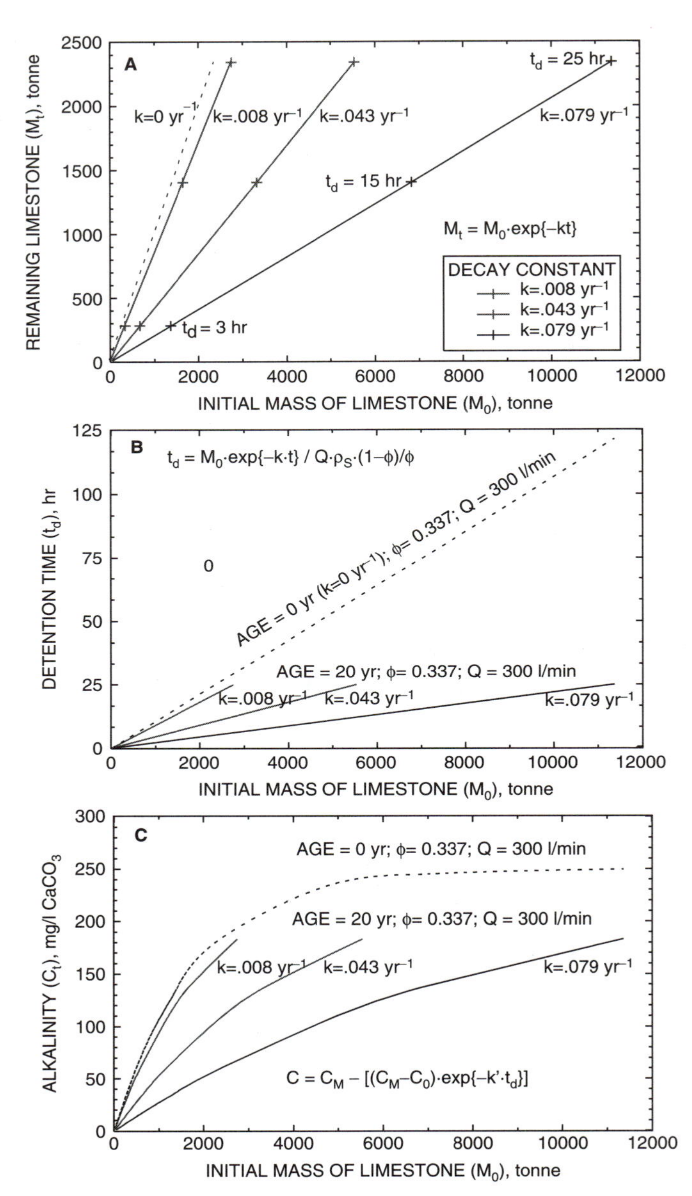

Figure 9 Predictions at 20 years elapsed time of continuous dissolution of an initial mass of limestone, assuming exponential decay at various rates: (A) mass remaining (M_t); (B) detention time (t_d); and C, alkalinity (C_t). Detention time estimated for porosity of 0.337 and flow rate of 300 liter/min. Alkalinity estimated for $C_0 = 0$, $C_M = 250$ mg/liter as $CaCO_3$, and $k' = 0.053\ h^{-1}$.

from a limestone drain or other porous reactive barrier can cause its failure due to the reduction of porosity and/or reduction of limestone surface area. Although the effect on alkalinity production due to reductions in porosity and detention time generally could be evaluated using the aforementioned equations, specific effects on limestone dissolution and alkalinity production rates require additional study.

All 13 of the limestone drain treatments reduced acidity (Table II). The reduction of acidity resulted from the increased pH and alkalinity and associated decreased concentrations of dissolved, readily hydrolyzable metals, including Al^{3+}, Fe^{3+}, and possibly Mn^{2+}, Fe^{2+}, and other divalent cations (e.g., Hedin *et al.*, 1994a; Rose and Cravotta, 1998).

All 13 drains considered for this chapter removed a substantial fraction of Al^{3+} in their influent (Table II). Decreased Al concentration with increased pH is consistent with its solubility control by Al-hydroxide and/or hydroxysulfate minerals (e.g., Nordstrom, 1982; Nordstrom and Ball, 1986). For example, as pH increased within the Orchard OLD, concentrations of Al declined (Fig. 5), whereas equilibrium with poorly crystalline $Al(OH)_3$ was maintained (Fig. 6D). Although the concentration and flux of Al at the Orchard drain were relatively small, where the flux of Al is large, clogging and failure of the drain due to the accumulation of precipitated Al can be rapid, occurring within months (Robbins *et al.*, 1999). The Jennings limestone drain and several others that were constructed to treat low pH, high Al (>5 mg/liter) influent have failed prematurely, primarily due to the precipitation of Al minerals within pores of the drains (Watzlaf *et al.*, 1994, 2000a; Robbins *et al.*, 1996). Although little information is available about the mineralogy of precipitated Al minerals within limestone drains, the formation of Al hydroxysulfate compounds is suspected to be the primary cause of fouling. Generally, Al hydroxysulfate minerals can form when AMD containing dissolved Al^{3+} and SO_4^{2-} attains pH > 4.3 by reaction with limestone or by mixing with near-neutral waters (Nordstrom, 1982; Nordstrom and Ball, 1986; Sterner *et al.*, 1998). For example, Robbins *et al.* (1996) determined an ALD in West Virginia was clogged by poorly crystalline aluminite [$Al_2(SO_4)(OH)_4 \cdot 7H_2O$] and silica. Robbins *et al.* (1999) showed that the Al hydroxysulfate can precipitate within calcite etch pits, interfering with dissolution, and suggested that these compounds can foul limestone drains because of their strong adhesion and high density compared to $Al(OH)_3$. Perforated pipes may not be adequate to flush Al hydroxysulfate minerals from drain pores because of their adhesive nature.

Although dissolved Fe in the influent to all 13 drains was predominantly Fe^{2+}, the influent at several sites also contained Fe^{3+} and/or dissolved O_2. Where anoxic conditions were maintained, Fe concentrations were not affected by the limestone treatments (Table II).[2] On the basis of equilibrium calculations, Hedin and Watzlaf (1994) suggested that siderite ($FeCO_3$) may have formed in some ALDs; however, the presence of siderite and the significance of Fe^{2+} removal by this mechanism have not been evaluated. Nevertheless, because stoichiometric oxidation of 1 mol of Fe^{2+} requires only 0.25 mol of O_2 (1 mg/liter O_2 for 7 mg/liter Fe^{2+}), losses of dissolved Fe^{3+} *and* Fe^{2+} within OLDs and some ALDs could result from the precipitation of Fe(III) hydroxide or hydroxysulfate minerals, such as ferrihydrite or schwertmannite (e.g., Bigham *et al.*, 1996; Cravotta and Trahan, 1999). The precipitation of Fe(III) compounds implies that O_2 and/or Fe^{3+} enters the drains at the inflow or elsewhere. The Orchard OLD contained 1–3 mg/liter dissolved O_2; despite undersaturation with siderite, Cravotta and Trahan (1999) reported complete removal of Fe^{2+} and Fe^{3+} (Figs. 5B, 5D, and 6C). The attenuation of Fe^{2+} was more efficient than expected on the basis of homogeneous oxidation and precipitation of $Fe(OH)_3$ (Williamson *et al.*, 1992; Kirby and Elder Brady, 1999). Hence, Cravotta and Trahan (1999) attributed the removal of Fe^{2+} to sorption and catalysis of oxidation by $Fe(OH)_3$ (e.g., Tamura *et al.*, 1976; Stumm and Sulzberger, 1992). Microorganisms that could catalyze the oxidation of Fe^{2+} and formation of $Fe(OH)_3$ are ubiquitous in AMD (e.g., Ferris *et al.*, 1989; Ehrlich, 1990; Robbins *et al.*, 1999).

Generally, on the basis of the average composition of influent and effluent, significant removal of Mn and other trace metals was not apparent for the drains evaluated (Table II). Nevertheless, the averages may not represent actual performace through time. For example, Cravotta and Trahan (1999) reported significant removal of Mn, Zn, Ni, and Co within the Orchard OLD after the first 6 months of operation (Figs. 10B and 10C). The hydrous-oxide precipitate that accumulated within the limestone bed had concentrations of Mn, Cu, Ni, and Zn that increased relative to Fe with distance downflow, hence with increased pH. Ratios of Mn/Fe and Mn/Al in the coatings increased by 2 orders of magnitude during the second 6-month monitoring interval relative to the first 6 months (Cravotta and Trahan, 1999). Watzlaf (1997) and Brandt and Ziemkiewicz (1997) also reported substantial Mn removal from AMD by limestone that had been coated by hydrous oxides after a time lag.

The removal of dissolved Mn, Zn, Ni, and Co from solution and a corresponding enrichment of the trace metals relative to Fe in particles and

[2]Apparent decreases in acidity, metals, and SO_4 concentrations in effluent relative to influent at Morrison, Schnepp, and REM-R sites (Table II) should be interpreted with caution because data for nearby seepage were used to represent the "influent" to the Morrison drain and historical data for the untreated AMD were used to represent the "influent" to the Schnepp and REM-R drains. Adjacent seeps rarely have identical chemistry, and water quality of untreated AMD generally improves with time (e.g., Wood, 1996).

coatings on limestone within the Orchard OLD resulted from sorption and coprecipitation reactions with the hydrous oxides of Fe(III), Mn(II–IV), and, to a lesser extent, Al (Cravotta and Trahan, 1999). Sorption of the trace metals generally can be enhanced by the incorporation of sulfate with hydrous Fe(III) oxides, where the sulfate can be part of the crystal lattice, as with schwertmannite (Bigham *et al.*, 1996; Webster *et al.*, 1998), or can be adsorbed, as with goethite or ferrihydrite (Ali and Dzombak, 1996; Rose and Ghazi, 1997). Uncoated limestone (calcite) generally is not an effective sorbent of the trace metals at pH < 7 (Zachara *et al.*, 1991). Similar trends have been reported for dissolved Mn^{2+} and trace metals to adsorb to hydrous Fe, Al, and Mn oxides, except that the Mn oxides are generally more effective sorbents at lower pH than Fe or Al oxides (Loganathan and Burau, 1973; McKenzie, 1980). Decreases in concentrations of both Al^{3+} and Zn^{2+} through many limestone drains (Table II) imply that some Zn^{2+} could be removed by sorption to $Al(OH)_3$ (e.g., Coston *et al.*, 1995).

The downgradient trends in pH and solute concentrations at the Orchard OLD became progressively more complex through time, particularly near the inflow (Fig. 5), and indicated deviation from piston-flow transport implied for the estimation of detention time as a function of distance downflow within the drain. The complex trends with respect to distance probably resulted from the localized dissolution of limestone and accumulation of hydrous oxides near the inflow that changed the porosity distribution and promoted the development of preferential flow paths. Thus, in addition to the predominant downflow transport and dispersion of hydrous oxide particles away from the inflow, turbulence could have promoted eddying and the recirculation of fluid and suspended particles locally toward the inflow. Similarly, complex trends and explanations for bromide tracer transport through the Howe Bridge 1 and Morrison ALDs were reported by Watzlaf *et al.* (2000a).

Although amorphous to crystalline Fe, Mn, and Al oxides had formed in the Orchard OLD, none of these phases was strongly adhesive. Nevertheless, the Orchard OLD was not equipped with a perforated pipe or other flushing system to prevent the accumulation of solids. However, some solids could be removed in slugs by backflushing, whereby the outflow was closed, causing the hydraulic head to increase inside the drains, and then opened. Hydrous Fe oxides were visible as loosely bound, rust-colored coatings on limestone samples near the inflow and as a gelatinous rusty floc in water samples. After 2 years of operation, numerous black flakes were also visible with the rusty floc. The mixed solids were enriched in Fe (Fe $>$ Al $>$ S $=$ Si $>$ Mn) and contained predominantly schwertmannite and goethite. The black flakes were enriched in Mn (Mn $>$ Fe $=$ Ca $>$ Al) and contained todorokite $[(Ca_{0.393}Mg_{0.473}Mn^{II}_{1.134})Mn^{III}{}_5O_{12} \cdot 2H_2O]$. Although equilibrium computations indicated undersaturation with pure solid phases of Mn(II–IV) and trace metals, formation of Mn oxides within the OLD may have been favored by near-neutral pH adjacent to limestone surfaces (Cravotta and Trahan, 1999; Robbins *et al.*, 1999). At pH ≥ 6, negatively charged surfaces of the $Fe(OH)_3$ and Mn oxides tend to attract cations, including Fe^{2+} and Mn^{2+}, which can be oxidized rapidly in an aerobic system (Tamura *et al.*, 1976; Hem, 1977, 1978). A variety of microorganisms could have promoted the oxidation of Fe and Mn in the OLD (Ehrlich, 1990; Robbins *et al.*, 1992, 1999). Despite significant concentrations of Al in the solids, the identities of crystalline Al phases could not be determined. Amorphous to poorly crystalline phases of Al, Fe, and Mn likely were present (e.g., Figs. 6C and 6D).

F. Design Considerations for Limestone Drains

The porosity and/or exposed limestone surface area and the flow rate and detention time for water in limestone drains are critical factors affecting their performance because of kinetic controls on dissolution, precipitation, and sorption reactions that control pH and dissolved ion concentrations. For a given flow rate and mass of limestone, detention time will increase with increased porosity; higher porosity can be achieved for uniform, large particles than for mixed size particles. Other factors being the same, alkalinity will increase with increased detention time. Hence, the limestone drain-sizing equation of Hedin and Watzlaf (1994) and Hedin *et al.* (1994a,b) [Eq. (17)] included porosity and detention time as specific variables; however, their equation assumes constant $CaCO_3$ flux and alkalinity over the specified lifetime of the drain. The exponential decay equations introduced in this chapter [Eqs. (10), (14), and (20)] incorporate the same variables as Eq. (17), but also account for corresponding decreases in limestone mass and alkalinity production rates over time on the basis of empirically derived $CaCO_3$ flux data and rate constants. As applied herein, Eqs. (10), (14), and (20) may be useful to extrapolate performance and/or determine the appropriate initial mass required for limestone beds. Their solution was simplified assuming constant rates of limestone dissolution and constant values for flow rate, porosity, and stone density.

The example computations did not consider variations in dissolution rate with water quality along the flow path nor other important chemical processes, such as the precipitation of secondary minerals. The precipitation and dissolution processes can affect the porosity distribution and/or the exposed surface area of the limestone bed. Generally, the formation of secondary minerals and the corresponding reduction of limestone surface area are likely to be associated with influent that does not meet criteria for an ALD (O_2, Fe^{3+}, or $Al^{3+} < 1$ mg/liter). Hence, the exponential equations [Eqs.

(10), (14), and (20)] using short-term test data would be applicable to estimate the mass of an ALD. Before their application for design of an OLD, additional tests may be appropriate. To evaluate mineral coatings on rate constants, parallel tests could be conducted with limestone that is uncoated and identical fragments coated with hydrous Fe, Mn, and/or Al oxides. Potential for clogging and/or flushing could also be evaluated considering measurements of porosity, permeability, and solids transport through a packed bed.

If possible, reducing detention time and maintaining calcite undersaturation in a limestone bed should be considered by splitting inflow to several parallel cells or along the length of a cell instead of directing flow through a single, elongate drain with inflow at one end and outflow at the other. For example, the Orchard OLD split inflow into three parallel troughs, and the Buck Mountain ALD intercepted multiple seeps along its length. The computed $CaCO_3$ fluxes and limestone dissolution rates for these drains were the highest among their respective counterpart OLDs and ALDs considered for this chapter (Table III). This multiple inflow or multiple cell arrangement has advantages and disadvantages. Splitting flow and reducing detention time takes advantage of the kinetics of limestone dissolution, where the rate of dissolution is expected to *decrease* with increased pH and alkalinity and where doubling detention time *will not* double alkalinity. Thus, treatment through parallel cells or along the length of a drain should promote greater dissolution and alkalinity production than a single cell or a series of cells containing the same mass of limestone. Furthermore, for two or more parallel cells, treatment can be conducted continuously through part of the system while flushing or other maintenance is conducted on the other part. The primary disadvantage would be that construction would be more complex and would involve greater area to install several trenches and plumbing for multiple cells than for a single cell. For example, to intercept seepage along its length, a continuous liner at the Buck Mountain. ALD had to be cut into pieces that were installed overlapping like shingles, with the top shingle at the upflow end. Although this arrangement worked to direct influent into and down the drain, it also complicated the monitoring and interpretation of water-quality variations with increased distance or detention time along the drain.

Hedin *et al.* (1994a) advised against the use of horizontally oriented limestone drains for the treatment of AMD containing >1 mg/liter of O_2, Fe^{3+}, or Al^{3+} due to the potential for armoring and clogging of the drains. The presence of any of these constituents can potentially reduce the effective longevity of a limestone drain. However, results of this chapter, combined with results of Watzlaf (1997), Sterner *et al.* (1998), and Cravotta and Trahan (1999), suggest that limestone treatment systems may effectively increase pH and remove dissolved metals, including Mn^{2+}, from AMD containing moderate concentrations of Fe^{3+} or Al^{3+} (1–5 mg/liter). Caution should be exercised for influent samples that have concentrations near this upper limit because the actual influent composition likely will vary temporally due to droughts, wet periods, and various other environmental factors. Designs should consider these extremes.

Horizontal or downflow systems constructed of coarse limestone could provide sufficient detention times to achieve neutralization of AMD and could be flushed periodically to remove accumulated solids. To enhance trace-metal attenuation by hydrous oxide particles within the drains, the cross section of the drain could be enlarged near the outflow to decrease flow velocities and increase detention time where pH is expected to be highest. To ensure against clogging from an excessive accumulation of hydrous oxides, perforated piping could be installed as subdrains for periodic flushing of excess sludge. Some of these conceptual designs have been constructed but have not been evaluated. For example, at the Hegins drain, valves were installed on perforated laterals that connected to a single longitudinal pipe that was not perforated. This 1-year old system can be flushed, but its long-term performance is uncertain. Additional studies are needed to evaluate such designs and to determine optimum criteria for utilization and construction of limestone drains for AMD remediation. In general, additional information is needed to understand potential for, and effects of, the dissolution of limestone and the formation, accumulation, and transport of secondary metal compounds in AMD treatment systems. Because of the wide range of water chemistry and hydrologic conditions at coal and metal mines (Rose and Cravotta, 1998; Nordstrom and Alpers, 1999), simple to complex remedial alternatives may be appropriate depending on site characteristics.

IV. SUMMARY AND CONCLUSIONS

Although numerous case studies have been reported, published criteria for the construction of limestone drains generally are imprecise and inadequate due to (1) the wide ranges in flow rates and compositions of mine drainage and (2) nonlinear and variable dissolution of limestone and production of alkalinity as functions of water chemistry, detention time, and limestone characteristics. Generally, chemical processes within a limestone drain can be characterized as functions of distance and time as water flows downgradient through the limestone bed. Immediately near the inflow, the pH of the treated water begins to increase as limestone dissolves, ultimately approaching neutrality and calcite saturation, provided that detention time within the drain is sufficient. If treated under anoxic conditions, dissolved

Fe, Mn, and various other metal contaminants such as Co, Ni, and, to a lesser extent, Zn in the AMD tend to be transported conservatively through the drain as pH and alkalinity increase; however, Al tends to be precipitated. Alternatively, if intercepted and treated under oxic conditions, most metals in the AMD can be attenuated by oxidation, precipitation, and/or adsorption processes within the drain. Accumulated metals must be flushed from the drain to prevent clogging, armoring, and failure. Nevertheless, some precipitated materials can be adhesive or encrusting, and hence practically impossible to flush.

The longevity of limestone drains and alkalinity concentration as a function of time can be evaluated as first-order decay, where the remaining mass of limestone at any time is determined as a function of the decay constant, k, and initial mass of limestone [Eqs. (14) and (20)], and the concentration of alkalinity for any detention time is determined as a function of the influent alkalinity, the maximum or steady-state alkalinity, and the rate constant, k' [Eq. (10)]. The application of these equations requires accurate information for the rate constants, porosity, and other variables. Generally, particle density can be assumed constant; however, porosity and bulk density can vary depending on the shape, sorting, and compaction of the particles. For the 13 limestone drains considered in this chapter, computed porosity ranged from 0.12 to 0.50. Computed values for the rate constant, k, ranged from 0.008 to 0.079 year^{-1} on the basis of long-term Ca flux rates. Comparable estimates for k were obtained on the basis of mass lost from immersed limestone samples and short-term cubitainer tests.

Cubitainer testing of the reaction between untreated AMD and limestone intended or available for construction of a limestone drain can provide estimates for maximum alkalinity and the rate constants, k' and k. These data for alkalinity production through time enable determination of the alkalinity concentration that will be produced in treating a specific mine water, the limestone consumption rates, the quantity of limestone needed for a desired design life, and whether the ALD or OLD will produce net alkaline effluent. Because the composition of AMD is a critical factor affecting the alkalinity produced by limestone of a given composition and particle size, cubitainer tests could be conducted repeatedly with different limestone(s) or through time to determine the range of expected conditions within a limestone drain. For an evaluation of mineral coatings on rate constants, tests could be conducted with uncoated limestone and that coated with hydrous Fe, Mn, and/or Al oxides. Potential for clogging and/or flushing should also be evaluated considering measurements of porosity, permeability, and solids transport through a packed bed.

Dimensions and mass of limestone need to be measured accurately to estimate porosity and, ultimately, field detention time. Accurate measurement of flow rate and chemistry of influent and effluent are needed to monitor and evaluate the performance of limestone drains. Additional data for intermediate sampling locations can be useful to evaluate spatial and temporal variations in water chemistry and limestone dissolution. Documentation of hydraulic head at sampling locations could be useful to characterize changes in porosity and permeability distributions within the drain. Variations in porosity within the drain due to the decay of limestone or the precipitation of minerals could be evaluated to refine estimates of detention time and longevity using the equations given previously.

In the Orchard OLD, the rate of limestone dissolution was minimally affected by the accumulation of hydrous Fe, Mn, and/or Al oxides as loosely bound coatings. Despite the ~ 1-mm-thick accumulation of hydrous oxides on limestone surfaces, pH increased most rapidly near the inflow as a result of aggressive dissolution of limestone by the influent AMD. The rate of limestone dissolution decreased with increased pH and concentrations of Ca^{2+} and HCO_3^- and decreased P_{CO_2}. Under the "closed system" conditions in which hydrolysis products including H^+ and CO_2 could not escape the drains and were retained as reactants, limestone surfaces dissolved rapidly, and armoring was avoided, despite oxygenated conditions. However, the computed dissolution rates for the field experiment were about an order of magnitude less than reported laboratory rates at comparable pH. The reason for slower rates of $CaCO_3$ dissolution by AMD (field) than by HCl (laboratory) was not determined but could be due to inhibition of dissolution by Mg^{2+} and/or $SO_4{}^{2-}$ and potential for various minerals and biofilms to grow on limestone surfaces. Additional studies are needed to determine the hydrochemical conditions and role of microorganisms in mineral precipitation reactions that promote or inhibit limestone dissolution and ultimately can cause failure of limestone treatment systems.

Different conceptual designs for ALDs and OLDs should be considered for promoting the transport of hydrous oxides and optimizing the long-term neutralization of AMD and removal of dissolved metals. For example, splitting inflow to several parallel treatment cells or distributing the influent along the length of a cell should be considered instead of a single, elongate drain with inflow at one end and outflow at the other. This multiple inflow or multiple cell arrangement takes advantage of the kinetics of limestone dissolution, and treatment can be conducted continuously through part of the system while flushing or other maintenance is conducted on another part. Before criteria can be refined for the construction of limestone drains to neutralize AMD, however, various innovative conceptual designs need to be tested, and available data for existing systems need to be evaluated. Long-term performance of OLDs has not been evaluated. Accurate information on variations in the water chemistry, detention time, rates of limestone dissolution, and effects of hydrolysis products on limestone dissolution, sorption of trace metals, and hydraulic properties is needed to optimize designs for OLDs and ALDs and to minimize costs for the effective implementation of these passive-treatment systems.

ACKNOWLEDGMENTS

The authors gratefully acknowledge long-term, significant contributions of data and interpretations by R. S. Hedin. Other individuals, who are too numerous to name and, in some cases, unknown to the authors, assisted with field work, laboratory analyses, and data compilations used for this chapter. The lead author also belatedly thanks A. C. Lasaga for his enthusiastic efforts at teaching and helpful writing on geochemical kinetics, which formed the basis for the analytical derivations in this chapter. Helpful comments on the early drafts of the manuscript were provided by several colleagues, including R. S. Hedin, R. Runkel, and two anonymous reviewers. However, the interpretations presented are solely the authors' responsibility.

REFERENCES

Ali, M. A., and Dzombak, D. A. (1996). Interactions of copper, organic acids, and sulfate in goethite suspensions. *Geochim. Cosmochim. Acta* **60**, 5045–5053.

Arakaki, T., and Mucci, A. (1995). A continuous and mechanistic representation of calcite reaction-controlled kinetics in dilute solutions at 25°C and 1 Atm total pressure. *Aquat. Geochem.* **1**, 105–130.

Arnold, D. E. (1991). Diversion wells: A low-cost approach to treatment of acid mine drainage. *In* "Proceedings of the 12th Annual West Virginia Surface Mine Drainage Task Force Symposium," pp. 39–50. West Virginia Mining and Reclamation Association, Charleston, WV.

Aschenbach, E. T. (1995). "An Examination of the Processes and Rates of Carbonate Dissolution in an Artificial Acid Mine Drainage Solution." M.S. thesis, Pennsylvania State University, University Park, PA.

Ball, J. W., and Nordstrom, D. K. (1991). User's manual for WAT EQ4F with revised data base: U.S. Geological Survey Open-File Report 91–183.

Bigham, J. M., Schwertmann, U., Traina, S. J., Winland, R. L., and Wolf, M. (1996). Schwertmannite and the chemical modeling of iron in acid sulfate waters. *Geochim. Cosmochim. Acta* **60**, 2111–2121.

Blowes, D. W., and Ptacek, C. J. (1994). Acid-neutralization mechanisms in inactive mine tailings. *In* "Environmental Geochemistry of Sulfide Mine-Wastes" (J. L. Jambor and D. W. Blowes, eds.), Vol. 22, pp. 271–292. Mineralogical Association of Canada, Short Course Handbook.

Brady, K. B. C., Hornberger, R. J., and Fleeger, G. (1998). Influence of geology on post-mining water quality—Northern Appalachian Basin. *In* "Coal Mine Drainage Prediction and Pollution Prevention in Pennsylvania" (K. B. C. Brady, M. W. Smith, and J. Schueck, eds.), pp. 8.1–8.92. Pennsylvania Department of Environmental Protection, 5600-BK-DEP2256, Harrisburg, PA.

Brady, K. B. C., Perry, E. F., Beam, R. L., Bisko, D. C., Gardner, M. D., and Tarantino, J. M. (1994). Evaluation of acid-base accounting to predict the quality of drainage at surface coal mines in Pennsylvania, U.S.A: U.S. Bureau of Mines Special Publication **SP 06A**, pp. 138–147.

Brant, D. L., and Ziemkiewicz, P. F. (1997). Passive removal of manganese from acid mine drainage. *In* "Proceedings of the 1997 National Meeting of the American Society for Surface Mining and Reclamation," pp. 741–744. American Society for Surface Mining and Reclamation, Princeton, WV.

Brodie, G. A., Britt, C. R., Tomaszewski, T. M., and Taylor, H. N. (1991). Use of passive anoxic limestone drains to enhance performance of acid drainage treatment wetlands. *In* "Proceedings of the 1991 National Meeting of the American Society for Surface Mining and Reclamation," pp. 211–228. American Society for Surface Mining and Reclamation, Princeton, WV.

Coston, J. A., Fuller, C. C., and Davis, J. A. (1995). Pb^{2+} and Zn^{2+} adsorption by a natural aluminum- and iron-bearing surface coating on an aquifer sand. *Geochim. Cosmochim. Acta* **59**, 3535–3547.

Cram, J. C. (1996). "Diversion Well Treatment of Acid Water, Lick Creek, Tioga County, PA". M. S. thesis, Pennsylvania State University, University Park, PA.

Cravotta, C. A., III (1994). Secondary iron-sulfate minerals as sources of sulfate and acidity: The geochemical evolution of acidic ground water at a reclaimed surface coal mine in Pennsylvania. *In* "Environmental Geochemistry of Sulfide Oxidation" (C. N. Alpers and D. W. Blowes, eds.), pp. 345–364. American Chemical Society Symposium Series 550, Washington, DC.

Cravotta, C. A., III, Brady, K. B. C., Rose, A. W., and Douds, J. B. (1999). Frequency distribution of the pH of coal-mine drainage in Pennsylvania. *In* "U.S. Geological Survey Toxic Substances Hydrology Program—Proceedings of the Technical Meeting, Charleston, South Carolina, March 8–12, 1999" (D. W. Morganwalp and H. Buxton, eds.), pp. 313–324. U.S. Geological Survey Water-Resources Investigations Report 99–4018A.

Cravotta, C. A., III, Breen, K. J., and Seal, R. (2001). Arsenic is ubiquitous but not elevated in abandoned coal-mine discharges in Pennsylvania. *In* "U.S. Geological Survey Appalachian Region Integrated Science Workshop Proceedings" p. 105. U.S. Geological Survey Open-File Report **01–406**.

Cravotta, C. A., III, and Trahan, M. K. (1999). Limestone drains to increase pH and remove dissolved metals from acidic mine drainage. *Appl. Geochem.* **14**, 581–606.

Cravotta, C. A., III, and Weitzel, J. B. (2001). Detecting change in water quality from implementation of limestone treatment systems in a coal-mined watershed, Pennsylvania. *In* "Eighth National Nonpoint Source Monitoring Workshop." U.S. Environmental Protection Agency EPA/905-R-01-008.

Ehrlich, H. L. (1990). "Geomicrobiology," 2nd Ed. Dekker, New York.

Ferris, F. G., Tazaki, K., and Fyfe, W. S. (1989). Iron oxides in acid mine drainage environments and their association with bacteria. *Chem. Geol.* **74**, 321–330.

Fishman, M. J., and Friedman, L. C. (eds.) (1989). Methods for determination of inorganic substances in water and fluvial sediments: U.S. Geological Survey Techniques of Water-Resources Investigations, Book 5, Chapter A1.

Freeze, R. A., and Cherry, J. A. (1979). "Groundwater." Prentice-Hall, Englewood Cliffs, NJ.

Greenberg, A. E., Clesceri, L. S., Eaton, A. D., and Franson, M. A. H., (eds.) (1992). "Standard Methods for the Examination of Water and Wastewater (18th)." American Public Health Association, Washington, DC.

Hedin, R. S., Nairn, R. W., and Kleinmann, R. L. P. (1994a). Passive treatment of coal mine drainage: U.S. Bureau of Mines Information Circular IC 9389.

Hedin, R. S., and Watzlaf, G. R. (1994). The effects of anoxic limestone drains on mine water chemistry: U.S. Bureau of Mines Special Publication SP 06A, pp. 185–194.

Hedin, R. S., Watzlaf, G. R., and Nairn, R. W. (1994b). Passive treatment of acid mine drainage with limestone. *J. Environ. Qual.* **23**, 1338–1345.

Hem, J. D. (1977). Reactions of metal ions at surfaces of hydrous iron oxide. *Geochim. Cosmochim. Acta* **41**, 527–538.

Hem, J. D. (1978). Redox processes at surfaces of manganese oxide and their effects on aqueous metal ions. *Chem. Geol.* **21**, 199–218.

Kepler, D. A., and McCleary, E. C. (1994). Successive alkalinity producing systems (SAP's) for the treatment of acidic mine drainage: U.S. Bureau of Mines Special Publication SP 06A, pp. 195–204.

Kirby, C. S., and Elder Brady, J. A. (1999). Field determination of Fe^{2+} oxidation rates in acid mine drainage using a continuously-stirred tank reactor. *Appl. Geochem.* **14**, 509–520.

Kooner, Z. S. (1993). Comparative study of adsorption behavior of copper, lead, and zinc onto goethite in aqueous systems. *Environ. Geol.* **21**, 342–250.

Lasaga, A. C. (1981). Rate laws of chemical reactions. *In* "Kinetics of Geochemical Processes" (A. C. Lasaga and R. J. Kirkpatrick, eds.), Vol. 8, pp. 1–68. Mineralogical Society of America Reviews in Mineralogy.

Loganathan, P., and Burau, R. G. (1973). Sorption of heavy metal ions by a hydrous manganese oxide. *Geochim. Cosmochim. Acta* **37**, 1277–1293.

McKenzie, R. M. (1980). The adsorption of lead and other heavy metals on oxides of manganese and iron. *Aust. J. Soil Res.* **18**, 61–73.

Morse, J. W. (1983). The kinetics of calcium carbonate dissolution and precipitation. *In* "Carbonates: Mineralogy and Chemistry" (R. J. Reeder, ed.), Vol. 11, pp. 227–264. Mineralogical Society of America Reviews in Mineralogy.

Murad, E., Schwertmann, U., Bigham, J. M., and Carlson, L. (1994). Mineralogical characteristics of poorly crystallized precipitates formed by oxidation of Fe^{2+} in acid sulfate waters. *In* "Environmental Geochemistry of Sulfide Oxidation" (C. N. Alpers and D. W. Blowes, eds.), pp. 190–200. American Chemical Society Symposium Series 550, Washington, DC.

Nelson, P. H. (1994). Permeability-porosity relationships in sedimentary rocks. *Log Analyst* **May-June**, 38–62.

Nordstrom, D. K. (1982). The effect of sulfate on aluminum concentrations in natural waters: Some stability relations in the system Al_2O_3-SO_3-H_2O. *Geochim. Cosmochim. Acta* **46**, 681–692.

Nordstrom, D. K., and Alpers, C. N. (1999). Geochemistry of acid mine waters. *Rev. Econ. Geol.* **6A**, 133–160.

Nordstrom, D. K., and Ball, J. W. (1986). The geochemical behavior of aluminum in acidified surface waters. *Science* **232**, 54–58.

Plummer, L. N., Parkhurst, D. L., and Wigley, M. L. (1979). Critical review of the kinetics of calcite dissolution and precipitation. *In* "Chemical Modeling in Aqueous Systems: Speciation, Sorption, Solubility, and Kinetics" (E. A. Jenne, ed.), pp. 537–573. American Chemical Society Symposium Series 93, Washington, DC.

Rice, P. A., Fontugne, D. J., Latini, R. G., and Barduhn, A. J. (1970). Anisotropic flow through porous media. *In* "Flow Through Porous Media," pp. 48–56. American Chemical Society, Washington, DC.

Robbins, E. I., Cravotta, C. A., III, Savela, C. E., and Nord, G. L., Jr. (1999). Hydrobiogeochemical interactions in "anoxic" limestone drains for neutralization of acidic mine drainage. *Fuel* **78**, 259–270.

Robbins, E. I., Cravotta, C. A., III, Savela, C. E., Nord, G. L., Jr., Balciauskas, K. A., and Belkin, H. E. (1997). Hydrobiogeochemical interactions on calcite and gypsum in "anoxic" limestone drains in West Virginia and Pennsylvania. *In* "1997 International Ash Utilization Symposium," pp. 546–559. University of Kentucky, Lexington, KY.

Robbins, E. I., D'Agostino, J. P., Ostwald, J., Fanning, D. S., Carter, V., and Van Hoven, R. L. (1992). Manganese nodules and microbial oxidation of manganese in the Huntley Meadows wetland, Virginia, USA. *In* "Biomineralization Processes of Iron and Manganese: Modern and Ancient Environments" (H. C. W. Skinner and R. W. Fitzpatrick, eds.), pp. 179–202. *Catena Supplement* **21**.

Robbins, E. I., Nord, G. L., Savela, C. E., Eddy, J. I., Livi, K. J. T., Gullett, C. D., Nordstrom, D. K., Chou, I.-M., and Briggs, K. M. (1996). Microbial and mineralogical analysis of aluminum-rich precipitates that occlude porosity in a failed anoxic limestone drain, Monongalia County, West Virginia. *In* "Coal Energy and the Environment" (S.-H. Chiang, ed.), Vol. 2, pp. 761–767. Proceedings Thirteenth Annual International Pittsburgh Coal Conference.

Rose, A. W., and Cravotta, C. A., III (1998). Geochemistry of coal-mine drainage. *In* "Coal Mine Drainage Prediction and Pollution Prevention in Pennsylvania" (K. B. C. Brady, M. W. Smith, and J. Schueck, eds.), pp. 1.1–1.22. Pennsylvania Department of Environmental Protection, 5600-BK-DEP2256, Harrisburg, PA.

Rose, S. and Ghazi, A. M. (1997). Release of sorbed sulfate from iron oxyhydroxides precipitated from acid mine drainage associated with coal mining. *Environ. Sci. Technol.* **31**, 2136–2140.

Skousen, J. G., Rose, A. W., Geidel, G., Foreman, J., Evans, R., Hellier, W., and others (1998). "Handbook of Technologies for Avoidance and Remediation of Acid Mine Drainage." National Mine Land Reclamation Center, Morgantown, WV.

Sterner, P. L., Skousen, J. G., and Donovan, J. J. (1998). Geochemistry of laboratory anoxic limestone drains. *In* "Proceedings of the 1998 National Meeting of the American Society for Surface Mining and Reclamation," pp. 214–234. American Society for Surface Mining and Reclamation, Princeton, WV.

Stumm, W., and Morgan, J. J. (1996). "Aquatic Chemistry: Chemical Equilibria and Rates in Natural Waters," 3rd Ed. Wiley-Interscience, New York.

Stumm, W., and Sulzberger, B. (1992). The cycling of iron in natural environments: Considerations based on laboratory studies of heterogeneous redox processes. *Geochim. Cosmochim. Acta* **56**, 3233–3257.

Tamura, H., Goto, K., and Nagayama, M. (1976). The effect of ferric hydroxide in the oxygenation of ferrous ions in neutral solutions. *Corrosion Sci.* **16**, 197–207.

Taylor, R. M., and Schwertmann, U. (1978). The influence of aluminum on iron oxides. I. The influence of Al on Fe oxide formation from the Fe(II) system. *Clays Clay Miner.* **26**, 373–383.

Turner, D., and McCoy, D. (1990). Anoxic alkaline drain treatment system, a low cost acid mine drainage treatment alternative. *In* "1990 Symposium on Surface Mining Hydrology, Sedimentology and Reclamation" (D. H. Graves and R. W. DeVore, eds.), pp. 73–75. University of Kentucky, Lexington, KY.

Watzlaf, G. R. (1997). Passive treatment of acid mine drainage in down-flow limestone systems. *In* "Proceedings of the 1997 National Meeting of the American Society for Surface Mining and Reclamation," pp. 611–622. American Society for Surface Mining and Reclamation, Princeton, WV.

Watzlaf, G. R., and Hedin, R. S. (1993). A method for predicting the alkalinity generated by anoxic limestone drains. *In* "Proceedings 14th Annual Meeting West Virginia Surface Mine Drainage Task Force." West Virginia University, Morgantown, WV.

Watzlaf, G. R., Kleinhenz, J. W., Odoski, J. R., and Hedin, R. S. (1994). The performance of the Jennings Environmental Center anoxic limestone drain: U.S. Bureau of Mines Special Publication SP 06B, p. 427.

Watzlaf, G. R., Schroeder, K. T., and Kairies, C. (2000a). Long-term performance of anoxic limestone drains for the treatment of mine drainage. *Mine Water Environ.* **19**, 98–110.

Watzlaf, G. R., Schroeder, K. T., and Kairies, C. (2000b). Long-term performance of alkalinity-producing passive systems for the treatment of mine drainage. *In* "Proceedings of the 2000 National Meeting of the American Society for Surface Mining and Reclamation," pp. 262–274. American Society for Surface Mining and Reclamation, Princeton, WV.

Webster, J. G., Swedlund, P. J., and Webster, K. S. (1998). Trace metal adsorption onto an acid mine drainage iron(III) oxy hydroxy sulfate. *Environ. Sci. Techno.* **32**, 1361–1368.

Wilde, F. D., Radtke, D. B., Gibs, J., and Iwatsubo, R. T. (1998). National field manual for the collection of water-quality data: U.S. Geological Survey Techniques of Water-Resources Investigations, Book 9, Handbooks for Water-Quality Investigations, variously paged.

Williamson, M. A., Kirby, C. S., and Rimstidt, J. D. (1992). The kinetics of iron oxidation in acid mine drainage. *In* "V. M. Goldschmidt Conference Program and Abstracts." The Geochemical Society, University Park, PA.

Wood, C. R. (1996). Water quality of large discharges from mines in the anthracite region of eastern Pennsylvania: U.S. Geological Survey Water-Resources Investigations Report **95–4243**.

Wood, W. W. (1976). Guidelines for the collection and field analysis of ground-water samples for selected unstable constituents: U.S. Geological Survey Techniques of Water-Resources Investigations, Book 1, Chap. D2.

Zachara, J. M., Cowan, C. E., and Resch, C. T. (1991). Sorption of divalent metals on calcite. *Geochim. Cosmochim. Acta* **55**, 1549–1562.

Ziemkiewicz, P. F., Skousen, J. G., Brant, D. L., Sterner, P. L., and Lovett, R. J. (1997). Acid mine drainage treatment with armored limestone in open limestone channels. *J. Environ. Qual.* **26**, 1017–1024.

Chapter 3

Preliminary Investigation into the Suitability of Permeable Reactive Barriers for the Treatment of Acid Sulfate Soils Discharge

T. David Waite,* Rosalind Desmier,*,† Michael Melville,† and Bennett Macdonald†

*Centre for Water and Waste Technology, School of Civil and Environmental Engineering, and

†School of Geography, The University of New South Wales, Sydney, NSW 2052, Australia

The generation and release of acidic water from acid sulfate soils is an environmental problem of international importance (global scale). Changes in land use and hydrological systems have promoted the oxidation of pyrite in acid sulfate soils with a concomitant generation of acidic discharge waters, which have significantly degraded coastal regions in many countries. Use of permeable reactive barriers packed with neutralizing agents such as calcite would appear to hold some potential as a treatment technology for assisting in the management of drainage from acid sulfate soils. A number of factors need to be resolved prior to installation of such a barrier in an acid sulfate soils region. Preliminary data necessary for design include the water chemistry and local hydrology of the region. It is particularly important to develop an understanding of the variability in flow and acid discharge through storm events, as these potentially constitute the times of greatest impact with respect to acid transport. While significant concern exists that armoring by iron and aluminum oxide precipitates may limit the reactivity and longevity of the permeable reactive barrier, maintenance of saturated conditions and installation of effective subdrainage and/or flushing procedures will be effective in overcoming problems and yield an effective, low maintenance solution to an increasingly troublesome problem.

I. INTRODUCTION

Most of the Holocene-age (<10,000 years B.P.) surfaces on coastal floodplains, lowlands, swamps, and estuary bottoms are underlain naturally by sediments that contain sulfide minerals (mostly iron pyrite, FeS_2). These sediments mainly formed during the sea level rise (until about 6500 years B.P.) after the last glacial maximum (at 20,000 years B.P.). The pyrite was formed from the reduction of dissolved sulfate in marine and brackish tidal water, particularly where sufficient organic detritus existed, such as in mangrove swamps and tidal marshes. Such accumulations constitute up to 5% of the sediments (Dent, 1986). The formation of pyrite is summarized by

$$\begin{aligned} &Fe_2O_{3(s)} + 4\,SO_4{}^{2-}{}_{(aq)} + 8\,CH_2O + 1/2O_{2(aq)} \rightarrow \\ &2\,FeS_{2(s)} + 8\,HCO_3{}^{-}{}_{(aq)} + 4\,H_2O \end{aligned} \tag{1}$$

This pyrite remains innocuous while saturated but problems arise following its oxidation after drainage or excavation. Natural processes can cause pyrite oxidation, such as evapotranspiration during droughts that can markedly lower water tables in floodplain backswamps, as can tectonic uplift of coastal margins. Human activities, such as land drainage and excavation or dredging of sulfidic sediments, tend to accelerate pyrite oxidation greatly. Even where no additional pyrite oxidation is caused, artificial drainage systems provide the conduit to increase the discharge of preexisting acidity created by natural processes, from back-swamp source areas, through the natural levees, into the estuary.

Pons (1973) used the term acid sulfate soil (ASS) generically for all sulfidic materials and soils in which, as a result of oxidative processes, sulfuric acid has been produced, is being produced, or will be produced in amounts that exceed the inherent buffering capacity of the material. ASS are sometimes further defined as actual acid sulfate soils (AASS) where porewaters exhibit $pH < 4$ after most pyrite has oxidized and as potential acid sulfate soil (PASS) where porewaters are not particularly acidic but which contain sufficient as yet unoxidized pyrite to become AASS. These two types of materials are formally defined as sulfuric and sulfidic horizons in soil taxonomy (Soil Survey Staff, 1999).

Within an ASS profile, a meter or so of AASS tends to overlie a considerable thickness of PASS. Variable thicknesses of fluviatile sediment may also bury ASS. This is particularly the case in regions of rapid catchment erosion and discharge or in regions of tectonic subsidence. Given that these coastal ASS are at elevations close to sea level, the general picture is one of a fairly fresh but very acidic, near-surface porewater that bathes the AASS. This overlies a more brackish groundwater in chemical equilibrium with the PASS. The water table is usually close to the surface in such regions with the actual level dependent on the variations in rainfall and evapotranspiration (Wilson *et al.*, 1999). The typically gelatinous PASS material exhibits a very low hydraulic conductivity (<1 m/day; White *et al.*, 1993). Therefore, little control of water table elevation comes from groundwaters of the surrounding bedrock landscapes into which the ASS were deposited. Discharge of acidic pyrite oxidation products occurs mainly by rain-fed upward movement of the water table and surface transport into natural or artificial drainage systems that penetrate the natural floodplain levee. During major rain events the entire ASS estuary floodplain can be flooded by runoff from the surrounding catchment. Such flooding temporarily dilutes the acidity discharged from the floodplain, but after the flood peak, acidic pyrite oxidation products continue to discharge into a fresh and unbuffered estuary. Most environmental impacts are then observed. Smaller rain events (< 50 mm) during generally dry times can also discharge acidity, but the buffering capacity of the estuary tends to reduce the impact.

Problems of acidic discharges also arise in the mining industry where acid rock drainage (ARD) requires careful management. There are, however, major differences between ARD and discharge from ASS. In particular, ARD is typically of high acidity but relatively low, constant flow from a single source (typically a tailings pile or waste rock dump), whereas ASS discharge arises from significantly more diffuse sources of large areal extent and extreme variability in volume and flow rate. In addition, ARD discharge often occurs in regions of considerable economic merit, whereas ASS typically occurs in regions of relatively low economic return. It is evident therefore that considerable scope exists for the development of innovative low cost and low maintenance treatment systems. There is a growing consensus (White *et al.*, 1997) that sustainable management of coastal regions affected by ASS requires the maximal retention of acidic pyrite oxidation products in or near to its diffuse floodplain stores. Management techniques should intercept the acidity, along the hydrological transfer pathway from within the ASS profile, up through the soil surface, and into the drainage systems, but before the acidic drain water enters the estuary. In light of this, permeable reactive barriers (PRB), inserted across the flow path of drainage waters, appear to be a particularly attractive passive treatment option. An example of the

and subtropics [e.g., for Thailand, see Dent and Pons (1995); for Vietnam, see Minh *et al.* (1997) and Husson *et al.* (2000); for Indonesia, see Bronswijk *et al.* (1995); for Australia, see Lin and Melville (1992) and White *et al.* (1996)]. In noncoastal or temperate regions, ASS are encountered less commonly [e.g., for Finland, see Astrom (1998); for Spain, see Bescansa and Roquero (1990); for New Zealand, see Metson *et al.* (1977); for the United States, see Carson and Dixon (1983) and Kraus (1998); for the United Kingdom, see Bloomfield (1972)]. ASS vary from one location to the next in the texture of the sediments, in the degree of PASS versus AASS, and in the extent of acidity stored in the profiles.

The estimates just given did not include any Australian ASS simply because the soil maps published at that time did not identify ASS in Australia. Even as late as 1989, there were less than five papers published on Australian ASS, although some other nonpublished technical reports existed. Naylor *et al.* (1995) published maps (scale 1:25,000) showing the probable distribution of ASS in coastal New South Wales (NSW). These maps, based on physiography and limited ground survey, showed a probable distribution in every estuary and all coastal lowlands up to about 5 m above mean sea level. The total extent of these materials in NSW was of the order of 0.5 million hectare. Lin and Melville (1992) estimated that there were at least 1 million hectares of ASS in Australia based on Galloway's (1982) mapping of coastal mangroves. White *et al.* (1997) proposed 3.5 million hectares of ASS in Australia.

C. Environmental Impacts

Acid waters generated from acid sulfate soils can be extremely toxic to plants, many invertebrates, and gilled organisms. The extent of the acid generation depends on a combination of weather, physiography, hydrology, and soil factors (permeability, composition, etc). Long periods of dry weather followed by large rainfall events tend to increase acid production within drainage waters (Callinan *et al.*, 1993). Such a sequence of events occurred in 1987 in the Tweed River on the far north coast of New South Wales, Australia. The result was a complete, massive kill of fish, crustaceans, and annelid worms along 23 km of the River (Easton,1989). There are approximately 10,000 hectares of ASS floodplain that could contribute to this acid discharge into the affected river section. Unfortunately, no water quality measurements are available for that devastating event. However, Easton (1989) observed that the affected section of the normally muddy flooded river was completely clarified. Clarification occurs because of the high concentrations of dissolved aluminium present and can be a feature of ASS discharges. Assuming the affected river volume as 2×10^4 megaliters and an alum dosing requirement of 100 mg/liter, we estimated that more than 2,000 tonnes of alum was likely to have been discharged in the 1987 Tweed River event. Wilson *et al.* (1999) reported that during the 1992 wet-season discharges into the Tweed River, only minor clarification and other impacts were observed, but an estimated discharge of about 1200 tonnes of sulfuric acid occurred.

Callinan *et al.* (1993) proposed that acid sulfate soil drainage waters with low pH and high concentrations of aluminum were responsible for the seasonal recurrence of fish kills and the ulcerative fish disease, epizootic ulcerative syndrome (EUS). Sammut *et al.* (1995) reported that fish kills by ASS discharge are usually caused by rapid damage to gills of the fish due to the exposure to elevated concentrations of acidity and aluminium. The affected fish asphyxiate even if there is sufficient dissolved oxygen. The increased aluminium and acidity cause skin lesions in fish after only minutes of exposure, and the EUS is caused by infection with the fungus *Aphanomyces invadans* (Sammut *et al.*, 1995). EUS, which was first reported in only 1973, is now globally the most important disease of fin fish in aquaculture and costs NSW commercial fishermen more than $5 million a year (Callinan *et al.*, 1993).

Acid sulfate soils have caused the removal of many native macrophytes and the invasion of acid tolerant plant species (Sammut *et al.*, 1994). Aquatic plants have been affected by the direct toxicity of acid and iron flocs, which have smothered and killed native vegetation (Sammut *et al.*, 1996). There is increasing evidence that elevated concentrations of dissolved iron due to ASS discharges into coastal waters of southeastern Queensland, Australia, are associated with blooms of the cyanobacterium, *Lyngbya majuscula* (Dennison *et al.*, 1997). Toxins from these blooms have been reported to result in dramatic reductions in fish populations and are purported to account for an increased incidence of tumors in dugongs and turtles. In addition, the toxins released by this organism represent a significant health risk to humans, causing severe dermatitis and respiratory problems.

In acid sulfate soil regions, engineered structures are also susceptible to subsidence due to the failure of unconsolidated estuarine sediments; corrosion and degradation of concrete and metal structures; and reduced water flow capabilities in wells, drains, and aquifers due to blocking by metal hydroxides (White *et al.*, 2001). Subsidence problems have caused severe difficulties and cost overruns in highway construction across acid sulfate soils underlain by highly gelatinous clays in northern NSW. This behavior is reminiscent of the Canadian "quick clays" (Lessard and Mitchell, 1985).

One area of investigation that has not yet been pursued is that of the mobilization of trace metals by the formation of AASS. Preliminary measurements show that elevated concentrations of Zn, B, As and sometimes Ni and Cu tend to be associated with the oxidation front of the ASS profile. These and other trace metals can be sourced from the parent sediment, particularly in dredge material from harbor bottoms or from agricultural chemicals.

III. CURRENT MANAGEMENT OPTIONS

Numerous guidelines have been established by government agencies in response to problems associated with the occurrence of acid sulfate soils. The government response in two states on the east coast of Australia has been to establish advisory committees that provide guidance on technical and policy matters [New South Wales Acid Sulfate Soils Management Advisory Committee (ASSMAC) and Queensland ASSMAC]. These committees have published guidelines for assessing acid sulfate soils (Ahern *et al.*, 1998; Stone *et al.*, 1998). On a smaller scale, local government bodies in NSW are now required to adopt local environment plans (LEPs) that address the best management for acid sulfate soils. In 1998, a National Working Party on Acid Sulfate Soils was formed, and strategies for the management of acid sulfate soils across Australia were developed (National Working Party on Acid Sulfate Soils, 1999).

The management practices that have been used for potential acid sulfate soils have typically focused on minimizing acidification by preventing or reducing the oxidation of sulfidic sediments (potential acid sulfate soils). To avoid the formation of actual acid sulfate soils (and the associated problems with acid drainage), it is important to recognize areas where these soils may be found and to conduct soil surveys to assess the nature of the soil prior to developments. The New South Wales government has undertaken this process by identifying all coastal acid sulfate soil areas and producing risk maps (scale 1:25,000) to show the acid sulfate soil distribution within the state, highlighting land uses that may cause an environmental risk by disturbing the soils (Naylor *et al.*, 1995).

Where avoidance is not an option, three common techniques are used to manage acid sulfate soils: preventing oxidation, neutralization, and leaching.

The selection of the appropriate management technique or combination of techniques is dependent on the relative economics and practicalities of the various options.

A. Preventing Oxidation

1. Burial of Acid Sulfate Soils

Anoxic conditions can be obtained by placing a thick layer of clean, nonacidic soil over the acid sulfate soils. This nonacidic soil cap should prevent oxidation of sulfidic sediments and slow the amount of acid being discharged from already oxidized soils. A continual water cover over the land can also be used to prevent further oxidation. To prevent the soil from oxidizing again, it is essential that the water cover remains on the land (Sammut, 1996). Nevertheless, in highly oxidized and very acidic acid sulfate soils, the presence of high concentrations of dissolved ferric iron can maintain some pyrite oxidation even after reflooding.

2. Design of Drainage Systems

Oxidation of sulfidic layers can be minimized by carefully designed drainage systems. White *et al.* (1997) suggested that soils with sulfidic layers at 0.5 m or less below soil surface or with acidic surface soils devoid of vegetation should be left undrained and returned to wetland conditions. However, soils with deeper sulfidic layers (0.5 to 2.0 m) should have carefully designed drainage systems, including treatment of any discharge. Soils with sulfidic layers greater than 2 m below soil surface can be drained with minimal oxidation of sulfides, provided drains are cut at least 1 m above the sulfidic layer.

White *et al.* (1997) also suggested that a series of broad shallow drains might improve the removal of excess surface water from agricultural land without oxidizing potential acid sulfate soils at depth. Laser leveling is commonly used on sugarcane and tea tree (*Melaleuca alternifolia*) farms (White *et al.*, 1999). Laser leveling allows efficient removal of surface water to drains. Instances of water logging are therefore reduced, without causing drawdown of the groundwater table. In clayey acid sulfate soils under sugarcane, elevation of the shallow groundwater table is controlled by evapotranspiration by the crop, not by the drains (White and Melville, 1993). If the groundwater table is maintained at a constantly high level in sandy ASS, the chance of pyrite oxidation in the soil and subsequent acid generation is reduced significantly (Blunden and Indraratna, 2000).

B. Neutralization

Neutralization is a reaction between acidic water and an alkaline agent. A number of materials can be applied to neutralize acid sulfate soils. Lime is the most common neutralization agent in use. However, other materials, such as seashells (Brandjes and Lauckner, 1997) and seawater-neutralized bauxite refinery residues (McConchie and Clark, 1999; discussed in more detail later), have been used or suggested. Other materials are available, but due to their cost or possible downstream impacts are not used commonly (see Section IV,C).

1. Addition of Lime to Soils

Liming is practiced widely throughout Europe, Australia, and Asia in the management of acid sulfate soils. On broad-acre agricultural land, large-scale liming treatment is expensive due to the enormous quantities of lime required for the neutralization of soils. Dent (1992) suggested that raising the pH of the soil above 5 should be sufficient to avoid most acid sulfate problems. White *et al.* (1997) found the cost of liming a 400-ha property to be $13 million dollars.

2. Addition of Lime to Surface Waters

Work in northern New South Wales has demonstrated that many acidic discharges occur from soils close to drains and are dependent on the level of the water table (Wilson *et al.*, 1999). Perhaps a more economical (yet still expensive) technique to treat acid soils is to lime the drains, rather than the whole drained area (White *et al.*, 1997).

The placement of coarse lime within a drainage channel or the scatter of fine lime to surface water is a common management technique for the treatment of acidic waters. These techniques use treatment structures referred to as open limestone channels (OLCs). They offer a short-term solution but require high amounts of maintenance. The main problems with OLCs as a treatment technique are (1) clogging with sediment, (2) dissolution of limestone during large rainfall events, and (3) armoring of limestone with iron hydroxides. This technique has been used recently in the treatment of acid waters generated during the construction of the Pacific Highway in northern NSW.

3. Dilution

Dilution of acid waters involves the addition of nonacidic water to raise the pH. Dilution can control the pH of acid water discharged into downstream waters. Dilution techniques are normally facilitated using a storage reservoir with a regulated discharge point. It is anticipated that by controlling the release of acid into the environment, there will be sufficient time for the acid water to be diluted further with other downstream waters, such as seawater, before there is an environmental impact. With seawater, a complementary strategy to dilution is the neutralization of acid water by the bicarbonate present in seawater.

C. Leaching

Natural processes can leach acid sulfate soils. However, faster leaching rates can be obtained using a "raised bed" technique. This technique involves

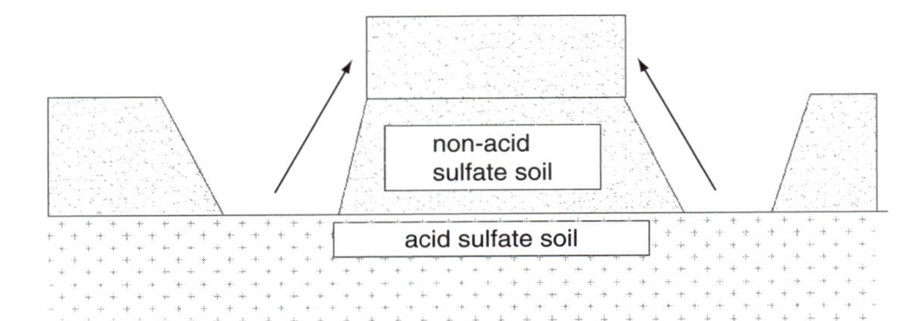

Figure 1 Cross section showing raised-bed treatment of acid sulfate soils.

controlled oxidation of sulfidic material within soils and careful leaching to remove the oxidation products. Often, leaching will involve the excavation of actual or potential acid sulfate soils into raised stockpiles, above the water table (Fig. 1). The excavated soil should be permeable so that acid water will be flushed from the stockpile when flooded. To prevent contamination of groundwater and streams, the stockpile should be located some distance away from any fresh water, and the base of the stockpile needs to be guarded with impermeable materials. Drains can be located at the base of the stockpile to promote oxidation and efficient removal of the effluent. The time required to complete oxidation and leaching is unpredictable and is influenced by factors such as rainfall, temperature, wind speed, and the size and shape of the stockpile (White and Melville, 1993). For success in this technique, it is important that oxidation, leaching, and effluent removal be accompanied by careful water monitoring and treatment.

This technique may be economically viable and/or environmentally acceptable for extensive agricultural developments or low-cost construction projects if the reserves of acidity are low and acid leachate containment and neutralization can be managed satisfactorily (Bowman, 1993). A range of crops such as yam, sugarcane, and cassava have been grown successfully on raised beds in the Plain of Reeds, Vietnam (Bowman, 1993).

IV. PERMEABLE REACTIVE BARRIER OPTIONS FOR ACID SULFATE SOILS DISCHARGE MANAGEMENT

Any system chosen to treat acid waters generated from acid sulfate soils regions will need to be thoughtful of existing and future land uses. Treatment technology will ultimately be used principally by farmers in the agricultural industry and developers in the construction industry. Therefore, systems that are inexpensive, need little maintenance, and utilize small amounts of land are essential.

The focus of the following section is on the application of permeable reactive barrier (PRB) technology, a technology commonly applied to resolve groundwater contamination and acid mine drainage problems but not used as yet to address more "broad-acre" problems. PRBs are passive, *in situ* water treatment constructions, and should not require continuous chemical inputs. PRB technology is not a new concept and has been used in the mining industry for treating specific waste streams since the early 1980s. Pearson and McDonnell (1975) conducted some of the early research that recognized its application in the management of acid mine drainage runoff.

The following sections describe the existing approaches to the treatment of acid mine drainage using PRB technology and provide an assessment, as well as some preliminary design possibilities, for application to acid sulfate soils drainage waters.

A. Calcareous Reactive Barriers

A range of chemicals could be considered for use within a PRB. However, the most commonly used material is limestone due to low cost and high availability. Four types of calcareous reactive barriers of potential use in acid sulfate soils regions are described briefly.

Protons and dissolved carbon dioxide [CO_2(aq)] in the water promote limestone dissolution [Eqs. (6) and (7)]. Bicarbonate alkalinity is thus added to the system and the pH is increased. The notation for dissolved CO_2 is H_2CO_3 [Eq. (8)]. This includes both CO_2(aq) and carbonic acid (Stumm and Morgan, 1996):

$$CaCO_{3(s)} + 2H^+ \rightarrow Ca^{2+} + H_2CO_3 \qquad (6)$$

$$CaCO_{3(s)} + H_2CO_3 \rightarrow Ca^{2+} + 2\,HCO_3^- \qquad (7)$$

$$CO_{2(aq)} + H_2O \leftrightarrow H_2CO_3 \qquad (8)$$

Major problems associated with use of calcite are its low solubility ($pK_{calcite} = 8.35$; Morel and Hering, 1993) and armoring effects. On exposure to O_2 the following reaction can occur:

$$Fe^{3+} + 3\,HCO_3^-(aq) \rightarrow Fe(OH)_3 + 3\,CO_{2(aq)}. \qquad (9)$$

As iron(III) oxyhydroxides precipitate, the limestone undergoes armoring. The rate of limestone dissolution and alkalinity production is therefore decreased as armoring increases (Sun *et al.*, 2000).

1. Open Limestone Channels

Open limestone channels (OLCs) are constructed by placing coarse limestone into a drainage channel. Problems with hindered reactivity and permeability of the limestone due to clogging with metal hydroxides are encountered commonly (Pearson and McDonnell, 1975; Ziemkiewicz *et al.*, 1997).

In a study of eight field drains and associated laboratory studies, Ziemkiewicz *et al.* (1997) recommended that OLCs should have a slope of greater than 20% to prevent armoring and to keep the limestone active. Rose and Laurenso (2000) found that channels with less gradient can be used effectively if the channel material is disturbed periodically (e.g., driving a small tractor with a cultivator attachment along the channel). The coatings could then be dislodged from the limestone and washed away, although attention should be given to the end fate of these washings as they will be high in iron, aluminium, and possibly other trace elements of potential deleterious biological impact. This procedure would depart from a passive system and the cost benefits would be large (Rose and Laurenso, 2000). This technique has been used in the treatment of drainage from acid sulfate soils (see Section III,B).

2. Anoxic Limestone Drains

Anoxic limestone drains (ALDs) are buried cells of limestone that treat anoxic waters (Turner and McCoy, 1990). The ALD should be buried underneath an impervious layer of material that prevents inputs of oxygen into the system and maximizes the accumulation of carbon dioxide (Fig. 2). At a pH less than 6 and under anoxic conditions, the limestone within the ALD will not be coated or armored with iron hydroxides because Fe^{2+} does not usually

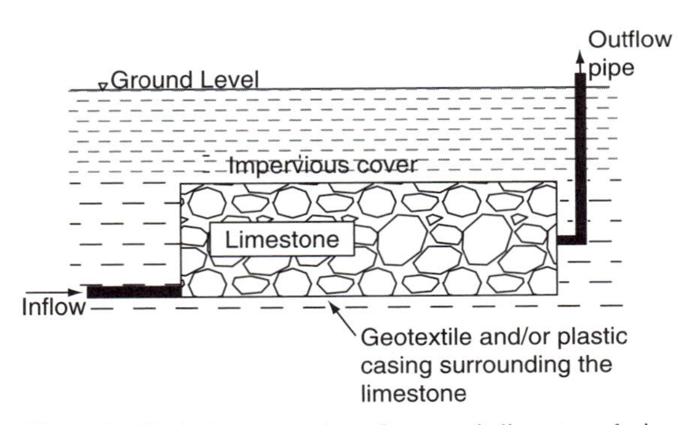

Figure 2 Typical cross section of an anoxic limestone drain.

precipitate [as $Fe(OH)_2(s)$, $FeCO_3(s)$, etc.] under such conditions (Skousen, 1997). Bicarbonate alkalinity is added to the water as it flows through the ALD and limestone dissolves. Subsequent aeration and ponding of the discharge from the drain result in the precipitation of metal oxyhydroxides [Eq. (9)].

Continued sustainability of an ALD system is a concern. Potential clogging of the ALD with metal hydroxides given sufficient concentrations of Fe^{3+}, Al^{3+}, and dissolved oxygen in the water poses a serious threat to the longevity of the system.

A number of ALDs have been operating successfully in the United States (Nairn *et al.*, 1992; Watzlaf *et al.*, 2000; Skousen *et al.*, 1999; Hedin and Watzlaf, 1994). There have also been ALD failures such as that located at the Jennings Environmental Center in Pennsylvania. Watzlaf *et al.* (2000) reported that flow through the ALD decreased within 6 months of operation. This was thought to be a result of iron and, more significantly, aluminum oxyhydroxides clogging the system.

Brodie *et al.* (1993) have suggested guidelines for ALD design, and results indicate that problems with armoring can be overcome if a drain is constructed properly. The main problem with using an anoxic drain for acid sulfate soils drainage waters is that water will be primarily oxic and a pretreatment will be required to remove the dissolved oxygen. This chapter describes two alternative methods.

3. Oxic Limestone Drains

An Oxic limestone drain (OLD) is similar to an ALD; however, it is designed to treat water that contains dissolved oxygen and ferric iron (iron that has been oxidized) in a one-stage operation. The partial pressure of $CO_2(P_{CO_2})$ is increased due to the drain being covered, hindering its escape. Consequently, there is higher limestone dissolution and alkalinity produced in this system compared to a limestone system open to the atmosphere.

Cravotta and Trahan (1999) undertook a field investigation into the treatment of acid mine drainage with OLDs in Pennsylvania and found that they were effective for neutralizing oxic, relatively dilute acid mine drainage (O_2 >1 mg/liter, acidity <90 mg/liter). Results of the study show that a residence time of 1–3 h was sufficient for nearly complete oxidation of Fe^{2+} in an OLD and for decreasing concentrations of dissolved Al^{3+}, Fe^{3+}, Fe^{2+}, Mn^{3+}, and trace metals (<5 mg/liter). It was found that a fraction of the hydrous oxides were transported as suspended particles as water flowed rapidly ($v = 0.1$ to 0.4 m/min and residence time <3.1 h) through the OLD.

An OLD system could be used in acid sulfate soils regions; however, careful design will be needed to prevent metal oxyhydroxide clogging. Perforated subdrains could be used to prevent clogging of an OLD with metal oxyhydroxides (Cravotta and Trahan, 1999).

4. Alkalinity Producing Systems

An Alkalinity producing system (APS) is a vertical flow pond where water flows from the surface of the APS through organic matter and limestone layers (Fig. 3). This configuration ensures that the incoming waters exhibit reducing conditions prior to entering the limestone, thus minimizing clogging as a result of metal oxyhydroxide formation. Water leaves the system via a drainage network located at the bottom of the pond. Variations of these systems have been designed with a series of limestone and drainage layers. A system known as the reverse alkalinity producing system (RAPS) is essentially a pond with an inflow at the bottom and an upflow through organic matter and limestone (Fig. 4) (Faulkner and Skousen, 1995). APSs are typically drained into an aerated pond where metals precipitate.

APS have been used successfully in the United States (Watzlaf *et al.*, 2000; Kepler and McCleary, 1994). Perhaps the most significant problems in using an APS in an acid sulfate soil environment is the large land area necessary, the gradient of the land (hydraulic head of water required to force water through the organic matter of a RAPS), and the exposure of more pyrite by constructing a pond.

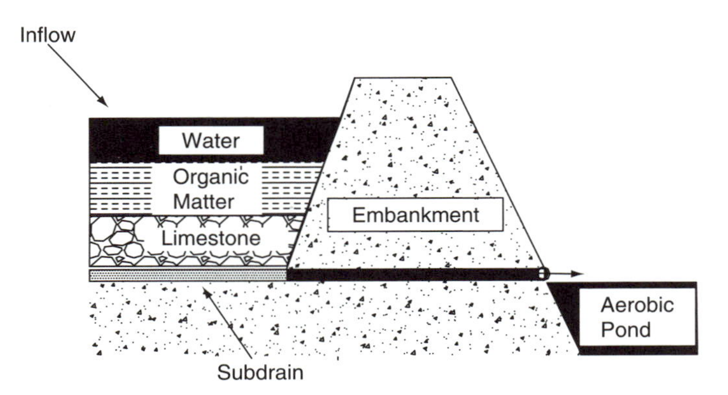

Figure 3 Typical cross section of an alkalinity producing system.

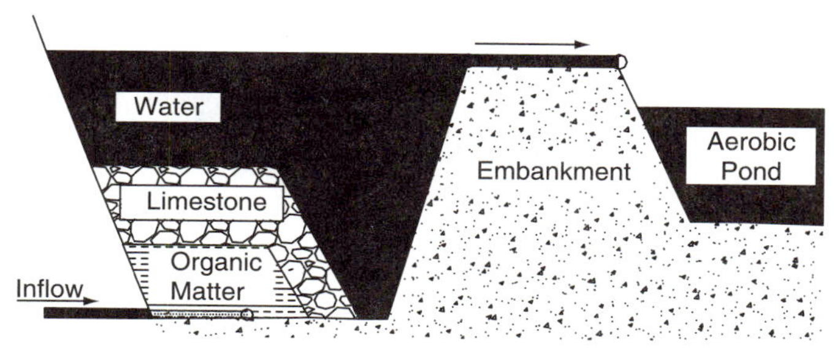

Figure 4 Typical cross section of a reverse alkalinity producing system.

B. Other Materials

Considerable scope for innovation exists when considering neutralizing agents other than calcium carbonate. Materials that have been used with varying success at mine sites include alkaline tailings liquor, fly ash (multiple metal oxides, carbonates), red mud from alumina operations, quicklime (CaO), hydrated lime [$Ca(OH)_2$], calcium peroxide (CaO_2), dolomite [$CaMg(CO_3)_2$], magnesite ($MgCO_3$), caustic magnesia (MgO), witherite ($BaCO_3$), hydroxyapatite [$Ca_5(PO_4)_3OH$], sodium orthosilicate (Na_4SiO_4), and alkaline paper-pulp residues (Taylor *et al.*, 1997). Parker *et al.* (1999) have considered the application of fast-weathering silicate minerals to the neutralization of acid drainage. Olivine [$(Mg, Fe)_2SiO_4$] is one of the more reactive silicate minerals, and its ability to neutralize acids is well known (Schuiling, 1998; Jonckbloedt, 1998). This mineral has almost four times the neutralizing capacity of carbonates, although its significantly slower rate of dissolution would appear to render it useful in only slow seep situations (Alpers and Nordstrom, 1990).

Application of red mud bauxite refinery residues to the treatment of wastewaters has been of particular interest for sometime, both because of the recognized high adsorptive capacity of its constituents (principally hematite, quartz, gibbsite, calcite, sodalite, and whewellite) and because of the attraction of using an abundant waste product for the treatment of other wastes (Couillard, 1982; Ho *et al.*, 1992). This waste product typically exhibits pH values in the vicinity of 13 and must be neutralized before it can be transported and applied. Various neutralizing agents have been proposed, including calcium sulfate (Lopez *et al.*, 1998) and seawater (McConchie *et al.*, 1996). The seawater-neutralized material stabilizes at a pH of around 8.5 and has been shown to have a high trace metal-trapping capacity ($>$1000 meq/kg) and a high acid neutralization capacity (3.5 mole/kg) (McConchie *et al.*, 1997) arising both from the original constituents and from the additional precipitates formed on neutralization (which include calcite, aragonite, whewellite, cancrinite, fluorite, portlandite, boehmite, gibbsite, hydrocalumite, hydrocalcite, and *p*-aluminohydrocalcite). McConchie *et al.* (1999) suggested that the seawater-neutralized red mud could be (1) blended with acid sulfate soil materials to prevent acid generation or (2) used to form reactive barriers to prevent acidic water entering drains or natural watercourses. The possibility also exists of pelletizing the material and using it in PRB construction. Of some concern is the possible mobilization of contaminant metals contained within this waste material on exposure to acidic drainage waters.

V. DESIGN CONCEPTS FOR A CALCAREOUS REACTIVE BARRIER

An oxic limestone drain (as described in Section IV,A,3) has been proposed as one means of mitigating acidic discharge from a sugarcane farming area affected by acid sulfate soils in far northern New South Wales, Australia (28°18'S and 153°31'E). The reactive barrier will treat water from the McLeods Creek catchment, which is a right bank tributary of the Tweed River (Fig. 5). The climate in this region is subtropical with an average annual rainfall of 1400 mm. As can be seen from Fig. 6, summers are warm to hot with high rainfall and winters are warm and dry.

A. Characteristics of ASS and their Near-Surface Groundwater at the Mcleods Creek Study Site

In order that a better understanding can be given of our trial application of PRB to remediation of ASS discharges, we now present some details of the ASS materials and the near-surface groundwater at our McLeods Creek study site.

McLeods Creek has been channelized and now discharges into the Tweed River anabranch, Stotts Channel, via a set of one-way (outward) flap gates that stop tidal inundation of the floodplain. A schematic of the McLeods Creek catchment is shown in Fig. 7. The "ASS Risk Maps" of Naylor *et al.* (1995) show that almost the entire Tweed River floodplain and coastal lowlands are acid sulfate soils [i.e., areas below 10 m relative to Australian Height Datum (AHD), where 0 m AHD is approximately mean sea level]. The

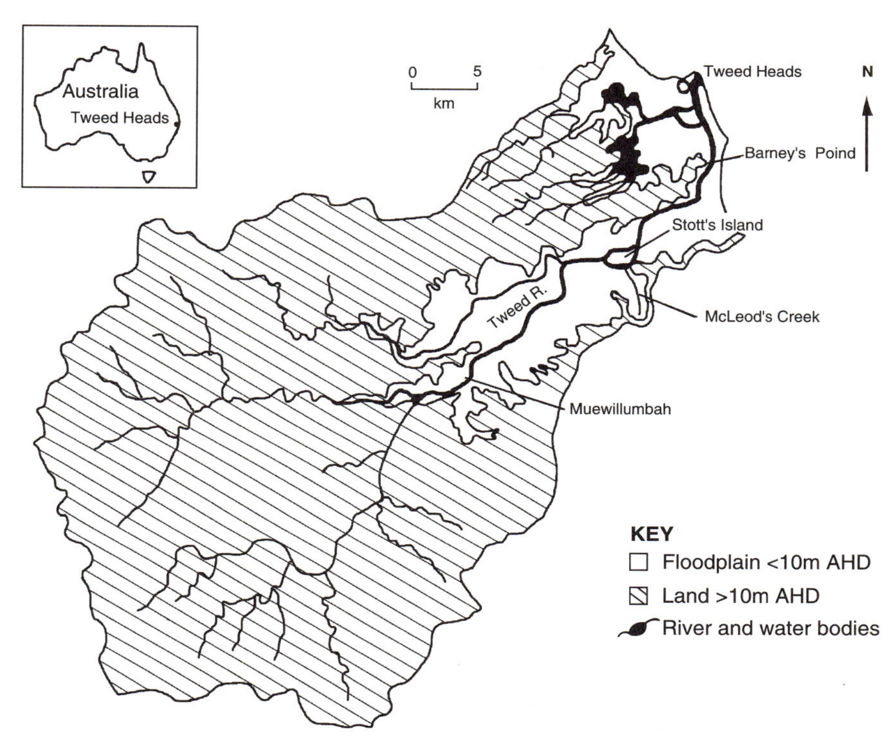

Figure 5 Location map of OLD study site: McLeods Creek catchment in northern New South Wales, Australia.

McLeods Creek floodplain backswamp and toe of the fluviatile levee along Stotts Channel are typical clayey ASS. Analysis of sediment along transects of deep drill holes across the Tweed valley has been completed by Roman (2000). These analyses show that most of the sediment is of marine origin, deposited in a tidal lake that formed after a coastal dune barrier formed from about 7000 years B.P.

The drill site material from McLeods Creek (Fig. 8) shows the typical clay gel PASS (layer C) moisture contents, bulk density, and pH from above 8.3 m depth (−7.7 m AHD) to 1.35 m depth (−0.75 m AHD). Above the PASS to the ground surface (0 m depth; + 0.6 m AHD) is the AASS (Layers A, B, and the transition zone). Below the Holocene-age AASS and PASS (below 8.3 m depth) is Pleistocene-age sediment. Figure 9 shows (from Leco-Furnace and XRF analyses) that the PASS material is fairly uniform with respect to sulfur content (about 2.5% Leco-S). Other Leco-S analyses show that nearly all of this is reduced sulfide minerals, and XRD analysis confirms this to be pyrite.

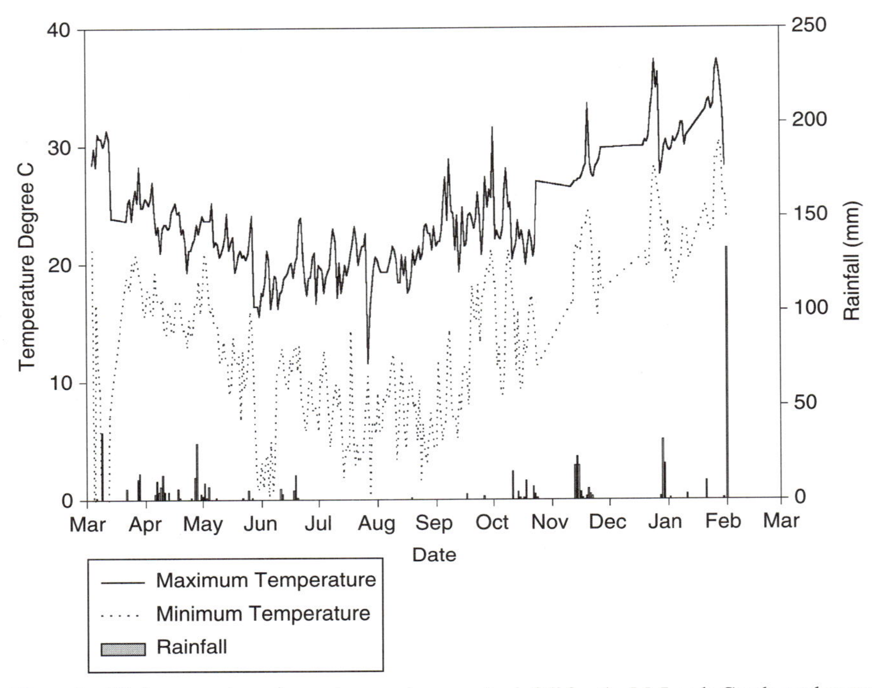

Figure 6 Minimum and maximum temperatures and rainfall for the McLeods Creek catchment for the period March 2000–March 2001.

PASS material has a very high water content and is highly mobile with very little load-bearing capacity. Indeed, the instability of this layer causes significant problems in developing these regions. Oxidation of the surface layers of this gel material appears to lead to rapid structure change, resulting in compaction. Such changes are possibly associated with the high concentration of ferrous iron that is generated at the surface of the gel layer as a result of oxygen penetration and which leads to compression of the double layers associated with the high surface charge clay particles making up the gel. Further studies of the link between geochemical transformations of this gel material and its geotechnical properties are underway.

The near-surface stratigraphy (about +1 m to −2 m AHD) of ASS at McLeods Creek is presented for transects of profiles (A–B, Fig. 10a; and C–D, Fig. 10b). Nearly 200 such ASS profiles at McLeods Creek have been measured for various purposes since the early 1990s. It is clear that AASS exist across the entire floodplain, except on the narrow fluviatile levee, such as close to C (Fig. 10b).

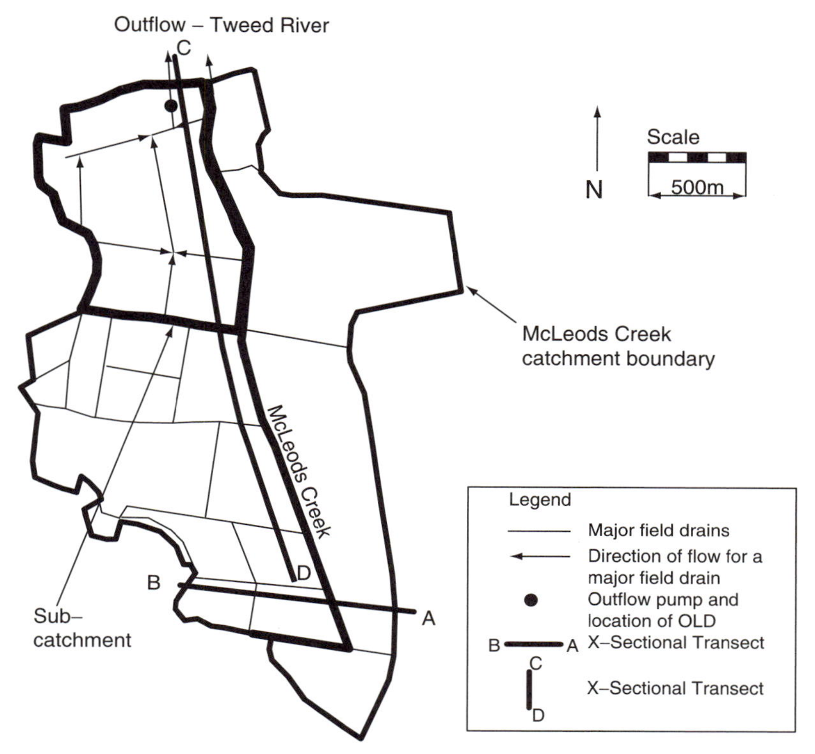

Figure 7 Schematic of the McLeods Creek catchment showing major field drains, location of the OLD test site, and location of transects (see Figs. 10a and b).

The floodplain at McLeods Creek was drained and cleared of its natural forest/swamp vegetation in the early 20th century to be used for dairying. Since about 1960 the area has been drained further for sugarcane production. In 1993, there was about 100 km of drains in the 450 ha of the McLeods Creek floodplain (Yang *et al.*, 2000). It is important to note that while this is entirely an ASS landscape, and has been associated with major acid discharges, it is one of the most productive sugarcane areas in NSW.

Detailed chemical analyses of the porewater (representing the near-surface groundwater) have been obtained by van Oploo (2000) at various McLeods Creek sites using dual-face, dialysis membrane (0.45 μm pore size) porewater samplers (also known as "peepers"). These devices, containing anoxic, Milli-Q water, are inserted into the ASS profile below the water table and are allowed to equilibrate for 7 days. After equilibration, the peeper is removed and the solution (about 8 ml) from each cell (at 20-mm depth increments) is

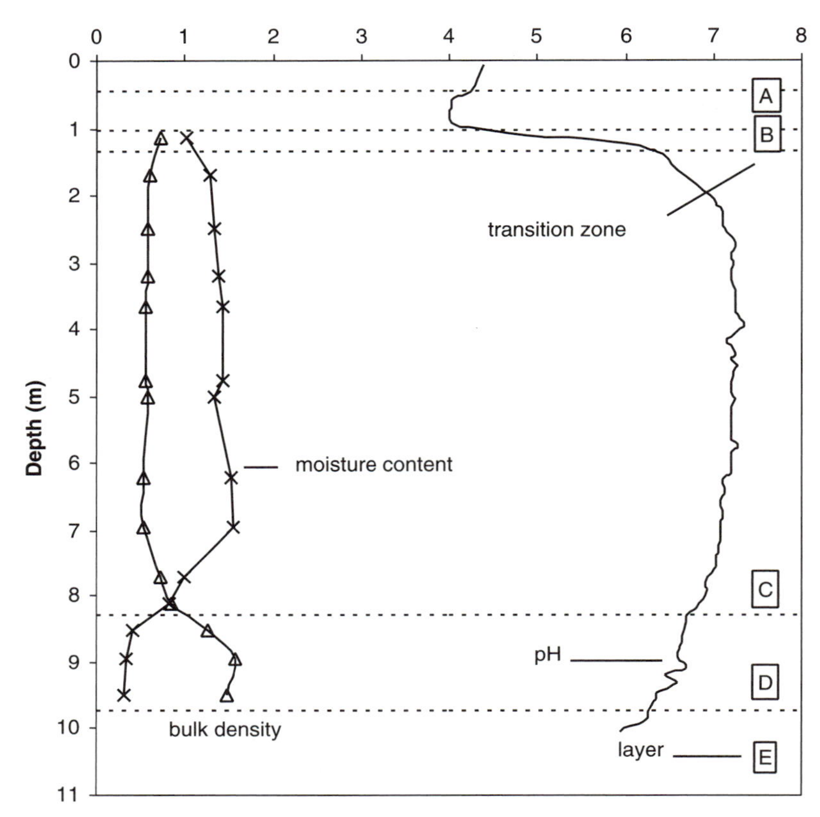

Figure 8 Bulk density, moisture content, and saturated soil pH as a function of depth for soils at the McLeods Creek test site.

withdrawn by syringe pipette, acidified with HNO_3, and quickly refrigerated or frozen prior to transport to the laboratory. Analyses by ICP-optical emission spectrophotometry were made for dissolved metal ions and dissolved sulfate.

Potentiometric chloride concentration, pH, and electrical conductivity were also measured on these samples. At the time of peeper removal in the field, an ASS core was taken adjacent to the peeper (within 1 m) using a bucket auger in the topsoil and a soil gouge below the water table. Redox potential and pH were measured immediately (Fig. 11) at elevations equivalent to the peeper using intermediate junction, glass spear electrodes inserted into the sediment (TPS field meters and Ionode electrodes). The usually sharp oxidation front, here at approximately −0.55 m AHD, is evident in Fig. 11.

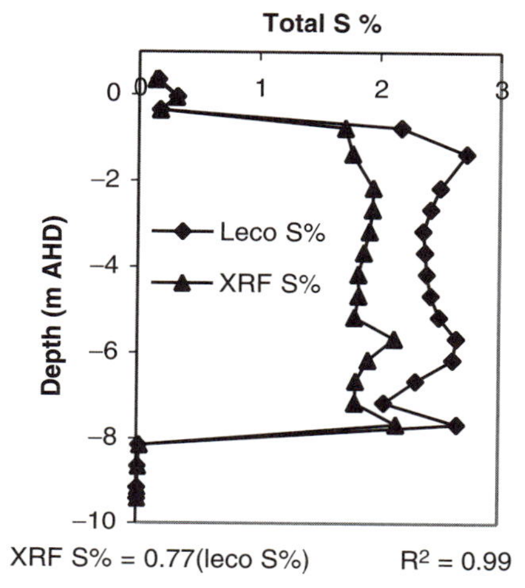

Figure 9 Sulfur content of sediments at McLeods Creek site as determined by Leco furnace and X-ray fluorescence (XRF).

Peeper devices are each only 1 m in length so that to obtain pore-water profiles spanning greater depths below the water table and providing some replication, nests of three peepers inserted to different depths were used. Such nested results (for soil profiles D, E, and F) for a range of ion species are presented in Fig. 12. The dissolved ion species within the concentration profiles are the direct products from pyrite oxidation (Fe and S) and the reaction products from acid attack on the host aluminosilicate clays and other mineral components of the estuarine sediments. Other peeper results obtained show that the magnitude and position of soluble species concentration profiles vary, depending on the antecedent hydrological conditions. For instance, prolonged dry periods followed by moderate rainfall give larger total concentrations at higher elevations in the ASS profile. A similar response occurs after harvesting the perennial sugarcane crop and small rain events because of the marked reduction in the evapotranspirative control on the water table elevation.

Van Oploo (2000) showed that these concentration profiles below the oxidation front could develop by downward diffusion over about 30 years, the approximate time since sugarcane was first grown at the site. Upward movement through the ground surface and into runoff, in response to rain-fed, near-surface water table dynamics, leaches out these soluble ion species, resulting in the presence of acidic oxidation products in the drainage water.

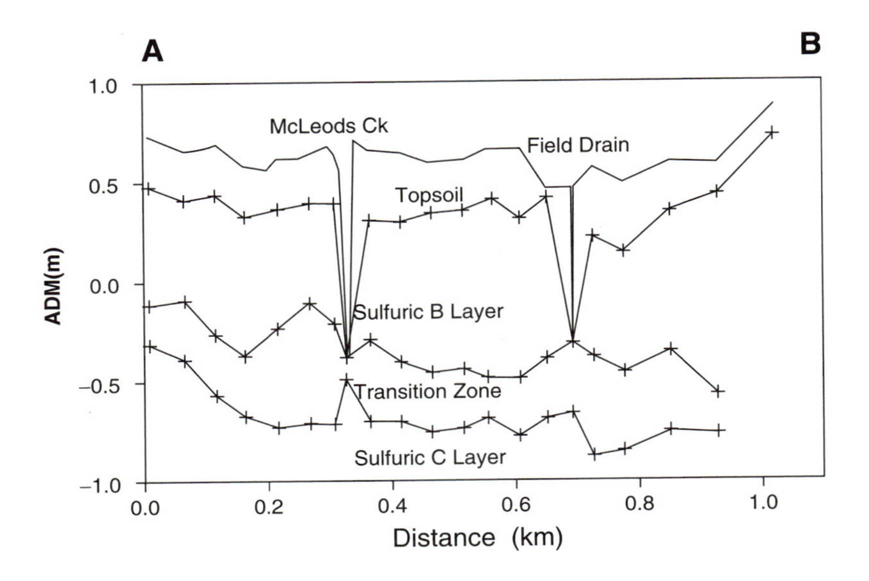

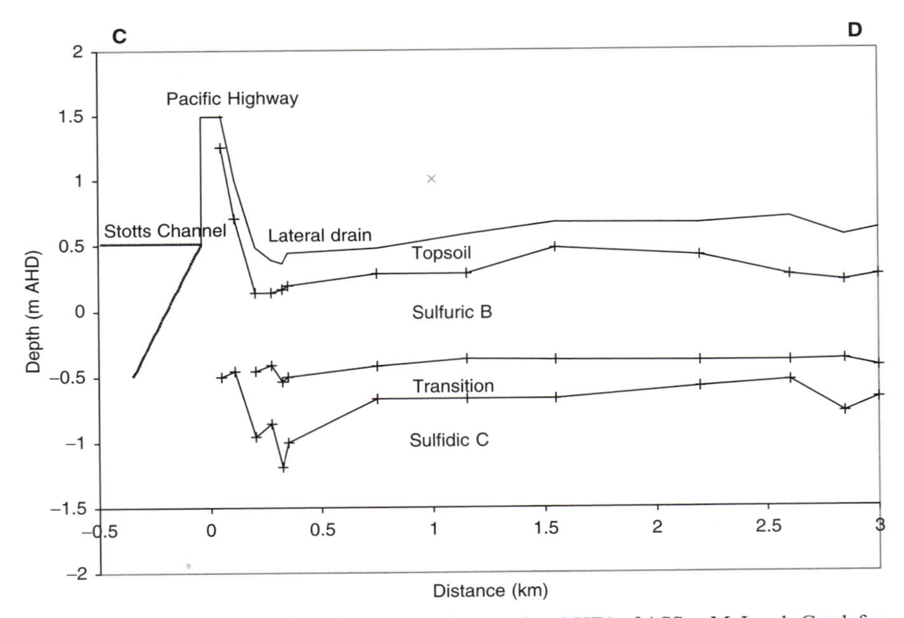

Figure 10 Near-surface stratigraphy (about +1 m to −2 m AHD) of ASS at McLeods Creek for transects (see Fig. 7) of profiles (a) A–B and (b) C–D.

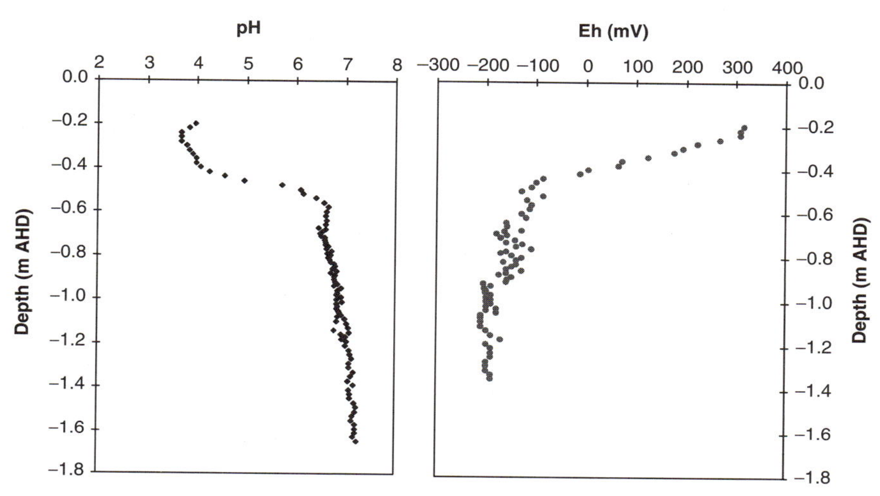

Figure 11 Field-measured profiles of pH (profiles D and F) and Eh (soil core) of acid sulfate soil at the McLeods Creek site.

The magnitude of the existing, leachable acidity (soluble plus exchangeable acidity species) can be measured by the depth-weighted "total actual acidity" (TAA; Konsten *et al.*, 1988) on representative samples from the ASS profile. For McLeods Creek canelands the mean value (from $n = 45$ profiles) is 47.4 tonnes of sulfuric acid equivalent per hectare. For ASS with predominantly pastureland in the adjacent Cudgen Lake catchment, the mean value ($n = 20$) is 44.1 tonne/ha. The overall mean value for these 65 ASS profiles is 46.4 tonnes H_2SO_4/ha (generalized to approximately 50 tonnes/ha). Estimates we have made (e.g., White *et al.*, 1997) suggest that acidity discharges from these sorts of sugarcane and pasture ASS landscapes are of the order 0.1 to 0.3 tonne H_2SO_4/ha/year. Given the large total areas of ASS involved with these coastal catchments (e.g., about 15,000 ha for the Tweed River catchment), there is a major need for the management of existing acidity discharge, as well as for avoidance in the creation of any new acidity from the huge potential acidity represented by the pyrite in the sulfidic sediment (Fig. 9).

B. Hydrology and Acidity Management of ASS Sugarcane Landscapes

The dynamics of acidity formation and discharge from ASS landscapes is a function of the hydrological behavior of the catchment system (e.g., Walker,

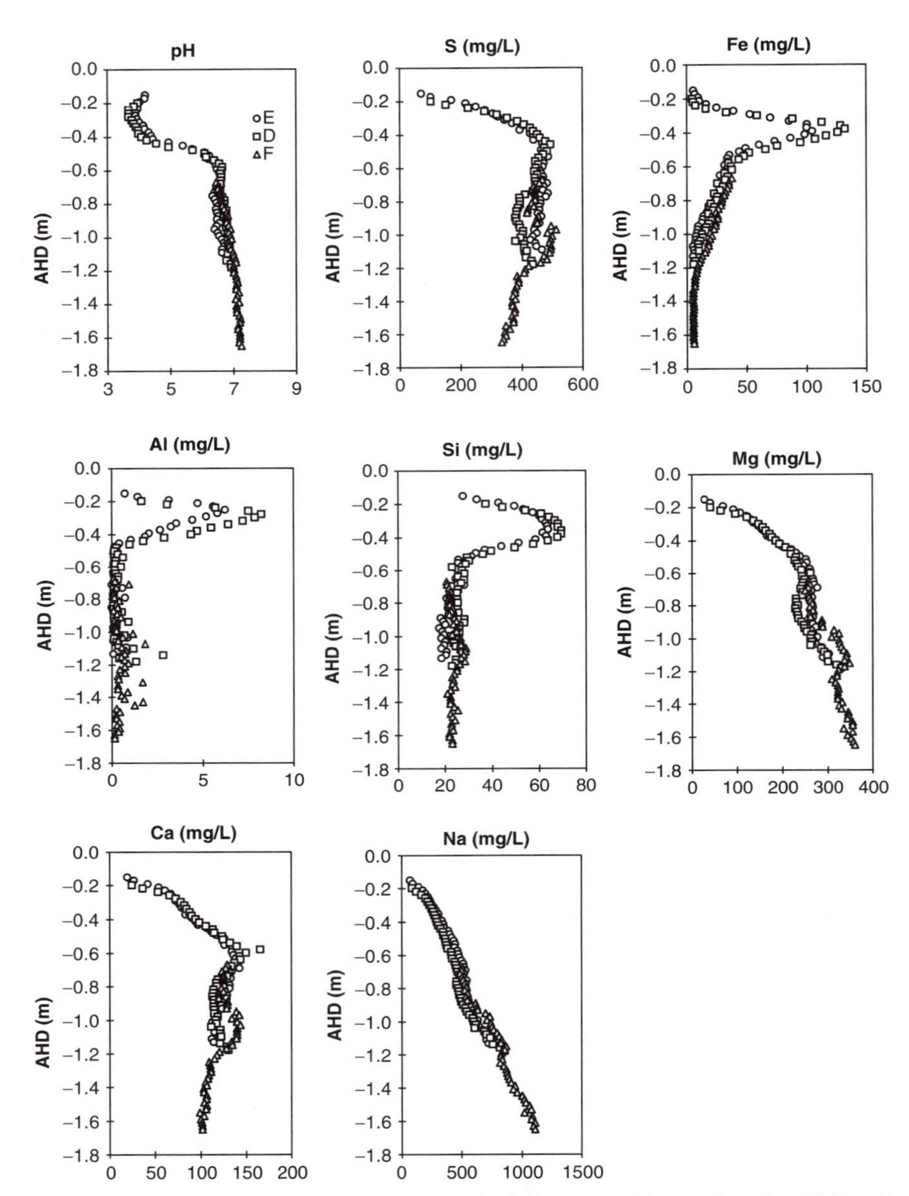

Figure 12 Concentrations of selected parameters in fluids extracted from soil profiles (D, E, and F) using "peepers."

1972; White and Melville, 1993; White *et al.*, 1997; Wilson *et al.* 1999; Cook *et al.*, 2000). Detailed hydrological studies at McLeods Creek reported by Wilson *et al.* (1999) show that acidity discharge is predominantly from very rapid, upward vertical movement of the acidity in the ASS profile in response to rain and its transfer into the caneland drains. Laser leveling of cane fields to minimize rain infiltration and incorporation of surface organic mulches after harvest to absorb acidic species are management options being trialed to maximize acidity retention in the landscape. The very small hydraulic conductivities and heads between the centers of cane paddocks and the drains ensure that the lateral flow of acidity through the ASS landscapes to drains is small. However, the exposed drain batters and contiguous ASS profiles (< 1 m from drains) do provide direct inputs of acidity so that in-filling of unnecessary drains, and ground limestone application to remaining drains, is a practicable management option now being applied in the cane industry (see White *et al.*, 1997). The acidity in the minor field drains is transferred to Stotts Channel via McLeods Creek, with partial neutralization in the lower creek section from brackish water that leaks through the tidal flap gates at the outlet. Within these sorts of sugarcane drainage systems, and other landscape sites such as wetlands, the principles of PRB have scope for development and application to control the slow leakage of any acidity generated by minor rain events. Major flows, such as in McLeods Creek, require rapid active acidity management such as can be provided by lime-dosing equipment.

C. Preliminary Design Considerations

As indicated earlier, OLDs represent potentially low-maintenance systems capable of treating small volumes of oxygenated discharge waters from acid sulfate soils. Such permeable reactive barriers can be installed with minimal disturbance to the existing land use. Important design issues in transferring OLD techniques from acid mine drainage applications to acid sulfate soils are addressed in the following sections.

1. Hydrology

The subcatchment in which the reactive barrier will be located is approximately 100 ha in area. The land is used for sugarcane farming, and numerous field drains are designed to remove flood waters to ground level within 24 h. Two pumps are located at the most downstream point to ensure quick removal of water from the OLD catchment to Stotts Channel, which runs into the Tweed River. There is minimal slope, thus large flow rates cannot be generated without substantial earthworks or pumps. While this is not an issue for flows through the OLD (as slow flow rates are required), there is concern that there will not be enough pressure head during flushing to agitate and remove the metal hydroxide from the drain.

Flow rates and volumes vary throughout the year on the cane field, depending on rainfall, soil moisture, cane development, and evaporation. While no gauging has been undertaken, estimates obtained from pumping rates used to minimize flooding at periods of high rainfall suggest that flows may range from zero to approximately 300 liter/s at the major outlet to Stotts Channel. If pumping were maintained for, say, 12 h over the wettest month of March, this would correspond to a total discharge of around 13 Megalitres for this month. Variable flow significantly complicates design of the OLD, as there are periods of excess water when water bypass will be required and periods of insufficient water supply when saturation of the OLD will not be maintained.

During dry periods it is important that the OLD remains saturated. The metal hydroxide flocs formed when the acid water reacts with limestone will remain suspended or loosely attached to the limestone surface as long as an aqueous phase is present; however, in the absence of a liquid phase, the iron oxides will form highly impermeable deposits on the limestone, which will hinder reactivity and permeability (Ziemkiewicz *et al.*, 1996).

During a large storm event, water will need to leave the site as sufficient land for storage will not be available. It is thus important to determine periods of a storm that need to be treated through the OLD (see Section VA2).

2. Water Quality

One of the major design requirements of a permeable reactive barrier for acid sulfate soils drainage treatment is the development of a system that will operate successfully over a range of flow volumes and acidities. It has been reported that major acid outbreaks occur after a long drought period followed by heavy rains (Callinan *et al.*, 1992). It is unclear, however, which portion of the hydrograph is most problematic. Is it necessary to treat the large flows associated with the hydrograph peak or is this component mainly surface runoff, which does not require treatment? Might the recession flows, when the extent of infiltration is greatest, be more active in transporting acid to the drainage channels and out of the system?

Wilson (1995) showed that large rises of the water table can be associated with relatively small rainfall events because of the low porosity of the soil. He found that a number of small rainfall events can gradually increase acidified water released into surface drains if prevailing conditions are suitable, while the large rainfall events did not always produce significant acidification of receiving waters, despite the larger flow volumes. It was shown that the position of the water table, available pore space, and store of acidic water behind floodgates can influence changes (Wilson, 1995). Sullivan and Bush (1999) also found that rapid acidification of drainage water could result from moderate rainfall. It has also been suggested by Sammut *et al.* (1996) that a delayed pulse of acid could be generated as a result of the mixing of monosulfides, formed in the base of drainage channels, with aerated runoff waters.

Scenarios such as these are being investigated intensively in predesign monitoring of the McLeods Creek catchment. Continuous monitors have been installed in the drainage channels, and data on water depth and quality (pH, EC, and temperature) are currently being collected. An example of data obtained from one of these continuous monitors through rainfall periods is shown in Fig. 13.

In this example, the pH is observed to drop from around 6.5 to approximately 4 on occurrence of a relatively small rainfall event. Interestingly, the pH remains low for a considerable time after the rainfall has ceased, suggesting the continued input of groundwater into the drainage channel.

Results from a second station are shown in Fig. 14. In this case, data for two time periods are reported with the first set (Fig. 14a) shown for a period of about 8 days while a second set lasting over 80 days is shown in Fig. 14b. For the shorter set, water depth in the drainage channel is also shown and reveals that the pH drops early in the hydrograph and is maintained low both during and well after peak flow. This result suggests that any PRB installed to

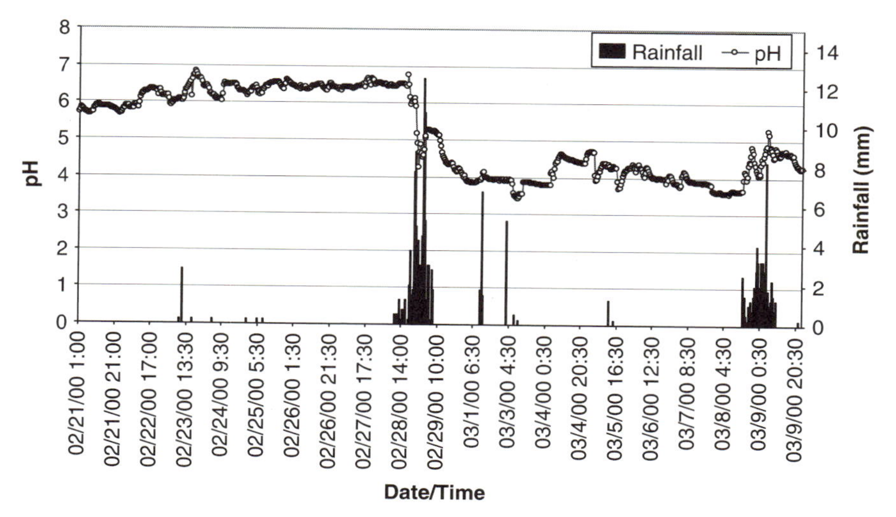

Figure 13 Continuous water quality record in a farm drain adjacent to McLeods Creek showing changes in pH through a rainfall event.

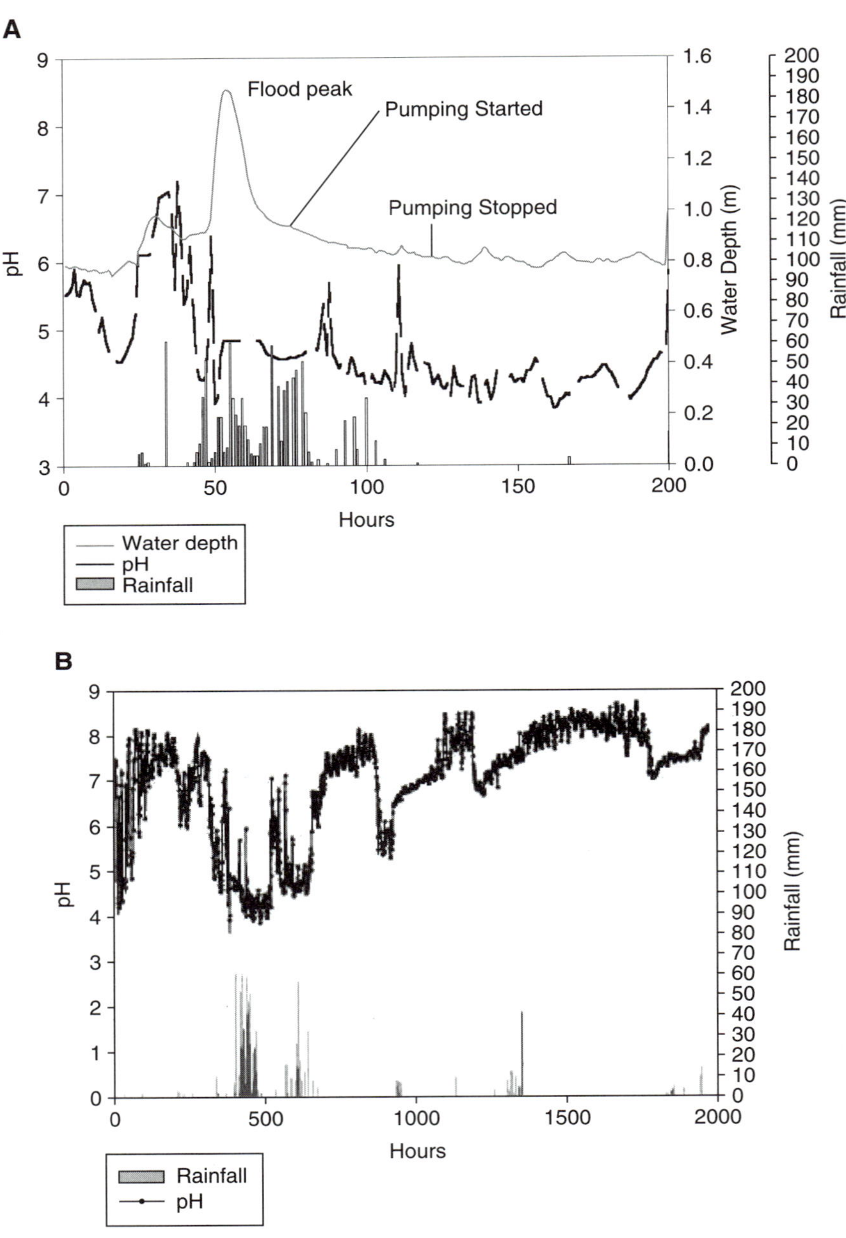

Figure 14 Results of continuous monitoring in field drains in the McLeods Creek catchment over (a) short and (b) long duration.

treat acid discharge would either have to cope with a very wide range of flows or be selective in which portion of the hydrograph is treated. This issue is discussed further in Section D.

In addition to continuous monitoring of a limited number of parameters, drainage waters in the McLeods Creek catchment have been obtained by near-surface grab sampling and analyzed for a wide range of chemical parameters. Electrical conductivity, pH, dissolved oxygen, redox potential, and temperature were measured at the site while samples were appropriately preserved for later analysis of major and minor anions and cations. In addition, acidity tests were performed on the samples in most cases within 1 h of collection. Titrations were conducted with NaOH and hot peroxide to end point 8.3, as stated in the American Public Health Association (1998) Standard Methods. Titrations were also conducted to end point 5.5, as this is the standard end point for acid sulfate soil acid tests.

The median, minimum, and maximum concentrations of selected parameters in samples obtained twice daily for 20 consecutive days in March and April 2001 are summarized in Table I.

Reasonably high concentrations of iron, aluminum, and silicon are observed at times with median concentrations of 2.4, 5.8, and 27.2 mg/liter, respectively. These median concentrations of iron and aluminum are well above the levels of 1 mg/liter for each element purported to cause armoring problems in anoxic limestone drains (Hedin *et al.*, 1994), although Cravotta and Watzlaf (Chapter 2) suggest that OLDs may remain effective for extended periods of time, even in the presence of Fe and Al in the 1- to 5-mg/liter range. These authors suggest that, provided the calcite surfaces remain moist, the iron and aluminum oxyhydroxide coatings remain relatively loosely attached and can be sloughed off. The high levels of silicon observed ,however, are of particular concern as polymeric iron silica and alumina silica coatings would be expected to be particularly troublesome.

The potential for serious armoring problems supports the proposal that an enclosed oxic limestone drain rather than an open limestone channel is preferred when treating acid sulfate soils discharge because (1) it will be easier to maintain saturated conditions and (2) the higher P_{CO_2} that will occur under enclosed conditions should result in more rapid calcite dissolution with a resultant "refreshing" of the surface of the neutralizing agent. Additional means of limiting armoring are also possible. For example, use of mixtures of calcite and magnesite (MgO) could limit the production of particularly troublesome aluminite [$Al_2(SO_4)(OH)_4 \cdot 7H_2O$] coatings due to the precipitation of $MgSO_4$. It is clear that a suitable flushing system will need to be incorporated into the drain design. In addition, consideration will need to be given to the disposal of any material generated in the flushing process, as it is certain to exhibit a high metal content. An alternate approach to the use of a

Table I
Water Quality Data from March and April 2001[a]

Parameter	N	Median	Minimum	Maximum
pH	41	3.9	3.3	5.2
DO (ppm)	37	6.1	3.3	11.0
EC (mS/cm)	40	3.6	0.6	6.0
Redox (mV)	25	363	149	486
Temperature (°C)	40	26.7	19.3	34.5
Acidity 5.5 (mg/liter $CaCO_3$)	38	35.8	<0	106
Acidity 8.3 (mg/liter $CaCO_3$)	41	53.8	<0	142
Cl (mg/liter)	41	776	114	1504
SO_4^{2-} (mg/liter)	41	619	152	938
S (mg/liter)	41	202	41.6	295
Fe^{2+} (mg/liter)	41	1.7	0.1	4.7
Fe (mg/liter)	41	2.4	0.1	6.6
Al (mg/liter)	41	5.8	0.1	11.0
Ca (mg/liter)	41	65.2	13.9	103
K (mg/liter)	41	14.3	2.1	32.0
Mg (mg/liter)	41	114	18.4	179
Mn (mg/liter)	41	1.5	0.2	2.7
Na (mg/liter)	41	473	64.3	920
Si (mg/liter)	41	27.2	11.0	32.9

[a]The location of sampling was at the site at which the OLD is being installed within the McLeods Creek catchment.

flushing system is to design the barrier in a modular manner such that segments can be replaced once fouled. Such an approach, while negating the passive treatment concept, minimizes the release of potentially troublesome iron and aluminum solids to receiving waters. Both options are being implemented in the current investigations.

D. Concept Design

Two OLD systems will be installed within the McLeods Creek catchment. One OLD is a modular system in which segments (particularly those that become clogged or inactive) can be removed periodically while the other is nonmodular (Fig. 15). The nonmodular system is a design that is typical for

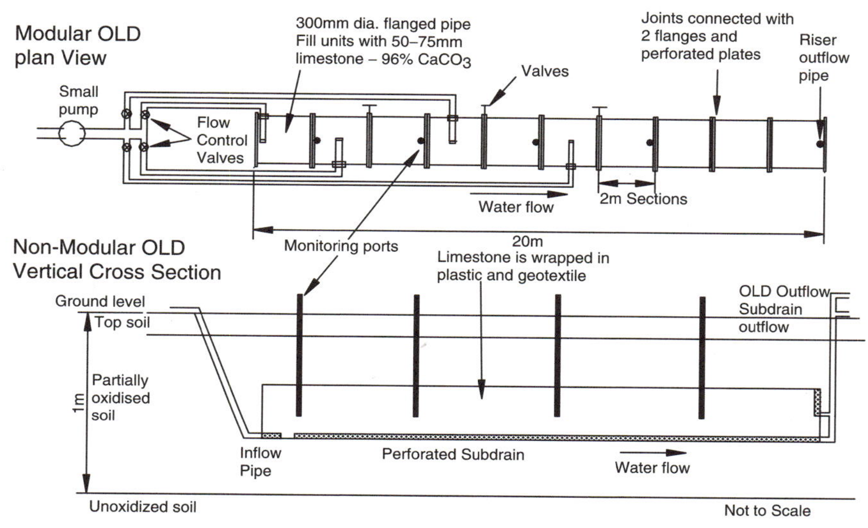

Figure 15 Schematics of modular and nonmodular permeable reactive barriers being trialed at the McLeods Creek test site for treatment of acid sulfate soils discharge.

OLDs and ALDs. Limestone will be buried within the soil and wrapped in plastic and geotextile to form a gas and water impermeable unit. The soil that will be displaced in burying the limestone is a clay that will be used to provide a soil cap over the OLD. This soil is typically partially oxidized and will need some lime treatment prior to use. A perforated subdrain will be installed at the base of the OLD, and trials using valves and/or pumps will be undertaken to examine the ease of flushing of metals oxyhydroxides from the system. Investigations into the use of different flushing techniques, such as building up pressure in the system by capping the outflow, pumping, mixing, pulsing, or baffling, will be undertaken. The addition of chemicals to the system to assist the flushing process is also an option requiring further investigation. As indicated earlier, the flushed material is anticipated to be high in metal content and, for this reason, should not be discharged into downstream waters. Treatment and disposal methods will thus need to be investigated.

The modular OLD will be constructed with flanged pipe sections that form 2-m modules. The limestone will be held within the module through the use of perforated PVC plates that attach to each flange. Typically, the inflow section of an OLD/ALD will accumulate metal hydroxides and limit flow through the system. This modular system has four inflow points along the drain in an attempt to distribute the accumulation of metal oxyhydroxides and to reduce the severity of flow blockage (see Chapter 12).

Cravotta and Watzlaf (see Chapter 2) provide detailed design information for limestone drains. Using their Eq. (12), a residence time of 3 h, a porosity of 50%, and a stone density of 2650 kg/m^3, the flow rate through each OLD is estimated to be on the order of 4 liter/min. After construction, the flow rate and porosity will be investigated further to determine the most suitable settings. Using the equation developed by Cravotta and Watlaf (Chapter 2), the total mass of limestone required is estimated to be on the order of 4 tons. Limestone fragments of 50–70 mm in size are used commonly in ALD/OLD installations and will be initially trialed here.

It should be pointed out that the OLDs under construction will not have the capacity for treating peak drainage flows following rainfall events. As indicated earlier, flows in the order of 300 liter/s are pumped from the site following rainfall events; flows that are far in excess of that which can be passed through the OLDs. At these times, more active acid neutralization procedures will need to be adopted with the OLDs used to treat base flow (which, as seen previously in Figs. 13 and 14, continue to be highly acidic following rainfall events). It should also be noted that a significant proportion of the peak flows will be made up of surface runoff, which is not particularly acidic and is best routed over or around the PRB.

VI. FUTURE POSSIBILITIES

Experience in the use of OLDs in the mining industry (U.S. Environment Protection Agency, 2000) indicates that failure occurs when

- the drain is clogged with sediment or hydrous oxides;
- the pressure of CO_2 cannot be maintained at a high level;
- the design residence time of the system is not maintained; e.g., short circuiting of the water flow in the system will not provide adequate contact time for limestone dissolution to occur;
- the acidity is higher than the reaction rate allowed by the drain; and
- the metal concentrations are greater than expected.

The OLD system described in Section V is not site specific and the concepts should be transferable to different localities. While still to be tested, the system should be practical in terms of capital and maintenance costs. It can be used in many acid sulfate soils problem areas once a number of design issues have been resolved.

The long-term success of the OLD is a concern. It is envisaged that the major problem associated with use of an OLD to treat acid sulfate soils discharge waters will be the potential armoring of the limestone used in the reactive barrier. Innovative cleaning and flushing techniques will need to be used to ensure longevity of an OLD in an acid sulfate soils region. The metal oxyhydroxides can be removed by carefully designed, perforated subdrainage systems (Cravotta and Trahan, 1999). This could involve pressurized or pumped agitation of the flocs.

Material within the OLD can also be varied to improve effectiveness, if required. Different combinations of packing material could be used, including compost, seawater-neutralized bauxite refinery residue (red mud), sugarcane fly ash, and so on.

Cravotta and Trahan (1999) found that $CaCO_3$ dissolution occurred more rapidly at the inflow of a horizontal OLD where the water is farthest away from calcite saturation. As the pH rises, the dissolution of limestone is less vigorous. This would be expected to lead to a higher deposition of iron and aluminum solid phases at the "front end" of the barrier. A gradation in barrier porosity could assist in normalizing reactivity through the barrier, although little attention has been given to this option as yet. Indeed, design guidelines for OLDs are yet to be developed that incorporate these factors, and there is considerable scope for innovation in this challenging area.

VII. CONCLUSION

The generation and release of acidic water from acid sulfate soils are environmental problems of international scale. Changes in land use and hydrological systems have promoted the oxidation of pyrite in acid sulfate soils with a concomitant generation of acidic discharge waters that have significantly degraded coastal regions of a number of countries.

Oxic limestone drain technology, used extensively in the remediation of acid mine drainage (with varying success), would appear to hold some potential as a treatment technology for assisting in the management of acid sulfate soils drainage waters. It should be stressed that OLD technology is a method to treat the acidic water once formed and does not attempt to prevent the generation of new acidity in the soils and, as such, can be used to prevent the transport of acid downstream. If the OLD is to be beneficial to downstream waterways, it will need to be installed at a number of discharge locations and used in conjunction with strategic approaches to avoid generation of new acidity.

A number of factors need to be resolved prior to installation of an OLD in an acid sulfate soils region. Among preliminary data necessary for design are the water chemistry and local hydrology of the region. It is particularly important to develop an understanding of the variability in flow and acid discharge through storm events as these potentially constitute the times of greatest impact. While significant concern exists that armoring by iron and aluminum oxide precipitates may limit the reactivity of the permeable barrier, it is believed that maintenance of saturated conditions and installation of effective subdrainage and/or flushing procedures will be effective in minimizing this problem.

REFERENCES

Ahern, C. R., Ahern, M. R., and Powell, B. (1998). Guidelines for sampling and analysis of lowland acid sulfate soils (ASS) in Queensland 1998. QASSIT, Department of Natural Resources, Resource Science Centre, Indooroopilly.

Alpers, C. N., and Nordstrom, D. N. (1990). Stoichiometry of mineral reactions from mass balance computations of acid mine waters, Iron Mountain, California. *In* "Acid Mine Drainage: Designing for Closure," pp. 23–33. Geological Association of Canada/ Mineralogical Association of Canada joint meeting, Vancouver.

American Public Health Association (1998). "Standard Methods for the Examination of Water and Wastewater," 20th Ed. APHA, Washinton, DC.

Astrom, M. (1998). Partitioning of transition metals in oxidized and reduced zones of sulfide-bearing fine-grained sediments. *Appl. Geochem.* **13**, 607–617.

Bescansa, P., and Roquero, C. (1990). Characterization and classification of tidal marsh soils and plant communities in north-west Spain. *Cremlingen* **17**, 347–355.

Bloomfield, C. (1972). The oxidation of iron sulphides in soils in relation to the formation of acid sulphate soils, and of ochre deposits in field drains. *J. Soil Sci.* **23**(1).

Blunden, B. G., and Indraratna, B. (2000). Evaluation of surface and groundwater management strategies for drained sulfidic soil using numerical simulation models. *Aust. J. Soil Res.* **38**, 569–590.

Bowman, G. M. (1993). Case studies of acid sulphate soil management. *In* "National Conference on Acid Sulphate Soils" (R. Bush, ed.), pp. 95–115. New South Wales Agriculture, Wollongbar, NSW.

Brandjes, P. J., and Lauckner, F. B. (1997). On-farm assessment of two liming materials in cabbage and hot pepper cultivation on acid sulphate soil in Guyana. *Exp. Agricul.* **33**(2), 225–235.

Brodie, G. A., Britt, C. R., Tomaszewski, T. M., and Taylor, H. N. (1993). Anoxic limestone drains to enhance performance of aerobic acid drainage treatment wetlands: Experiences of the Tennessee Valley Authority. *In* "Constructed Wetlands for Water Quality Improvement" (G. A. Moshiri, ed), pp. 129–138.

Bronswijk, J. J. B., Groenenberg, J. E., Ritsema, C. J., van Wijk, A. L. M., and Nugroho, K. (1995). *Agricult. Water Manage.* **27**, 125–142.

Callinan, R. B., Fraser, G. C., and Melville, M. D. (1993). Seasonally recurrent fish mortalities and ulcerative disease outbreaks associated with acid sulphate soils in Australian estuaries. *In* "Selected Papers of the Ho Chi Minh City Symposium on Acid Sulphate Soils" (D. L. Dent. and M. E. F. van Mensvoort, eds.), pp. 403–410. ILRI Publication 53, International Institute for Land Reclamation and Improvement, Wageningen.

Carson, C. D., and Dixon, J. B. (1983). Mineralogy and acidity of an inland acid sulfate soil of Texas. *Soil Sci. Soc. Am. J.* **47**, 828–833.

Christensen, L. B., and Olesen, S. E. (1985). Leaching of ferrous iron after drainage of pyrite-rich soils and means of preventing pollution of streams. *In* "Proceedings of a Symposium on

Agricultural Water Management" (A. L. M. Van Wijk and J. Wesseling, eds.), pp. 279–289. A. A. Balkema, Rotterdam, Boston, 1986, Arnhem, Netherlands.

Cook, F. J., Hicks, W., Gardner, E. A., Carlin, G. D., and Froggatt, D. W. (2000). Export of acidity in drainage water from acid sulphate soils. *Mar. Pollut. Bull.* **41**(7–12), 319–326.

Couillard, D. (1982). Use of red mud, a residue of aluina production by the Bayer process, in water treatment. *Sci. Total Environ.* **25**, 181–191.

Cravotta, C. A., and Trahan, M. K. (1999). Limestone drains to increase pH and remove dissolved metals from acidic mine drainage. *Appl. Geochem.* **14**. 581–606.

Dent, D. (1986). Acid sulphate soils: A baseline for research and development. ILRI Publication 39, International Institute for Land Reclamation and Improvement, Wageningen.

Dent, D. (1992). "Reclamation of Acid Sulphate Soils" (Rattan Lal and B. A. Stewart, eds.). Springer-Verlag, New York.

Dent, D. L., and Pons, L. J. (1995). A world perspective on acid sulphate soils. *Geoderma* **67**, 263–276.

Easton, C. (1989). The trouble with the Tweed. *Fish. World* (March), 58.

Faulkner, B., and Skousen, J. (1995). Effects of land reclamation and passive treatment systems on improving water quality. *Green Lands* **25**(4), 34–40.

Hedin, R. S., Nairn, R. W., and Kleinmann, R. L. (1994). Passive treatment of coal mine drainage. U.S. Bureau of Mines Information Circular IC 9389.

Hedin, R. S., and Watzlaf, G. R. (1994). The effects of anoxic limestone drains on mine water chemistry. *In* "Proceedings of International Land Reclamation and Mine Drainage Conference and Third International Conference on the Abatement of Acidic Drainage," pp. 185–194. United States Department of the Interior.

Ho, G. E., Gibbs, R. A. and Mathew, K. (1992). Bacteria and virus removal from secondary effluent in sand and red mud columns. *Water Sci. Technol.* **23**, 261–270.

Husson, O., Verburg, P. H., Phung, Mai Thanh, and van Mensvoort, M. E. F. (2000). Spatial variability of acid sulfate soils in the Plain of Reeds, Mekong delta, Vietnam. *Geoderma* **97**, 1–19.

Jonckbloedt, R. C. L. (1998). Olivine dissolution in sulphuric acid at elevated temperatures: Implications for the Olivine process, an alternative waste acid neutralizing process. *J. Geochem. Explor.* **62**, 337–346.

Kepler, D. A., and McCleary, E. C. (1994). Successive alkalinity producing systems (SAPS) for the treatment of acidic mine drainage. *In* "Proceedings of International Land Reclamation and Mine Drainage Conference and Third International Conference on the Abatement of Acidic Drainage," pp. 195–204. United States Department of the Interior.

Konsten, C. J. M., Brinkman, R., and Anderson, W. (1988). A field laboratory method to determine total potential and actual acidity in acid sulphate soils. *In* "Selected Papers of the Dakar Symposium on Acid Sulphate Soils" (H. Dost, ed.), pp. 106–116. IRLI Publication 44, International Institute for Land Reclamation and Improvement, Wageningen.

Kraus, M. (1998). Development of potential acid sulfate paleosols in Paleocene floodplains, Bighorn Basin, Wyoming, USA. *Paleogr. Palaeochem. Palaeoecol.* **144**, 203–224.

Lessard, G., and Mitchell, J. K. (1985). The causes of aging in quick clays. *Can. Geotech. J.* **22**, 335–346.

Lin, C., and Melville, M. D. (1992). Mangrove soil: A potential contamination source to estuarine ecosystems of Australia. *Wetlands (Aust.)* **11**, 68–74.

Lin, C., Melville, M. D., White, I., and Wilson, B. P. (1995). Human and natural controls on the accumulation, acidification and drainage of pyritic sediments: Pearl River Delta, China and coastal New South Wales. *Aust. Geograph. Stud.* **33**(1), 77–88.

Lopez, E., Soto, B., Arias, M., Nunez, A., Rubinos, D. and Barral, M. T. (1998). Adsorbent properties of red mud and its use for wastewater treatment. *Water Res.* **32**, 1314–1322.

McConchie, D., and Clark, M. (1999). Acid sulfate soil neutralisation techniques. *In* "Proceedings of the Workshop on Remediation and Assessment of Broadacre Acid Sulfate Soils" (P. Slavich, ed.), pp 88–93. Acid Sulfate Soil Management Advisory Committee (ASSMAC), Southern Cross University, Lismore.

McConchie, D. M., Clark, M. W., and Hanahan, C. (1997). The use of seawater neutralised red mud from bauxite refineries to control acid mine drainage and heavy metal leachates. *In* "Geological Society of Australia Abstracts No. 49," p. 298. Geological Society of Australia, Townsville.

McConchie, D. M., Saenger, P., and Fawkes, R. (1996). An environmental assessment of the use of seawater to neutralise bauxite refinery wastes. *In* "Proceedings of 2nd International Symposium on Extraction and Processing for the Treatment and Minimisation of Wastes" (V. Ramachandran and C. C. Nesbitt, eds.), pp. 407–416. Scottsdale, AZ.

Metson, A. J., Gibson, E. J., Cox, J. E., and Gibbs, D. B. (1977). The problem of acid sulphate soils, with examples from North Auckland, New Zealand. *New Zeal. J. Sci.* **20**, 371–394.

Minh, L. Q., Tuong, T. P, van Mensvoort, M. E. F., and Bouma, J. (1997). Contamination of surface water as affected by land use in acid sulfate soils in the Mekong River Delta, Vietnam. *Agricult. Ecosyst. Environ.* **61**, 19–27.

Morel, F. M. M., and Hering, J. G. (1993). "Principles and Applications of Aquatic Chemistry." Wiley, New York.

Nairn, R. W., Hedin, R. S., and Watzlaf, G. R. (1992). Generation of alkalinity in an anoxic limestone drain. *In* "Achieving Land Use Potential through Reclamation: Proceedings of the 9th Annual National Meeting of the American Society for Surface Mining and Reclamation," pp. 206–219, American Society for Surface Mining and Reclamation, Duluth, MN.

National Working Party on Acid Sulfate Soils (1999). "National Strategy for the Management of Coastal Acid Sulfate Soils." NSW Agriculture, Wollongbar Agricultural Institute.

Naylor, S. D., Chapman, G. A., Atkinson, G., Murphy, C. L., Tulau, M. J., Flewin, T. C., Milford, H. B., and Morand, D. T. (1995). "Guidelines for the Use of Acid Sulfate Soil Risk Maps." Soil Conservation Service of N.S.W., Department of Land and Water Conservation, Sydney.

Palko, J., and Yli-Halla, M. (1993). Assessment and management of acidity release upon drainage of acid sulphate soils in Finland. *In* "Selected Papers of the Ho Chi Minh City Symposium" (D. L. Dent and M. E. F. van Mensvoort, eds.), pp. 411–418. ILRI Publication 53 International Institute for Land Reclamation and Improvement, Wageningen.

Parker G. K., Noller B. N., and Waite T. D. (1999) Assessment of the use of fast-weathering silicate minerals to buffer AMD in surface waters in tropical Australia. *In* "Proceedings of Sudbury '99: Mining and the Environment," pp. 1251–1260. Sudbury, Ontario.

Pearson, F. H., and McDonnell, A. J. (1975). Limestone barriers to neutralise acidic streams. *J. Environ. Engin. Divi. ASCE* **101**, (No. EE3), 426–440.

Pons, L. J. (1973). Outline of the genesis, characteristics, classification and improvement of acid sulphate soils. *In* "Proceedings of the International Symposium on Acid Sulphate Soils" (H. Dost, ed.), Vol I, pp. 3–27. ILRI Publication 18,International Institute for Land Reclamation and Improvement, Wageningen.

Pons, L. J., and van Breemen, N. (1982). Factors influencing the formation of potential acidity in tidal swamps. *In* "Proceedings of the Bangkok Symposium on Acid Sulphate Soils" (H. Dost and N. van Breemen, eds.), pp. 37–49. ILRI Publication 31, International Institute for Land Reclamation and Improvement, Wageningen.

Robbins, E. I., Cravotta, C. A., Savela, C. E. and Nord, G. L. Jr. (1999). Hydrobiogeo-chemical interactions in "anoxic" limestone drains for neutralization of acidic mine drainage. *Fuel* **78**, 259–270.

Rose, A. W., and Laurenso, F. J. (2000). Evaluation of two open limestone channels for treating acid mine drainage. *In* "Proceedings 2000 National Meeting of the American Society for

Surface Mining and Reclamation," pp. 236–247. American Society for Surface Mining and Reclamation, Tampa, FL.

Sammut, J. (1996). "An Introduction to Acid Sulphate Soils." Acid Sulfate Soils Management Advisory Committee, The Department of the Environment, Sports and Territories and the Australian Seafood Industry Council.

Sammut, J., White, I., and Melville, M. D. (1994). Stratification in acidified coastal floodplain drains. *Wetlands (Aust.)* **13**, 49–64.

Sammut, J., White, I., and Melville, M. D. (1996). Acidification of an estuarine tributary in eastern Australia due to drainage of acid sulfate soils. *Mar. Freshwater Res.* **47**(5), 669–684.

Schuiling, R. D. (1998). Geochemical engineering; taking stock. *J. Geochem. Explor.* **62**, 1–28.

Skousen, J. (1997). Overview of passive systems for treating acid mine drainage. *Green Lands* **27**(4), 34–43.

Skousen, J., Sexstone, A., Sterner, P., Calabrese, J., and Ziemkiewicz, P. (1999). Acid mine drainage treatment with a combined wetland/anoxic limestone drain: Greenhouse and field systems. *In* "Proceedings 1999 National Meeting of the American Society for Surface Mining and Reclamation," pp. 621–633. American Society for Surface Mining and Reclamation, Scottsdale, AZ.

Smith, J., and, Melville, M. D. (2000). An assessment of spatial variations in actual acidity at McLeods Creek, Northeast NSW, Australia. *In* "Soil 2000: New Horizons for a New Century" (J. A. Adams and A. K. Metherell, eds.), Vol. 2, pp. 281–282. Lincoln University.

Soil Survey Staff (1999). "Soil Taxonomy." U.S. Dept. Agric. Handbook 436, 2nd Ed. U.S. Gov't Printing Office, Washington, DC.

Stone, Y., Ahern, C. R., and Blunden, B. (1998). "Acid sulfate soils manual 1998." Acid Sulfate Soil Management Advisory Committee, Wollongbar, NSW, Australia.

Stumm, W., and Morgan, J. J. (1996). "Aquatic Chemistry, Chemical Equilibria and Rates in Natural Waters." Wiley, New York.

Sullivan, L., and Bush, R. (1999). The behaviour of drain sludge in acid sulfate soil areas: Some implications for acidification of waterways and drain maintenance. *In* "Proceedings of the Workshop on Remediation and Assessment of Broadacre Acid Sulfate Soils" (P. Slavich, ed.), pp.43–48. Acid Sulfate Soil Management Advisory Committee (ASSMAC), Southern Cross University, Lismore.

Sun, Q., McDonald, L. M., Jr., and Skousen, J. G. (2000). Effects of armouring on limestone neutralization of AMD. *In* "West Virginia Surface Mine Drainage Task Force Symposium." Morgantown, WV.

Taylor, J. R., Waring, C. L., Murphy, N. C., and Leake, M. J. (1997). An overview of acid mine drainage control and treatment options including recent advances. *In* "Proceedings of the 3rd Australian Workshop on Acid Mine Drainage," pp. 147–160. Darwin, Australia.

Turner, D., and McCoy, D. (1990). Anoxic alkaline drain treatment system; a low cost acid mine drainage treatment alternative. *In* "Proceedings 1990 National Symposium on Mining," pp. 73–75. American Society for Surface Mining and Reclamation, University of Kentucky, Lexington, KY.

U.S. Environment Protection Agency (2000). "Coal Remining Best Management Practice Guidance Manual." 821-R-00-007, Environmental Protection Agency, Engineering and Analysis Division, Washington, DC.

Waite, T. D., and Parker, G. K. (1999). Overview of acid mine drainage management options: potential application in control of discharges from acid sulfate soils. *In* "Proceedings of the Workshop on Remediation and Assessment of Broadacre Acid Sulfate Soils" (P. Slavich, ed.), pp. 94–102. Acid Sulfate Soil Management Advisory Committee (ASSMAC), Southern Cross University, Lismore.

Walker, P. H. (1972). Seasonal and stratigraphic controls in coastal floodplain soils. *Aust. J. Soil Res.* **10**, 127–142.

Watzlaf, G. R., Schroeder, K. T., and Kairies, C. L. (2000a). Long term performance of anoxic limestone drains. *Mine Water Environ.* **19**, 98–110.

Watzlaf, G. R., Schroeder, K. T., and Kairies, C. L. (2000b). Long-term performance of alkalinity-producing passive systems for the treatment of mine drainage. *In* "Proceedings 2000 National Meeting of the American Society for Surface Mining and Reclamation," pp. 262–274. American Society for Surface Mining and Reclamation, Tampa, FL.

White, I., Heath, L., and Melville, M. (1999). Ecological impacts of flood mitigation and drainage in coastal lowlands. *Aust. J. Emerg. Manage.* **14**(3), 9–15.

White, I., and Melville, M. D. (1993). Treatment and containment of potential acid sulphate soils: Formation, distribution, properties and management of potential acid sulphate soils. Technical Report No. 53, Centre for Environmental Mechanics, Report to The Roads and Traffic Authority.

White, I., Melville, M. D., Sammut, J., van Oploo, P., Wilson, B.P., and Yang, X. (1996). "Acid Sulphate Soils: Facing the Challenges," Monograph No. 1. Earth Foundation of Australia, Sydney.

White, I., Melville, M. D., Wilson, B. P and Sammut, J. (1997). Reducing acidic discharge from coastal wetlands in eastern Australia. *Wetlands Ecol. Manage.* **5**, 55–72.

Wilson, B. P. (1995). "Soil Hydrological Relations to Drainage from Sugarcane on Acid Sulphate Soils." Ph.D. University of New South Wales, Sydney.

Wilson, B. P., White, I., and Melville, M. D. (1999). Floodplain hydrology, acid discharge and change in water quality associated with a drained acid sulphate soil. *Mar. Freshwater Res.* **50**(2), 149–157.

Yang, X., Zhou, Q., and Melville, M. D. (2000). An integrated drainage network analysis system for agricultural drainage management. 2. The applications. *Agricult. Water Manage.* **45**, 87–100.

Ziemkiewicz, P. F., Brant, D. L., and Skousen, J. G. (1996). Acid mine drainage treatment with open limestone channels. *In* "Proceedings of the Thirteenth Annual Meeting of the American Society for Surface Mining and Reclamation, Successes and Failures: Applying Research Results to Insure Reclamation Success" (W. L. Daniels, J. A. Burger, and C. E. Zipper, eds.), pp. 367–374. ASSMR and the Powell River Project of Virginia Tech, Knoxville, TN.

Ziemkiewicz, P. F., Skousen, J. G., Brant, D. L., Sterner, P. L., and Lovett, R. J. (1997). Acid mine drainage treatment with armored limestone in open limestone channel. *J. Environ. Qual.* **26**, 1017–1024.

Chapter 4

Permeable Reactive Barrier/ GeoSiphon Treatment for Metals-Contaminated Groundwater

W. E. Jones, M. E. Denham, M. A. Phifer, F. C. Sappington, and F. A. Washburn
Savannah River Technology Center, Westinghouse Savannah River Company, Aiken, South Carolina 29808

A combination of *in situ* (permeable reactive barrier) treatment, an innovative, relatively passive extraction system (the GeoSiphon), and secondary *ex situ* treatment has been used to treat low-pH, metal-contaminated groundwater. The GeoSiphon system uses natural hydraulic head to induce essentially passive groundwater flow for remediation. In this case, flow was induced through a limestone–gravel-filled permeable reactive barrier to increase pH to near the optimum for aluminum hydroxide and chromium precipitation. *Ex situ* treatment included aluminum hydroxide and chromium precipitate settling, followed by increasing the groundwater redox potential to convert iron from the ferrous-to-ferric state. Ferric iron and nickel precipitates were then allowed to settle prior to discharge of the treated groundwater. The GeoSiphon moved water effectively through the permeable reactive barrier to the *ex situ* treatment system. It is adaptable to a variety of single or multiple *in situ* and/or *ex situ* treatment systems.

I. INTRODUCTION

Groundwater remediation science and engineering have advanced since the early 1970s through a quest for more effective, faster, and inexpensive cleanup methods. In general, *in situ* groundwater remediation methods have received increasing attention (compared to pump and treat) due to the expectation of lower lifecycle costs. This chapter describes an innovative combination of *in situ* (permeable reactive barrier; PRB) treatment, a recently patented, relatively passive extraction system (the GeoSiphon), and secondary *ex situ* treatment to address multiple contaminants. The advantages of incorporating a relatively passive extraction system into remedial strategies can include reduced initial and long-term costs, and minimization or elimination of secondary waste streams.

The GeoSiphon system uses natural hydraulic head differences to induce flow for remediation through a passive or essentially passive operation (i.e., no external power is required to extract the water). A location with higher pressure head can be within the contaminated portion of an aquifer or a surface-water body. A location with lower pressure head can be within the same aquifer, another aquifer, the unsaturated (vadose) zone, the same surface-water body, another surface-water body, or the ground surface. The two points of hydraulic head difference do not have to be along the same natural groundwater or surface-water flow path (i.e., they may not be naturally hydraulically connected). GeoSiphon is used for relatively shallow groundwater recovery, having a maximum practical lift of 25 ft (7.6 m) at standard temperature and pressure.

The GeoSiphon method can significantly lower operating and maintenance cost compared to traditional pump-and-treat methods. The equipment itself is simple and typically available off the shelf. The passive operation of the GeoSiphon is similar to funnel-and-gate and continuous permeable-wall treatment systems. Depending on the type of GeoSiphon configuration, it can potentially accelerate clean up, be applied to a range of site conditions and contaminants, and increase treatment medium longevity and effectiveness (i.e., minimize medium armoring and/or cementation and formation pluggage by removing precipitates from the PRB). Other benefits may include no significant secondary waste production (as in the case of ion exchange for resin regeneration) and pH adjustment to minimize the formation of metal complexes that interfere with metals removal.

Another unique aspect of this technology is its broad-range applicability in a number of configurations. Depending on the configuration, it can be used to treat groundwater or surface water. By using appropriate treatment media and configuration, it can be used to treat radionuclides, metal, and/or volatile organic compounds (VOCs). The GeoSiphon treatment technology can also be configured *in situ* or *ex situ*, depending on the overall remedial strategy.

This pilot-scale project was funded by the U.S. Department of Energy's (DOE's) Subsurface Contaminants Focus Area to evaluate a relatively passive, low-cost treatment method for low-pH, metal-contaminated groundwater that links a PRB with an aboveground treatment system through a siphon line. This concept was the focus of the demonstration. Although an aboveground secondary treatment system was demonstrated, there was less emphasis on quantifying its effectiveness because it can consist of any number of technologies that can be designed to be as effective as required by the chemistry of the water, the regulatory limits, and cost.

GeoSiphon History

GeoSiphon systems have been demonstrated for the treatment of groundwater VOCs at the TNX Area and for dissolved groundwater metals at D Area (this study) at the DOE Savannah River Site (SRS) in South Carolina. The GeoSiphon configuration at the TNX Area is significantly different from the D-Area work described in this study. The TNX configuration includes two cells, one installed in 1997 and a second installed in 1998. The cells comprise 2.4-m (8-ft)-diameter wells installed via auger and removable caisson. The cells contain a granular cast iron treatment medium. Abiotic degradation of the trichloroethylene occurs as groundwater is drawn through the iron by siphon through a vertical 0.3-m (12-in.)-diameter screen located in the center of the granular iron. The treated groundwater is discharged to a permitted outfall ditch (Phifer *et al.*, 1999).

II. SITE DESCRIPTION

The SRS is a 842-km^2 (325 mi^2) DOE facility located near Aiken, South Carolina (Fig. 1). Steam and electricity are produced at the SRS in a moderate-to-low sulfur coal-burning power plant located in D Area (Fig. 2). Most of the coal is stored in an open, 3.6-ha (8.9-acre) pile adjacent to the 484-D PowerHouse. Runoff from the coal pile is discharged into the 5.1-ha (12.5-acre) coal pile runoff basin (CPRB). The CPRB was designed as a sedimentation/seepage basin for removal of suspended solids and for minimization of direct discharge of coal pile runoff to a nearby creek.

Long-term chemical and biological oxidation of coal-associated sulfur compounds produces sulfuric acid. Rain water and the sulfuric acid leach

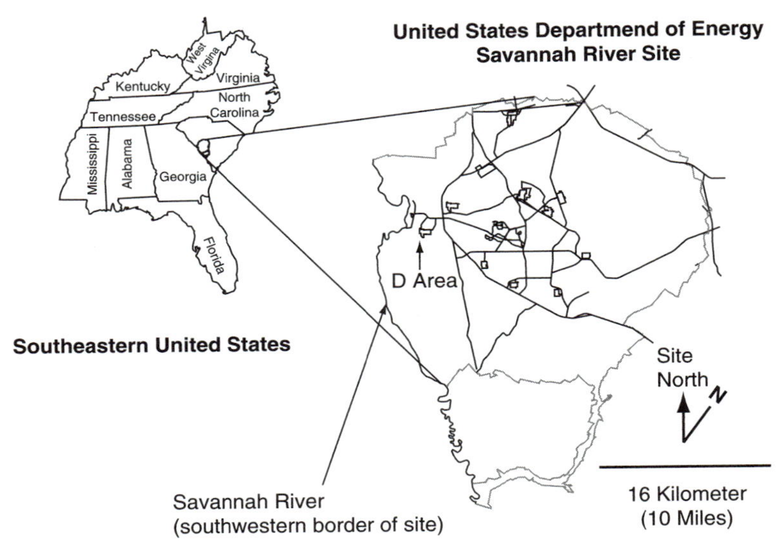

Figure 1 Location of the Savannah River site, Aiken, South Carolina.

other impurities from the coal. This produces an acidic runoff with substantial metals contamination that is collected and discharged via gravity drainage to the CPRB. This runoff predominantly seeps into the water table aquifer through the bottom of the CPRB. The chemistry of D-Area CPRB groundwater is acidic (pH ranging from 2 to 4), with typical constituent concentrations of 60 mg/liter total iron, 200 mg/liter aluminum, 1300 mg/liter sulfate, <1 mg/liter nickel and chromium(III), and minor concentrations of lead, cadmium, and trichloroethylene. A detailed description of the D-Area permeable reactive Barrier/GeoSiphon system construction and testing is in by Washburn *et al.* (1999) and is available electronically from the DOE's Office of Scientific and Technical Information "Information Bridge" internet site (http://www.osti.gov/bridge/product. biblio.jsp?osti_id=7506).

The D-Area Permeable Reactive Barrier/GeoSiphon pilot-scale treatment system consisted of a primary treatment trench filled with limestone, which functioned as a PRB, a siphon to transport contaminated groundwater, and a precipitate settling and secondary treatment system (Fig. 2 and 3). The PRB was designed to raise pH sufficiently to precipitate aluminum and chromium. The secondary treatment system, located aboveground at the GeoSiphon discharge point, was designed to allow the predominantly aluminum hydroxide flocculate to settle out and then allow addition of an oxidizer and a base for precipitation and removal (by settling) of the remaining contaminants, primarily iron and nickel hydroxide (Fig. 4).

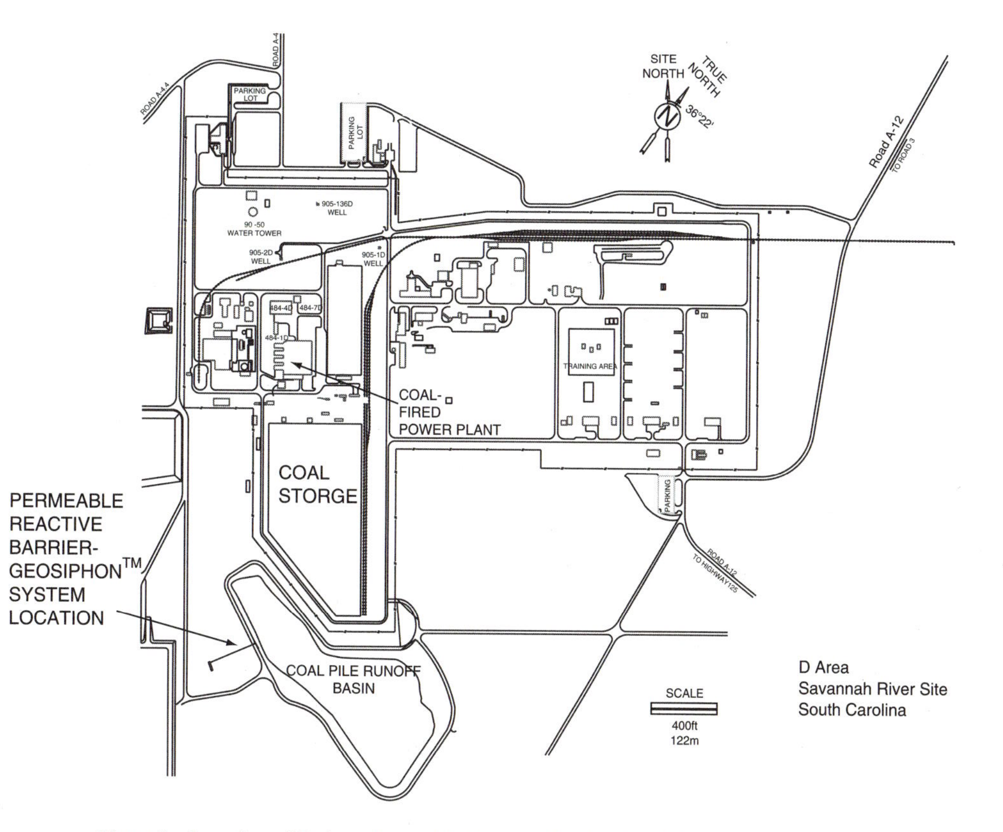

Figure 2 Location of D-Area Permeable Reactive Barrier/GeoSiphon treatment system.

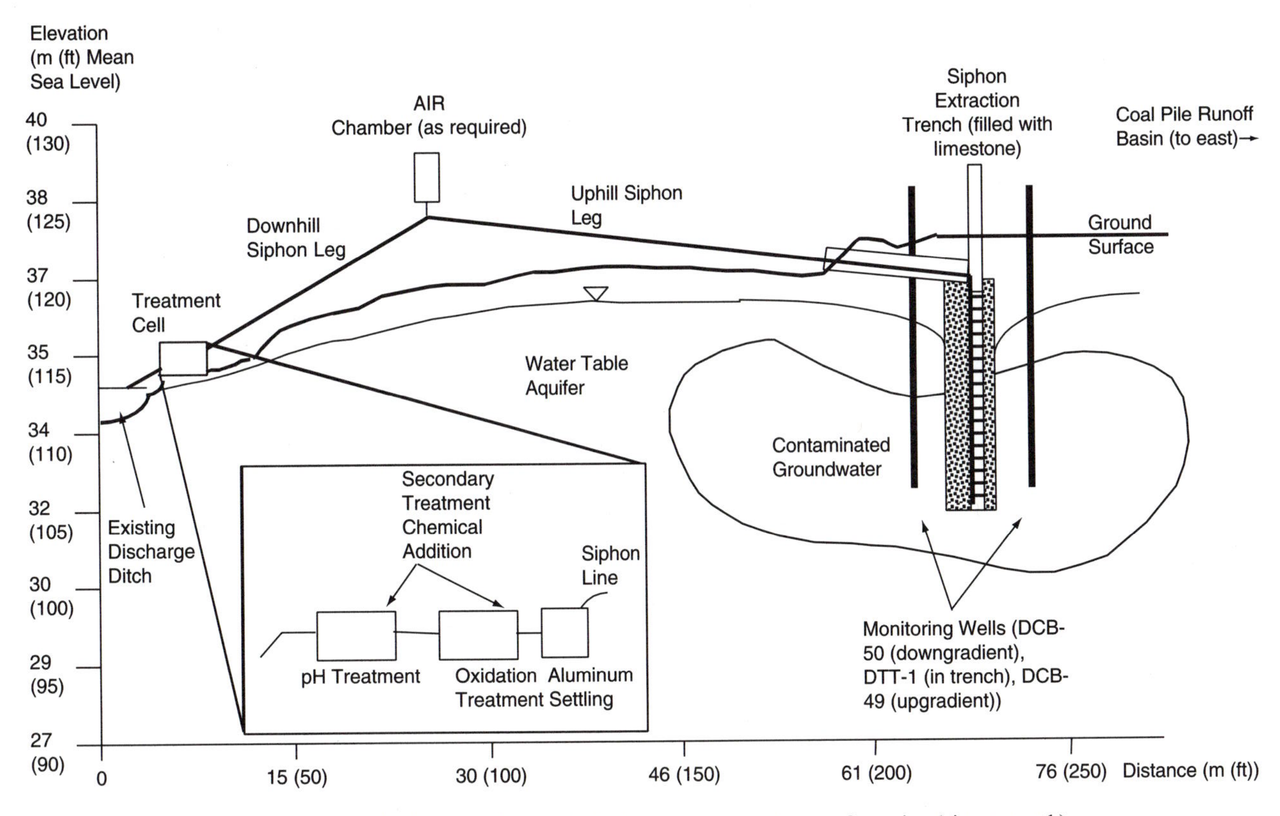

Figure 3 Permeable Reactive Barrier/GeoSiphon treatment system configuration (view to north).

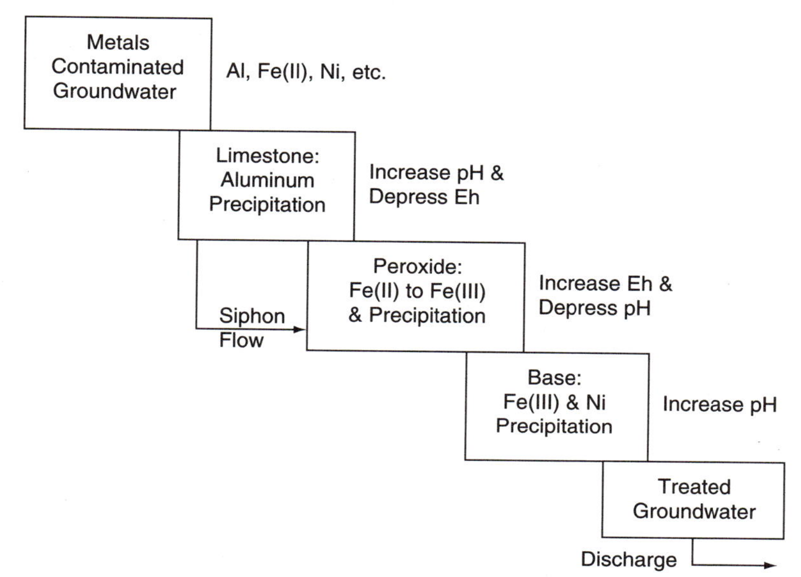

Figure 4 Diagram of D-Area Permeable Reactive Barrier/GeoSiphon treatment process.

III. HYDRAULIC DESIGN

Contaminated groundwater is collected by a limestone-filled trench (PRB) near the CPRB. The GeoSiphon transfers the water from the PRB to the precipitate settling and secondary treatment area (Fig. 3). Dissolved metals concentrations associated with the CPRB are highest in the uppermost 10 ft (3.0 m) of groundwater. The Permeable Reactive Barrier/GeoSiphon trench depth was designed to extract groundwater from this shallow zone comprising relatively fine grained, low hydraulic conductivity (2×10^{-3} cm/s to about 2×10^{-7} cm/s), interbedded sands, silts, and clays. Hydrogeologic characteristics for this interval (ground surface to about 5.2 m (17 ft) below ground surface) were used for PRB and siphon piping design.

The PRB design was 0.6 m (2 ft) wide by 12.2 m (40 ft) long by 4.9 m (16 ft) deep, with an approximate 3.0-m (10-ft) vertical saturated thickness. Within the PRB, 10.2-cm (4-in.)-diameter Schedule 40 polyvinylchloride (PVC) piping was configured in three vertical risers connected by two perforated horizontal pipes spaced 1.2 m (4 ft) apart (Fig. 5). Groundwater was extracted from the central riser, designated DTT-1. Both perforated horizontal pipes had four rows of 0.95-cm (3/8-in.)-diameter holes spaced about 10.16 cm (4 in.) apart to allow groundwater into the piping. All three risers extended to

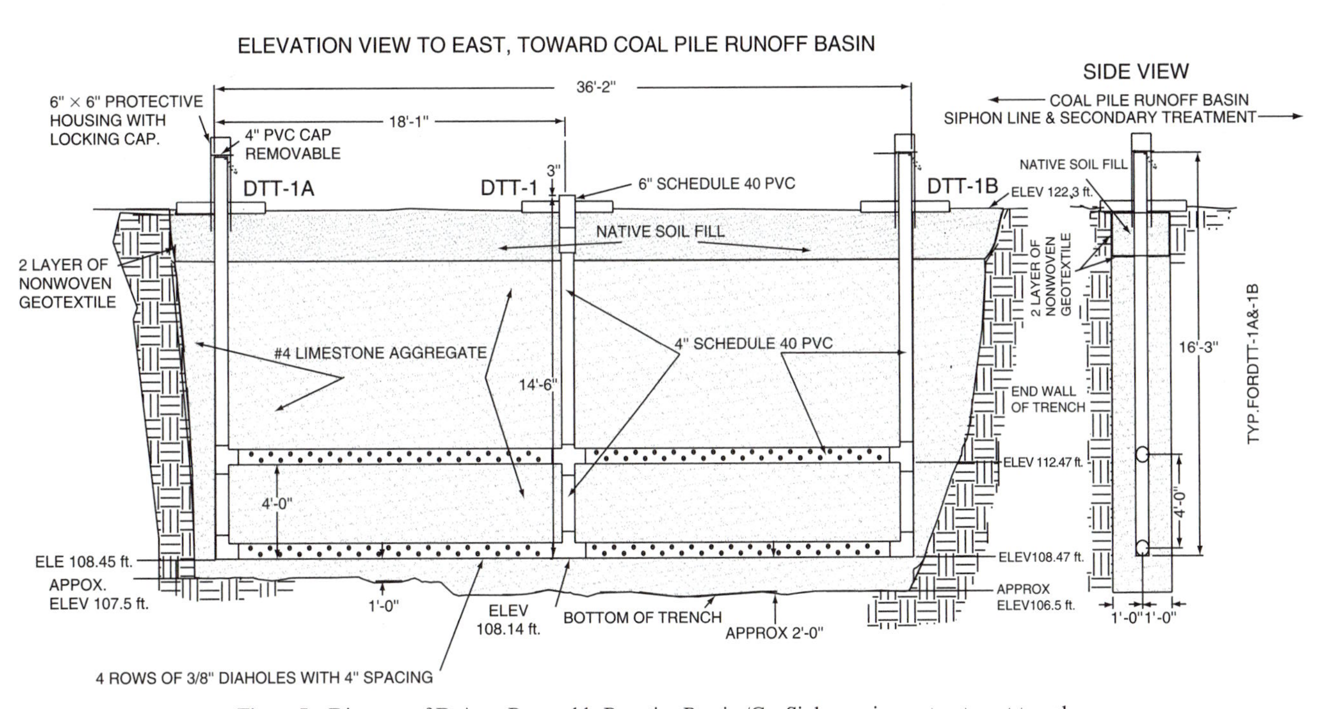

Figure 5 Diagram of D-Area Permeable Reactive Barrier/GeoSiphon primary treatment trench.

above the ground surface for access to the PRB. The trench was backfilled around the piping with Number Four size [i.e., 1.91 cm (3/4 in.) to 3.81 cm (1.5 in.)] limestone cobble aggregate on May 4, 1999.

An estimate of the specific capacity of the PRB was required for siphon system design and flow rate estimation. Transmissivity of the alluvium was estimated at 0.00053 m^2/min (0.0057 ft^2/min) from a nearby well screened in roughly the same interval as the PRB (Washburn *et al.*, 1999). A specific capacity of the PRB was estimated at 0.00754 liter/s/m (0.000873 ft^3/s/ft), assuming its hydraulic behavior is similar to a rectangular system of closely spaced wells (Powers, 1992). The groundwater extraction piping was sized considering both the specific capacity of the trench and the piping head loss, as outlined in Phifer *et al.* (1999).

The siphon line was approximately 79.2 m (260 ft) in total length, with the upward leg about 55.5 m (182 ft) and the downward leg about 23.5 m (77 ft). Four different siphon-line configurations were tested:

1. A 2.54-cm (1-in.)-diameter upward leg to crest and a 1.27-cm (0.5-in.)-diameter downward leg to discharge, siphon on an engineered grade, no air chamber.
2. Continuous 2.54-cm (1-in.)-diameter siphon with an air chamber at the crest and a valved discharge (restricted flow) on an engineered grade.
3. Continuous 2.54-cm (1-in.)-diameter siphon with an air chamber at the crest (unrestricted flow) on an engineered grade.
4. Continuous 1.27-cm (0.5-in.)-diameter siphon on a nonengineered grade, no air chamber (simplest possible GeoSiphon configuration).

Siphons require priming (initial filling of line) to initiate flow. This can be accomplished by gravity filling from the high point in the line (crest) with the inlet and outlet valved off or by using a vacuum pump at the crest with the inlet and outlet submerged and open. After priming, the siphon will convey liquid from the point of higher hydraulic head to the point of lower hydraulic head as long as head differential is maintained and prime is not lost (Gibson, 1961; Loitz *et al.*, 1990; Phifer *et al.*, 1998, 1999).

Air accumulation can break the siphon. This can be avoided by use of submerged inlets and outlets to prevent air from being drawn into the siphon and by maintenance of full flow in the siphon. Maintenance of full flow is achieved by various methods. One method is maintenance of the minimum flow velocity (MFV) required to transport gasses (that have degassed from the liquid) through and out the siphon. This method is employed most effectively when significant head differential results in adequate flow velocity.

Alternative methods are also available for the removal of accumulating gasses from the siphon (e.g, an air chamber located at the siphon line crest). For this project, very little head differential was available [from about 0.8 m

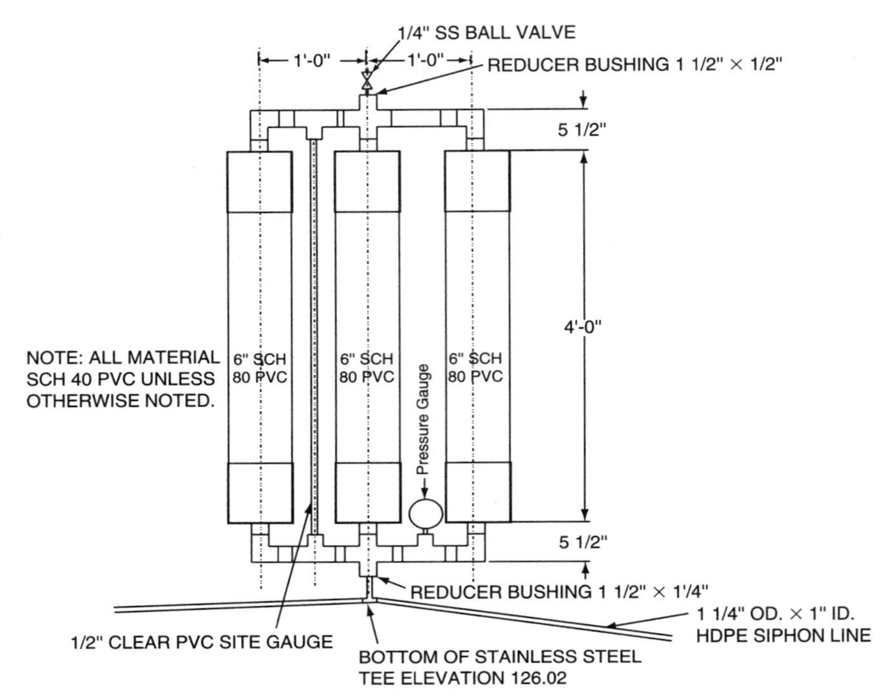

Figure 6 Diagram of air chamber configuration for the D-Area Permeable Reactive Barrier/GeoSiphon treatment system.

(2.5 ft) to about 1.8 m (6 ft)]. Although extended flow without maintaining an air chamber was possible, particularly with the optimally sized piping, the air chamber did enhance system performance. A schematic diagram of the air chamber is presented in Fig. 6.

IV. CHEMISTRY DESIGN

The D-Area CPRB groundwater is acidic (test location pH is approximately 3.5) and is contaminated primarily with ferrous iron, aluminum, nickel, and sulfate (with minor chromium). Determination of iron speciation was necessary for optimal treatment system design. The dominant metals (ferrous iron and aluminum in this case) were considered in designing a two-stage treatment system. For example, if ferric iron were present, ferric hydroxide would likely precipitate within the PRB, armoring the limestone and reducing PRB efficiency.

Prior studies at SRS demonstrated that limestone is an effective treatment for aluminum and chromium in this groundwater (Washburn *et al.*, 1998). The elevated pH caused by contact of the groundwater with limestone induces precipitation of these metals. However, as it does in an anoxic limestone drain, ferrous iron passes through the limestone and must be treated by an additional step. The solubility vs pH curves in Fig. 7, calculated using the equilibrium geochemistry program PHREEQC (Parkhurst, 1995), show that precipitation of iron in D-Area CPRB groundwater requires a large increase in pH to precipitate ferrous iron or oxidation of the ferrous iron to ferric iron. Oxidation was chosen as the additional precipitation step for the D-Area GeoSiphon. One consideration for the use of oxidation is that chromium(III) may be oxidized to more toxic chromate species. The two-stage treatment is designed to prevent this by removing chromium(III) from groundwater in the first stage, prior to oxidation.

The resulting two-stage treatment design was as follows.

- Primary treatment in PRB: Increase pH to near the optimal pH for aluminum hydroxide precipitation by flow through the limestone-filled trench. During this stage, aluminum and chromium precipitate, and iron remains in solution as the ferrous (iron II) form. The aluminum precipitates are then removed in the trench and by settling at the GeoSiphon discharge point prior to the second stage.
- Secondary treatment: Increase the redox potential to convert iron from ferrous (iron II) to ferric (iron III). During this stage, nickel

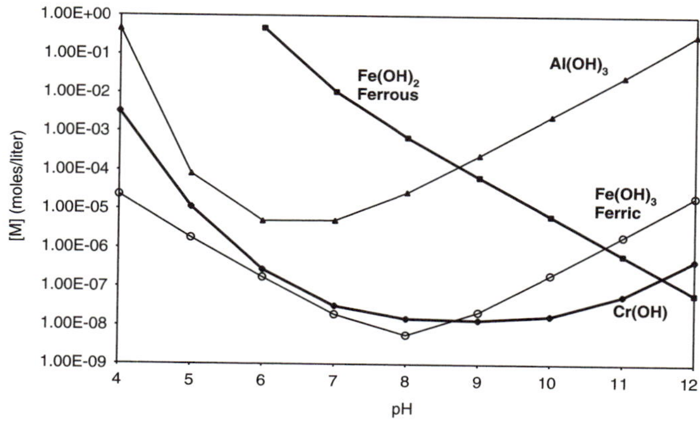

Figure 7 Solubility of hydroxides of aluminum, ferric iron, and chromium vs pH in D-Area Coal Pile Runoff Basin groundwater. Calculated using the equilibrium geochemistry program PHREEQC (Parkhurst, 1995).

precipitates along with ferric iron as a coprecipitate. The precipitates are removed by settling.

Successive treatment campaigns were conducted to test different second-stage treatments. These were as follows:

- Limestone treatment/settling alone.
- Limestone treatment/settling followed by hydrogen peroxide.
- Limestone treatment/settling followed by calcium peroxide.
- Limestone treatment/settling followed by hydrogen peroxide and sodium carbonate for additional alkalinity.

Traditional wastewater practices for primary sedimentation and calculated flow rates for the trench output were used to design the settling tanks (Metcalf and Eddy, 1991). Water depth in the tanks was limited to 0.30 m (12 in.) to ensure proper siphon operation, as the head difference between extraction and discharge points ranged from about 0.61 m (2 ft) to 1.83 m (6 ft). Tank placement and general working space were limited to an area 7.31 m (24 ft) long by 3.66 m (12 ft) wide to minimize construction impacts on area wetlands. The treatment tank volume calculation was based on flow rate and a typical residence time of 2 h. The resulting tank dimensions were 1.22 m (4 ft) wide by 1.52 m (5 ft) long by 0.53 m (1.75 ft) high. A smaller, 0.91 m (3 ft) wide by 1.22 m (4 ft) long by 0.53 m (1.75 ft) high, tank was also constructed for placement of additional limestone in the event of trench material armoring during testing.

V. HYDRAULIC PERFORMANCE

A. PRB/Trench Hydraulics

Groundwater extraction from the trench was initiated on May 26, 1999. Graphical data for the siphon-line operation are provided in Fig. 8. During the initial test campaigns (or runs), May 26, 1999 through July 30, 1999, both with and without the air chamber, the maximum flow rate allowed by the state regulatory permit was 3.8 liter/min (1 gpm).

During run 1 (the 2.54-cm (1-in.)-diameter to 1.27-cm (0.5-in.)-diameter siphon-line configuration), unrestricted flow was typically about 1.9 liter/min (0.5 gpm). The air chamber configuration (runs 2 and 3) used 2.54-cm (1-in.)-diameter high-density polyethylene (HDPE) tubing for both the uphill and downhill legs. During the first air chamber configuration (run 2), flow control [reduction to 1.27-cm (0.5-in.)-diameter line at the discharge end along with a valve] was required to maintain a flow at or less than the regulatory limit of

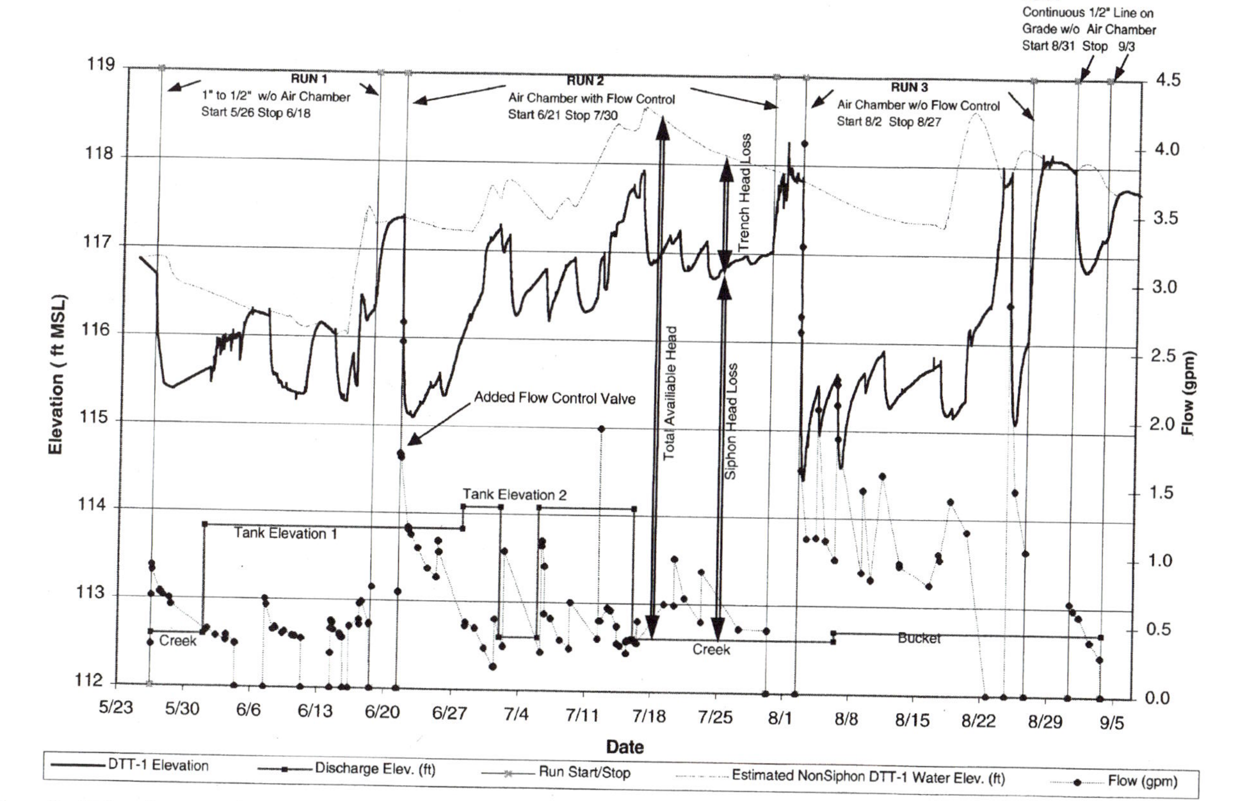

Figure 8 Siphon line operation for runs 1–4 showing waterlevel elevation and flow, D-Area Permeable Reactive Barrier/GeoSiphon treatment system.

3.8 liter/min (1 gpm). With state regulatory permission granted July 29, 1999, the system was allowed to operate with unrestricted flow (runs 3 and 4) after the treatment tests were completed. The highest rates of unrestricted flow were achieved during run 3 using the 2.54-cm (1-in.)-diameter line with the air chamber on the elevated, engineered grade. The purpose of operating with unrestricted flow was to determine the maximum sustainable flow rate of the trench/siphon. Unrestricted flow during run 4 was also used for the continuous 1.27-cm (0.5-in.)-diameter line lying on a nonengineered grade, without the air chamber, the simplest possible GeoSiphon configuration.

Maximum sustainable flow using the 2.54-cm (1-in.)-diameter line with air chamber varied from about 3.8 liter/min (1 gpm) to just over 1.5 gpm. The variation in sustainable flow rate is related to air chamber recharge (vacuum renewal) episodes. Flow rates increased to near or above 5.7 liter/min (1.5 gpm) with air chamber recharge. Between recharge events, flow rates decreased to at or just below 3.8 liter/min (1 gpm). The maximum sustainable flow rate for the continuous 1.27-cm (0.5-in.)-diameter line lying on nonengineered grade (without air chamber) was about 1.9 liter/min (0.5 gpm), similar to the flow rate for the 2.54-cm (1-in.)-diameter to 1.27-cm (0.5-in.)-diameter line.

Flow and drawdown measurements from an electronic data logger were used to determine a range of specific capacities for the trench during the field-testing period. Specific capacity estimates were derived by dividing the discharge rate by the trench drawdown measured at DTT-1. Specific capacity values from operational data were calculated for approximately 24 and 48 h following the initial extraction of groundwater from the trench during one or more extraction episodes for each of the line configurations.

Typical drawdown within the trench was 0.61 m (2 ft) or less. Therefore, a representative minimum volume to be removed to achieve equilibrium extraction conditions would be a 0.61-m (2-ft) drawdown by a 0.61-m (2-ft) width by a 12.2-m (40 ft) length, or 4.53 m^3 (160 ft^3), multiplied by 0.3 (approximate aggregate porosity), to yield 1359 liters (48 ft^3 or 360 gallons). Dividing by a conservative extraction rate of 1.5 liter/min (0.4 gpm) yields a minimum dewatering period of 898 min, or 15 h. A drawdown of 0.91 m (3 ft) would require a minimum extraction period of 22 h. The 15-h period corresponds well with the approximate dewatering period (time from inception of withdrawal to extraction equilibrium) of about 16 h. This minimum period was extended to 24 h as a conservative measure to account for potential variables, such as noninstantaneous release of water from surrounding soils. The period was not extended beyond 48 h to minimize variables such as natural water table fluctuation. Specific capacity measurements were taken during periods having a reliable static initial water level, lasting at least 48 h with no, or relatively little, extraction disturbance, and having no, or relatively little, rainfall during the extraction period.

Specific capacity estimates for the various line configurations and flow rates are within a relatively narrow range, from 0.0827 liter/m (0.00089 ft^3/ft) to 0.1161 liter/m (0.00125 ft^3/ft), with a mean overall specific capacity of 0.1013 liter/m (0.00109 ft^3/ft). The measured specific capacity range corresponds reasonably with the conservative theoretical specific capacity of 0.0811 liter/m (0.000873 ft^3/ft). The lower-end specific capacity measurement (0.0827 liter/m (0.00089 ft^3/ft)] matches the conservative theoretical value quite closely.

Overall, specific capacity values do not decrease significantly from project beginning to end, suggesting that a obvious flow rate decrease into and within the trench due to precipitate accumulation had not yet occurred; however, verification of a decrease in specific capacity over time would require drawdown measurements for identical flow rates. A recommendation would be to extract water from DTT-1 at 3- or 6-month intervals for at least a 24-h period at the same flow rate [perhaps 1.89 or 3.79 liter/min (0.5 or 1.0 gpm)] for each test. A decrease in specific capacity over time would likely indicate a reduction in flow due to precipitate accumulation. Another method for investigating the specific capacity of the trench would be to extract water at greater than natural siphon rates as a step test with incrementally increasing flow rates [perhaps increasing 3.79 l (1 gpm) every 24 h until maximum withdrawal rate is reached].

B. Siphon Hydraulics

Siphon line gas removal and passive operation of the GeoSiphon system is possible. All four GeoSiphon configurations were able to maintain siphon flow for extended periods. Degassing was accomplished by either collecting the gas phase within an air chamber or transporting it out the siphon end. Flow rates for the 2.54-cm (1-in.)-to 1.27-cm (0.5-in.)-diameter line averaged about 1.67 liter/min (0.44 gpm). Flow rates for the restricted flow continuous 2.54-cm (1-in.)-diameter line typically ranged from 1.51 to 2.27 liter/min (0.4 to 0.6 gpm). Flow rates for the unrestricted flow continuous 2.54-cm (1-in.)-diameter line averaged about 4.16 liter/min (1.1 gpm). Flow rates for the continuous 1.27-cm (0.5-in.)-diameter line were typically less than 1.89 liter/min (0.5 gpm). The actual flow rates associated with each configuration were somewhat less than the calculated theoretical single-phase flow rates. The unrestricted flow, with a continuous 2.54-cm (1-in.)-diameter line, configuration flow rates deviated most from that calculated. The actual flow rates may have deviated from calculated values due to one or a combination of the following:

- Calculations were based on the assumption of single-phase flow. If the actual flow was dual phase, the calculations would produce overestimates.
- During air chamber use, the periodic air-chamber recharge produced significant variations in flow and head within the PRB. Calculated flow based on instantaneous measurements during these periods could be expected to deviate from actual flows.
- The density of the fluid within the siphon may have been greater than that of water due to the significant aluminum precipitate being transported. This would reduce flow compared to calculated values.
- The aluminum precipitate could have coated the inside of the line, reducing the effective inside-line diameter.

The head differential required to drive water through each GeoSiphon configuration increased in the following order: unrestricted flow with a continuous 2.54-cm (1-in.)-diameter line, restricted flow with a continuous 2.54-cm (1-in.)-diameter line, a 2.54-cm (1-in.)- to 1.27-cm (0.5-in.)-diameter line, and a continuous 1.27-cm (0.5-in.)-diameter line. Configurations with air chambers required periodic air-chamber recharge. The groundwater transported in the GeoSiphon line had the following major dissolved gasses in descending order of concentration: carbon dioxide, nitrogen, oxygen, and argon. Transport of this water within the GeoSiphon results in degassing of a gas phase composed of (in descending order of concentration) nitrogen, oxygen, carbon dioxide, water vapor, and argon. Degassing was shown to increase with increasing temperature and increasing flow rate. It should also increase with increasing vacuum.

VI. PERMEABLE REACTIVE BARRIER AND SECONDARY TREATMENT PERFORMANCE

A. PRB Performance

The PRB increased the pH of the acidic groundwater and caused aluminum to precipitate—the two functions for which it was designed. The chromium concentrations from well DCB-49 were too low (many below detection limits) to allow any examination of chromium removal by the PRB. The effects of the PRB were measured by inferring the chemistry of water entering the trench from the chemistry of water exiting the trench through the siphon line. However, the spatial and temporal chemical heterogeneity of water entering the PRB made accurate correlation between measurements of water entering the trench and water exiting the trench impossible. Monitoring well DCB-49 was installed upgradient from the PRB to provide general information on the groundwater entering the PRB. However, because of chemical heterogeneity with PRB depth and width, groundwater sampled from well DCB-49 was not necessarily correlative to water exiting the trench through the siphon line. The siphon line obtains water from a discrete depth, whereas well DCB-49 provides a composite of groundwater compositions over several feet of vertical extent. The cause of such heterogeneity is likely the heterogeneous geology combined with close proximity to the shallow basin.

Groundwater chemical heterogeneity in the PRB was reflected by the lower aluminum concentrations in water from the PRB sampling location (DTT-1, the central trench-piping riser) than in water from the siphon-line intake. Sampling from three different depths in the PRB showed a variation in aluminum concentration from <2 to 135 mg/liter and in iron concentration from 284 to 667 mg/liter. To measure the performance of the limestone PRB, the mass balance of calcium and magnesium was used to estimate the aluminum concentration and pH of untreated groundwater based on the composition of the siphon-line output.

A linear relationship exists between calcium and magnesium in groundwater from well DCB-49 (inflowing groundwater) (Fig. 9). It is assumed here that congruent dissolution of the calcite in the limestone also causes a linear relationship between calcium and magnesium, as depicted in Fig. 9. This trend is defined by the origin and the concentrations of calcium and magnesium in PRB limestone dissolved in the laboratory. Calcium and magnesium concentrations from the siphon-line water plot between these lines, and the mass balance of calcium and magnesium dictates that the siphon-line points are intersections of lines parallel to the limestone-dissolution line and the inflowing groundwater line. The line through point "A" that is parallel to the limestone-dissolution line intersects the inflowing groundwater line at a magnesium concentration of 41 mg/liter and a calcium concentration of 50 mg/ liter (Fig. 10). These are the original calcium and magnesium concentrations of groundwater that ultimately evolved to the concentrations at point "A" by limestone dissolution in the trench. The composition of this water evolved by dissolving 319 mg/liter calcium and 39 mg/liter magnesium from the limestone.

Groundwater chemistry data also show that aluminum is correlated linearly ($r^2 = 0.95$) to calcium in groundwater from well DCB-49 (Washburn *et al.*, 1999). Thus, the original aluminum concentration of siphon-line waters can be calculated from the linear relation to calcium and the original estimated calcium concentration. Comparison of the estimated original aluminum concentrations with aluminum concentrations measured at the siphon-line exit shows that the trench was effective in removing from 10% to near 100% of dissolved aluminum, with a median removal value of 61% (Fig. 11).

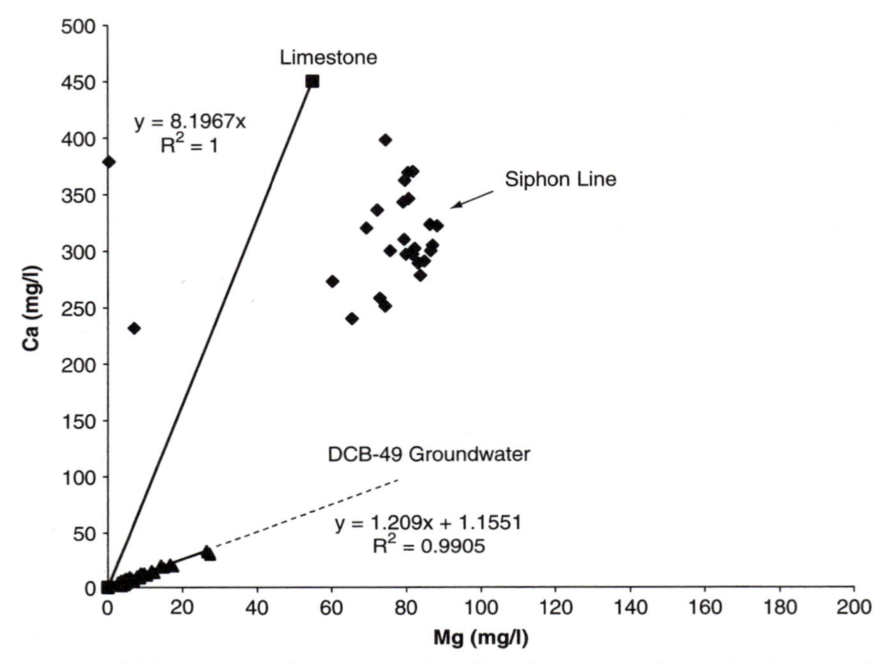

Figure 9 Calcium vs magnesium concentrations for inflowing groundwater (monitoring well DCB-49) compared to D-Area permeable reactive barrier limestone.

Variation in removal effectiveness was most likely due to the horizontal and vertical stratification of the contaminants entering the trench and the variable residence times within the PRB due to varying flow rates. Less contaminated areas of the plume entering the PRB were generally treated to less than detectable aluminum concentrations, as indicated by the DTT-1 aluminum concentrations. Areas of higher contamination in the plume were treated to various levels of aluminum removal. Areas of the plume with lowest original pH and highest original aluminum concentration require longer residence times in the PRB to achieve complete aluminum removal.

The original pH of groundwater entering the trench at the elevation of the siphon-line intake was estimated assuming congruent dissolution of calcite and the reaction:

$$XCO_3 + 2H^+ = X^{+2} + H_2CO_3 \text{ (where X = Ca or Mg).} \quad (1)$$

The contribution of acid from precipitation of aluminum was ignored. This is justified based on the low mass of aluminum precipitated relative to the mass of calcite dissolved. In addition, ideal behavior in solution was

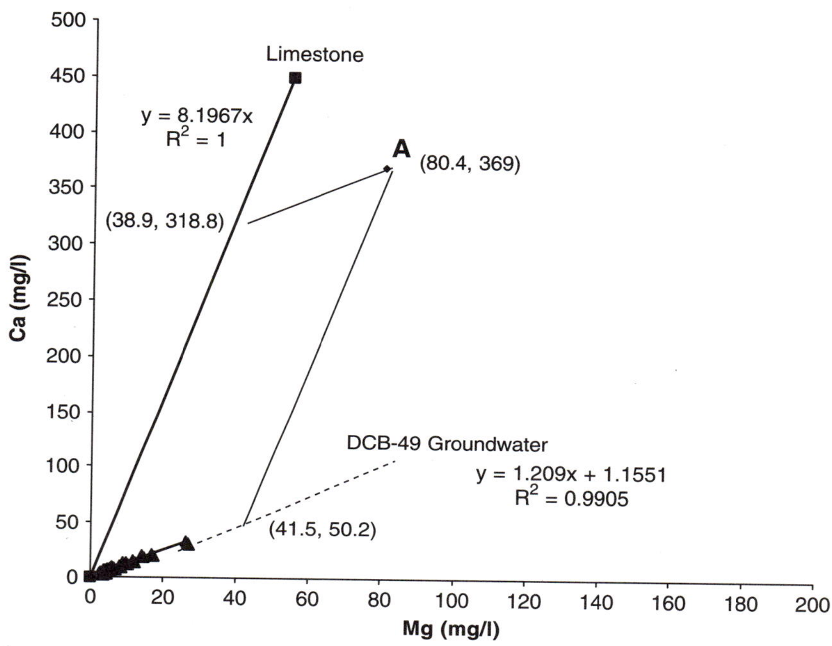

Figure 10 Siphon-line water determination of original calcium and magnesium concentrations of groundwater from monitoring well DCB-49, upgradient from the treatment system.

assumed. It is expected that the logarithmic nature of the pH scale would minimize any effects of these assumptions.

The estimated original pH of siphon-line water ranged from 1.9 to 2.5, with a median of 2.2. Estimated original pH and measured siphon-line exit pH are shown in Fig. 12. The PRB effectively raised groundwater pH to 4.5 to 6.5. The sampling and analysis plan for the D-Area Permeable Reactive Barrier/GeoSiphon treatment system is presented in Table 1.

B. Secondary Treatment Performance

Secondary treatment tests used hydrogen peroxide, calcium peroxide, and hydrogen peroxide/sodium carbonate for the oxidation of ferrous iron to ferric iron. The delivery systems used for secondary treatment included a drip system for application of solutions (hydrogen peroxide and sodium carbonate) and nylon sock and batch application for the introduction of solids (calcium peroxide). The nylon sock was meant to provide a continuous

Table 1

Sampling and Analysis Plan for the D-Area Permeable ReactiveBarrier/GeoSiphon Treatment System

Analytes	Sample bottle	Sample volume	Preservative	Analytical method	Frequency	Location
RCRA metals: aluminum, calcium, chromium, iron, magnesium, manganese, nickel, phosphorus, and silica	20 ml plastic	Fill bottle – 20 ml	1-ml 33.5 to 38% reagent-grade HCl	Inductively Coupled Plasma Emission Spectroscopy	Initial	Siphon line, treatment tanks, and effluent exit
		No overflow	Filtered			
		No headspace	Refrigeration		Daily for treatment system, monitoring wells	DCB-49 (upgradient of trench) DCB-50 (down-gradient of trench)
					Trench frequency as specified during operation	DTT-1 (trench)
Chloride, nitrate, nitrite, phosphate, and sulfate	15 ml plastic	Fill bottle	Filtered	Ion Chromatography	Initial	Siphon line, treatment tanks, and effluent exit
			Refrigeration			
					Daily for treatment system, monitoring wells	DCB-49 DCB-50
					Trench frequency as specified during operation	DTT-1
pH, Eh, DO, and temperature	Field measurement	Field measurement	None	Field Meter	Initial	Siphon line, treatment tanks, and effluent exit
					Daily for treatment system, monitoring wells	DCB-49 DCB-50
					Trench frequency as specified during operation	DTT-1

(*continues*)

Table 1 (*continued*)

Analytes	Sample bottle	Sample volume	Preservative	Analytical method	Frequency	Location
Sulfide	Field measurement	Field measurement	None	Chemetrics Colorimetric	Initial	
					Within first campaign during siphon downtime	DTT-1
VOCs	22-ml headspace	7.5 ml	Refrigeration	Gas Chromatography	Initial	DCB-49
					As needed	
Gases	As provided by Microseeps	As required by Microseeps	None	Microseeps, Inc. Gas Chromatography Mass Spectrometry	Once during air chamber usage	DCB-49, DTT-1, siphon, air chamber

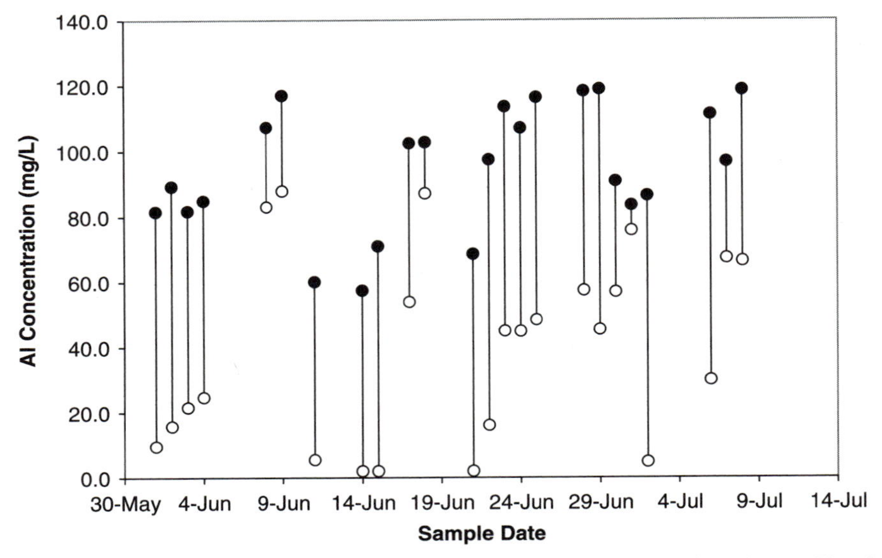

Figure 11 Aluminum Treatment in the D-Area Permeable Reactive Barrier/GeoSiphon. Closed circles are estimated initial aluminum concentrations. Open circles are aluminum concentrations at siphon-line exit.

supply of calcium peroxide to the secondary treatment tank, but failed to achieve this objective. The batch method of calcium peroxide addition was more effective than the nylon sock method. The batch method consisted of the addition of reagent in a single application each day.

Both hydrogen peroxide and calcium peroxide were effective in oxidizing ferrous iron (Fe II) to ferric iron (Fe III). However, the oxidation and subsequent precipitation of ferric hydroxide is an acid-producing reaction:

$$4Fe^{+2} + O_2 + 10\,H_2O = 4\,Fe(OH)_3 + 8\,H^+. \tag{2}$$

To maximize ferric iron precipitation, the acid produced by this reaction must be neutralized by an acid-consuming reaction or addition of alkalinity. Calcium peroxide has an advantage over hydrogen peroxide alone because the net reaction releases less acid:

$$2\,Fe^{+2} + 4\,H_2O + CaO_2 = 2\,Fe(OH)_3 + Ca^{+2} + 2\,H^+. \tag{3}$$

The net reaction of ferrous iron with hydrogen peroxide produces twice the number of moles of acid per mole of iron oxidized:

$$2\,Fe^{2+} + 4\,H_2O + H_2O_2 = 2\,Fe(OH)_3 + 4\,H^+. \tag{4}$$

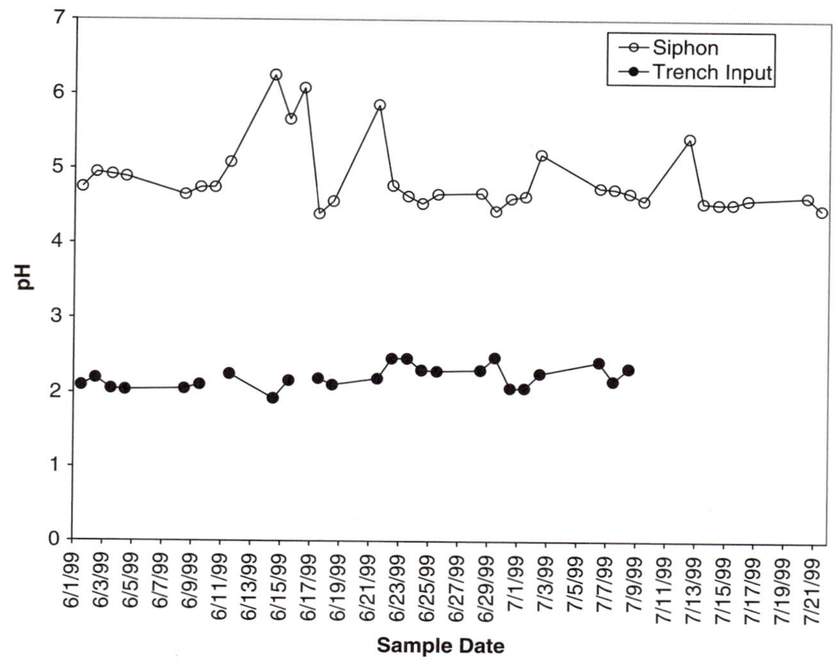

Figure 12 Change in pH between estimated original pH (closed circles) and measured pH at siphon-line exit (open circles), D-Area Permeable Reactive Barrier/GeoSiphon treatment system.

This is reflected in a comparison of the discharge pH of hydrogen peroxide-treated water (pH 3 to 4) with that of calcium peroxide-treated water (pH 4.5 to 6.5). The higher pH produced by calcium peroxide resulted in better metals removal. However, the liquid hydrogen peroxide was applied more easily (through the drip application system) than the solid calcium peroxide. Thus, hydrogen peroxide with subsequent addition of sodium carbonate for alkalinity was investigated.

The hydrogen peroxide/sodium carbonate addition effectively oxidized iron and raised pH. Toward the end of this secondary treatment, essentially complete conversion of ferrous to ferric iron was achieved, and pH increased to the 4.5 to 7.0 range. More effective metals precipitation and removal were achieved at the higher pH. However, maintenance of pH in the desired range was not consistent due to variable flow from the passive delivery system. The flow rate varied in proportion to the head differential in the delivery system.

The three secondary treatment tanks did not provide adequate settling capacity for the limestone and hydrogen peroxide treatments due to their respective settling velocities being significantly less than the critical settling velocities allowed by the system configurations and flow rates. Adequate settling capacity also was not provided for the nylon sock application of calcium peroxide. It appears that near adequate settling capacity was provided for the batch application of the calcium peroxide treatment and also for the hydrogen peroxide/sodium carbonate treatment. Laboratory-determined settling velocities for hydrogen peroxide/sodium-hydroxide flocculant indicate that it would settle faster than the hydrogen-peroxide/sodium carbonate flocculant. Therefore, if exit pH can be controlled, sodium hydroxide may be preferable to sodium carbonate.

Calcium peroxide and hydrogen peroxide followed by sodium carbonate proved to be the most effective secondary treatment for metals removal. Their effectiveness could be improved by increasing pH beyond that achieved with these reagents. This could be accomplished by adding sodium hydroxide as a solution by drip application. Iron and nickel removal are presented in Fig. 13

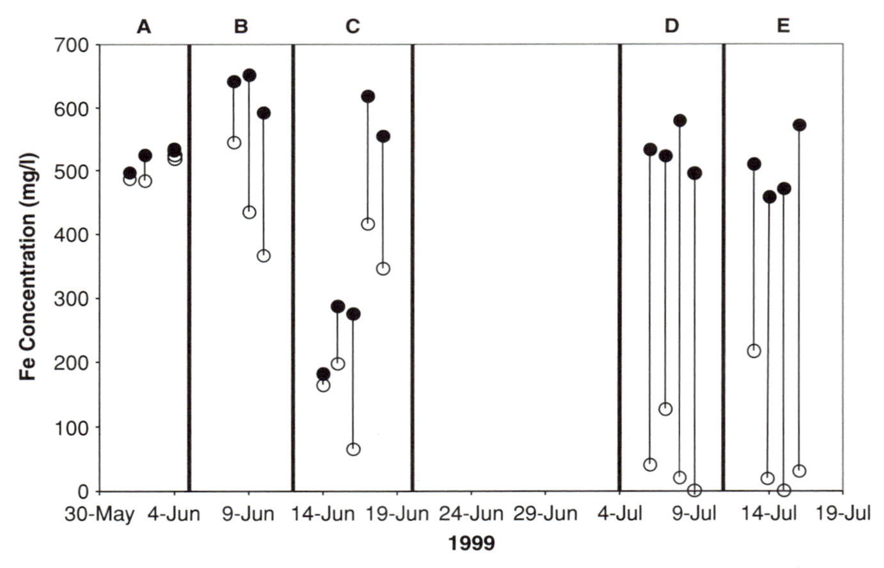

Figure 13 Iron treatment in the D-Area Permeable Reactive Barrier/GeoSiphon system. Closed circles are iron concentrations before secondary treatment; open circles are iron concentrations at exit of secondary treatment. Secondary treatments shown are (A) no secondary treatment; (B) H_2O_2 treatment (10% solution at 6 ml/min drip rate); (C) H_2O_2 treatment (10% solution at 9 ml/min drip rate); (D) CaO_2 batch treatment (1 kg/day); (E) H_2O_2 with Na_2CO_3 (10% H_2O_2 solution at 9 ml/min drip rate with 50 g/liter solution of Na_2CO_3 at a drip rate of 80 ml/min per gpm of groundwater flow). Note that data from CaO_2 and lime treatments using nylon socks for delivery are not shown. This delivery system did not disperse reagents sufficiently and treatment was minimal.

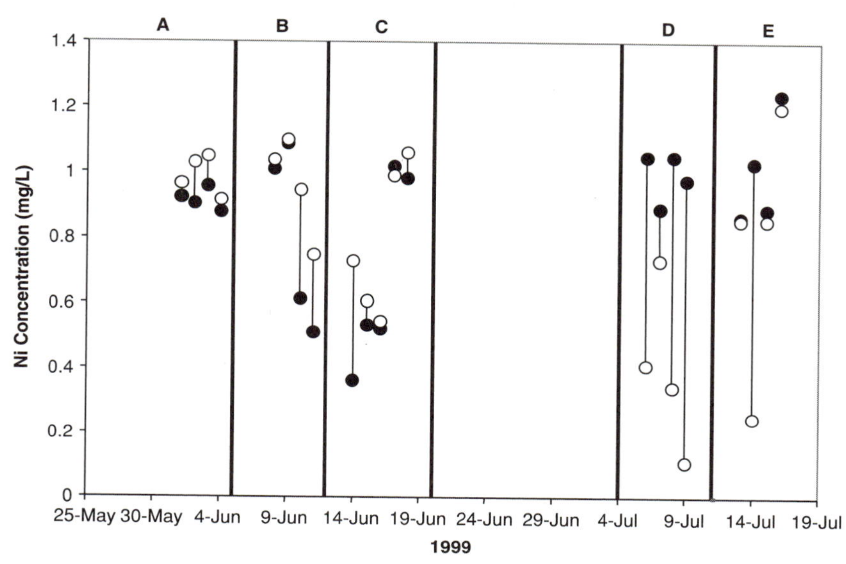

Figure 14 Nickel treatment in the D-Area Permeable Reactive Barrier/GeoSiphon treatment system. Closed circles are nickel concentrations before secondary treatment. Open circles are nickel concentrations at exit of secondary treatment. Secondary treatments shown are (A) no secondary treatment; (B) H_2O_2 treatment (10% solution at 6 ml/min drip rate); (C) H_2O_2 treatment (10% solution at 9 ml/min drip rate); (D) CaO_2 batch treatment (1 kg/day); (E) H_2O_2 with Na_2CO_3 (10% H_2O_2 solution at 9 ml/min drip rate with 50 g/liter solution of Na_2CO_3 at a drip rate of 80 ml/min per gpm of groundwater flow). Note that data from CaO_2 and lime treatments using nylon socks for delivery are not shown. This delivery system did not disperse reagents sufficiently and treatment was minimal.

and 14, respectively. Values are presented for removal efficiencies between each tank and for the overall treatment system. Maximum overall removal of aluminum, iron, and nickel is 96.8, 99.8, and 75.8%, respectively. These results emphasize the effectiveness of treatments that are able to oxidize the iron, maintain an elevated pH, and operate at the proper settling velocities.

VII. SUMMARY

This pilot-scale project was designed to evaluate a relatively passive, low-cost treatment method for low pH, metal-contaminated groundwater that links a PRB with an aboveground treatment system through a siphon line. This concept was the focus of the demonstration. Although an aboveground secondary treatment system was demonstrated, there was less emphasis on quantifying its effectiveness because it can consist of any number of technologies that can be designed to be as effective as required by the chemistry of the water, regulatory limits, and cost.

Groundwater extraction by siphon from the PRB was initiated on May 26, 1999; secondary treatment system testing was conducted from June 1 through July 16, 1999; and all field testing was completed on September 3, 1999. Operation of the primary limestone treatment trench was evaluated mainly by the measurement of specific capacity and aluminum removal efficiency. Significant treatment of ferrous iron (Fe II) within the limestone-filled PRB was not anticipated nor did it occur. A conservative theoretical specific capacity of 0.0811 liter/m (0.000873 ft^3/ft) was calculated prior to operation. Average specific capacity during operation was 0.1013 liter/m (0.00109 ft^3/ft). Specific capacity values over the 3-month test period did not decrease significantly, suggesting that substantial precipitate pluggage did not occur within the PRB or along PRB walls. The PRB was effective at raising the pH and removing aluminum from contaminated groundwater. The effectiveness varied from 100% to less than 10% (median value 61%), with no discernible temporal pattern. The variation in effectiveness was most likely due to the horizontal and vertical stratification of the contaminants entering the PRB and the variable residence times within the PRB.

Passive operation of the D-Area PRB/GeoSiphon system siphon line is feasible. All four siphon-line configurations were able to maintain siphon flow for extended time periods. However, the actual flow rates associated with each configuration were somewhat less than the calculated theoretical single-phase flow rates. The configurations with air chambers required periodic air chamber recharge. The limestone-treated groundwater transported in the siphon line had the following major dissolved gasses in descending order of concentration: carbon dioxide, nitrogen, oxygen, and argon. Transport of this water within the siphon results in degassing of a gas phase composed of nitrogen, oxygen, carbon dioxide, water vapor, and argon, in descending order. This gas phase was either collected within an air chamber or transported out the end of the siphon.

Secondary treatment tests were conducted utilizing hydrogen peroxide, calcium peroxide, and hydrogen peroxide/sodium carbonate. The delivery systems utilized for secondary treatment included a drip system for the application of solutions (hydrogen peroxide and sodium carbonate) and nylon sock and batch application for the application of solids (calcium peroxide). Summaries of treatment results for aluminum, iron, and nickel are presented in Fig. 11, 13, and 14, respectively. A detailed description of the D-Area PRB/GeoSiphon system is presented in Washburn *et al.*, (1999), which is available electronically from the DOE's Office of Scientific and

Technical Information "Information Bridge" internet site (http://www.osti.gov/bridge/product. biblio.jsp?osti_id=7506 Key observations associated with the secondary treatment system are as follows.

- The settling times in the treatment tanks were insufficient for complete settling of the precipitate produced. Additional settling capacity could be provided to complete settling.
- Consistent addition of solid reagents was a problem. Passive delivery systems exist that would be beneficial on a larger scale, but were not appropriate for the pilot-scale system.
- Both hydrogen peroxide and calcium peroxide are effective oxidizing agents for the conversion of ferrous (Fe II) to ferric (Fe III) iron.
- Calcium peroxide has an advantage over hydrogen peroxide because the reaction of calcium peroxide with ferrous iron produces less acid:

$$2\,Fe^{+2} + 4\,H_2O + CaO_2 = 2\,Fe(OH)_3 + Ca^{+2} + 2\,H^+.$$

- The aluminum precipitate produced was a very fine flocculent, some of which was transported through the entire treatment system.
- The optimal pH for ferric iron (Fe III) removal, approximately 8, was not achieved in the field. Use of a sodium hydroxide solution or lime delivered appropriately could have achieved this pH.
- Calcium peroxide and hydrogen peroxide/sodium carbonate secondary treatment tests produced the most effective metals removal.
- While sodium hydroxide has the advantage over sodium carbonate of producing a better-settling sludge, its use in a passive system produces more difficulty in controlling the exit pH levels for regulatory constraints.
- Treatments that combined oxidation with elevated pH (pH of 8 would be optimal) were most effective.

VIII. LESSONS LEARNED/CONCLUSIONS

Following are the major conclusions drawn from this demonstration:

- Determination of iron speciation is necessary for optimal treatment system design. The dominant metals (ferrous iron and aluminum in this case) must be considered. For example, if ferric iron were present, ferric hydroxide would have precipitated within the PRB, likely armoring the limestone and reducing PRB efficiency.
- *In situ* system design must account for vertical and horizontal plume heterogeneity. Groundwater extraction through the PRB trench and

Geosiphon system prior to secondary treatment eliminates the impact of plume heterogeneity, essentially averaging out the contaminant concentrations for secondary treatment.

- For a larger scale system, delivery of secondary treatment solutions or powders should be designed carefully. For this pilot-scale project, the drip application was relatively effective, the batch application method was less effective, and the nylon sock application method was relatively ineffective.
- Sufficient observation wells should be installed at various distances and depths from the PRB to monitor incoming groundwater chemistry (including stratification), primary contaminant removal, and hydraulic performance. Continuous electronic monitoring/recording of flow and water levels is beneficial for operational evaluation, including potential reduction in PRB specific capacity over time due to pluggage. A small, removable treatment medium cell within the PRB would aid the evaluation of medium armoring and pluggage.

REFERENCES

Gibson, A. H. (1961). "Hydraulics and Its Applications," 3rd Ed. Constable & Company, London.

Loitz, D. B., de Steiguer, A. L., and Broz, W. R. (1990). "CRC Handbook of Chemistry and Physics," 79th Ed. CRC Press, New York.

Metcalf and Eddy, Inc. (1991). "Wastewater Engineering: Treatment, Disposal, Reuse," McGraw Hill, New York.

Parkhurst, D. L., (1995). "User's Guide to PHREEQC: A Computer Program for Speciation, Reaction-Path, Advective Transport, and Inverse Geochemical Calculations." Water-Resources Investigations Report 95–4227, United States Geological Survey, Lakewood, CO.

Phifer, M. A., Sappington, F. C., and Denham, M. E. (1998). TNX Geosiphon Cell (TGSC-1) Phase I Deployment/Demonstration Final Report (U), WSRC-TR-98-00032, Rev. 0, Savannah River Site, Aiken, SC.

Phifer, M. A., Sappington, F. C., Nichols, R. L., and Dixon, K. L. (1999). TNX GeoSiphon Cell (TGSC-1) Phase II Deployment/Demonstration Final Report (U), WSRC-TR-99-00432, Rev. 0, Savannah River Site, Aiken, SC.

Phifer, M. A., Sappington, F. C., Nichols, R. L, Ellis, W. N., and Cardoso-Neto, J. E. (1999). GeosSiphon/Geoflow Treatment Systems, Waste Management '99 Symposium, February 28 to March 4, 1999, Tucson, AZ.

Powers, J. P. (1992). "Construction Dewatering, New Methods and Applications," 2nd Ed. Wiley, New York.

Washburn, F. A., Denham, M. E., Jones, W. E., Phifer, M. A., and Sappington, F. C. (1999). Permeable Reactive Barrier/GeoSiphon Treatment System for Metals Contaminated Groundwater (U), WSRC-RP-99-01063, Rev. 0, Savannah River Site, Aiken, SC.

Washburn, F. A., Phifer, M. A., Denham, M. E., and Sappington, F. C. (1998). D-Area Coal Pile Runoff Basin Treatability Study Preliminary Report, WSRC-RP-98-01138, Rev. 0, Savannah River Site, Aiken, SC.

Chapter 5

Deep Aquifer Remediation Tools: Theory, Design, and Performance Modeling

Geoff W. Freethey,* David L. Naftz,* Ryan C. Rowland,* and James A. Davis†

*U.S. Geological Survey, Salt Lake City, Utah 84119
†U.S. Geological Survey, Menlo Park, California 94025

The majority of commercial permeable reactive barriers (PRBs) have been constructed as either funnel and gate or continuous wall systems and have been limited to relatively shallow treatment depths (less than 21 m). Both of these installation techniques preclude the cost-effective replacement and removal of spent reactive material. Deep aquifer remediation tools (DARTs) have been developed for the emplacement of PRBs through arrays of nonpumping wells and offer a number of advantages compared to classic trench installation methods, including easy replacement of spent reactive material and deeper installation depths. A DART is composed of a rigid polyvinyl chloride shell with high-capacity flow channels that contains the permeable reactive material and flexible wings to direct the flow of groundwater into the reactive material. Three items should be considered when evaluating the feasibility of using DARTs for PRB emplacement: (1) the hydrology of the system, (2) the construction engineering of the DART itself, and (3) the residence time needed for contaminant removal reactions. Computer simulation indicates that the reactive material used in a DART should be engineered to have a hydraulic conductivity (*K*) 50 to 200 times greater than the *K* of the aquifer material. This difference in *K* will capture groundwater from an upgradient width of an aquifer equal to about 1.8 to 1.9 times the diameter of the nonpumping well. Simulated groundwater residence time in a 10-cm. diameter

DART ranged from 1.4 h in flow paths in the outer edges of the DART to a maximum of 29.3 h in flow paths in the center of the DART. These modeled residence times compared favorably with groundwater residence times needed for 99% uranium removal in a ZVI PRB installed at Fry Canyon, Utah.

I. INTRODUCTION

Starting in the mid-1990s, permeable reactive barriers (PRBs) began to be used in full-scale groundwater remediation projects (U.S. Environmental Protection Agency, 1998). The majority of commercial PRBs are built in two basic configurations: (1) funnel and gate and (2) continuous wall (U.S. Environmental Protection Agency, 1998). Both of these configurations require some degree of excavation and have been limited to fairly shallow depths (< 21 m) (U.S. Environmental Protection Agency, 1998). Depending on the conditions, standard excavators can reach to depths of about 11 m and modified excavators can reach maximum depths of about 21 m.

Although the majority of PRB installations have used conventional excavation techniques, other reactive material emplacement technologies exist for deeper (> 11 m) groundwater contaminant plumes (Gavaskar *et al.*, 1998; U.S. Environmental Protection Agency, 1998). Deeper emplacement methods include (1) tremie tube/mandrel, (2) deep soil mixing (U.S. Navy, 2001), (3) high-pressure jetting, (4) vertical hydraulic fracturing (Hocking *et al.*, 2000; Slack *et al.*, 2000), and (5) deep well injection (Fruchter *et al.*, 2000). Because of the high equipment and mobilization costs associated with many of these techniques, they generally are not cost effective for small and or remote site installations. Verification of reactive material placement and integrity with deep emplacement methods could be of concern; however, active resistivity monitoring methods are being used for the onsite verification of material placement (Hocking *et al.*, 2000).

Other limitations of PRB remediation technology are the finite longevity of the reactive material and the long-term fate of the inorganic contaminants that accumulate on the reactive material over time (Gavaskar *et al.*, 1998). Replacement or rejuvenation of reactive materials may be cost prohibitive and should be factored into long-term cost models for PRB technologies (U.S. Environmental Protection Agency, 1998). For example, the buildup of mineral precipitates in zero-valent iron (ZVI) PRBs (Naftz *et al.*, 2000; Morrison *et al.*, 2001) could cause a substantial reduction in permeability, possibly resulting in a net decrease in contaminant treatment efficiencies.

After the groundwater has been remediated, the long-term fate of the spent reactive material is of concern. For example, the long-term accumulation of inorganic contaminants such as uranium (U) on the reactive media will

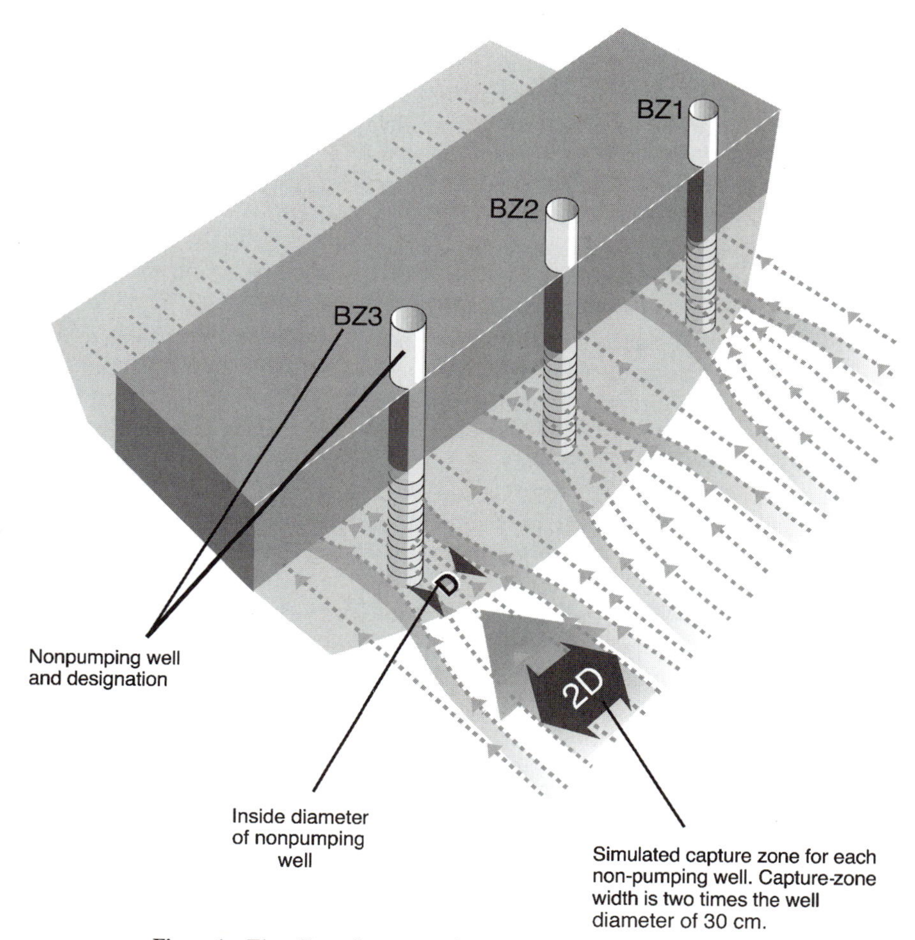

Figure 1 The effect of nonpumping wells on simulated flow paths.

probably require extraction and containment in an approved disposal facility or *in situ* stabilization. If left *in situ*, the spent reactive material could release the inorganic contaminants that accumulated on the material during site cleanup.

The construction of arrays of nonpumping wells may provide an alternative PRB emplacement method for treating contaminant plumes, especially at depth. Wilson and Mackay (1997a,b) and Wilson *et al.* (1997) noted that groundwater converges to arrays of nonpumping wells in response to the difference in hydraulic conductivity (K) between the well and the aquifer (Fig. 1). Because well-drilling techniques are well established and drilling equipment is readily available, the use of nonpumping wells as conduits to emplace PRBs represents a potentially cost-effective emplacement method, especially in remote areas where the mobilization costs of specialized equipment would be cost prohibitive. Specific advantages afforded by the use of nonpumping wells for the emplacement of PRBs include (1) materials, equipment, and contractors are readily available; (2) well installation under most conditions is straightforward; (3) deep installations (> 50 m) are possible; and (4) because of the limited access points, only a fraction of the treatment materials are needed relative to the same coverage with a continuous wall PRB (Wilson and Mackay, 1997a).

Puls *et al.* (1999) placed iron fillings in a series of nonpumping wells to test the removal of chromate from contaminated water. Although this trial emplacement resulted in high chromate removal efficiencies, the design did not allow for future removal and or replacement of the spent reactive material. Deep aquifer remediation tools (DARTs) were developed for use with arrays of nonpumping wells (Naftz and Davis, 1999; Naftz *et al.*, 1999; see Chapter 7) and allow for easy retrieval, replacement, and disposal of reactive material after chemical breakthrough, potentially decreasing the long-term operation and maintenance costs for PRBs.

This chapter describes how DART technology works in combination with arrays of nonpumping wells. Detailed information will be presented on DART design elements, simulated groundwater movement within DARTs, and model-computed contaminant residence times within DARTs in a variety of hydrologic conditions. The Technology Enterprise Office of the U.S. Geological Survey in Reston, Virginia, provided funding for this work.

The movement of groundwater through a well containing a DART was simulated using MODFLOW96 (Harbaugh and McDonald, 1996a,b), a U.S. Geological Survey computer code for simulating flow through porous media. Visualization of simulated flow lines was done using MODPATH/PLOT (Pollock, 1994), U.S. Geological Survey computer codes for generating a plotting path line of groundwater flow simulated by MODFLOW96.

Boundary conditions and hydraulic properties for simulations were assigned to closely emulate boundary conditions and hydraulic gradients at field sites where DARTs were being tested. Hydraulic conductivity and effective porosity values measured for actual reactive material and aquifers were used. The open space between a perforated well casing and a DART was assigned a hydraulic conductivity value of 1500 m/day and an effective porosity of 1. The fins and unperforated portions of the DART housing were assigned as inactive cells in the simulations.

II. DEEP AQUIFER REMEDIATION TOOLS

A DART is composed of three basic components (Fig. 2): (1) a rigid PVC shell with high-capacity flow channels that contains the permeable reactive material, (2) flexible wings to direct the flow of groundwater into the permeable reactive material, and (3) passive samplers to determine the quality of the treated water. Multiple DARTs can be joined together for the treatment of thicker contaminant plumes. DARTs also allow for "vertical stacking" of different reactive materials for the treatment of chemically segregated contaminant plumes. The reactive material within each DART can be removed for subsequent disposal or replaced when spent. The DART can be (1) designed to accommodate a variety of aquifer geometries and well diameters, (2) used for shallow or deep contamination problems, and (3) modified for a variety of plume thicknesses.

In 1997, the U.S. Geological Survey installed DARTs in two deep wells (>135 m) below land surface at a Wyoming mining facility to test the effectiveness of removing U from groundwater (see Chapter 7). In 1998, DARTs were deployed in a line of shallow wells in southern Utah where U had invaded a shallow alluvial aquifer near an abandoned U upgrader facility (Naftz and Davis, 1999; Naftz *et al.*, 1999; see Chapter 7). Chemical analysis of the reactive material inside the DARTs and treated water samples verified that DARTs can be used to emplace PRBs through arrays of nonpumping wells (Naftz *et al.*, 1999; see Chapter 7).

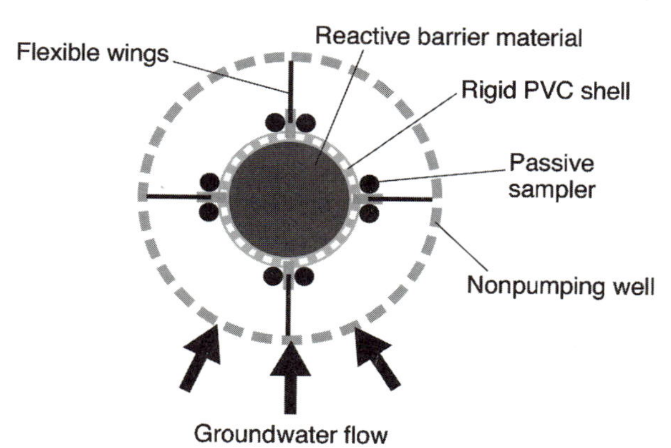

Figure 2 Schematic diagram of a deep aquifer remediation tool.

A. Dart Design Considerations

Three items should be considered to determine how a DART will function in a selected environment: (1) the hydrology of the system, (2) the construction engineering of the DART itself, and (3) the residence time needed for contaminant removal reactions.

1. Hydrology

The system hydrology governs the velocity and direction of contaminated groundwater flow. The hydraulic properties of an aquifer that is homogeneous and isotropic can be predicted with great confidence; however, such an aquifer is rare. The properties of the more common heterogeneous, anisotropic aquifer create varying flow velocities in different parts of the aquifer, and in different directions. Flow paths will deviate from the direction defined by the hydraulic gradient when fingers, layers, or areas of greater *K* lie across a flow path (Fig. 3). Once a particle of water enters a more permeable zone

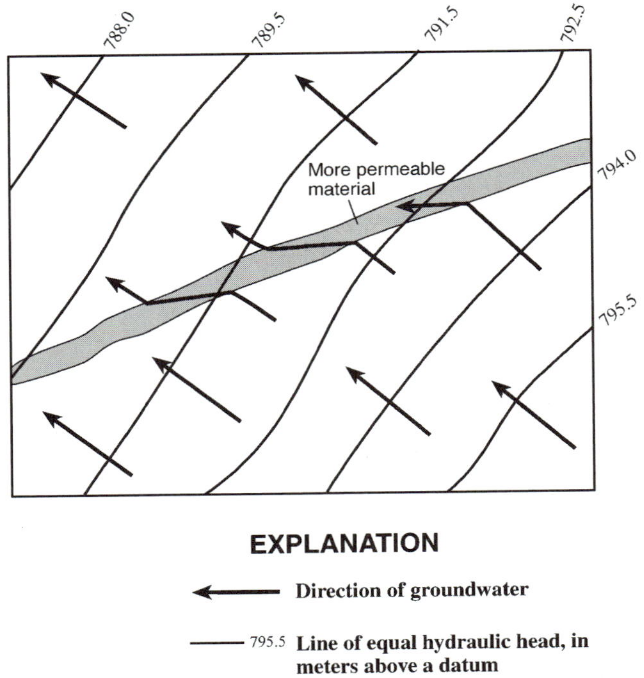

Figure 3 Modeled effect of an area of more permeable material on groundwater flow paths.

that is not parallel to the dominant direction of flow, it moves preferentially in that environment until the hydraulic gradient forces it to exit on the opposite side.

The average linear velocity of groundwater will increase when a more permeable aquifer zone is encountered (Fig. 4). Flow paths for groundwater will turn toward a more permeable aquifer zone, and when a particle enters that zone, it will move more rapidly under the same hydraulic gradient. Particle velocity increases for two reasons: (1) average linear velocity (V) is governed by the product of the hydraulic gradient (dh/dl) and K divided by the effective porosity (n_e) [Eq. (1)] and (2) longitudinal dispersion is greater when pore spaces are larger and farther apart:

$$V = \frac{K dh}{n_e dl}. \tag{1}$$

Heterogeneity in aquifers is difficult to characterize, even at shallow depths. It occurs at all scales, from regional (10 to 100 km) down to site

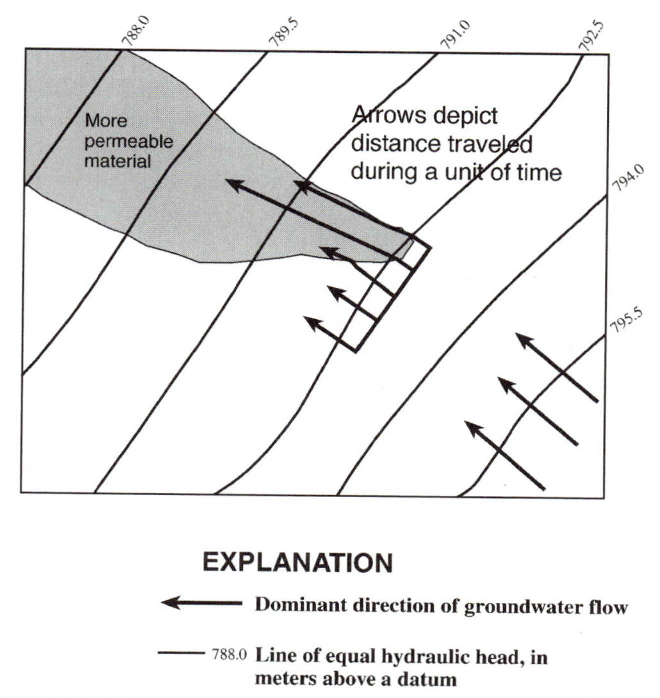

Figure 4 Modeled effect of an area of more permeable material on groundwater velocity.

specific (submeter). Because PRB installations are primarily site specific, knowing the small-scale heterogeneity is important to the success or failure of most types of passive treatments. Eykholt *et al.*, (1999) demonstrated how a heterogeneous three-dimensional aquifer, generated to mimic a braided stream-channel deposit, would affect groundwater flow directions and travel times around and through a trench-type PRB. Results showed that (1) the randomly generated flow field contributed to variable travel times and three-dimensional directions of flow within the homogeneous PRB material and (2) residence times within the barrier may not always be long enough to remove a simulated vinyl chloride contaminant. Especially notable was the vertical movement of particles within the barrier caused by zones of higher hydraulic conductivity occurring at different elevations adjacent to the PRB.

These considerations also affect the performance of a DART. If variations in aquifer hydraulic conductivity can be discerned through field measurements, DART construction and emplacement should be designed to accommodate these characteristics as closely as possible.

Inducing groundwater to move through a well that contains a DART is based on the configuration of wells that form the "permeable reactive barrier" and the difference between the K of the aquifer and that of the reactive material in the DARTs. Computer simulation indicates that the reactive material used in a DART should be engineered to have a K 50 to 200 times greater than the K of the aquifer material adjacent to the DART (Fig. 5). In a homogeneous isotropic system, this difference will capture groundwater from an upgradient width of aquifer equal to about 1.8 to 1.9 times the diameter of the well (Fig. 5). Heterogeneities in the system and the increased K in a sand pack placed on the outside of a perforated or screened interval could increase or decrease this capture width.

The large contrast in a K needed for optimum plume capture by the DART indicates that the K of material used in DARTs must be as large as possible and DART capture efficiency will be optimized in aquifers with lower K. For example, the K values in the colluvial aquifer at Fry Canyon, Utah, range from 17 to 26 m/day, and reactive material K values at the deployed site range from 122 to $>$ 915 m/day (see Chapter 14). Modeling results indicate that a minimum K value of 850 m/day would be needed for efficient plume capture at the Fry Canyon site.

The optimum orientation for the well configuration used to create a DART treatment zone is a line of wells perpendicular to the dominant direction of groundwater flow (Fig. 6). The wells should be spaced close enough and perforated at appropriate intervals to intercept all contaminated groundwater passing through the zone. Because the practical aspects of well drilling generally preclude placing wells close together, two or more lines of wells perpendicular to flow will be required. These lines can be placed far

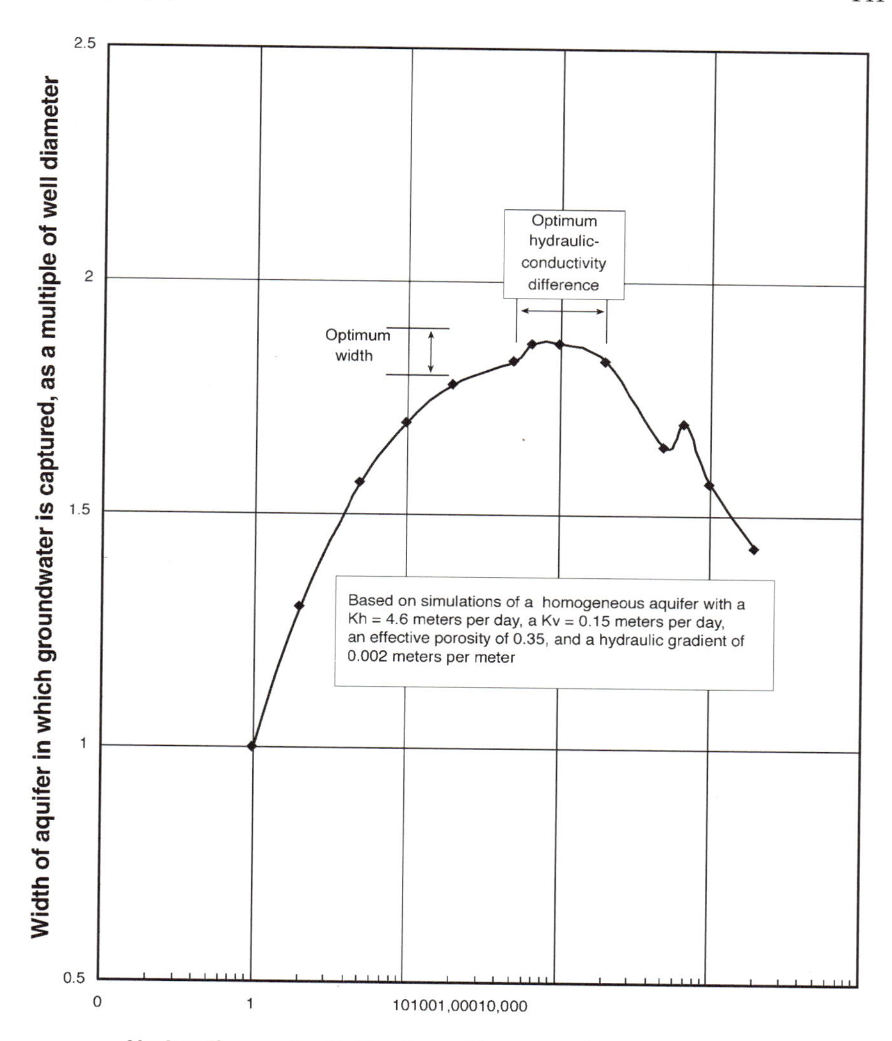

Figure 5 Relation between the width of aquifer treated and the difference between hydraulic conductivity of the aquifer and reactive material in a DART.

enough apart to reduce the difficulty associated with drilling wells in proximity (Fig. 7A). Because wells typically have a greater vertical deviation with depth, three or more lines of wells may be considered in deeper applications to

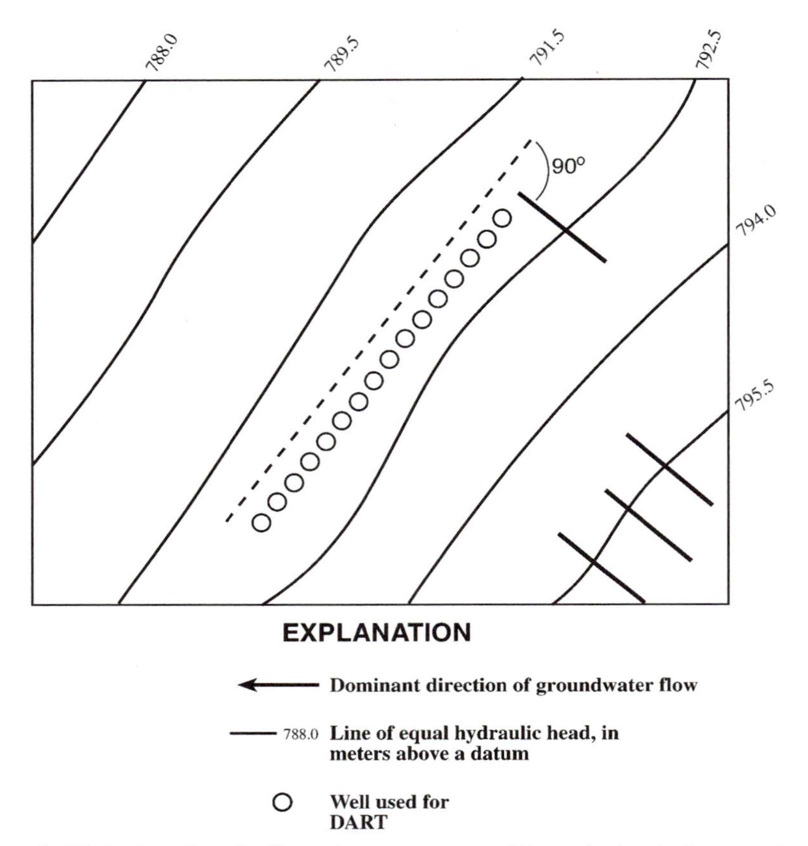

Figure 6 Ideal orientation of wells used to create a permeable reactive barrier for groundwater remediation.

allow a larger separation at the surface (Fig. 7B). Any configuration that provides for the optimum lateral separation that is perpendicular to flow direction will result in all groundwater moving through some part of a reactive material. Wells placed in the most permeable parts of the aquifer will intercept the largest volume of groundwater with the fewest number of wells (Fig. 7C). If the direction of groundwater flow changes in response to changes in recharge or discharge boundaries, some groundwater may not pass through reactive material because the optimum lateral separation between wells might be exceeded. Thus, prior knowledge of variations in flow directions is needed to plan for such a contingency.

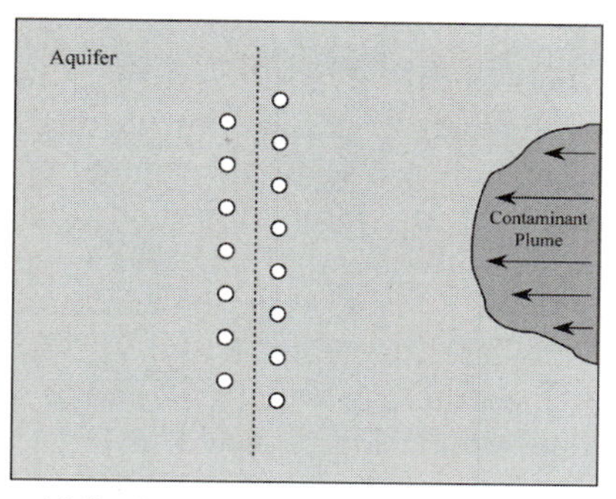

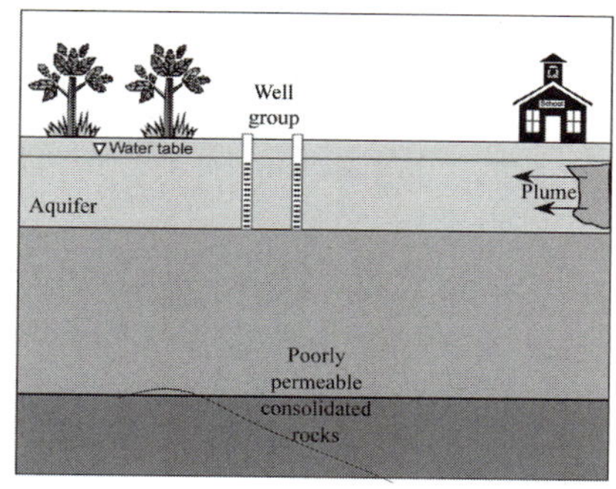

(a) Configuration for shallow contaminant remediation—no vertical deviation of wells expected.

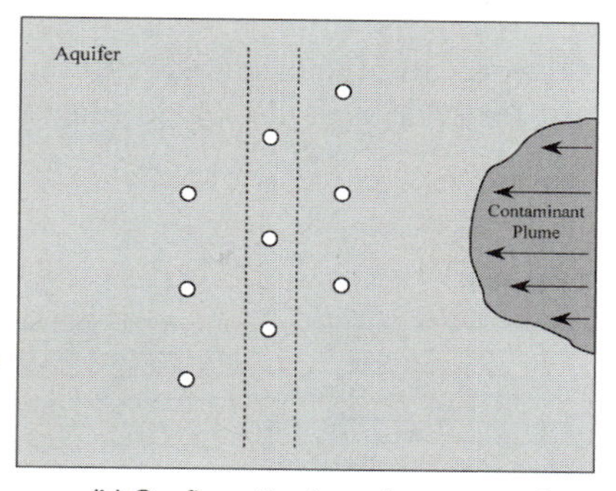

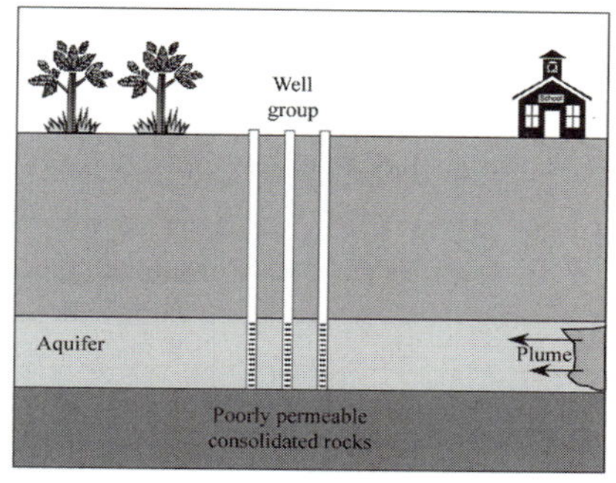

(b) Configuration for a deep contaminant remediation—vertical deviation of wells likely.

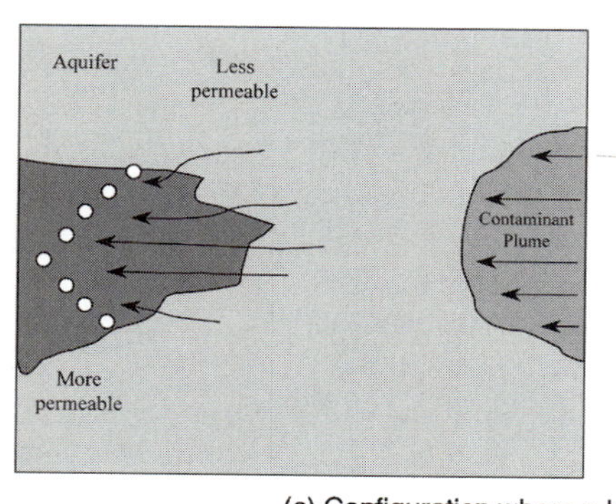

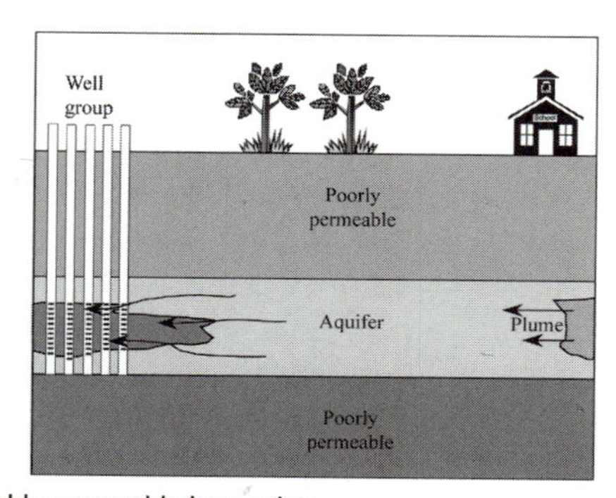

(c) Configuration where a highly permeable lens exists.

Figure 7 Plan and section sketches of three typical well configurations used for DART emplacement.

The flow characteristics inside a well housing a DART are governed by the flow characteristics of the annular space between the DART and the well casing, by the openings in the solid body of the DART, and by flow through the reactive material inside the DART. Generalized sectional simulation through a well containing a DART indicates minor diversion of groundwater flow through areas difficult to seal, such as the bottom end of a DART (Fig. 8), and that trying to divert flow into the DART by restricting the length of perforations in the outer casing is marginally effective (Figs. 8 and 9). The simulation depicted in Fig. 9 shows 11 flow lines tracking around the end of a DART inside a fully perforated casing. The simulation depicted in Fig. 8 shows that eight flow lines still track around the end of the DART, even though unperforated casing is adjacent to the DART bottom. The benefit gained by specific well construction techniques is rarely cost effective.

The wings on a DART are the principal feature that diverts groundwater flow through the reactive material within the core, and the orientation of the wings relative to flow direction is important for creating a flow field that is

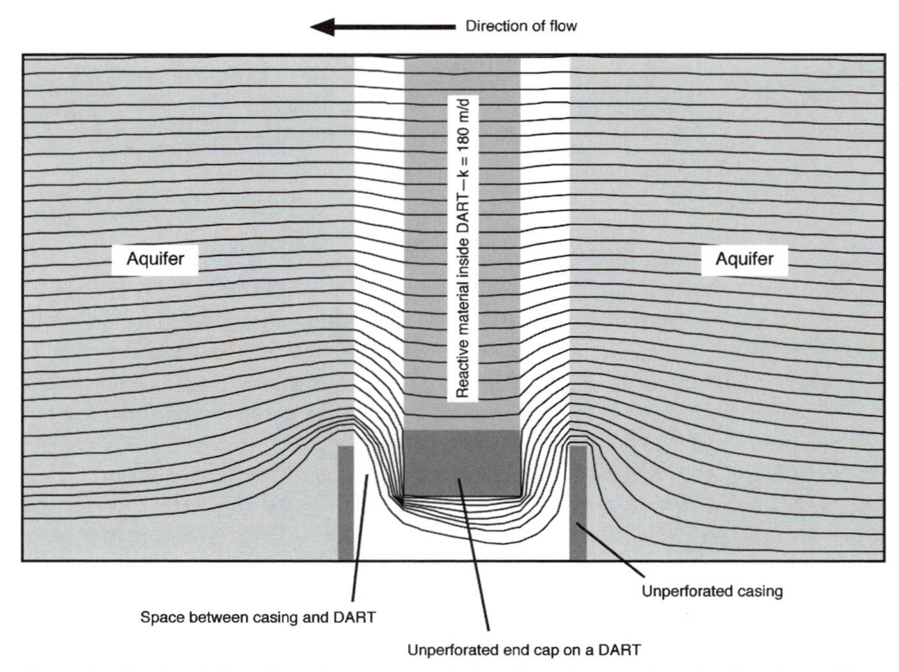

Figure 8 Simulated flow lines through a vertical well section showing the diversion of flow around the end of a DART.

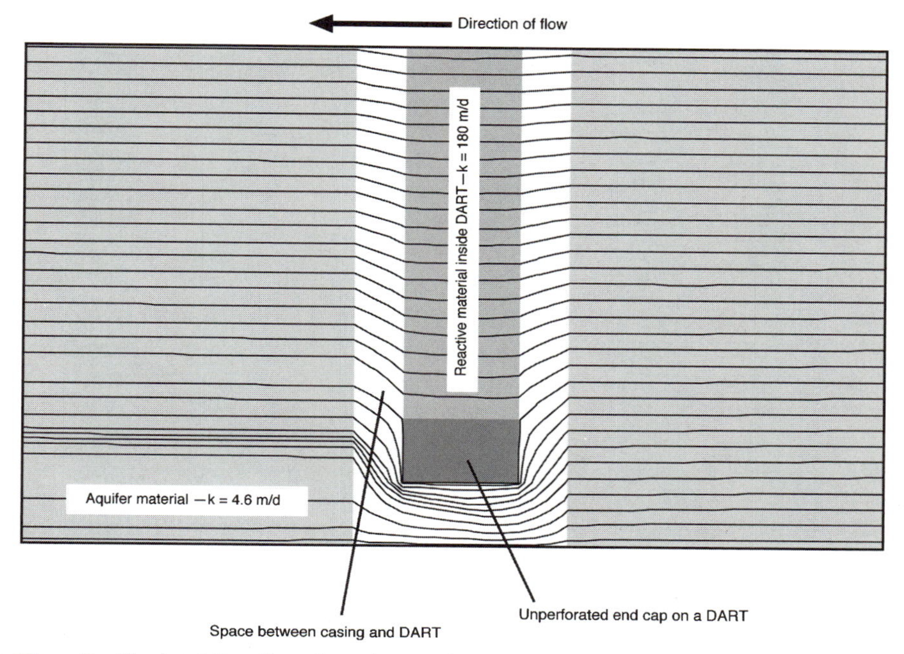

Figure 9 Simulated flow lines through a vertical well section showing the diversion of flow around the end of a DART in a fully perforated casing.

of maximum width. Figure 10 illustrates the significance of wings on a DART. The first illustration demonstrates how groundwater mostly moves around the DART core, which has a much lower K than the open space between the well casing and the DART housing (Fig. 10a). Only about 20% of the flow entering the well would pass through the reactive material. The addition of wings, as illustrated in Fig. 10b, perpendicular to flow direction causes all of the flow entering the well to pass through some part of the reactive material in the DART. These diversion wings create a flow field nearly as efficient as filling the entire well casing with reactive material (Fig. 10c). The simulated wells shown in Figs. 10b and 10c would treat about the same volume of groundwater.

2. Engineering

General criteria used in the design of the DARTs were to (1) maximize the open surface area of the DART housing while maintaining structural

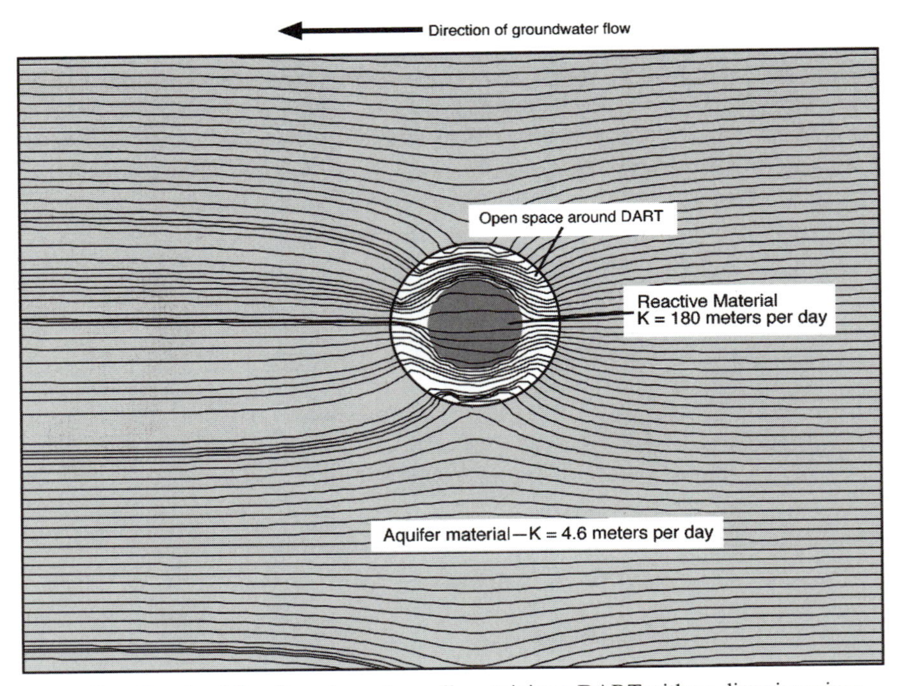

Figure 10a Simulated flow lines through a well containing a DART with no diversion wings.

strength, (2) maximize the volume of groundwater that could flow through the reactive material(s) inside the DART, (3) allow for easy production of DARTs with various diameters and lengths, (4) allow for deployment of a DART by one or two people using easily transportable equipment such as ropes and pulleys, (5) allow access to the reactive material so that it can be easily removed and replaced with fresh material, and (6) minimize the cost of production. A typical DART consists of schedule 80, threaded polyvinyl chloride (PVC) casing with 0.6-cm slots, two threaded PVC end caps, four rigid PVC fins with plastic or rubber extensions, a swiveling eyebolt on the top end cap for rope or cable attachment, and a fiberglass screen lining (Fig. 11). Schedule 80 PVC casing is strong enough for the large 0.6-cm slots and thick enough for grooves that are machined along the vertical length of the DART for fin attachment (Fig. 12).

Rigid PVC fins are glued into the grooves, and thin, flexible plastic is attached along the entire length of each rigid fin to fit snugly against the inside of a cased well. The fins maximize the amount of water that will flow

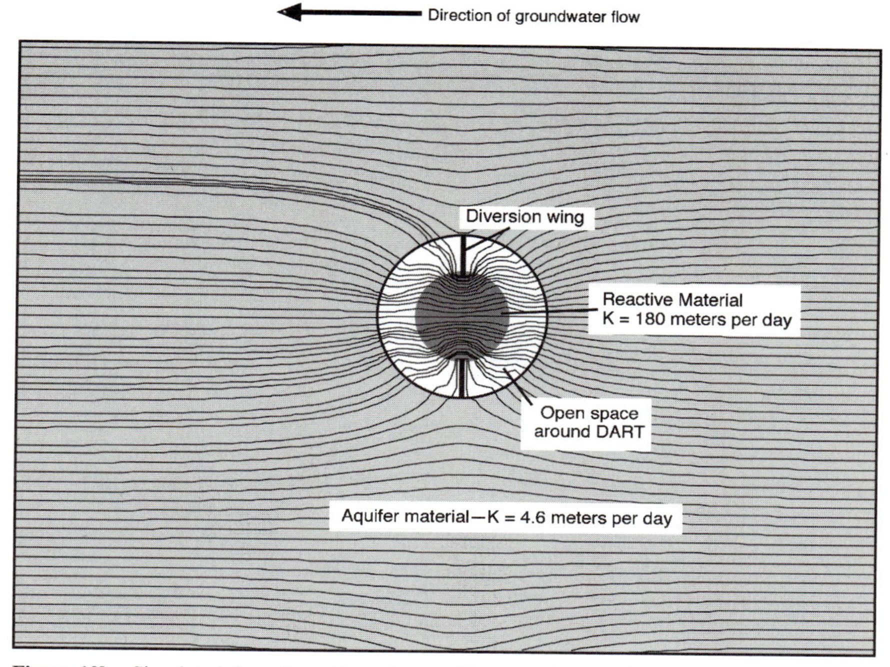

Figure 10b Simulated flow lines through a well containing a DART with diversion wings perpendicular to flow direction.

through the DART by largely eliminating flow paths between the DART and the inside of the well casing. Four fins are used to maximize the surface area of inflow and to largely negate having to orient a DART such that two fins are always perpendicular to the direction of groundwater flow. With four fins, a DART placed randomly in a well will range from about 70 to 100% of a maximum efficiency (Figs. 13a and 13b). Six wings would narrow this range from about 82.5 to 100% of maximum but would decrease the inflow surface area substantially (Fig. 13c). Given the heterogeneities in any aquifer, placing the fins exactly perpendicular to flow probably does not guarantee maximum capture width in the aquifer.

The use of commercially available threaded PVC casing allows DARTs to be made any length, in segments, and at various diameters at a minimal cost. DARTs with this basic design have been placed successfully in shallow (< 6.1 m) 20-cm-diameter wells and deep (>137 m) 10-cm-diameter wells.

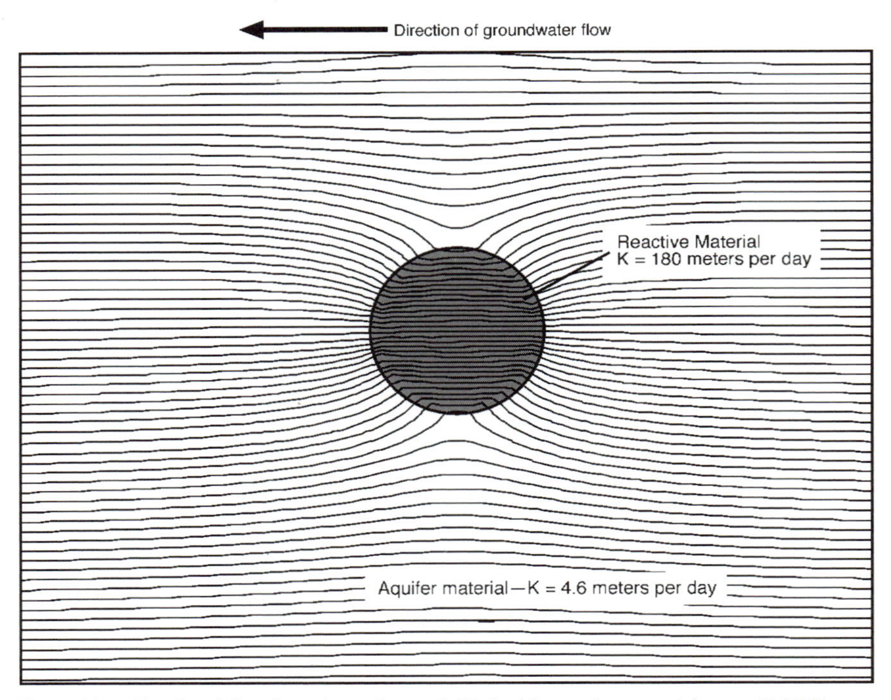

Figure 10c Simulated flow lines through a well filled with reactive material — no DART.

3. Kinetics of Contaminant Removal

It is clear from the previous discussion that DARTs in nonpumping wells will capture contaminated groundwater from upgradient parts of the aquifer. However, the residence time of the contaminated groundwater in contact with the reactive material contained in each DART must be long enough for high contaminant removal efficiencies. Because of the small widths of DARTS (e.g., 15-cm diameters), shorter groundwater residence times will generally occur relative to trench installations of PRBs. Most PRBs installed with trenching methods will be 30 to 90 cm in thickness.

To address this concern, the residence time of groundwater was simulated in detail for a DART placed in a nonpumping well. Hydraulic properties used in this simulation were (1) aquifer $K = 4.6\,\mathrm{m/day}$, (2) aquifer effective porosity $= 0.20$, (3) reactive material $K = 180\,\mathrm{m/day}$, (4) reactive material

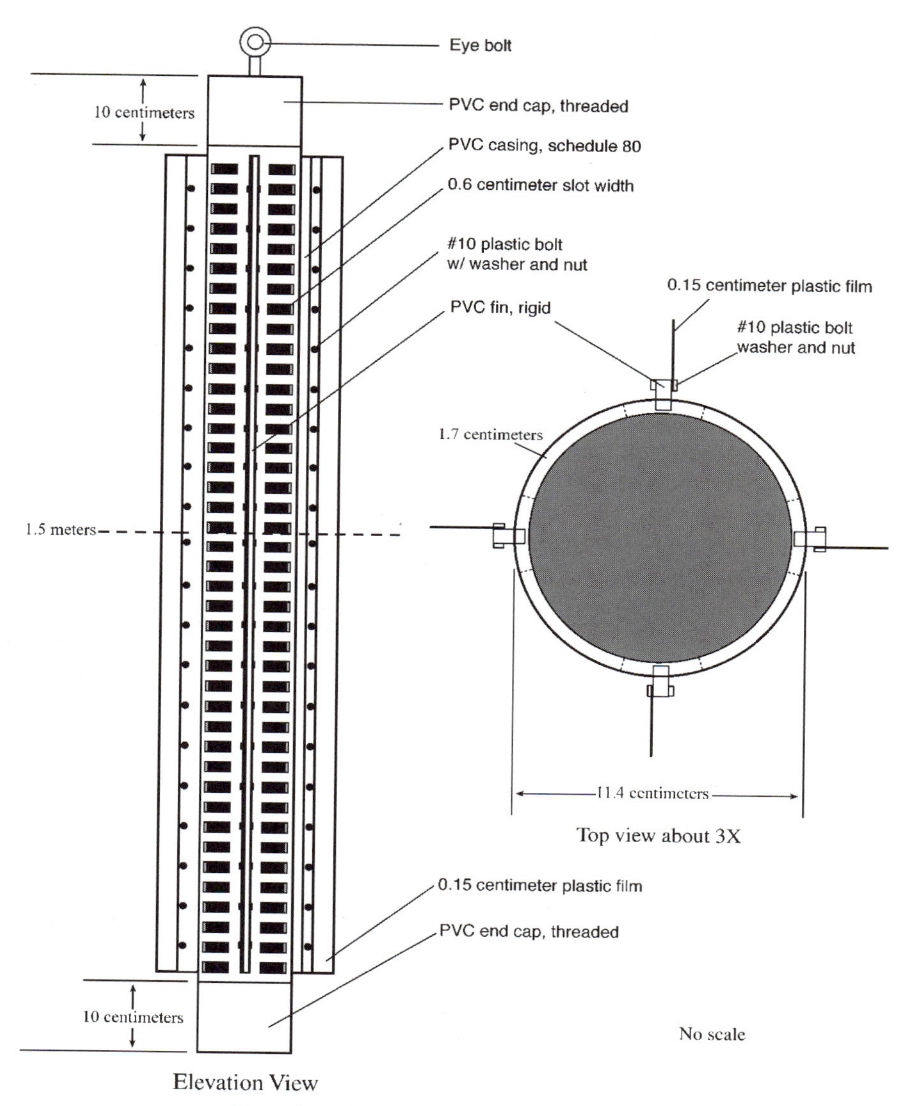

Figure 11 Sketch of a typical DART assembly.

effective porosity = 0.56, and (5) hydraulic gradient = 0.002 m/m (Fig. 14). The diameter of the DART used in the simulation was 10 cm, and the diameter of the nonpumping well was 15 cm. Two flow wings were placed on the

Figure 12 DART deployment during September 1998, Fry Canyon, Utah.

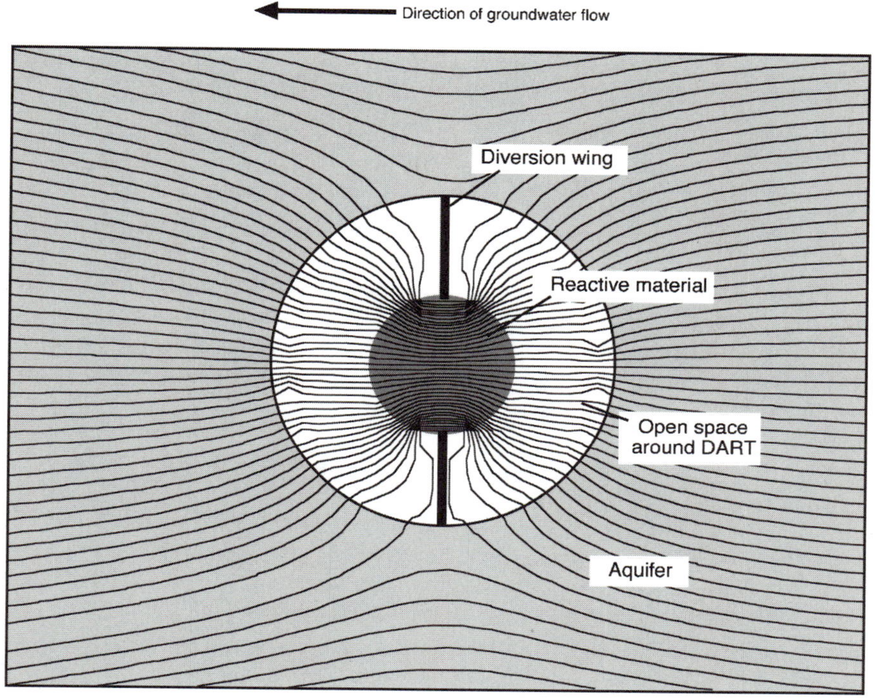

Figure 13a Simulated flow lines through a DART with at least two diversion wings perpendicular to groundwater flow direction.

DART and were oriented perpendicular to the simulated direction of groundwater flow (Fig. 14).

The results of the simulations are shown in Fig. 14. The longest groundwater flow paths and residence times occurred in the center of the DART, and the shortest flow paths and residence times occurred along the outer edges of the DART, adjacent to the flow wings. Simulated residence time of the groundwater ranged from a minimum of 1.4 h in flow paths in the outer edges of the DART to a maximum of 29.3 h in flow paths in the center of the DART (Fig. 14). The median residence time of groundwater during the simulation was 15.4 h.

Residence times for 90 and 99% U removal efficiencies in a ZVI PRB (trench installation) at Fry Canyon, Utah (see Chapter 14), were calculated and are shown in Table I. These calculated residence times provide insight into potential U removal efficiencies that might be expected using DARTs emplaced with ZVI-reactive material under the hydrologic assumptions

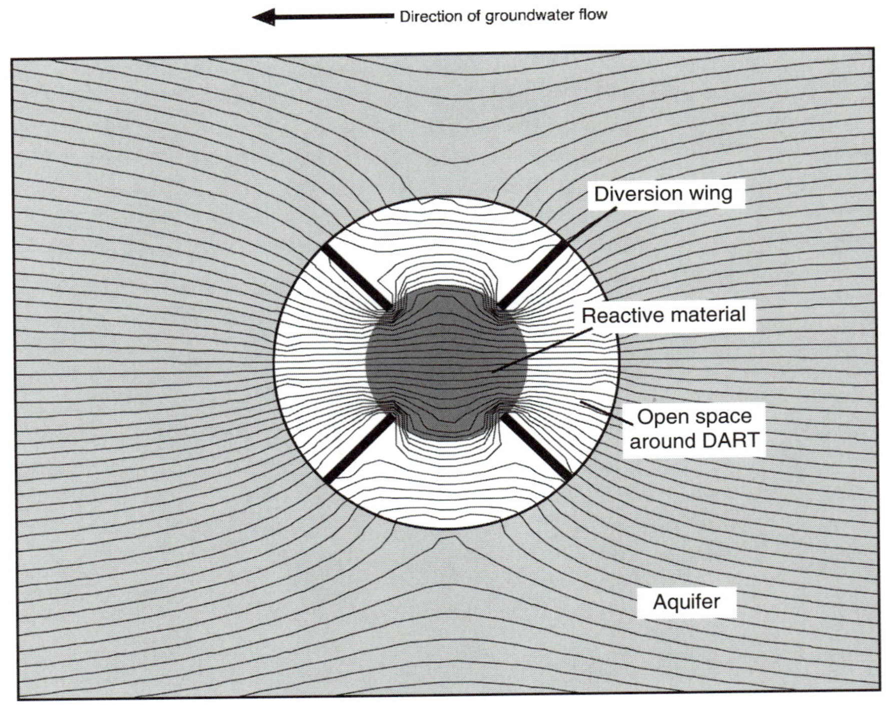

Figure 13b Simulated flow lines through a DART with four diversion wings, none of which are perpendicular to groundwater flow direction.

presented previously. Input U concentrations to the ZVI PRB at the Fry Canyon demonstration site ranged from 5330 to 10,760 μg/liter) during May 1999 (Table I). A 2.2-h residence time was required for 90% U removal, and a 2.5-h residence time was required for 99% U removal in the ZVI PRB. These results are consistent along three separate flow paths within the ZVI PRB at Fry Canyon (Table I).

Simulated groundwater residence times in the DART appear to be sufficient for high U removal efficiencies. The observed groundwater residence times needed for 90 and 99% U removal at Fry Canyon are substantially less than the maximum (29.3 h) and median (15.4 h) simulated residence times for the DART (Fig. 14). The minimum simulated residence time of 1.4 h (Fig. 14) is slightly less than the observed residence time of 2.2 h needed for 90% U

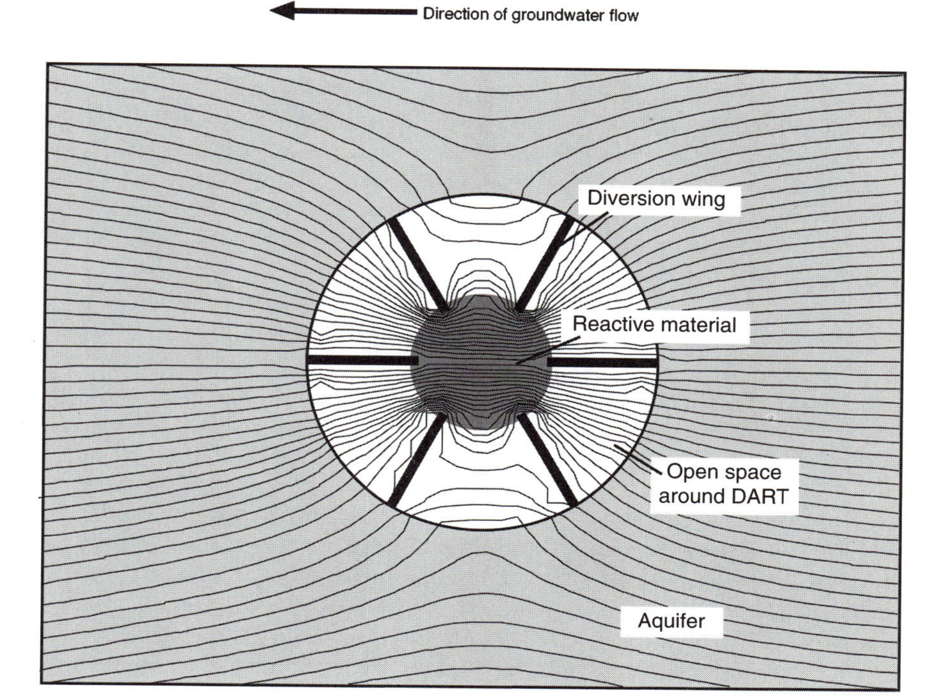

Figure 13c Simulated flow lines through a DART with six diversion wings, none of which are perpendicular to groundwater flow direction.

removal (Table I). Under the conditions specified for these simulations, groundwater residence times in all but the very outer flow paths will be sufficient for 99% U removal efficiencies.

Clogging of the DARTs by mineral precipitates could occur and would depend on the type of reactive material and geochemical conditions in the groundwater. Carbonate mineral precipitation has been documented in trench installations of ZVI reactive material (see Chapters 13 and 14). Because of the limited treatment thickness, mineral precipitation in a DART may substantially impact the contaminant treatment efficiencies relative to the effects of clogging reactions in a trench-type deployment. Clogging in a DART could result in the diversion of groundwater flow to regions outside the treatment zone.

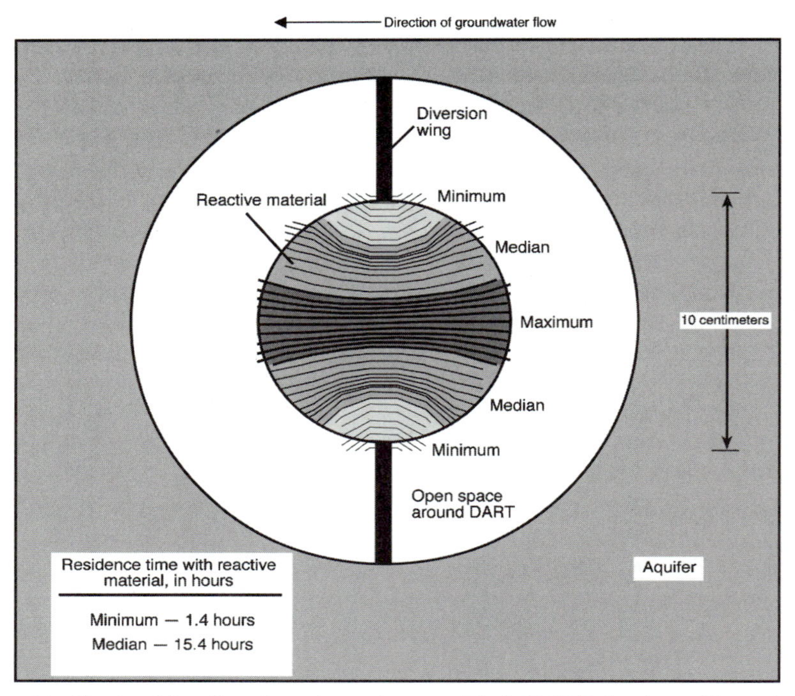

Figure 14 Simulated flow lines through reactive material of a DART showing residence time of contaminated groundwater.

In contrast, initial plugging in a trench installation may simply divert groundwater flow to another area within the PRB where contaminant removal can still occur. Unlike other PRB installation techniques, the extent of PRB plugging can be inspected visually by simply removing the DART from the nonpumping well. If substantial plugging has occurred, replacement of the reactive material within the DART and redeployment may be relatively simple tasks in contrast to reactive material replacement in a trench PRB installation.

B. Dart Deployment Considerations

The practical aspects considered here are largely based on economics. If passive *in situ* treatments are to be practical, they must be cost effective for installation and long-term maintenance. Their ability to work in a wide variety of hydrologic and geologic environments without additional cost is also important.

Table I

Calculated Residence Times Needed for 90 and 99% Removal of Input Uranium Concentrations during May 1999 in the ZVI PRB (Trench Installation), Fry Canyon, Utah[a]

Flow path	Input U concentration, in micrograms per liter, during May 1999	Treated U concentration in micrograms per liter, after traveling 15.2 cm into the ZVI PRB, May 1999	Mean groundwater velocity, in meters per day, measured in the ZVI PRB, April 1999	Calculated flow path distance needed to remove 90% of the input U concentration in cm	Calculated flow path distance needed to remove 99% of the input U concentration in cm	Calculated residence time needed to remove 90% of the input U concentration in h	Calculated residence time needed to remove 99% of the input U concentration in h
ZVIT1 to ZVIR2-2	10,760	<0.06	1.5	13.2	15.0	2.2	2.5
AVG[b] to ZVIT3	8045[c]	3	1.5	13.2	15.0	2.2	2.5
ZVIT2 to ZVIR2-2	5330	10	1.5	13.2	15.0	2.2	2.5

[a] See Chapter 14 for a discussion and illustration of the PRBs installed near Fry Canyon, Utah.
[b] No well, designates a hypothetical monitoring well between wells ZVIT1 and ZVIT2.
[c] Input concentration calculated from average uranium concentration in water from wells ZVIT1 and ZVIT2.

DARTs are designed to be placed in a well inside the perforated casing or well screen. The construction is such that the design can be changed to accommodate any depth and any diameter of casing by using commonly available materials. Maximizing the diameter of a DART allows contaminated groundwater to contact more of the reactive material. Thus, a compromise between this maximum diameter and a diameter that allows minimum effort for insertion and removal must be achieved. Diversion wings are necessary for directing flow through the reactive material, but if plastic fins are too short, their flexibility is decreased to a point where they are too tight against the well casing and vertical movement becomes difficult. DART lengths can be constructed to match any configuration of screened intervals in an existing well. Emplacement in a well with more than three perforated intervals where contaminants are present might require a string connected by intervening lengths of pipe. One to three DARTS, all less than 1.5 m long, could be placed individually at different depths with wire lines. Longer strings would most easily be placed by using a boom on a truck. Placement in horizontal wells would require pushing the assembly into place and thus would require a solid connection, such as a threaded PVC pipe.

If exact orientation of the diversion wings is required, the DART assembly must be connected to a solid string of pipe and wing orientation must be marked on each string segment as they are assembled at the well head. Rotation for desired orientation can only be clockwise unless a positive locking device is used at the string joints.

The form of a DART can be altered for special applications. For example, if removal of contaminants from a single low-volume pumping well were desired, wings could be placed around the circumference of a DART below a pump to force water flowing up inside a well to move through reactive material before reaching the pump intakes (Fig. 15). This would protect the pump and attached plumbing from contacting contaminated water.

Another method of removing contaminants from water flowing to a pumping well is to place a well array between the pumping-well screens and the plume of contaminated groundwater. Because flow lines converge on a pumping well, the array could be in a crescent shape to intercept groundwater being diverted toward the well by its lower hydraulic head. In some cases, the array might have to surround the pumping well to assure that all contaminated water was passed through the reactive material in the DARTs (Fig. 16)

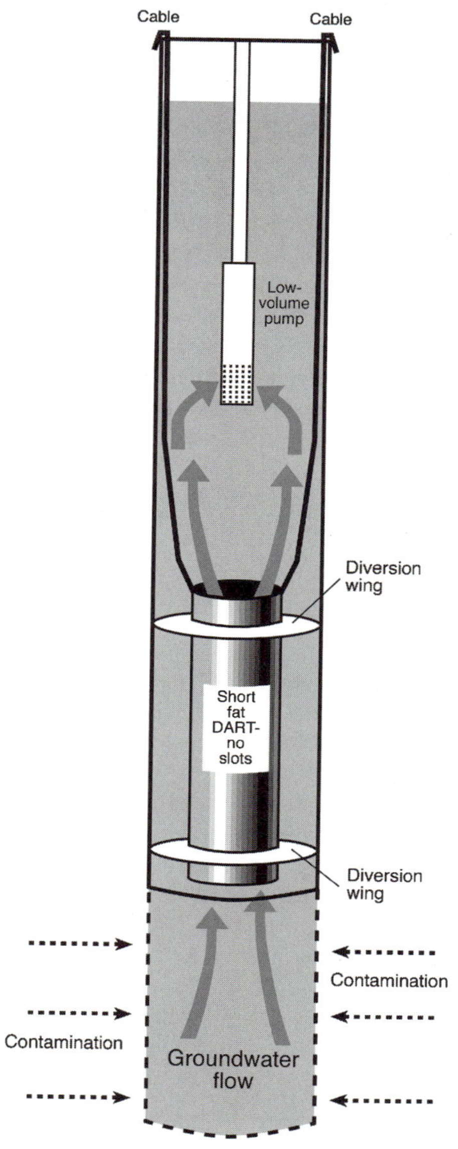

Figure 15 Diagram of a DART placed below a pump in a low-volume well.

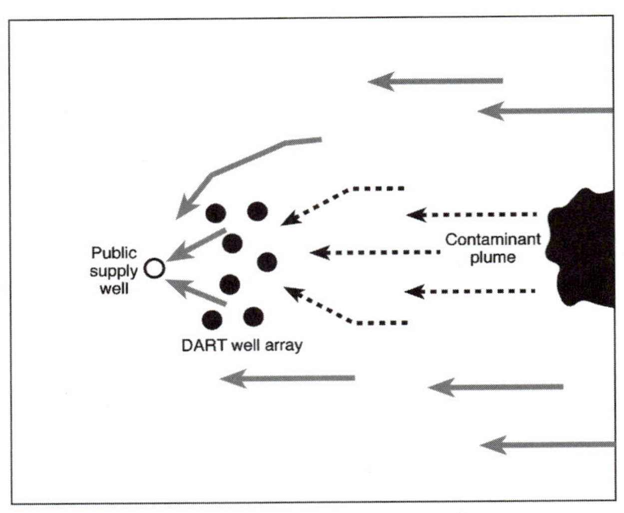

(a) Narrow plume

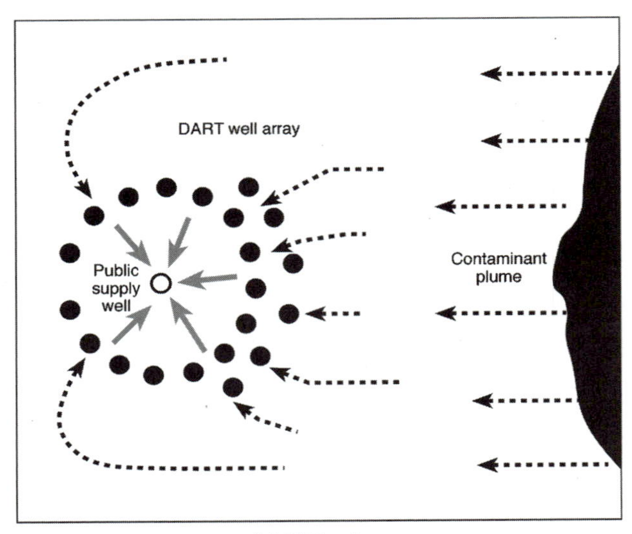

(b) Wide plume

Figure 16 Possible DART well arrays for removal of contaminants from groundwater resulting from a (a) narrow contaminant plume and (b) wide contaminant plume.

III. SUMMARY

Since the mid-1990s, commercial PRBs have been installed in two basic configurations: (1) funnel and gate and (2) continuous wall. The majority of PRB installations have been limited to shallow depths (< 21 m). After PRB installation, the replacement of spent reactive material and eventual removal of the PRB will result in additional construction costs that could be substantial.

DARTs have been developed for the emplacement of PRBs through arrays of nonpumping wells. Specific advantages afforded by the use of nonpumping wells in combination with DARTs include (1) materials, equipment, and contractors are readily available; (2) well installation under most conditions is straightforward; (3) deep installations (> 50 m) are possible; (4) only a fraction of the treatment materials are needed relative to the same coverage with a continuous-wall PRB; and (5) replacement and removal of spent reactive material are cost effective. The DART is composed of a rigid PVC shell with high-capacity flow channels that contains the permeable reactive material and flexible wings to direct the flow of groundwater into the permeable reactive material. Three items should be considered when evaluating the feasibility of using DARTs for PRB emplacement: (1) the hydrology of the system, including the degree of heterogeneity; (2) the construction engineering of the DART itself; and (3) the residence time needed for contaminant removal reactions.

Computer simulation indicates that the reactive material used in a DART should be engineered to have a K value 50 to 200 times greater than the K value of the aquifer material. This difference in K will capture groundwater from an upgradient width of aquifer equal to about 1.8 to 1.9 times the diameter of the nonpumping well. Important design features for DARTs include (1) maximizing the open surface area of the DART housing, (2) maximizing the volume of groundwater that can flow through the reactive materials, (3) allowance for easy production of DARTs with various diameters and lengths, (4) easy emplacement of a DART, (5) easy access to the reactive material, and (6) minimizing the cost of production. Simulated groundwater residence time in a 10-cm-diameter DART ranged from 1.4 h in flow paths in the outer edges of the DART to a maximum of 29.3 h in flow paths in the center of the DART. These modeled residence times compared favorably with groundwater residence times needed for 99% U removal in a ZVI PRB installed at Fry Canyon, Utah.

In addition to classic PRB emplacement scenarios, alternative applications for DART emplacements are possible. These include emplacement in and protection of single wells used for individual and small community water supplies.

REFERENCES

Eykholt, G. R., Elder, C. R., and Benson, C. H. (1999). Effects of aquifer heterogeneity and reaction mechanism uncertainty on a reactive barrier. *J. Hazard. Mater.* **68**, 73–96.

Fruchter, J. S., Cole, C. R., Williams, M. D., Vermeul, V. R., Amonette, J. E., Szecsody, J. E., Istok, J. D., and Humphrey, M. D. (2000). Creation of a subsurface permeable treatment zone for aqueous chromate contamination using in situ redox manipulation. *Groundwater Monitor. Remediat.* **20**, 66–77.

Gavaskar, A. R., Gupta, N., Sass, B. M., Janosy, R. J., and O'Sullivan, D. (1998). "Permeable Barriers for Groundwater Remediation." Battelle Press, Columbus, OH.

Harbaugh, A. W., and McDonald, M. G. (1996a). User's documentation for MODFLOW-96, an update to the U.S. Geological Survey modular finite-difference ground-water flow model. U.S. Geological Survey Open-File Report 96–485.

Harbaugh, A. W., and McDonald, M. G. (1996b). Programmer's documentation for MODFLOW-96, an update to the U.S. Geological Survey modular finite-difference ground-water flow model. U.S. Geological Survey Open-File Report 96–486.

Hocking, G., Wells, S. L., and Ospina, R. I. (2000). Deep reactive barriers for remediation of VOCs and heavy metals. *In* "Chemical Oxidation and Reactive Barriers: Remediation of Chlorinated and Recalcitrant Compounds" (G. B. Wickramanayake, A. R. Gavaskar, and A. Chen, eds.), Vol. 6, pp. 307–314. Battelle Press, Columbus, OH.

Morrison, S. J., Metzler, D. R., and Carpenter, C. E. (2001). Uranium precipitation in a permeable reactive barrier by progressive irreversible dissolution of zero-valent iron. *Environ. Sci. Technol.* **35**, 385–390.

Naftz, D. L., and Davis, J. A. (1999). Deep Aquifer Remediation Tools (DARTs): A new technology for ground-water remediation. U.S. Geological Survey Fact Sheet 156–99.

Naftz, D. L., Davis, J. A., Fuller, C. C., Morrison, S. J., Freethey, G. W., Feltcorn, E. M., Wilhelm, R. G., Piana, M. J., Joye, J. L., and Rowland, R. C. (1999). Field demonstration of permeable reactive barriers to control radionuclide and trace-element contamination in groundwater from abandoned mine lands. *In* "U.S. Geological Survey Toxic Substances Hydrology Program Proceedings of the Technical Meeting, Charleston, South Carolina, March 8–12, 1999 Contamination from Hardrock Mining: U.S. Geological Survey Water-Resources Investigations Report 99–4018A" (D. W. Morganwalp, and H. T. Buxton, eds.), Vol. 1, 281–288.

Naftz, D. L., Fuller, C. C., Morrison, S. J., Davis, J. A., Piana, M. J., Freethey, G. W., and Rowland, R. C. (2000). Field demonstration of permeable reactive barriers to control uranium contamination in groundwater, *In* "Chemical Oxidation and Reactive Barriers: Remediation of Chlorinated and Recalcitrant Compounds" (G. B. Wickramanayake, A. R. Gavaskar, and A. Chen, eds.), Vol. C2-6, pp. 281–290. Battelle Press, Columbus, OH.

Pollock, D. W. (1994). User's guide for MODPATH/MODPATH-PLOT, Version 3: A particle tracking post-processing package for MODFLOW, the U.S. Geological Survey finite-difference ground-water flow model. U.S. Geological Survey Open-File Report 94–464.

Puls, R. W., Paul, C. J., and Powell, R. M. (1999). The application of in situ permeable reactive (zero-valent iron) barrier technology for remediation of chromate-contaminated groundwater: A field test. *App. Geochem.* **14**, 989–1000.

Slack, W. W., Murdoch, L. C., and Meiggs, T. (2000). Hydraulic fracturing to improve in situ remediation. *In* "Chemical Oxidation and Reactive Barriers: Remediation of Chlorinated and Recalcitrant Compounds" (G. B. Wickramanayake, A. R. Gavaskar, and A. Chen, eds.), Vol. 6, pp. 291–298. Battelle Press, Columbus, OH.

U.S. Environmental Protection Agency (1998). Permeable reactive barrier technologies for contaminant remediation. U.S. Environmental Protection Agency Report EPA/600/R-98/125.

U.S. Navy (2001). Deep soil mixing: accessed June 26, 2001, at URL http://erb.nfesc.navy.mil/Default.htm.

Wilson, R. D., and Mackay, D. M. (1997a). Arrays of unpumped wells: An alternative to permeable walls for in situ treatment. *In* "International Contaminant Technology Conference Proceedings", pp. 888–894. St. Petersburg, FL.

Wilson, R. D., and Mackay, D. M. (1997b). Arrays of unpumped wells for plume migration control or enhanced intrinsic remediation. *Bioremediation* **4**, 187–191.

Wilson, R. D., Mackay, D. M., and Cherry, J. A. (1997). Arrays of unpumped wells for plume migration control by semi-passive in situ remediation. *Ground Water Monit. Remedia.* **17**, 185–193.

Chapter 6

Creation of a Subsurface Permeable Reactive Barrier Using *in Situ* Redox Manipulation

V. R. Vermeul, M. D. Williams, J. E. Szecsody, J. S. Fruchter, C. R. Cole, and J. E. Amonette
Battelle, Pacific Northwest National Laboratory, Richland, Washington 99352

I. INTRODUCTION

Many groundwater plumes requiring treatment are dispersed over large areas. Contamination of this type is difficult and expensive to treat using extraction wells and surface-based treatment (pump-and-treat) methods. Mackay and Cherry (1989) have shown that many of the problems encountered during pump-and-treat operations are the result of the pumping phase rather than the treatment phase. As a result, a variety of *in situ* treatment technologies that avoid the pumping phase are under investigation. One promising group of *in situ* technologies involves use of permeable reactive barriers. The typical permeable reactive barrier approach consists of trenches excavated downgradient of the contaminant plume to a depth sufficient to intercept the migrating plume. These trenches are backfilled with one or more reactive materials selected to promote a particular treatment process.

The *in situ* redox manipulation (ISRM) concept proposed by Fruchter *et al.* (1992, 1994) provides for creation of permeable treatment zones *in situ* without excavation and extends the applicability of *in situ* methods to deeper aquifers. ISRM provides an improvement over the baseline pump and treat technology because it places the treatment capacity in the most permeable regions of the subsurface, where the bulk of the contamination resides. This treatment capacity remains in the subsurface, where it is available to treat contaminants that seep slowly out of less permeable regions. Cost estimates by Cummings and Booth (1997) indicate ISRM technology is expected to be less expensive than pump and treat technology. ISRM installation costs are similar to pump and treat, but there are no ongoing operation and maintenance costs. Additional advantages of this technology are that (1) the barrier can be left in place, (2) human exposure is diminished greatly, (3) it is unobtrusive, and (4) the barrier is renewable if monitoring determines it is necessary.

Potential secondary effects associated with ISRM technology include metals mobilization, residuals concentrations, hydraulic performance (i.e., aquifer plugging), and dissolved oxygen depletion. Initial bench-scale laboratory testing of ISRM did not identify any secondary effects that would preclude field scale testing the technology. Field-scale experiments were designed to quantify these effects and determine if they could potentially limit the full-scale application of this technology for the *in situ* remediation of contaminated groundwater.

The objective of the ISRM method is to create a permeable reactive barrier (PRB) in the subsurface to remediate redox-sensitive contaminants (Fig. 1). As illustrated, redox-sensitive contaminants in the plume are immobilized or destroyed as they migrate through the manipulated zone. The permeable treatment zone is created by reducing the ferric iron to ferrous iron within the aquifer sediments (Fig. 2). In the ISRM concept, a permeable treatment zone is placed within the contaminant plume downgradient of the contaminant source. The treatment zone is created by injecting appropriate reagents to chemically reduce the structural iron in the sediments. The reduced Fe(II) appears to be present in at least two different phases: adsorbed Fe(II) and Fe(II)–carbonate (siderite). Adsorbed Fe(II) appears to be the dominant phase.

Amonette *et al.* (1994; 1998), Fruchter *et al.* (1994, 2000), Szecsody *et al.* (2000a,b), Vermeul *et al.* (2000), and Williams *et al.* (2000) have shown that if the redox potential of an aquifer can be made reducing, a variety of redox-sensitive contaminants can be treated. Redox-sensitive contaminants that are candidates for treatment by this technology include chromate (CrO_4^{2-}), uranium, technetium, TNT, RDX, and some chlorinated solvents [e.g., TCE, CCl_4, TCA, PCE]. Chromate is immobilized (Fig. 2) by the reduction of Cr(VI) to Cr(III), which forms a highly insoluble chromium hydroxide or iron chromium hydroxide solid solution (Rai *et al.*, 1987; Sass and Rai, 1987). This case is particularly favorable, as Cr(III) is less toxic and not easily reoxidized under ambient environmental conditions. Uranium and technetium can also be reduced to less soluble forms [although, unlike Cr(III), they may

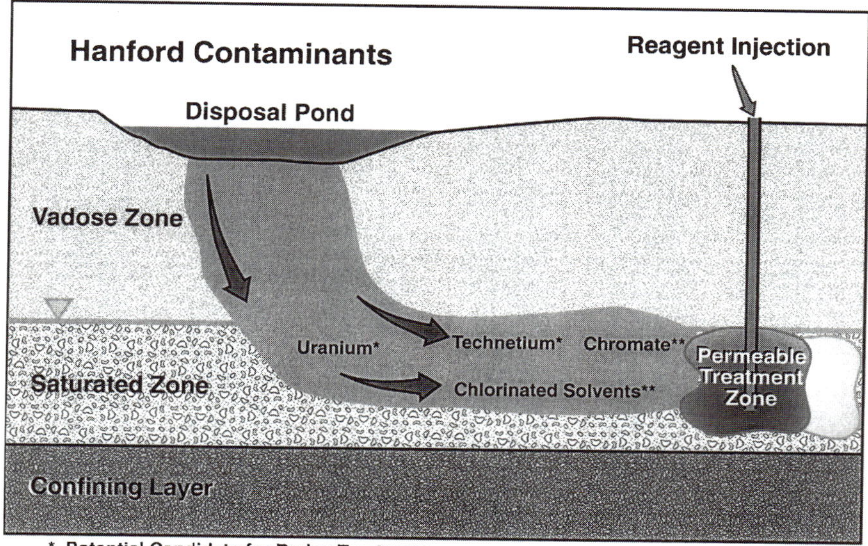

Figure 1 *In situ* permeable treatment zone concept.

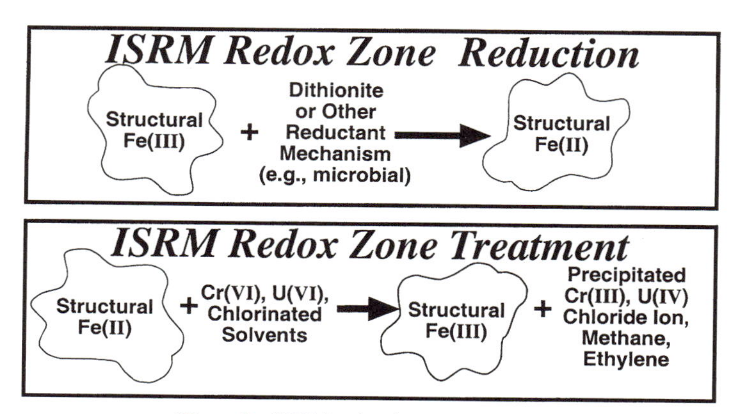

Figure 2 ISRM reduction and treatment.

reoxidize/remobilize once the treatment zone is reoxidized], explosives (i.e., TNT, RDX) can be destroyed by reductive mineralization, and chlorinated solvents can be destroyed by reductive dechlorination.

A multiple-scale experimental approach, as described in this chapter, was used to develop and test ISRM remediation technology. Initial bench-scale batch and column experiments conducted on Hanford sediments evaluated a variety of potential reducing reagents (Amonette *et al.*, 1994; 1998; Fruchter *et al.*, 1994). Once sodium dithionite was selected as a preferred reagent, a variety of batch and column experiments with sediment and dithionite were used to develop an understanding of the important reactions, final reaction products (i.e., residuals), and nature and fate of any ions released from the sediments and sediment surface coatings under reducing conditions (e.g., mobilization of trace metals). The bench-scale laboratory experiments were followed by a series of intermediate-scale experiments conducted at Oregon State University (Istok *et al.*, 1999).

Mathematical models were developed to integrate the results from the bench-scale experiments with the site characterization information to define nominal specifications for the field experiments (Williams *et al.*, 1994). This process then proceeded in an iterative fashion with designs being checked against experiments at the intermediate and each successively larger scale and improved as necessary before the next test was conducted. Important design factors include well and corehole placement, reagent and buffer concentrations, injection and withdrawal rates, and the duration of each phase of the experiment (injection, reaction, and withdrawal). RAFT, a reactive transport numerical model (Chilakapati, 1995), was also used to interpret the results of these experiments to determine reaction kinetics at larger scales (i.e., intermediate and field).

Three successively larger scale field demonstrations of the ISRM technology were conducted at the U. S. Department of Energy's Hanford site in southeast Washington State, including a single injection well proof-of-principle test (Fruchter *et al.*, 1996, 2000), a multi-injection well-treatability test (Williams *et al.*, 2000), and a full-scale deployment of the technology as a remedy for a 700-m (2300-ft) wide Cr(VI) plume impacting the Columbia River. ISRM treatment zone performance data presented in this chapter extend through March 2001 for the proof-of-principle field test ($\sim$ 5.5 years after treatment) and April 1999 for the treatability test. Additional performance assessment monitoring at the treatability test site has been limited by construction activities associated with expansion of the barrier for full-scale deployment, which is scheduled for completion in 2003.

Emplacement of an ISRM treatment zone involves three distinct phases, including injection, reaction, and withdrawal (i.e., push/pull test). During the injection (or push) phase, contaminated waters of the targeted treatment zone are replaced by one pore volume of the reagent solution, injected into the treatment zone through the central injection well. The injection phase is followed by a stagnant reaction (or residence) phase in which the injected reagent is allowed to react with the aquifer sediments. During the withdrawal phase, unreacted reagent, other reaction products, and any mobilized trace metals are removed through the central injection well. This single well, push/pull approach was chosen because it provides for better residual recovery than any other technique. The ISRM treatment zone concept can be extended to emplacement of an ISRM permeable reactive barrier by coalescing a series of cylindrical treatment zones, created through a series of single well reagent injections, across the downgradient path of a migrating contaminant plume.

II. BENCH-SCALE BATCH AND COLUMN EXPERIMENTS

The primary objectives of the bench-scale experiments were to identify an appropriate chemical reducing agent for ISRM and quantify the important reactions (and rates) that occur when iron-bearing sediments are reduced. Results of these experiments were used to design the larger scale tests and to determine the required reagent mixture and concentrations. Bench-scale tests were also used to demonstrate the treatment of specific contaminants with reduced sediments and to estimate the treatment capacities (i.e., longevity) of the ISRM-treated sediments.

During development of the ISRM technology, sodium dithionite was shown to be an effective chemical reducing agent for reducing structural ferric iron to ferrous iron in Hanford sediments (Amonette *et al.*, 1994). The reduction and disproportionation reaction rates for dithionite and Hanford soils were adequate for the reduction of solids in the aquifer, while ensuring that dithionite would not remain as a contaminant in the groundwater for extended periods of time. Because stability is enhanced greatly at high pH, the injected solution consisted of sodium dithionite with a potassium carbonate/potassium bicarbonate buffer (pH 11). The potassium-based buffer was chosen instead of its less expensive sodium-based analog to decrease the possibility of clay dispersal and mobilization during and after the barrier emplacement process.

Dithionite reacts with iron in the aquifer sediments, reducing Fe(III) surface phases to Fe(II) according to the overall reaction described by

$$S_2O_4^{2-}(aq) + 2\,Fe(III)(s) + 2\,H_2O \rightarrow 2\,SO_3^{2-}(aq) + 2\,Fe(II)(s) + 4\,H^+ \quad (1)$$

The reduced sediments in the treatment zone can remove redox-sensitive contaminants from groundwater flowing through the zone. For example, the chromate ion (CrO_4^{2-}) is a common groundwater contaminant at many sites, but it is only significantly water soluble under oxidizing conditions. Within the zone of dithionite-reduced sediments, aqueous chromate reacts with Fe(II) produced by the dithionite reaction [Eq. (1)] and is precipitated as a solid hydroxide [e.g., $Cr(OH)_3$] according to the example reaction described by

$$CrO_4^{2-}(aq) + 3\,Fe(II)(s) + 5\,H^+ \rightarrow Cr(OH)_3(s) + 3\,Fe(III) + H_2O \qquad (2)$$

Similar precipitation reactions will occur for other oxidized redox-sensitive metal species.

III. INTERMEDIATE-SCALE EXPERIMENTS

Following completion of the initial bench-scale experiments, three intermediate-scale injection experiments were conducted in a 7-m-long, 0.20-m-thick, 11° wedge-shaped flow cell (Istok *et al.*, 1999). The wedge shape approximated the radial flow field associated with injection well hydraulics. The cell was constructed at Oregon State University and packed with sands from a quarry near the Hanford site that were similar to the Hanford formation sands encountered at the 100-H area field site. The intermediate-scale experiments, designed to mimic planned field-scale experiments, were based on numerical simulations that incorporated hydrogeologic data collected during site specific characterization of the field test site. The primary objective of intermediate-scale experimentation was to gain experience and make improvements in the operational aspects of each experiment (e.g., location and frequency of sampling, measurement methods, sample and reagent handling, refinement of analytical techniques, assessment of metals mobilization) before conducting the field-scale experiment. Three intermediate-scale experiments, designed to mimic planned field tests, were conducted, including a full-scale tracer test, a minidithionite injection/withdrawal test, and a full-scale dithionite injection/withdrawal test.

IV. FIELD-SCALE PROOF-OF-PRINCIPLE EXPERIMENT

The primary objective of the pilot-scale proof-of-principle field experiment was to determine the feasibility of creating a reduced zone in an aquifer using the ISRM method and to estimate the longevity of the reduced zone in a natural environment. Additional objectives of this test were to evaluate secondary effects of the process (e.g., reagent recovery and residual chemicals in the aquifer, trace metal mobilization, and effects on aquifer permeability). The limited scale of this test made it di;cult to evaluate downgradient effects due to uncertainties in the groundwater flow direction.

The ISRM proof-of-principle test site is located in the Hanford 100-H area (Fig. 3) southwest of the 100-H reactor, about 730 m from the Columbia

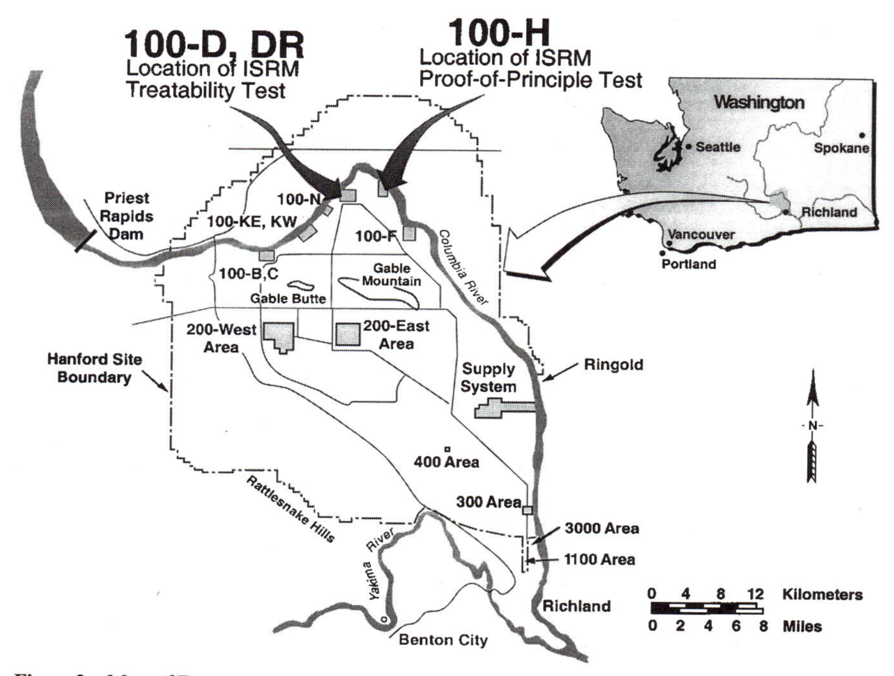

Figure 3 Map of Department of Energy's Hanford site, Washington State, showing the location of the 100-H area ISRM proof-of-principle test site and the 100-D area ISRM treatability test site.

River. The site is relatively uncontaminated, with only low levels (<0.100 mg/liter) of hexavalent chromium in the groundwater. Sixteen wells were drilled and completed, including the central injection well, the upper and lower zone monitoring wells, and the up- and downgradient monitoring wells (Fig. 4). Prior to conducting the full-scale dithionite injection/withdrawal experiment, a series of site characterization activities were completed. These included hydraulic tests, a full-scale bromide tracer injection/withdrawal test that mimicked all three phases planned for the full-scale dithionite injection/withdrawal experiment, and a minidithionite injection/withdrawal test. These activities are fully described in Fruchter *et al.* (1996) and Vermeul *et al.* (1995).

A. Experiment Operation

The full-scale dithionite injection/withdrawal experiment (initiated on September 7, 1995) was designed to produce a 15-m (50-ft)-diameter reduced

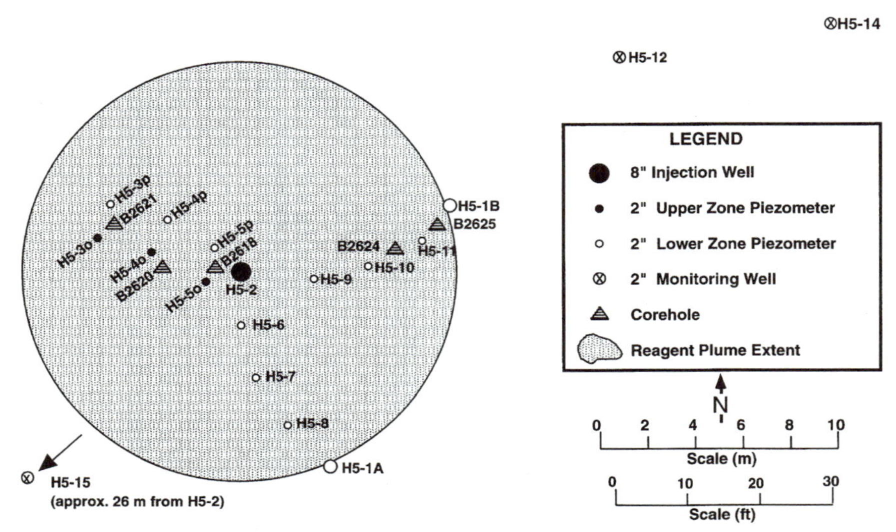

Figure 4a Well layout and corehole locations of the 100-H area ISRM proof-of-principle test site.

zone around the central injection/withdrawal well H5-2 (Fig. 4). Approximately 77,000 liters (20,500 gal) of buffered sodium dithionite solution was injected into the aquifer during the 17.1-h injection phase. The injection design incorporated three successively lower concentrations of reagent (Table I) to minimize cost and to decrease the amount of derived waste and aquifer residuals. The injection phase was followed by an 18.5-h reaction (or residence) phase that provided time for the reagent to react with the ferric iron within the sediments. The final 83-h withdrawal phase of the experiment involved pumping 4.9 injection volumes of groundwater from the central injection well to remove any unreacted reagent, reaction products (e.g., thiosulfate, sulfate, sulfite), and mobilized metals.

During the injection phase, reagent was injected from three 27,000-liter (7100-gal) mixer tanks. These tanks were prefilled to an appropriate level with site groundwater pumped from the injection well. Concentrated liquid reagent (delivered to the site in a refrigerated tanker truck) was added to each tank to achieve desired injection concentration and volume. Time-dependent adjustments in the injection concentration of reagent were used

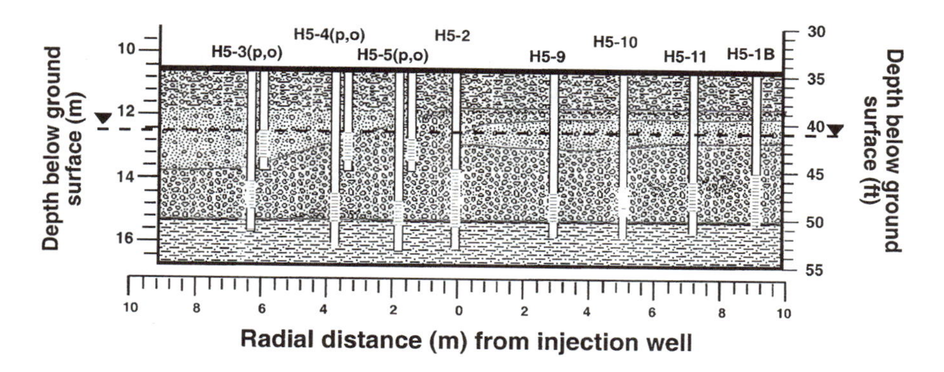

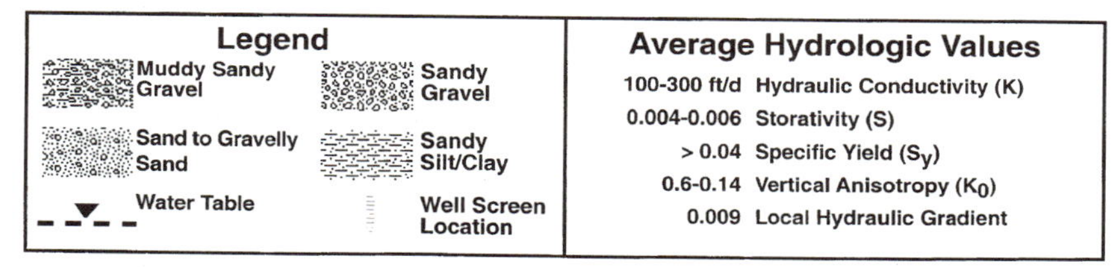

Figure 4b Cross section of aquifer/wells at 100-H. Wells outside the treatment zone (H5-1A, H5-12, H5-13, H5-14, and H5-15) are screened across the entire unconfined aquifer.

Table I

Concentrations used during Injection

Tank	Sodium dithionite (*m*)	Potassium carbonate (*m*)	Potassium bicarbonate (*m*)	Sodium bromide (*ppm*)
A	0.1	0.4	0.04	90
B	0.065	0.26	0.026	90
C	0.033	0.13	0.013	90

to compensate for differences in average residence times and reaction path lengths that occur during the injection phase. For example, solution injected at the leading edge of the injection plume will be in the aquifer for 18 h (i.e., more than three reduction reaction half-lives), whereas solution injected near the end of the injection phase will encounter sediments that have already been continuously exposed to full reagent for 18 h.

Following the residence phase, which provided sufficient time for the reagent to react with the aquifer sediments, the reagent, any mobile reaction products, and by-products (e.g., mobilized trace metals) were withdrawn from the aquifer through the central injection well and stored in a 380,000-liter (100,000 gal) Modutank installed at the site. The withdrawal volume was 4.9 times greater than the injection volume. The larger withdrawal volume is a result of groundwater drift and dispersive effects. Samples of the withdrawn water were collected and submitted for analyses to provide data for mass balance calculations to determine (1) the amount of reagent and reaction products recovered and (2) whether concentrations of sulfate and trace metals were below the purge-water collection criteria. Analytical results indicated that all Hanford purge-water collection criteria were met, and the stored withdrawal water was subsequently discharged to ground.

Dithionite movement and reactivity during the injection, residence, and withdrawal phases of the field test were characterized by monitoring geochemical changes in the groundwater (oxygen, pH, electrical conductivity) and by direct measurement of dithionite. Groundwater geochemical measurements and samples were taken at intervals ranging from 0.25 to 6.0 h from the injection/withdrawal well and 12 monitoring wells. This resulted in the collection of approximately 2500 aqueous samples, which were analyzed to characterize dithionite distribution and reactivity within the aquifer. More details on the injection, residence, and withdrawal phases are presented in Fruchter *et al.* (1996, 2000).

Mass balance calculations compared the mass of injection constituents with the mass of withdrawn constituents to determine what percentage of reagents and reaction products was recovered during the withdrawal phase. Samples collected from the extraction stream during the withdrawal phase of the experiment were analyzed for thiosulfate, sulfate, sulfite, and bromide using ion chromatography (IC). Integration of the extraction stream total sulfur species and bromide concentration data indicated that 87% of the injected sulfur mass and 90% of the injected bromide mass were recovered during the withdrawal phase.

B. Reductive Capacity and Barrier Longevity

Indirect evidence of the reduction of aquifer sediments was obtained through the collection and analysis of aqueous samples from the site (e.g., dis-solved oxygen and total chromium/hexavalent chromium concentrations). Direct evidence of reduction of aquifer sediment was achieved through analysis of sediment samples collected from seven borings drilled at the ISRM field experiment site following the dithionite injection/withdrawal experiment (Fig. 4).

The primary methodology for determining the reductive capacity of core samples collected from the treatment zone following the ISRM field experiment was to determine the oxygen reduction capacity (ORC) of the sediment samples [for a detailed discussion, see Istok *et al.*, (1999), Fruchter *et al.*, (1996), and Szecsody *et al.*, (2000a)]. Samples were packed in columns, and the ORC of the sediment in the column was then measured by flowing a solution of oxygen-saturated water (~8 mg/liter) through the column until effluent dissolved oxygen concentrations approached influent concentrations. The O_2 reoxidation method was selected as the primary methodology because the processes involved in the methodology were automated easily, which increased significantly the number of core samples that could be analyzed.

Results of the ORC analysis for the 13 sediment core samples collected from the treatment zone following the dithionite injection/withdrawal test are summarized Table II. These data show that significant treatment capacity was created by the test and a high percentage of reduction of the available Fe(III) (i.e., that reducible by dithionite) was achieved within the treatment zone. The ORC measurements for core samples that were fully reduced by dithionite in the laboratory resulted in an average Fe(III) value of 0.059% by weight for Hanford formation sediments. Estimates for the lifetime of the treatment zone were calculated using the results from the analysis of these core samples. The treatment capacity of the targeted treatment zone was estimated to be between 51 and 85 pore volumes based on (1) the average Fe(III) based on ORC measurements in Table II; (2) effective porosity and bulk density estimates of 0.2 and 1.9 g/cm^3, respectively, and assumed dissolved oxygen and hexavalent chromium concentrations of 9 and 1 mg/liter, respectively; and (3) core data indicating that 60 to 100% of the available reactive iron [from core samples collected up to a 7.5-m (25-ft) radial distance] was reduced during the dithionite injection/withdrawal experiment. Assuming a 15-m (50-ft)-wide treatment zone and a groundwater velocity of 0.3 m/day (1 ft/day), which is typical of this area of the Hanford site, the estimated longevity of the ISRM treatment zone at the proof-of-principle test site is between 7 and 12 years.

Spatial trends in the effectiveness of the treatment zone emplacement are evident when comparisons of the percentage of available Fe(III) that was reduced by the treatment are made between core samples (Fig. 5). There is a general trend of increasing percent reduced with depth; this is consistent with the experimental design, which targeted the bottom 1.5 m (5 ft) of the aquifer. There is also a general trend of decreasing percent reduced with increasing radial distance from the injection/withdrawal well. This trend is also expected, as the dithionite concentrations measured in samples collected from the wells during the injection/withdrawal test were lower at greater radial distances due to reaction losses along the longer path lengths and from dispersion.

Table II

Oxygen Reductive Capacity (ORC) Measurements from Sediment Samples Collected Following the ISRM Proof-of-Principle Test at the 100-H Area[a]

Boring ID	Depth (m)	Radial distance (m)	% of sample < 4.75 mm	Sample ORC (meq/kg)	Pore volumes of treatment[b]	Maximum sample ORC (meq/kg)	% of available Fe(III) reduced
B2618	12.0	1.2	79	9.9	63	23.4	42
B2618	13.0	1.2	37	12.6	38	22.4	56
B2618	14.2	1.2	51	42.7	176	39.1	100
B2618	14.9	1.2	43	6.5	22	35.9	18
B2620	13.0	3.1	79	1.0	6	18.5	5
B2620	14.5	3.1	34	10.0	28	19.6	51
B2620	15.1	3.1	30	32.5	78	32.4	100
B2621	15.1	5.8	29	10.7	25	17.8	60
B2624	12.7	6.4	79	13.9	88	39.8	35
B2624	13.6	6.4	42	25.1	85	32.9	76
B2624	14.8	6.4	38	21.0	64	37.1	57
B2624	14.5 (int)	6.4	60	17.9	86	21.5	83
	14.5 (ext)	6.4	60	13.8	66	35.9	38
B2625	14.6	8.2	21	1.3	2	16.5	8

[a]Borehole locations are shown in Fig. 4a.
[b]Assumes bulk density = 1.9 g/cm^3; porosity = 20%; 9 mg/liter dissolved oxygen; 1 mg/liter hexavalent chromium.

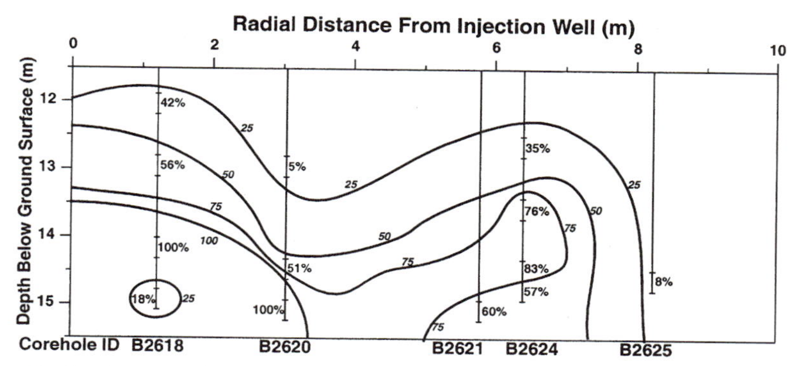

Figure 5 Contour plot of percentage Fe reduced based on ORC measurements of core samples collected after the 100-H area ISRM proof-of-principle test. Corehole locations are shown as radial distance from the injection/withdrawal well (H5-2). ORC data are from Table II. See Fig. 4a for plan-view locations of coreholes.

C. Performance Assessment Groundwater Monitoring

Groundwater monitoring to assess treatment zone performance at the ISRM site was conducted for approximately 5 years following the proof-of-principle experiment conducted in September 1995. Dissolved oxygen and hexavalent chromium, which can be used as an indirect measure of the performance of the reduced zone, were monitored along with other standard geochemical field parameters (pH, EC, temperature). pH, conductivity, and dissolved oxygen were measured with electrodes. A portable Hach DR/2000 spectrophotometer was used to measure hexavalent chromium and for corroboration with dissolved oxygen electrode measurements. Samples were also collected for IC analysis of sulfur species, and filtered samples were acidified with nitric acid for trace metals analysis by ICP/MS. To prevent reoxidation through contact with atmospheric oxygen within the well bore, argon gas is fed continuously to the wells from a liquid argon dewar stored at the site. A permanent solution would be to install packers just above the well screen or high-quality well seals at the surface to minimize the diffusion of atmospheric oxygen at this interface. Argon blanketing was used to permit easy access to the wells during sampling and hydraulic testing. The concentration of oxygen in the well bore was checked with a hand-held oxygen meter before each sampling event.

1. Chromium Monitoring

Baseline and selected posttest chromium concentration results for wells at the 100-H area ISRM site are provided in Table III, and in a generally

Table III

Pre- and Posttest Summary of Dissolved Oxygen and Chromium Measurements from Wells at 100-H Area ISRM Test Site[a]

Well ID	Baseline (August 1995)		Posttest (November 1997)			Posttest (April 1999)			Posttest (March 2001)	
	Dissolved oxygen (mg/liter)	Chromium (ICP/MS) (μg/liter)	Dissolved oxygen (mg/liter)	Chromium (ICP/MS) (μg/liter)	Hexavalent chromium (mg/liter)	Dissolved oxygen (mg/liter)	Chromium (ICP/MS)[b] (μg/liter)	Hexavalent chromium[b] (mg/liter)	Dissolved oxygen (mg/liter)	Hexavalent chromium[b] (mg/liter)
Wells within treatment zone										
H5-2	—	—	0.00	13.2	< 0.008	0.17	<1	< 0.008	0.4	< 0.008
H5-3p	4.40	50.8	3.50	37.0	0.03	6.00	13.2	0.01	7.9	0.01
H5-3o	5.70	71.0	11.47	35.0	0.03	9.74	6.0	< 0.008	9.7	< 0.008
H5-4p	2.39	47.1	0.00	3.4	< 0.008	0.00	±0.2	< 0.008	0.00	< 0.008
H5-4o	6.07	68.9	8.41	12.6	0.01	8.80	5.5	< 0.008	9.5	< 0.008
H5-5p	4.27	64.8	0.00	< 3	< 0.008	0.00	1.6	< 0.008	0.00	< 0.008
H5-5o	5.95	67.4	0.00	< 3	< 0.008	5.68	2.7	< 0.008	9.6	< 0.008
H5-6	5.36	60.1	0.00	4.1	< 0.008	0.00	<1	< 0.008	0.00	< 0.008
H5-7	5.14	67.4	0.00	4.5	< 0.008	0.00	1.5	< 0.008	0.00	< 0.008
H5-8	3.90	61.5	0.00	4.1	< 0.008	1.92	8.9	0.01	3.45	< 0.008
H5-9	6.93	59.8	0.00	< 3	< 0.008	0.00	<1	< 0.008	0.00	< 0.008
H5-10	5.57	45.8	0.00	3.4	< 0.008	0.00	<1	< 0.008	0.00	< 0.008
H5-11	6.61	59.5	0.00	< 3	< 0.008	0.00	<1	< 0.008	0.00	< 0.008

(continues)

Table III *(continued)*

Well ID	Baseline (August 1995)		Posttest (November 1997)			Posttest (April 1999)			Posttest (March 2001)	
	Dissolved oxygen (mg/liter)	Chromium (ICP/MS) (μg/liter)	Dissolved oxygen (mg/liter)	Chromium (ICP/MS) (μg/liter)	Hexavalent chromium (mg/liter)	Dissolved oxygen (mg/liter)	Chromium (ICP/MS)[b] (μg/liter)	Hexavalent chromium[b] (mg/liter)	Dissolved oxygen (mg/liter)	Hexavalent chromium[b] (mg/liter)
Wells outside treatment zone										
H5-12	1.88	12.9	6.46	33.0	0.03	3.18	9.5	0.01	2.4	< 0.008
H5-13	3.48	34.0	8.31	59.4	0.05	8.17	16.7	0.01	na	0.01
H5-14	—	—	6.89	53.1	0.06	6.50	13.3	0.01	na	0.01
H5-15	Not drilled	Not drilled	9.20	46.5	0.04	8.74	9.5	0.01	9.8	0.01

[a]Well locations are shown in Fig. 4a.

[b]Chromium concentrations within the plume in the area of the ISRM test site were significantly lower by the 1999 sampling event due to the injection of large volumes of treated groundwater nearby from a pump-and-treat system, which started in mid-1997.

west-to-east transect of wells screened in the lower portion of the aquifer (i.e., the targeted treatment zone) in Fig. 6. Total chromium concentrations before the experiment ranged from 12.9 to 64.8 μg/liter at the site. Following the dithionite injection/withdrawal experiment, total chromium concentrations declined to less than 10 μg/liter and near the detection limit of the analytical method (ICP/MS; 1 to 3 μg/liter) in wells located within the targeted treatment zone. Hexavalent chromium concentrations within the reduced zone following the injection/withdrawal experiment were mostly below the detection limit of the field measurement method (8 μg/liter). Hexavalent chromium was not analyzed before the field experiment, but the total chromium measurements after the experiment had concentrations and trends similar to the postexperiment hexavalent chromium concentrations (see Table III). Hexavalent chromium concentrations within the reduced zone also remained below detection limits during the most recent sampling event (March 2001 in Table III), ~ 5.5 years after the injection/withdrawal test. Samples collected from wells outside the reduced zone also showed significantly lower chromium concentrations during the 2001 sampling event, but the concentrations

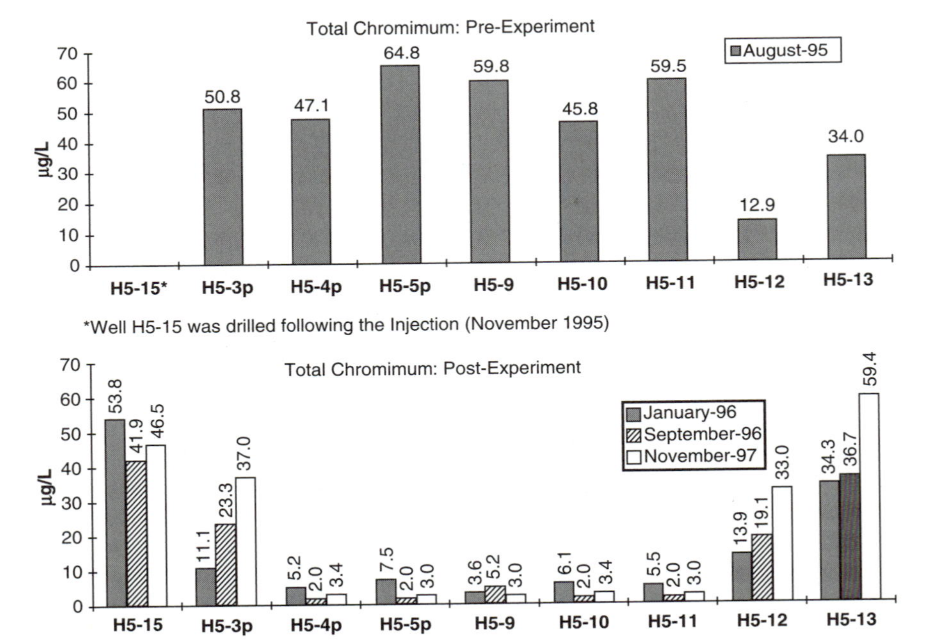

Figure 6 ICP/MS total chromium measurements along a west-to-east transect showing baseline and selected post-ISRM injection/withdrawal test values.

in the plume around the test site have been impacted from the injection of treated groundwater from a pump- and-treat system that began operation in mid-1997.

Chromium concentrations for samples collected from wells located on the edges of the reduced zone (e.g., H5-3p and H5-8) and upper portions of the aquifer (H5-3o, H5-4o, and H5-5o) have been increasing since the early times after the test. Wells located on the outer fringe of the reduced zone encountered lower concentrations of dithionite during the injection/withdrawal test and are expected to oxidize first from the approximate regional east–northeast groundwater flow direction. Wells screened in the upper portion of the aquifer encountered significantly lower dithionite concentrations than the corresponding lower zone wells because the test targeted the lower portion of the aquifer [i.e., the injection well was screened only in the lower 1.5 m (5 ft) of the aquifer].

2. Dissolved Oxygen Monitoring

Selected dissolved oxygen concentrations measured at the site are shown in Table III and Fig. 7. The trends seen in dissolved oxygen are similar to those

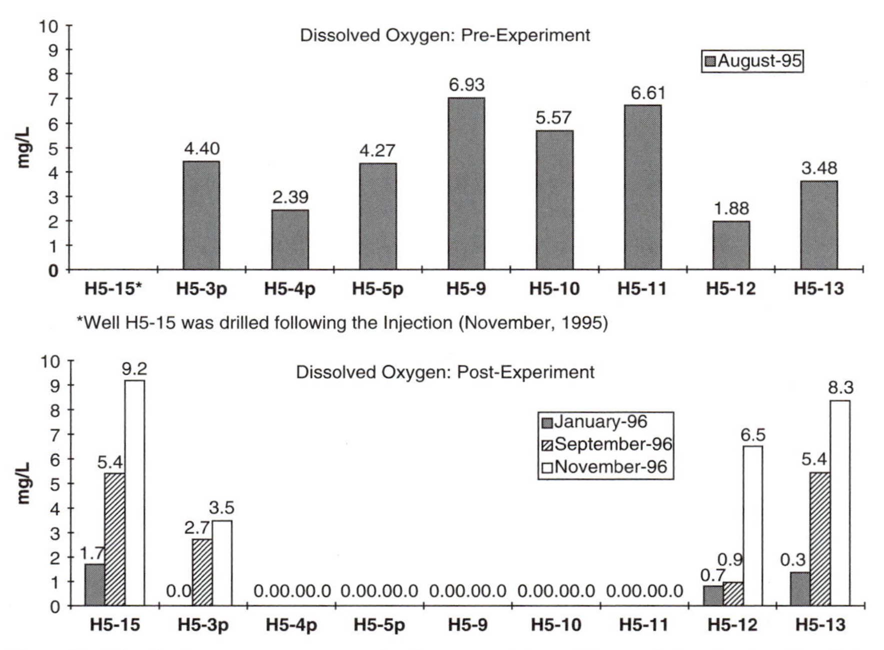

Figure 7 Dissolved oxygen measurements along a west-to-east transect showing baseline data and selected post-ISRM dithionite injection/withdrawal test data.

discussed in the previous section on chromium. Preexperiment dissolved oxygen concentrations ranged between 1.88 and 6.93 mg/liter. Following the injection/withdrawal test, the dissolved oxygen concentrations in samples collected from most of the wells within the reduced zone were below detection limits, as shown in Fig. 7. With the exception of the wells on the fringes of the reduced zone, the treatment zone remains anoxic 5.5 years following the test.

3. Residual Chemicals and Trace Metals Mobilization

Potentially adverse impacts to groundwater quality include residual chemicals (i.e., injected reagent or reaction products that were not recovered during the withdrawal) and naturally occurring trace metals mobilized under changing Eh/pH conditions or from dissolution by the reagent during injection. Residual chemicals in the aquifer were potassium, sodium, sulfur species (which eventually oxidize to sulfate), and carbonate. The residuals also affect the pH and electrical conductivity within the treatment zone. One year after the injection/withdrawal test, wells within the reduced zone remained elevated in potassium, sodium, and sulfate compared with preexperiment (baseline) values (Table IV). Samples collected from wells outside the treatment zone indicate they have not been altered significantly from preexperiment (baseline) measurements.

Baseline trace metals concentrations and concentrations 1 year after the dithionite injection/withdrawal test are compared to primary and secondary drinking water standards in Tables V and VI, respectively. Relative to baseline data, iron and manganese within the reduced zone were elevated significantly. These trace metals are not expected to migrate a significant distance downgradient of the reduced zone because of their high sorption in the reduced oxidation state [Fe(II), Mn(II)]. These reduced species are also expected to reoxidize into less soluble species [Fe(III), Mn(III), Mn(IV)] when in contact with unreduced sediment downgradient of the treatment zone. Zinc was also elevated slightly above the baseline values 1 year after the test. Other trace metals within the reduced zone (e.g., As and Pb), although elevated in the withdrawal stream, were not above baseline data 1 year after the experiment. The increase in these trace metals was due to the dissolution of oxides within the sediment by dithionite. The postexperiment trace metals concentrations in the wells outside the treatment zone were similar to baseline values.

D. Hydraulic Performance

Pre- and postexperiment hydraulic test responses were compared for selected observation wells at the ISRM test site to assess the impact of the

Table IV

Residual Chemicals (Na, K, SO_4), pH, and Electrical Conductivity

			Preexperiment (August 1995) all wells		Postexperiment (August 1996) treatment zone		Postexperiment (August 1996) H5-12 H5-13 H5-14	
Quality parameter	Secondary standard	Units	Mean	Standard deviation	Mean	Standard deviation	Mean	Standard deviation
K^+	None	ppm	6	1	116	71	7	0.4
Na^+	None	ppm	25	3	55	19	35	7
SO_4^{2-}	250	ppm	—	—	324	213	95	23
pH	—	pH	7.55	0.12	8.19	0.49	7.51	0.01
Conductivity	—	μS/cm	452	42	871	302	559	32

Table V

Comparison of Trace Metals at ISRM Site with Primary Drinking Water Standards

			Preexperiment (August 1995) all wells		Postexperiment (August 1996) treatment zone		Postexperiment (August 1996) H5-12 H5-13 H5-14	
Quality parameter	Primary Standard	Units	Mean	Standard deviation	Mean	Standard deviation	Mean	Standard deviation
Sb	6	ppb	<1	—	<1	—	<1	—
As	50	ppb	<2	—	<1	—	<1	—
Cd	5	ppb	<1	—	<1	—	<1	—
Cu	50	ppb	5	1	8	1	9	2
Pb	15	ppb	0.4	2	<1	—	<1	—
Ni	1000	ppb	4	3	4	3	5	1
Se	50	ppb	0.3	1	<6	—	<6	—

Table VI

Comparison of Trace Metals at ISRM Site with Secondary Drinking Water Standards

			Preexperiment (August 1995) all wells		Postexperiment (August 1996) treatment zone		Postexperiment (August 1996) H5-12 H5-13 H5-14	
Quality parameter	Secondary standard	Units	Mean	Standard deviation	Mean	Standard deviation	Mean	Standard deviation
Al	50–200	ppb	32	13	<10	—	<10	—
Fe	3000	ppb	5	13	132	162	20	34
Mn	50	ppb	32	59	170	237	14	12
Zn	5000	ppb	25	18	51	5	54	11

field experiment on existing *in situ* hydraulic properties. A direct comparison of hydraulic responses was not possible, as the aquifer thickness increased from 3 to 4 m for the pre- and postexperiment tests, respectively, due to the aquifer response to high water levels in the Columbia River in 1996 and 1997. However, analysis of available data, using methods to account for these differences, indicated that the dithionite injection did not result in any measurable degradation in formation permeability.

A comparison of preinjection and postinjection hydraulic test data did indicate a near-well decrease in permeability at the injection/withdrawal well following the injection. This small zone of reduced permeability (i.e., skin effect) may be attributed to the entrapment of suspended or colloidal material (or mineralization associated with the carbonate buffer) in the well screen or sandpack zone immediately outside the well screen during emplacement operations. This near-well reduction in permeability caused no adverse effects during the injection or withdrawal phases of the demonstration and is not expected to result in any significant degradation in the overall hydraulic performance of the treatment zone. A complete description of hydraulic analysis methods is included in Fruchter *et al.* (1996).

V. CERCLA TREATABILITY TEST

Following the successful demonstration of the ISRM technology at the laboratory and field pilot scale, a larger scale treatability test (Fig. 8) was conducted at a chromium-contaminated site located within the Hanford 100-D area (Williams *et al.*, 2000). The test used five injection wells to create a linear ISRM permeable reactive barrier that could be used to assess issues related to full-scale deployment and downgradient effects that were not addressed by the pilot-scale field testing. As shown in Fig. 8, these barrier emplacement operations created a reduced zone in the aquifer approximately 46 m (150 ft) long (perpendicular to groundwater flow), 15 m (50 ft) wide, and $\sim$ 5 m (15 ft) thick, extending over the thickness of the unconfined aquifer. Depth to groundwater in this area is about 26 m (85 ft) below ground surface, and the average groundwater velocity, based on measurements of hydraulic gradient, hydraulic conductivity, and porosity, is $\sim$ 0.3 m/day (1 ft/day). Characterization activities conducted during well installation included sampling the sediment, aquifer testing, establishing baseline aqueous geochemistry, and conducting a bromide tracer test. The first dithionite injection/withdrawal was conducted during September and October 1997; groundwater monitoring was conducted from the fall of 1997 to the spring of 1998. The remaining four dithionite injection/withdrawal tests were conducted from May to July 1998.

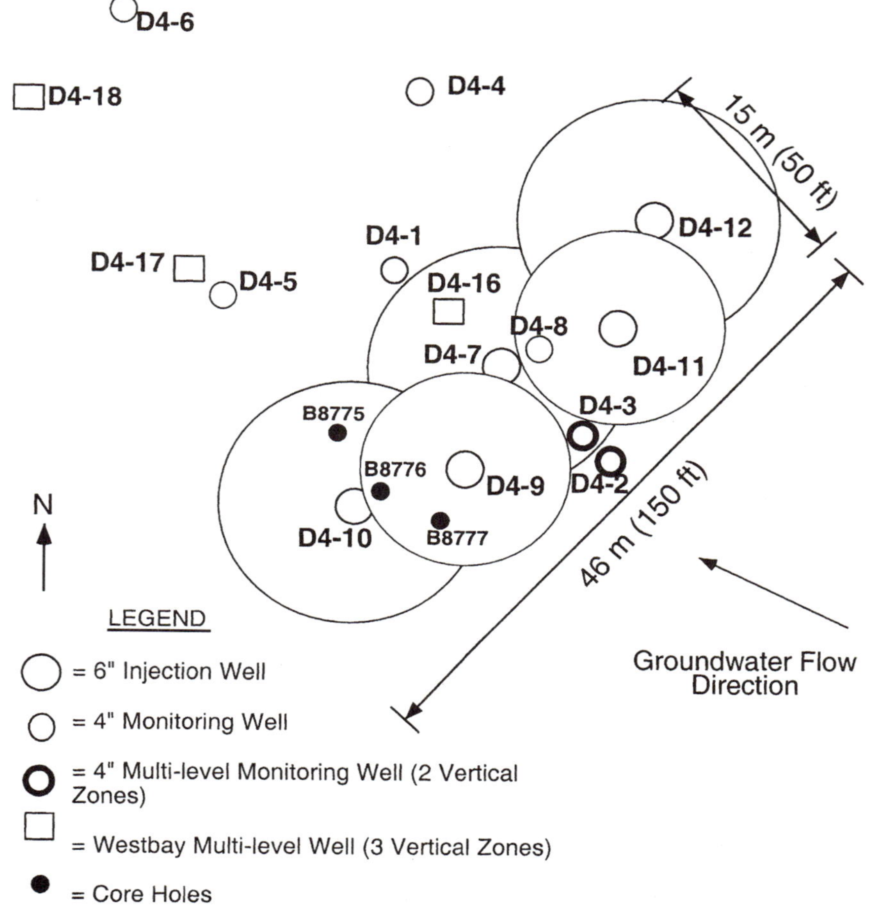

Figure 8 Well emplacement layout for the ISRM treatability test at the Hanford 100-D area.

Groundwater monitoring was conducted at the site on a monthly to bimonthly basis during 1998 and roughly quarterly during 1999. Analytes consisted of field parameters (EC, pH, and dissolved oxygen), chromate, anions, and trace metals. Results from the most recent complete sampling event (April 6, 1999) are shown in Table VII. Average values and ranges of selected parameters for this sampling event are given for samples collected from wells within the treatment zone and from downgradient wells (Table VII). Hexavalent chromium [Cr(VI)] results are shown in plan view in Figs. 9 and 10.

Table VII

Groundwater Measurement Summary

Parameter	Units	Range	Mean
Within the treatment zone (4/6/99)			
pH		7.71–8.78	8.38
Electrical conductivity	$\mu S/cm$	792–1891	1206
Dissolved oxygen[a]	mg/liter	0	0
Sulfate	mg/liter	197–886	455
Cr(VI)	$\mu g/liter$	<8	<8
Downgradient of the treatment zone (4/6/99)			
pH		7.50–7.82	7.61
Electrical conductivity	$\mu S/cm$	728–1180	956
Dissolved oxygen	mg/liter	0.00–4.45	2.66
Sulfate	mg/liter	249–535	403
Cr(VI)	$\mu g/liter$	20–210	75

[a]Westbay Well D4-16 DO measurements were not included.

Analysis of postemplacement groundwater sampling at the site showed Cr(VI) concentrations below detection limits (8 μg/liter) within the treatment zone compared with average baseline concentrations of 1000 μg/liter. Cr(VI) concentrations also declined in the downgradient wells to as low as 20 μg/liter by the April 1999 sampling event, approximately 1 year after emplacement of the barrier. Dissolved oxygen concentrations were also significantly lower within the treatment zone and downgradient wells than the baseline values. Because the ISRM treatability test site is located near the Columbia River [~150 m (500 ft)] and potential salmon-spawning habitat, regulatory and stakeholder concerns regarding attenuation of the anoxic plume downgradient of the barrier were addressed in a modeling study that simulated the near-river system and investigated important attenuation mechanisms. Predictions from numerical modeling resulted in dissolved oxygen concentrations from 75 to 95% of saturation by the time the groundwater discharges to the river (Williams *et al.*, 1999; Williams and Oostrom, 2000).

The secondary effects observed following the ISRM treatability test were similar in magnitude to those observed following the proof-of-principle field test. Sulfate, sodium, and potassium concentrations were elevated at the site from the dithionite residuals. Although residuals remain elevated, multilevel monitoring data collected from the site indicated that residuals were being transported downgradient of the treatment zone and that the primary source

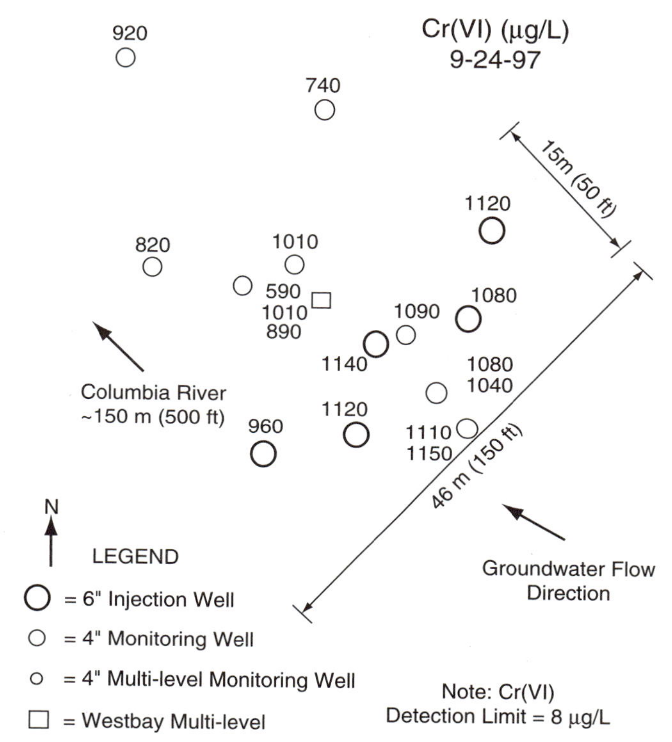

Figure 9 Cr(VI) concentrations (μg/liter) at the Hanford 100-D treatability test site before dithionite injection (9/24/97).

of the elevated residuals was continued seepage of reagent trapped in the vadose zone during emplacement of the treatment zone. Major trace metals involved in the redox process (Fe, Mn) were elevated above baseline values in the reduced zone due to dissolution of Fe and Mn oxides by the dithionite solution and the enhanced solubility of these naturally occurring oxides in the aquifer sediments under reducing conditions. As discussed previously, although these metals are elevated in the treatment zone, they are not likely to be mobilized a substantial distance downgradient from the zone due to their high retardation factors and reprecipitation once they contact oxidizing sediments outside the zone. The pH within the treatment zone was slightly above the baseline value of 7.5 as shown in Table VII. This elevated pH is associated with potassium carbonate/potassium bicarbonate pH buffer added to the reagent to enhance dithionite stability. The electrical conductivity was elevated in the reduced zone above baseline values due to the residual

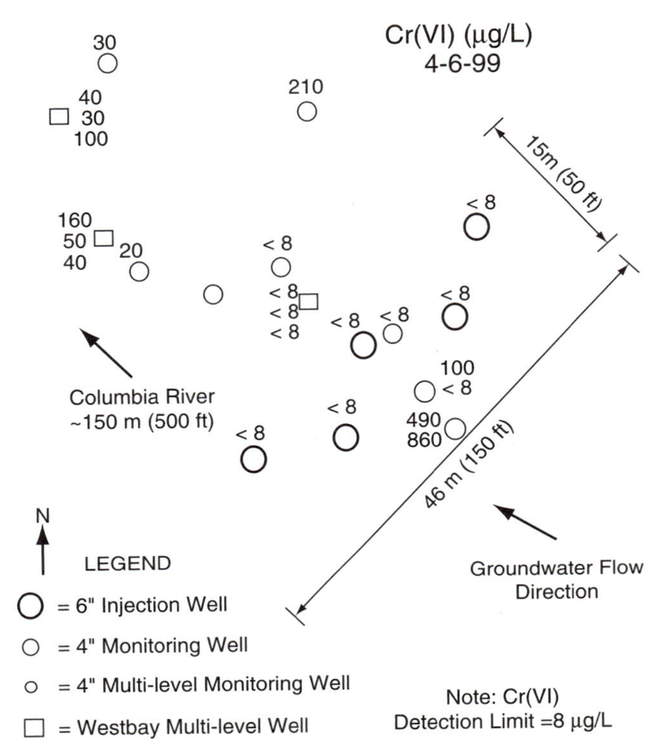

Figure 10 Cr(VI) concentrations (μg/liter) at the Hanford 100-D treatability test site after dithionite injection (4/6/99).

chemicals left in the aquifer from unrecovered reagents (Na, K, carbonate/bicarbonate, and sulfate). Unreacted dithionite and the sulfate/thiosulfate reaction products ultimately oxidize to sulfate. Comparison of preinjection and postinjection pressure response data indicated that injection of the sodium dithionite reagent resulted in no significant degradation in the overall hydraulic performance of the treatment zone (i.e., no aquifer plugging).

Laboratory analysis of sediment collected from the aquifer during the initial site characterization activities indicated that the average potential reductive capacity of the sediment was 11.0 $\pm$ 3.0 μmol Fe(III) per gram of sediment. Following treatment zone creation, sediment samples were collected from coreholes within the reduced zone to measure the reductive capacity of the sediment that was achieved by the treatability test dithionite injection/withdrawal operations. The average reductive capacity of the field-reduced sediment (11.2 $\pm$ 7.4 μmol/g) was similar to the average potential reductive capacity determined for the site sediments, indicating that the dithionite injection strategy was effective at reducing the targeted aquifer sediments for creation of the ISRM treatment zone. The range in the results of these analyses, as measured in breakthrough curves during field tracer and dithionite injection tests, is a function of the physical and chemical heterogeneities at the site.

Barrier longevity was estimated from the reductive capacity of the field-reduced sediments. Using 11.2 μmol Fe(II) per gram of sediment, the treatment zone should reduce chromate for an estimated 174 pore volumes [using 14% porosity, 2.3 g/cm^3 dry bulk density, and groundwater concentrations of 8 mg/liter dissolved oxygen and 1 mg/liter Cr(VI)]. Assuming this average reductive capacity was attained for a 15-m (50-ft) width at the 100-D area site, the predicted longevity of the 100-D area ISRM permeable treatment zone is 23 years [(using the estimated average groundwater velocity of 0.3 m/day (1 ft/day)]. This estimate neglects the flux of oxygen into the treatment zone at the interface between the vadose zone and unconfined aquifer and the possibility of preferential flow paths through the treatment zone that could significantly affect barrier longevity.

VI. SUMMARY AND CONCLUSIONS

An *in Situ* redox manipulation method for creating a permeable reactive barrier in the subsurface has been developed, and a field-scale demonstration has been conducted at a chromate-contaminated site on the U. S. Department of Energy's Hanford site in southeastern Washington State. The ISRM treatment zone is created by reducing the ferric iron [Fe(III)] phases naturally present in the aquifer sediments to ferrous iron phases [mainly adsorbed Fe(II)] with a chemical reducing agent using an injection/withdrawal (i.e., push/pull) emplacement strategy. Sodium dithionite ($Na_2S_2O_4$) is injected into the aquifer, provided a residence time sufficient to react the sediment, and any remaining unreacted reagent and reaction products are withdrawn from the aquifer. Standard groundwater wells are used, allowing treatment of contaminants too deep below the ground surface for conventional trench-and-fill technologies. Once in place, redox-sensitive contaminants migrating through this manipulated zone are destroyed (organic solvents) or immobilized (metals). A multiscale approach was utilized during development of the technology with experiments ranging from bench-scale laboratory tests to emplacement of a full field-scale treatability test barrier. Each successively larger scale experiment permitted isolation of effects related to each scale of experiment and enabled better design and regulatory approval of the next larger scale experiment.

Performance data obtained from field-scale demonstrations of the ISRM technology indicate that the method is effective for the remediation of Cr(VI)-contaminated groundwater. In both pilot-scale and treatability test-scale field tests, Cr(VI) concentrations within the treatment zone were reduced from background levels as high as 1000 μg/liter to below the detection limits of the analytical method (8 μg/liter), and reducing conditions continued to persist 5.5 years after the initial pilot-scale treatment, which is currently the longest monitoring period available for either of the sites. At the treatability test site, which was the only field experiment sized adequately to assess downgradient effects of the ISRM treatment, Cr(VI) concentrations downgradient of the barrier have also declined to as low as 20 μg/liter. No secondary effects were identified during laboratory- or field-scale testing of the technology that would preclude deployment of the ISRM technology for full-scale remediation of Cr(VI)-contaminated groundwater.

Based on the successful results of the ISRM treatability study and a cost analysis by the Hanford site environmental restoration contractor, the ISRM barrier concept is currently being deployed at Hanford as a remedy for a 700-m (2300-ft) wide Cr(VI) plume impacting the Columbia River.

ACKNOWLEDGMENTS

The authors acknowledge the efforts of Dr. J. Istok and his students at Oregon State University for assistance in conducting a number of the field tests. Additionally, we to thank B.N. Bjornstad, D.E. Hollingsworth, D.C. Lanigan, J.C. Evans, and T.L. Likala for support during field activities. This work was prepared with the support of the following contributors: Office of Science and Technology (headquarters), Skip Chamberlain; Subsurface Contaminants Focus Area (focus area/program), James Wright; Richland Operations Office, Science and Technology Programs Division, Craig Richins, technical program officer; and Pacific Northwest National Laboratory, Environmental Science and Technology, Environmental Technology Division (contractor), Walt Apley, manager.

REFERENCES

Amonette, J. E., Szecsody, J. E., Schaef, H. T., Templeton, J. C., Gorby, Y. A., and Fruchter, J. S. (1994). Abiotic reduction of aquifer materials by dithionite: A promising in-situ remediation technology. In "Proceedings of the 33rd Hanford Symposium on Health and the Environment – In Situ Remediation: Scientific Basis for Current and Future Technologies", (GW Gee and NR Wing, eds.), pp. 851–882. Pacific Northwest National Laboratory, Richland, WA.

Amonette, J. E., Fruchter, J. S., Gorby, Y. A., Cole, C. R., Cantrell, K. J., and Kaplan, D. I. (1998). "Method of Removing Oxidized Contaminants from Water." U. S. Patent 5,783,088.
Chilakapati, A. (1995). "A Simulator for Reactive Flow and Transport of Groundwater Contaminants." PNNL-10636/UC-602, Pacific Northwest National Laboratory, Richland, WA.
Cummings, M. A., and Booth, S. R. (1997). "Cost Effectiveness of in Situ Redox Manipulation for Remediation of Chromium-Contaminated Groundwater." LA-UR-97–165, Los Alamos National Laboratory, Los Alamos, NM.
Fruchter, J. S., Amonette, J. E., Cole, C. R., Gorby, Y. A., Humphrey, M. D., Istok, J. D., Spane, F. A., Szecsody, J. E., Teel, S. S., Vermeul, V. R., Williams, M. D., and Yabusaki, S. B. (1996). "In Situ Redox Manipulation Field Injection Test Report – Hanford 100H Area." PNNL-11372, Pacific Northwest National Laboratory, Richland, WA.
Fruchter, J. S., Spane, F. A., Fredrickson, J. K., Cole, C. R., Amonette, J. E., Templeton, J. C., Stevens, T. O., Holford, D. J., Eary, L. E., Bjornstad, B. N., Black, G. D., Zachara, J. M., and Vermeul, V. R. (1994). "Manipulation of Natural Subsurface Processes: Field Research and Validation". PNL-10123, Pacific Northwest National Laboratory, Richland, WA.
Fruchter, J. S., Cole, C. R., Williams, M. D., Vermeul, V. R., Amonette, J. E., Szecsody, J. E., Istok, J. D., and Humphrey, M. D. (2000). Creation of a subsurface permeable treatment barrier using in situ redox manipulation. *Groundwater Monitor. Remediat. Rev.* Spring 2002.
Fruchter, J. S., Zachara, J. M., Fredrickson, J. K., Cole, C. R., Amonette, J. E., Stevens, T. O., Holford, D. J., Eary, L. E., Black, G. D., and Vermeul, V. R. (1992). "Manipulation of Natural Subsurface Processes: Field Research and Validation." pp. 88–106. Pacific Northwest Laboratory Annual Report for 1991 to the DOE Office of Energy Research, Part 2: Environmental Sciences. PNL-8000 Pt. 2, Pacific Northwest Laboratory, Richland, WA.
Istok, J. D., Amonette, J. E., Cole, C. R., Fruchter, J. S., Humphrey, M. D., Szecsody, J. E., Teel, S. S., Vermeul, V. R., Williams, M. D., and Yabusaki, S. B. (1999). In situ redox manipulation by dithionite injection: Intermediate-scale laboratory experiments. *Ground Water* **37**, 884–889.
Mackay, D. M., and Cherry, J. A. (1989). Groundwater contamination: Pump and treat remediation. *Environ. Sci. Technol.*, **23**(6), 630–636.
Rai, D., Sass, B. M., and Moore, D. A. (1987). Chromium(III) hydrolysis constants and solubility of chromium(III) hydroxide. *Inorg. Chem.* **26**, 345–349.
Sass, B. M., and Rai, D. (1987). Solubility of amorphous chromium(III)–iron(III) hydroxide solid-solutions. *Inorg. Chem.* **26**, 2228–2232.
Szecsody, J. E., Fruchter, J. S., McKinley, M. A., Resch, C. T., and Gilmore, T. J., (2001). "Feasibility of the *In Situ* Redox Manipulation of Subsurface Sediments for RDX Remediation at Pantex." PNNL-13746, Pacific Northwest National Laboratory, Richland, WA.
Szecsody, J. E., Fruchter, J. S., Sklarew, D. S., and Evans, J. C. (2000a). "In Situ Redox Manipulation of Subsurface Sediments from Fort Lewis, Washington: Iron Reduction and TCE Dechlorination Mechanisms". PNNL-13178, Pacific Northwest National Laboratory, Richland, WA.
Szecsody, J. E., Williams, M. D., Fruchter, J. S., Vermeul, V. R., Evans, J. C., and Sklarew, D. S. (2000b). Influence of sediment reduction on TCE degradation. In "Remediation of Chlorinated and Recalcitrant Compounds," (G. Wickramanayake, ed.), Proceedings of Monterey 2000 conference.
Vermeul, V. R., Teel, S. S., Amonette, J. E., Cole, C. R., Fruchter, J. S., Gorby, Y. A., Spane, F. A., Szecsody, J. E., Williams, M. D., and Yabusaki, S. B. (1995). "Geologic, Geochemical, Microbiologic, and Hydrologic Characterization at the in Situ Redox Manipulation Test Site". PNL-10633, Pacific Northwest Laboratory, Richland, WA.
Vermeul, V. R., Williams, M. D., Evans, J. C., Szecsody, J. E., Bjornstad, B. N., and Liikala, T. L. (2000). "In Situ Redox Manipulation Proof-of-Principle Test at the Fort Lewis Logistics

Center: Final Report", PNNL-13357, Pacific Northwest National Laboratory, Richland, WA.

Williams, M. D., and Oostrom, M. (2000). Oxygenation of anoxic water in a fluctuating water table system. *J Hydrol.*, **230**, 70–85.

Williams, M. D., Vermeul, V. R., Oostrom, M., Evans, J. C., Fruchter, J. S., Istok, J. D., Humphrey, M. D., Lanigan, D. C., Szecsody, J. E., White, M. D., Wietsma, T. W., and Cole, C. R. (1999). "Anoxic Plume Attenuation in a Fluctuating Water Table System: Impact of 100–D Area in Situ Redox Manipulation on Downgradient Dissolved Oxygen Concentrations." PNNL-12192. Pacific Northwest National Laboratory, Richland, WA.

Williams, M. D., Vermeul, V. R., Szecsody, J. E., and Fruchter, J. S. (2000). "100-D Area in Situ Redox Treatability Test for Chromate-Contaminated Groundwater." PNNL-13349, Pacific Northwest National Laboratory, Richland, WA.

Williams, M. D., Yabusaki, S. B., and Cole, C. R. (1994). In situ redox manipulation field experiment: Design analysis. In "Proceedings of the 33rd Hanford Symposium on Health and the Environment – In Situ Remediation: Scientific Basis for Current and Future Technologies" (G. W. Gee and N. R. Wing, eds.). Pacific Northwest Laboratory, Richland, WA.

Part II

Development of Reactive Materials

Chapter 7

Development and Performance of an Iron Oxide/Phosphate Reactive Barrier for the Remediation of Uranium-Contaminated Groundwater

Jennifer L. Joye,* David L. Naftz,† James A. Davis,* Geoff W. Freethey,† and Ryan C. Rowland†

*U.S. Geological Survey, Menlo Park, California 94025

†U.S. Geological Survey, Salt Lake City, Utah 84119

This chapter reports on the laboratory and field testing of engineered solid mixtures for use with deep aquifer remediation tool (DART) technology for the removal of groundwater metal contamination resulting from mining and ore processing. In laboratory testing mixtures of pelletized bone charcoal (bone char apatite) and hematite pellets coated with amorphous iron oxide have been studied in various ratios in both batch and column experiments to evaluate the effectiveness of the mixture in removing uranium from groundwater. Variables considered include phosphate levels in solution and on the iron oxide surface, pellet ratios, and aging of the solids. Laboratory batch experiments indicated phosphorus adsorption by the iron oxide pellets, but there was no apparent increase in uranium removal in mixtures compared to bone char pellets alone. In contrast, column experiments showed greater uranium removal by the mixture of materials, possibly as a result of longer aging times and/or higher adsorbed phosphorus concentrations. No evidence of autonite formation was found. Field tests using DART deployment technology also compared different pellet ratios and included (1) shallow DART deployment in an unpumped well array in Fry Canyon, Utah, the site of an abandoned uranium upgrader operation, and (2) deep DART emplacement 146 m below land surface during active mining at the Christensen Ranch *in situ* uranium mine in northeastern Wyoming. An additional field test at Fry Canyon also included DART testing with zerovalent iron.

I. BACKGROUND / OBJECTIVES

A wide range of mechanisms exists through which inorganic contaminants can be removed from groundwater, including adsorption (Morrison *et al.*, 1996), precipitation (Blowes *et al.*, 1998), and biologically mediated transformations (Robertson and Cherry, 1995). Inorganic contaminants in groundwater are often managed by operating expensive, aboveground pump-and-treat systems. There is a need for a cost-effective means for preventing the release of contaminants from abandoned mine drainage or for immobilizing contaminants that have already become mobile. The deep aquifer remediation tool (DART) is used in conjunction with nonpumping wells and offers a low-cost and low-maintenance alternative to *ex situ* treatment methods and a deeper installation depth than trenching installations of permeable reactive barriers (PRBs) (see Chapter 5). A DART is composed of a rigid polyvinyl chloride shell with high-capacity flow channels that contains permeable reactive material and flexible wings to direct the flow of groundwater into the reactive material. The new DART technology, developed by the U.S. Geological Survey, can be deployed at a range of depths, takes advantage of existing well structures, and the reactive materials can be removed and replaced easily. This study focuses primarily on the behavior of mixed solids designed to immobilize one or more contaminants with a combination of both adsorption and precipitation reactions, as well as the use of these solids with DART technology. Contaminant removal through chemical reduction is also investigated in field testing with DARTs.

The two materials tested in mixtures with the DARTs were bone charcoal (bone char apatite) and an iron oxide material. Apatites are by far the most abundant phosphates found in nature (Van Wazer, 1973) and have been tested for use in reactive barriers (Fuller *et al.*, 2002) because of their stability and their natural occurrence in halos around uranium (U) ore deposits as a final-stage weathering mineral. Due to dissolved phosphate (PO_4) release during apatite dissolution, it was expected that one or more uranyl phosphate phases, e.g., autunite [$Ca(UO_2)_2(PO_4)_2 \cdot 10\text{–}12\,H_2O$], a naturally occurring stable hydrous calcium–uranyl phosphate, would precipitate within the mixtures of reactive materials. Fired bone material, or apatite, bound with an Al–Si binder make up the bone char pellets and provide the PO_4 source for precipitation of a uranyl phosphate phase. In nature, buried bone will take up to as much as 0.1% by weight of U, with uptake increasing with length of time buried and local uranium [U(VI)] groundwater concentration (Millard and Hedges, 1996).

The iron oxide solid provides a highly reactive surface to adsorb both U(VI) and phosphorus. Iron oxides are not only ubiquitous in the environment but have been used in a myriad of metal removal studies (e.g., Payne and Waite, 1996; Coston *et al.*, 1995). Payne *et al.* (1996) showed that amorphous iron oxide with adsorbed PO_4 can enhance U(VI) uptake in solution. The proposed mechanism for dissolved U(VI) immobilization in the DART mixtures, therefore, involves both adsorption of PO_4 and U(VI) by the reactive iron oxide surface, followed by nucleation of a uranyl phosphate-phase formation on the iron oxide surface. In addition, a uranyl phosphate phase could form on the bone char pellet surface or by homogeneous nucleation in solution when PO_4 concentrations are high. However, Fuller and colleagues (Chapter 9) have shown that the removal mechanism is likely U(VI) surface complexation rather than uranyl phosphate precipitation under the conditions typical for Fry Canyon groundwater. Thus, enhanced U(VI) uptake by a mixture of bone char and iron oxide solids, as a result of the PO_4 influence on U(VI) uptake by the iron oxide, may occur.

Although the majority of research concerning PRB materials has focused on single material use, we consider here the possibility that a mixture of materials may improve PRB performance. There are toxic metals other than U associated with the Christensen Ranch *insitu* uranium mine site, and it is possible that a mixed solid reactive material may offer efficient removal of multiple contaminants. At the Christensen site, there are significant concentrations of selenium (Se), vanadium (V), and arsenic (As), as a result of the highly alkaline and oxidizing lixiviant used to mobilize U from the ore body. The amorphous iron oxides in the mixture could act to scavenge these other metals as well as U.

In this study, mixtures of bone charcoal and iron oxide pellets at various ratios were tested in both batch and column experiments to evaluate the effectiveness of the mixtures in removing U(VI) from groundwater, relative to bone char alone. The DART mixtures were field tested in a shallow-contaminated aquifer as an alternative to passive trenching and for deep contamination scenarios not accessible by trench techniques. Field tests included (1) shallow DART deployment in a unpumped well array in Fry Canyon (southeastern), Utah, the site of an abandoned U upgrader operation (see Chapter 14), and (2) deep DART emplacement approximately 146 m below land surface during active *insitu* U mining in northeastern Wyoming.

Zero valent iron filings (ZVI), not in mixture but alone, were tested to investigate the reductive precipitation of U(VI) (Naftz *et al.*, 1999). ZVI is available commercially in several forms, which have good structural and hydrodynamic properties. In addition, U(VI) reduction by ZVI is less sensireactive barrier materials: high surface area and high permeability. Both types of pellets possess the high surface area required so that surface reactions can occur more readily per unit volume of material. However, the high surface area needed cannot be achieved with small particles, as these would typically pack into a bed with low permeability. Millimeter-sized pellets allow for high permeability so that flow through the reactive material is enhanced when DARTs with reactive materials are placed in aquifers (see Chapter 5).

C. Characterization of Barrier Materials

Scanning electron microscopy (SEM) was used to image the texture and porosity of the solid phases before experiments. Specific surface areas were determined at atmospheric pressure using the flow-through method on a Micrometrics Flow Sorb II (Model 2300). This method allows accurate determination of specific surface areas on granular samples with values greater than $0.1 m^2/g$ to be determined using adsorption of N_2 gas (Lowell, 1979). Laboratory batch and column experiments were performed to study the U(VI) uptake capacity of the solids mixed in different ratios, along with effects of aging and PO_4 concentration.

SEM-EDS and electron backscatter imaging were used to identify the presence of U after the solids were used in field deployments. Selected samples from the Christensen site were analyzed using XRD to test for the presence of crystalline U phases. Twenty-four-hour nitric acid (2N) extractions were used to determine the amount of U that reacted with the materials after deployment in the field. As, V, and Se were also measured in the extractions from the Christensen DARTs and are reported for reference. To prepare for the extractions, the samples were mixed into 1-kg composites and then 50 g of each composite was weighed wet and subsequently dried in air. After drying, the sample was separated into 10-g samples, 2 g of which was used for extraction and 5 g of which was crushed with a mortar and pestle. Two grams of each crushed sample was also used for extraction. Field-reacted samples from both field sites were also extracted for 2 and 9 h with the artificial groundwater at pH 7.2 (same composition as in batch experiments) to determine the amount of U(VI) that could be desorbed under the aquifer conditions at the Fry Canyon site. Field-reacted ZVI solids were observed using SEM and extracted using 4N HNO_3.

D. U(VI) Adsorption Batch Experiments

In field deployments, DARTs with mixtures of the bone char apatite and iron oxide pellets were packed with approximately known dry volume ratios.

In order to compare the laboratory results with the field deployments, dry volume ratios were used in mixtures of these two materials in both column and batch experiments. Artificial groundwater (AGW) for all batch experiments was similar to Fry Canyon groundwater conditions (see Chapter 14), except that a lower calcium concentration was used to avoid the precipitation of carbonate phases (Ca=36 mg/liter, K=5.4 mg/liter, Mg= 43.8 mg/liter, Na=380 mg/liter, SO_4 = 610 mg/liter, Cl= 90 mg/liter, HCO_3 = 400 mg/liter). Under these conditions, dissolved uranium exists in the +6 oxidation state and is denoted as U(VI). Wyoming groundwater conditions were not used because they were difficult to mimic in the laboratory as a result of the high partial pressure of CO_2 present at the average 146-m Wyoming deployment depth and the sodium bicarbonate lixiviant used in U removal from the ore (alkalinity = 1020 mg/liter as $CaCO_3$).

1. Ratioed Pellet Batch Tests

Four ratios of the bone char apatite and iron oxide pellets (100:0, 75:25, 50:50, 0:100) were tested in batch experiments to select the dry volume ratios to be tested in DART field deployments and to determine if there was an increase in U(VI) removal with a mixture of the materials compared to U(VI) removal by the end member materials. The solids were weighed in polycarbonate centrifuge tubes and preequilibrated with the Fry Canyon AGW for 24 h at pH 7.2. The amount of AGW added to the mixture of pellets was sufficient to yield a total solid/liquid ratio of 10 g/liter; batch experiments were done in triplicate. The experimental volume was generally 200 ml, but larger volumes were sometimes used to decrease the variability in results caused by a variation in pellet size and acidity. The sample solutions were adjusted to pH 7.2 using dilute acid and/or base, and the pH was maintained using a 2.5% CO_2/97.5% air mixture. After the 24-h preequilibration period, samples were spiked with U(VI) and equilibrated for 48 h to allow time for diffusion into the porous pellets. The initial U(VI) concentration in the batch experiments was approximately 23.8 mg/liter (10^{-4} *M*), in the range of field groundwater U(VI) concentrations. Supernatants were separated from the solids using centrifugation. U(VI) and PO_4 were measured in the supernatants, and uptake was calculated by difference.

2. Batch Experiments with Aged Materials

Aged samples were prepared in the same way as unaged samples except that the aged samples were preequilibrated for 1 week in AGW prior to U(VI) addition as opposed to a preequilibration time of 24 h. "Aging" or "aged samples" refer to this longer preequilibration time. After being spiked with

U(VI), the samples were equilibrated on a shaker table for 48 h. The pH of the solutions was measured every 8 h and was maintained at pH 7.2 ± 0.02 by bubbling with a 2.5% CO_2 mixture. Every effort was made to keep samples in a constant 2.5% CO_2 atmosphere.

3. Phosphorus influence

Iron oxide pellets were equilibrated with different dissolved PO_4 concentrations to determine the effects on U(VI) uptake of increased PO_4 adsorption on the iron oxide pellets. Three volumes of iron oxide pellets were each added to 200 ml of AGW and were spiked with an appropriate aliquot of a 0.01 *M* KH_2PO_4 solution to attain approximate PO_4 adsorption levels of 0.03, 0.25, and 0.5 mg PO_4/g. These samples were allowed to equilibrate for 48 h. Supernatants were measured for PO_4 using inductively coupled plasma-optical emission spectroscopy (ICP-OES) to determine PO_4 uptake levels onto the iron oxide pellets (data not shown). U(VI) uptake of the PO_4-sorbed iron oxide pellets was then tested alone and with bone char in a 50:50 volume mixture. Twice the total solid volume was used in this experiment in order to obtain a more representative sample of the PO_4-rich iron oxide pellets, as we were not familiar with the extent of PO_4 sorption coverage. Iron oxide pellets with no PO_4 adsorbed were equilibrated with an AGW solution containing U(VI) for comparison. Iron oxide pellets with no PO_4 and mixed with bone char were also equilibrated with AGW containing U(VI) for comparison with mixtures containing increased PO_4 levels.

E. Column Experiment

U(VI) uptake by a 50% bone char/50% iron oxide pellet by dry volume mixture was compared to that of a 100% bone char volume in a column experiment to determine if there is increased U(VI) uptake with the presence of the iron oxide pellets. The column experimental apparatus is described in Chapter 9.

Generally, dry solids were packed into 2.5-cm diameter × 30-cm-long glass columns and were gravity fed. Columns were flushed with CO_2 and then flushed with deionized water to dissolve CO_2, removing gas pockets. Reactive materials were pre-equilibrated with artificial groundwater until the column outlet pH stabilized. pH was measured both in line and at the effluent. At that point, the inlet flow was changed to AGW containing U(VI). U(VI) influent concentrations were 5.2 and 4.5 mg/liter in the 100% bone char and 50/50 mixture columns, respectively. Columns, feed lines, and carboy reservoirs were covered to discourage photosynthesis.

In order to calculate total column volume (*cv*), intraparticle volumes were determined from weight loss upon drying of water-saturated pellets, and the general pellet, density of the water-saturated pellets was determined by water displacement. The ratio of effluent to influent U(VI) concentration (C/C_o) is plotted versus *cv* of AGW passed. Total sorption capacity is defined as the maximum cumulative sorption of U(VI) at complete breakthrough ($C/C_o = 1$). After 100% breakthrough, the potential for U(VI) release was tested by passing U(VI)-free AGW through the columns.

III. MATERIAL/METHODS: FIELD STUDIES

Fry Canyon, Utah, the site of an abandoned U tailings pile (see Chapter 14) and the Christensen *in situ* U mine in Wyoming, north of Casper, were the two field sites at which the DARTS were tested. The materials tested for use in the DARTs at these locations are described.

A. Dart Design

Specific details on the DART design can be found in Naftz and Davis (1999) and in Chapter 5. DARTs use TIMCO high-flow, schedule 80, threaded polyvinyl chloride casing with 0.6-cm slots to provide a rigid package for placement of the reactive materials down existing wells and are designed for easy removal. Flow wings, attached to the outside of the DART, are flexible and extend to the well casing to direct water flow into the material to improve treatment efficiency. Flow modeling predicts a treatment area of the up gradient aquifer twice the diameter of the nonpumping well for each DART package (see Chapter 5). In the initial deployment of DARTs, pumping was used to collect annular space between the well casing and the DART. A subsequent DART design used point-of-compliance (POC) samplers made of dialysis membrane that are attached to the outside surface of the DART between the flow wings (Fig. 1) and at three different depths as an alternative to the pumped sample tubes. POC samplers show the groundwater U(VI) concentrations that existed during the last week or so prior to sampling, which can be compared to the influent concentration for an estimation of "breakthrough." Figure 1 shows a schematic of the DARTs that were deployed at both Fry Canyon and the Christensen site.

At the Christensen site, DARTs were deployed in the borehole beneath the well casing, whereas at the site in Fry Canyon, DARTs were deployed within

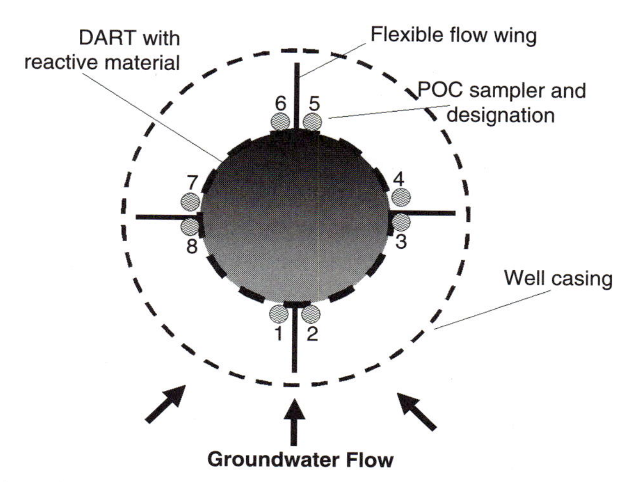

Figure 1 Schematic diagram (plan view) of DART deployment. A second set of POC samplers (numbers 9–16) was located 76 cm below the first set of samplers.

the well casing. Water samples collected from the POC samplers were filtered and acidified with nitric acid to pH 2 immediately after collection in the field. Background samples were taken from the deployment wells for comparison with POC U(VI) levels in order to estimate average U(VI) removal at time of sampling.

B. Christensen *In Situ* U Mine

At the Christensen site, the average ore depth is 146 m below land surface. The groundwater is oxidizing (200 mg/liter), has a high alkalinity (1020 mg/liter as $CaCO_3$), and has a pH of 6.4, likely because of the lixiviant used to dissolve U at the site. The dissolved U(VI) concentration is between 9 and 11 mg/liter, and Se concentrations range from 1.5 to 5 mg/liter. Measured V and As concentrations were as high as 0.5 mg/liter. DARTs were deployed in an actively pumped area with a forced gradient and, therefore, may have experienced high flow rates. Three DARTs, COG1, COG2, and COG3, were deployed at the Christensen site on May 20, 1999 (Table I). The Christensen DARTs were deployed in different mining modules and over a distance of approximately 620 m.

Table I

Summary of Christensen Mine, Wyoming, DART Deployment

DART designation	Contents	Date deployed	Well U concentration (mg/liter)	Date retrieved
Phase II (POCs)				
COG 1	75% bone char, 25% iron pellet (intermixed)	May 20, 1999	8.7[a]	August 8, 1999
COG 2	100% bone char	May 20, 1999	11.9[b]	January 5, 2000
COG 3	50% bone char, 50% iron pellet (intermixed)	May 20, 1999	11.1[a]	January 5, 2000

[a]Measured on September 14, 1998.
[b]Measured on April 26, 1999.

C. Fry Canyon, Utah

The Fry Canyon demonstration site is an abandoned U upgrading operation located in southeastern Utah. The alkalinity at Fry Canyon averages 450 mg/liter as $CaCO_3$, the pH is 7.4, and the U(VI) concentration averages 668 μg/liter. Here, the target treatment zone is relatively shallow at 5–6 m below the surface, which, along with the different chemistry relative to Christensen, may provide more information regarding the extent of DART application.

Two sets of DARTs were deployed at Fry Canyon (Table II). The first set was an early deployment that utilized pumping tubes to sample groundwater from the annular space surrounding the DART to measure treated water quality. Although water samples were collected for these DARTs, only solid extraction results are reported here. Pumping the annular space included untreated, up gradient water as well as treated water. As a result, these water samples could not be used to evaluate efficiency. The second set of Fry Canyon DARTs (Table II), deployed in September 1999, had the POC samplers. Both water samples and solid extraction results for these DARTs are included.

To test reversibility of U(VI) uptake, the reactive materials deployed in the DARTs at Fry Canyon were reacted in batch experiments with an artificial groundwater at pH 7.2 for 9 h. pH was maintained with the 2.5% CO_2 gas mixture.

Table II

Summary of Fry Canyon DART Deployment

DART designation	Contents	Date deployed	Well U concentration (mg/liter)	Date retrieved
Deployment I (pumping)				
BZ1	50% bone char, 50% iron pellet (intermixed)	October 20, 1998	0.64[a]	September 1, 1999
BZ2	100% bone char	October 20, 1998	0.67[a]	September 1, 1999
BZ3	50% bone char, 50% iron pellet (vertically layered 50:50)	October 20, 1998	0.73[a]	September 1, 1999
Deployment II (POCs)				
BZ4	100% bone char	September 1, 1999	0.43[b]	November 14, 2000
BZ5	100% ZVI	September 1, 1999	0.42[b]	November 14, 2000
BZ6	50% bone char, 50% iron pellet (intermixed)	September 1, 1999	0.44[b]	November 14, 2000

[a]Measured after DART removal in September 1999.
[b]Measured after DART removal in November 2000.

IV. LABORATORY RESULTS

A. Characterization of Solids

The iron oxide and bone char pellets were highly porous materials as can be observed in the SEM micrographs in Figs. 2 and 3. Intraparticle volumes

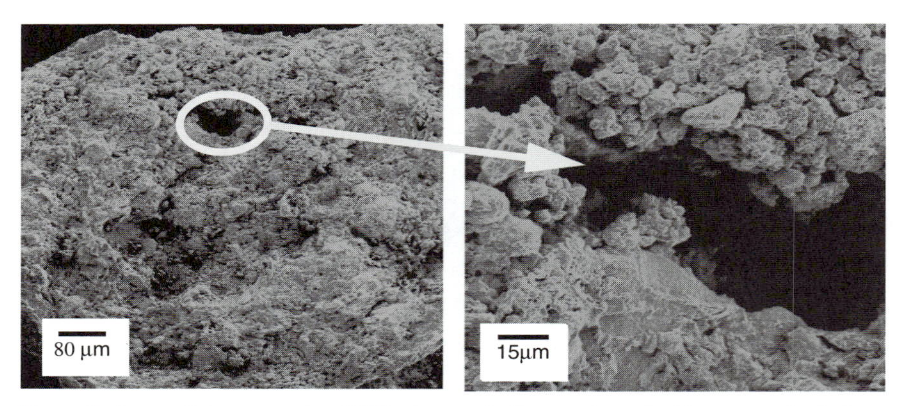

Figure 2 Iron pellet cross-section SEM micrograph. Images show significant porosity in iron pellets with larger pores measuring 80 μm. The characteristic elevated porosity of the engineered pellets is necessary in order to increase permeability and therefore create preferred flow through the material packages when deployed in an aquifer.

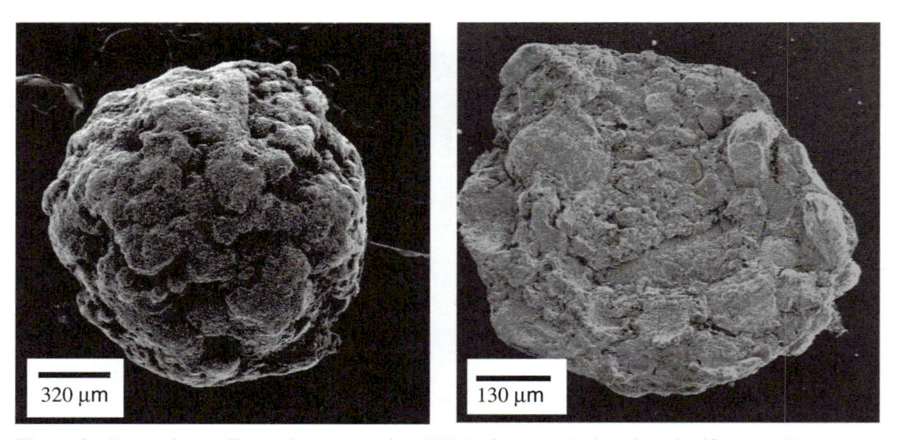

Figure 3 Bone char pellet and cross-section SEM micrograph showing significant pellet porosity. Bone char pellets can be made in bulk by firing bone meal in the absence of air and binding it on a porous supporting material.

were 0.591 cm^3/g for the bone char and 0.328 cm^3/g for the iron oxide pellets.

The bone char pellets were 2 to 4 mm in diameter and the iron oxide pellets ranged from 3 to 4 mm. General particle densities of 0.7 and 1.2 g/cm^3 were measured for the bone char pellets and iron oxide pellets, respectively. Surface areas, determined by BET N2 gas adsorption, were 44 m^2/g for the bone char (see Chapter 9) and 15 m^2/g for the iron oxide pellets.

B. Batch Experiments

Figures 4–7 show the U(VI) uptake results for different mixtures of iron oxide pellets and bone char, along with the effects of aging the mixtures. U(VI) uptake increased with the amount of bone char present (Fig. 4). Upon aging the same solids with the AGW for 1 week, U(VI) uptake increased with aging time. U(VI) uptake on the aged solids also increased with the amount of bone char present, with the highest U removal being by the 100% bone char

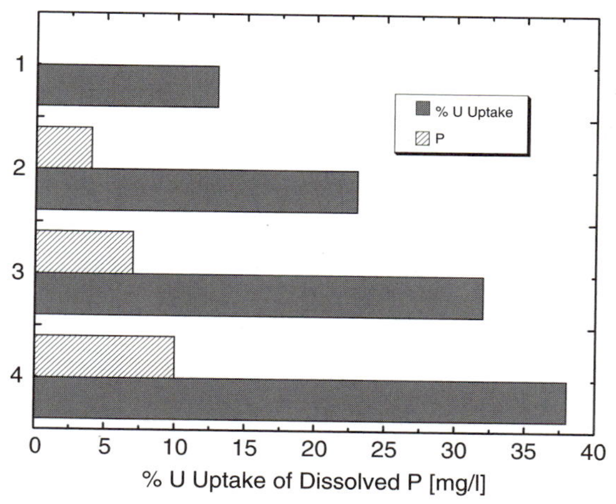

Figure 4 U(VI) uptake results for the following pellet mixtures: (1) 100% iron oxide pellets, (2) 50% iron oxide pellets: 50% bone char mixture, (3) 75% bone char:25% iron oxide pellet mixture, and (4) 100% bone char. Samples were preequilibrated with Fry Canyon artificial groundwater for 24 h prior to spiking. Total U(VI) concentration was 23.8 mg/liter and a pH of 7.2 was maintained with a 2.5% CO_2 gas mixture. Samples with U(VI) were allowed to equilibrate for 48 h. There was no enhancement of U removal by the mixtures. Also shown is dissolved PO_4 measured in solution at the end of the experiment, which increases with increasing percentage of bone char.

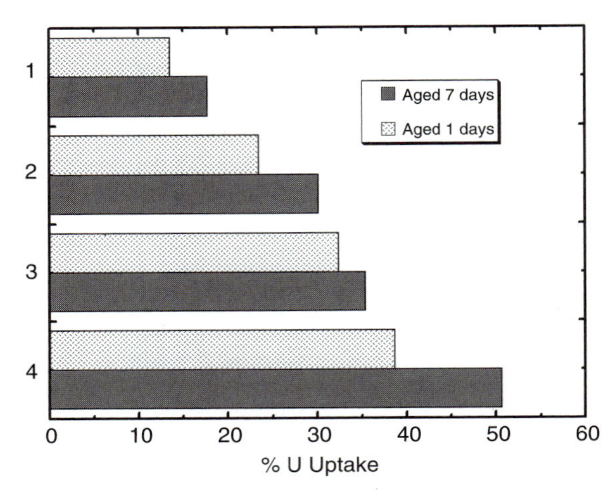

Figure 5 Effects of aging on U(VI) uptake in the same mixtures 1–4. Experimental conditions were the same [total U(VI) concentration = 23.8 mg/liter pH = 7.2 maintained with a 2.5% CO_2 gas mixture, equilibration time = 48 h], except that these samples were aged 7 days in addition to the 24-h preequilibration period. U(VI) uptake increases with aging possibly as a result of more PO_4 dissolution, but the trend still suggests that there is no benefit in a mixture.

samples (Fig. 5). Dissolved PO_4 concentrations measured in solution also increased with aging time. In the 75:25 and 50:50 aged mixtures, dissolved PO_4 decreased as compared with the 100% bone char alone, likely as a result of the presence of the iron oxide pellets (Fig. 6). XRD analysis of the aged samples showed no evidence of autunite or other crystalline U structure.

To test PO_4 influence during aging, aliquots of iron oxide pellets were reacted with three different concentrations of PO_4: 0.4, 3.0, and 6.0 mg/liter. Dissolved PO_4 measured in solution showed sorption levels of 0.03, 0.17, and 0.43 mg PO_4/g iron oxide pellets, respectively. Iron oxide pellets with 0.43 mg PO_4/g sorbed showed 5% greater U(VI) uptake compared to the same weight of iron pellets with no sorbed PO_4 (14%) (Fig. 7). Iron oxide pellets with sorbed PO_4 from this experiment were then mixed with bone char pellets in a 50:50 volume mixture and reacted with U(VI). Figure 7 shows an increase in U(VI) adsorption in samples with bone char and the PO_4-sorbed iron pellets.

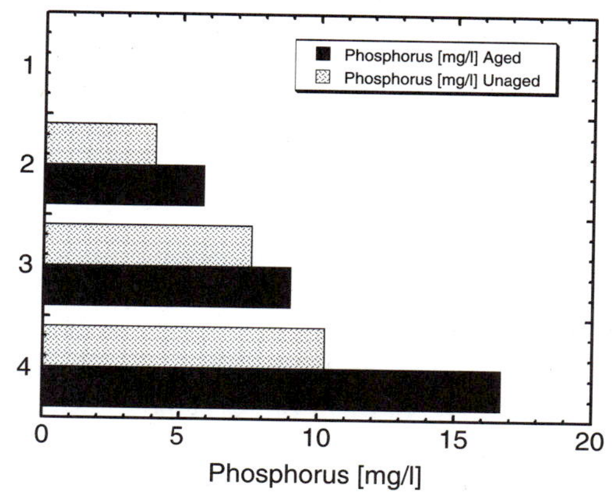

Figure 6 Dissolved PO_4 release with aging. Experimental conditions are the same as described in Fig. 5. P values are dissolved PO_4 levels measured in solution at the end of the U(VI) uptake experiment described in Fig. 5. More PO_4 is released with increased aging time, and there is decreased PO_4 in the presence of iron oxide pellets, but percentage PO_4 removal was not large.

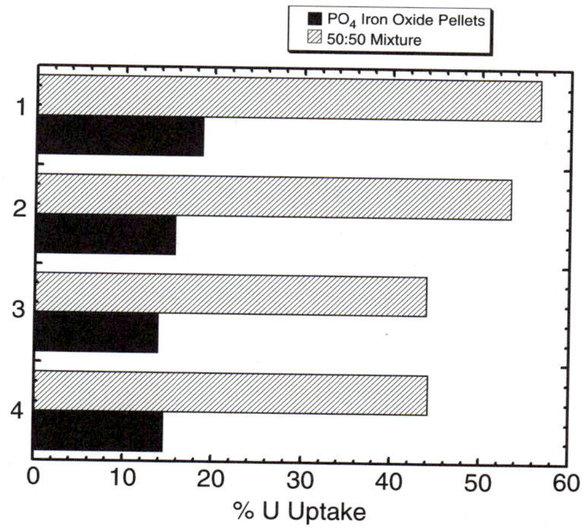

Figure 7 U(VI) uptake on iron oxide pellets with the following amounts of sorbed PO_4: (1) 16 μmol of PO_4/g iron oxide pellets, (2) 8 μmol PO_4/g, (3) 1 μmol PO_4/g and (4) 0 μmol PO_4/g. Also shown is the uranium uptake for 50:50 mixtures of the PO_4-sorbed iron oxide pellets with bone char. Amount of solid used was twice that of previous U (VI) uptake experiments to provide a more representative sample, and the total U (VI) concentration was increased to 34 mg/liter to avoid 100% uptake. All other experimental conditions were the same. Trend shows increasing uranium uptake with increasing PO_4 adsorption onto iron pellets, up to 0.5 mg PO_4/g. An increase in U(VI) uptake is also seen in mixtures of the PO_4- rich iron pellets and bone chare.

C. Column Experiments

Figure 8 shows a comparison of U(VI) breakthrough in columns packed with 100% bone char pellets or a 50/50 mixture by volume of bone char and iron pellets. Although there was significant scatter in data at breakthrough, the column packed with bone char pellets achieved U(VI) breakthrough at the inlet concentration at roughly 2500 column pore volumes. The total sorption capacity for the 100% bone char column was 1500 μg U/g solid. The 50/50 mixture column did not achieve full breakthrough until about 9000 column pore volumes and had a total sorption capacity of 5500 μg U/g solid. Upon elution with U(VI)-free AGW, desorption was characterized by an initial fast release in the 50/50 mixture column to a C/C_o of about 0.2 followed by a slower release continuing over many column pore volumes. The 100% bone char column also showed an initial fast release of U(VI), but was terminated at a C/C_o of approximately 0.3.

Figure 9 shows dissolved PO_4 concentrations in the column eluant early during the same experiments. Dissolved PO_4 concentrations were higher in the bone char-only column during the first 20 column volumes, indicating a

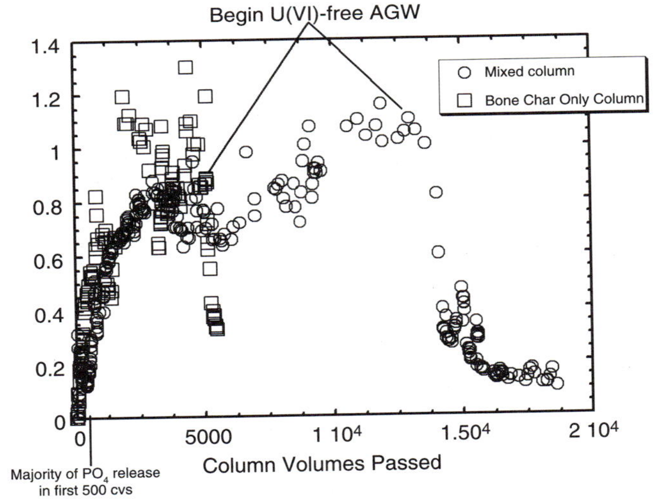

Figure 8 U(VI) uptake comparison of a 100% bone char-only column and a 50/50 mixture column, including subsequent elution with U-free groundwater. Inlet U(VI) concentrations were 5.2 and 4.5 mg/liter for the 100% bone char and 50/50 mixed column, respectively. U(VI) appears to be breaking through the 100% bone char column faster than the 50/50 mixture column, contrary to batch results. Highest effluent concentrations of dissolved PO_4 measured in first 500 cvs (Fig. 9).

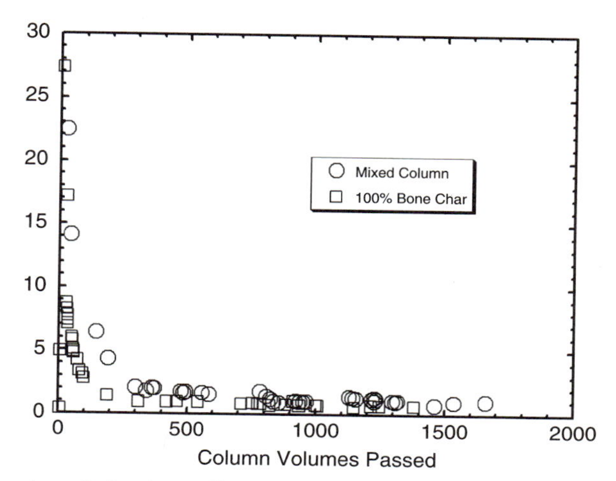

Figure 9 Phosphate elution (normalized to volume of bone char pellets) comparison between 100% bone char and 50/50 mixture columns described in Fig. 8. The majority of PO_4 released is within the first 500 column pore volumes. Less PO_4 is released per cm^3 of bone char pellet in the mixed column possibly as a result of sorption of PO_4 by the iron oxide pellets.

rapid flushing of available PO_4 from the bone char. PO_4 concentrations were near 30 mg/liter during the first flush and decreased rapidly until reaching a steady-state value of about 1 mg/liter after 300 column volumes. In the column packed with the mixture of pellets, PO_4 concentrations were lower than observed for the bone char-only column during the first 20 column volumes, but the PO_4 effluent curve exhibited more tailing at higher pore volumes than bone char alone, indicating likely retardation in PO_4 transport due to sorption on the iron oxide pellets.

V. FIELD STUDY RESULTS

Tables I and II give a summary of the DART deployments at the Fry Canyon and Christensen sites.

A. Christensen *in situ* Uranium Mine, Wyoming

Table III includes all U(VI) data for the POC samplers deployed at the Christensen site. The 75:25 mixture was deployed for only 75 days as a result of technical difficulties and thus there is only one sampling. According to the results of the POC samplers at the time of sampling, the 50/50 mixture was

Table III

Summary of Christensen Site POC U Analyses

COG1[a] (75:25 bone char:iron pellet)			COG2[b] (100% bone char)			COG3[c] (50:50 bone char:iron pellet)		
POC sampler #	Sept. 99 U[d]	Oct. 99 U	Sept. 99 U	Oct. 99 U	Jan. 00 U	Sept. 99 U	Oct. 99 U	Jan. 99 U
1	5.0	Removed in September	11.0	7.6	7.5	1.1	0.8	0.9
2	5.1		10.6	7.6	7.4	1.0	0.9	NS
3	4.5		10.6	7.7	7.2	1.1	0.8	NS
4	4.6		10.9	7.6	7.4	1.0	0.8	0.9
5	5.1		10.8	7.6	NS[e]	1.0	0.8	NS
6	5.2		10.7	7.5	7.2	NS	0.8	NS
7	5.0		10.9	7.6	7.3	1.1	0.9	0.8
8	5.0		10.6	7.6	7.3	1.0	0.5	0.8
9	5.2		10.9	7.5	7.2	1.0	0.9	0.9
10	5.1		11.0	7.6	7.3	1.1	0.5	0.9
11	4.6		10.7	7.6	NS	1.0	0.8	NS
12	4.5		10.7	7.7	7.3	1.0	0.2	NS
13	5.8		10.6	7.6	7.2	1.0	0.9	2.0
14	5.3		10.8	7.6	7.3	0.9	0.8	NS
15	5.7		10.7	7.6	NS	0.9	0.9	0.9
16	5.6		10.8	7.7	7.3	0.9	0.8	0.9

[a] Starting [U(VI)] = 8.7 mg/liter.
[b] Starting [U(VI)] = 11.9 mg/liter.
[c] Starting [U(VI)] = 11.1 mg/liter.
[d] All U(VI) values given for POCs, in mg/liter.
[e] Point of compliance sampler damaged; no sample taken.

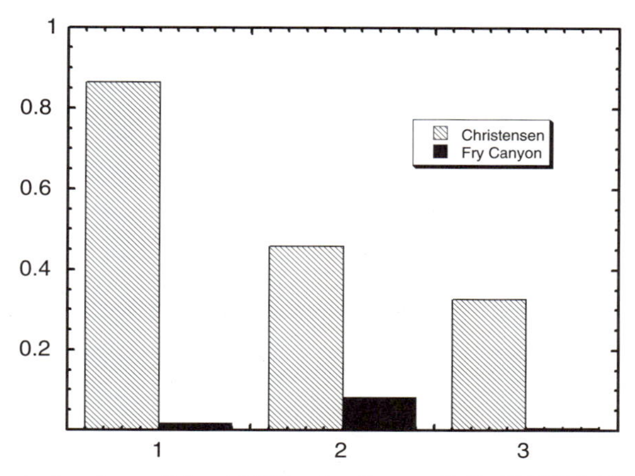

Figure 10 Twenty-four-hour nitric acid (2N) extraction results for DART field-reacted solids at Christensen and Fry Canyon: (1) 100% bone char, (2) 50/50 bone char, iron oxide pellet mixture, and (3) 75/25 bone char, iron oxide pellet mixture (only at Christensen Mine). The 100% bone char package shows the greatest amount of U extracted in composite samples taken from COGEMA packages. In the Fry Canyon field solids, two times more U was extracted from the 50:50 mixed packages than from the bone char alone.

removing 90% or greater of dissolved U(VI) over the three sampling periods, whereas the 100% bone char and the 75/25 mixture were removing less than 50%. Results from extractions of the reactive pellet materials (Fig. 10, Table IV) showed that the 100% bone char material had the highest U content, the 75:25 mixture the second highest, and the least amount of U was extracted from the 50:50 mixture. Even when the extraction results are normalized to the mass of bone char present, the 100% bone char pellets had the highest U content, whereas the 50/50 mixture had the least. SEM and SEM-EDS analyses of the Christensen DART solids were inconclusive because U was not detected.

B. Fry Canyon, Utah

At Fry Canyon, the 100% bone char package had the lowest U(VI) concentrations in the POC samplers, in contrast to the results observed for the Christensen site (Table V). Extractions of the reactive pellets showed that the 50:50 mixture solid had the greatest amount of U (Fig. 10), also in contrast to the results observed for the Christensen site. Based on the extraction results, the ZVI and 100% bone char materials had similar U contents, about 50% of that observed for the 50:50 mixture of bone char/iron oxide pellets (Table VI).

Table IV

Summary of Christensen Site DART Extract Data

DART designation	Contents	Days deployed	Well U concentration (mg/liter)	Extracted trace metals (mg/g solid)			
				U	As	Se	V
Phase II (POCs)							
COG1	75:25 bone char:iron pellet	79	10.2	0.21	—[a]	NQ[b]	0.02
COG2	100% bone char	230	14.0	0.84	—[a]	0.03	0.04
COG3	50:50 bone char: iron pellet	230	13.0	0.10	—[a]	NQ	0.05

[a]None detected.
[b]Not quantifiable.

Table V

Summary of Fry Canyon Deployment II POC U(VI) Analyses

POC sampler #	BZ4[a] 100% bone char		BZ5[b] 100% ZVI		BZ6[c] 50:50 bone char:iron pellet (intermixed)	
	Oct. 1999 U[d]	Dec. 1999 U	Oct. 1999 U	Dec. 2000 U	Oct. 1999 U	Dec. 1999 U
1	0.19	0.05	0.23	0.36	0.37	0.42
2	0.26	—[e]	0.35	0.37	0.43	0.38
3	0.17	0.05	0.43	0.33	0.29	0.21
4	0.26	NS[f]	0.51	0.33	0.20	0.32
5	0.26	0.02	0.69	0.28	0.23	0.44
6	0.50	0.03	0.68	0.19	0.34	0.48
7	0.22	0.02	NS	0.33	0.44	0.47
8	0.04	0.01	0.65	0.43	NS	0.49

[a] Starting [U(VI)] = 0.43 mg/liter.
[b] Starting [U(VI)] = 0.42 mg/liter.
[c] Starting [U(VI)] = 0.44 mg/liter.
[d] All U(VI) POC data, given in mg/liter.
[e] U(VI) concentration too low for quantification.
[f] Point of compliance sampler damaged; no sample taken.

Table VI

Fry Canyon Deployment-Solid Extracts

DART designation	Contents	Days deployed	Well U concentration (mg/liter)	Extracted U (mg/g solid)
Phase I (pumping)				
BZ1	50:50 bone char: iron pellet intermixed	327	0.64	0.04
BZ2	100% bone char	327	0.67	0.03
BZ3	50:50 bone char: iron pellet, layered	327	0.73	0.04, 0.02[a]
Phase II (POCs)				
BZ1	100% bone char	441	0.43	0.02
BZ2	100% ZVI	441	0.42	0.01
BZ3	50:50 bone char:iron pellet	441	0.44	0.04

[a] U extracted for iron pellets and bone char, respectively, from the layered package.

The reversibility of sorption test showed that U(VI) desorbed from the ZVI and 50:50 mixture materials was only 2 to 3% of uranium present on the reactive materials (by extraction); U(VI) desorbed from the 100% bone char material was an order of magnitude less.

VI. DISCUSSION

In batch experiments with bone char and iron oxide pellets, there was no evidence that a mixture of these materials would be beneficial for the removal of U(VI) compared to the bone char pellets alone. In contrast, results from the column experiments showed increased U(VI) removal by the mixture of materials, possibly as a result of longer reaction times and/or higher adsorbed PO_4 concentrations. As indicated in the batch experiments, aging of mixtures and increased sorption of PO_4 increased U(VI) uptake.

XRD and SEM-EDS analyses showed no evidence of crystalline uranyl phases in laboratory batch samples. In addition, backscatter SEM and SEM-EDS analyses of the reactive materials used in the DARTs at the Christensen site also revealed that U was below detection. Uranyl phosphate phase formation is likely limited by low dissolved U(VI) and PO_4 concentrations relative to that required for nucleation; Fuller *et al.* (2002) have shown that (1) a sorbed uranium concentration of at least 6000 μg/g was necessary for uranyl phosphate formation in the presence of hydroxyapatite and (2) U(VI) bonding at lower concentrations was consistent with U(VI) sorption. Thus, it is likely that any U(VI) retained by batch, column, and reactive materials in the field was due to a sorption process. However, it is possible, especially in the case of the ZVI material, that some of the U retained by the reactive materials in the field was present as U(IV).

Extracts of the reactive materials deployed in DARTs at the Christensen site indicated that the 100% bone char package retained the most U, whereas at the Fry Canyon site, results indicated that the 50/50 mixture of bone char and iron oxide pellets retained the most U. However, the results were likely affected by differing flow rates and U(VI) concentrations in the wells at each DART location. The Christensen DARTs were placed in an actively mined area with high flow rates and with much higher U(VI) concentrations. Due to a lack of data for the pumping rate history and influent U(VI) concentrations, it is not possible to calculate quantitative U(VI) fluxes through the DARTs deployed at the Christensen site. It is known that the flow rates fluctuated substantially due to mining operations. In addition, the Christensen DARTs were separated from each other by fairly large distances (hundreds of meters), resulting in considerable variability in flow characteristics and U(VI) concentrations at each DART. Thus, the fact that the 100% bone char package retained the most U does not mean that this material was the most efficient, as it may have had a much higher flux of U(VI) than the other DARTs deployed at the site.

At the Fry Canyon site, groundwater flow rates during the second DART deployment varied from 0.21 to 0.61 m/day, and the input U(VI) concentration is known to vary seasonally (see Chapter 14). However, because the DARTs were located very close to each other, it is very likely that each DART received the same inlet U(VI) flux. Thus, the fact that the 50/50 mixture of bone char and iron oxide pellets had the highest U content (by extraction) at the Fry Canyon site is significant, as it suggests that the efficiency of U(VI) removal was greater for the mixture of reactive materials than was observed for either the bone char pellets alone or the ZVI material.

Although the ZVI material can be very efficient in the removal of U(VI) (see Chapter 14), it was not known whether the thickness of ZVI material that can be achieved in DARTs of the diameter deployed here (8 in) was sufficient to achieve the necessary reducing conditions for efficient U(VI) removal. Results from the single ZVI DART deployed at Fry Canyon suggest that it may not have produced sufficient reducing conditions to achieve the same degree of U(VI) removal observed in the much thicker trench emplacement of ZVI at the same site (see Chapter 14).

Together with the extraction results, the POC sampler results can provide an indication of how quickly the reactive materials were approaching their capacity for remediation. At the Christensen site, the DART with the 100% bone char material was likely approaching its U uptake capacity at the very high U(VI) concentrations and high flow rates present during the deployment. Thus, when the POC samplers were sampled at the end of the deployment, high U(VI) concentrations were observed, suggesting that a roughly 50% "breakthrough" of U(VI) was occurring. The hypothesis that the 100% bone char material was near its capacity is supported by the fact that extractions showed 800 μg U/g solid, which is nearing the capacity calculated from column results (1500 μg U/g). The 75:25 bone char/iron oxide mixture, which was only in the ground for 79 days, was already at half the uptake of the 100% bone char package that was deployed for 279 days. According to the POC samplers, the 50/50 mixture DART at Christensen was removing over 90% of U(VI) during the three samplings and was, according to column results, only around 15% capacity. It would appear that the 50/50 DART would have a longer lifetime and provide more efficient remediation; however, without constraining the input U(VI) concentrations and well fluxes there can be no efficiency comparison.

This study demonstrated that DARTs can effectively remove contaminants. Both laboratory column and the Fry Canyon field results suggest that a mixture of materials can be more effective at contaminant removal than a single reactive barrier material, although efficiencies of removal in the field tests could not be determined accurately due to a lack of knowledge of the input fluxes of U(VI) to the DARTs. A future study with DARTs is needed in which input flow rates and contaminant concentrations are measured accurately in order to determine the contaminant flux to the reactive materials. Preliminary results using ZVI as a reactive material suggest that the DART diameter should be larger than 8 in. in order to increase the effective treatment zone with reducing conditions.

REFERENCES

Blowes, D. W., Ptacek, C. J., Benner, S. G., McRae, C. W. T., and Puls, R. W. (1998). Treatment of dissolved metals using permeable reactive barriers. *International Association of Hydrological Sciences-AISH Publication.* **250**, 483–490.

Coston, J. A., Fuller, C. C., and Davis, J. A. (1995). Pb^{2+} and Zn^{2+} adsorption by a natural Al- and Fe-bearing surface coating on an aquifer sand. *Geochem. Cosmochim. Acta* **59**, 3535–3548.

Fuller, C. C., Bargar, J. R., Davis, J. A., and Piana, M. J. (2002). Mechanisms of uranium interactions with hydroxyapatite: Implications for groundwater remediation. *Environm. Sci. Technol.* **36**, 158–165.

Lowell, S. (1979). "Introdroduction to Powder Surface Area". Wiley-Interscience, New York.

Millard, A. R., and Hedges, R. E. M.(1996). A diffusion- adsorption model of uranium uptake by archaelogical bone. *Geochem. Cosmochim. Acta.* **60**, 2139–2152.

Morrison, S. J., Spangler, R. R., and Morris, S. A. (1996). Subsurface injection of dissolved ferric chloride to form a chemical barrier; laboratory investigations. *Ground Water* **34**(1), 75–83.

Naftz, D. L., Davis, J. A., Fuller, C. C., Morrison, S. J., Freethey, G. W., Felthorn, E. M., Wilhelm, R. G., Piana, M. J., Joye, J. L., and Rowland, R. C. (1999). Field demonstration of permeable reactive barriers to control radionuclide and trace-element contamination in ground water from abandoned mine lands, in Morganwalp. *In* "Proceedings of the Technical Meeting, Charleston, South Carolina, March 8–12, 1999: U. S. Geological Survey" (D. W. and H. T. Buxton, eds.), p. 281–288. Water Resources Investigations Report, 99–4018A.

Payne, T. E., Davis, J. A., and Waite, T. D. (1996). Uranium adsorption on ferrihydrite: Effects of phosphate and humic acid. *Radiochim. Acta*, **74**, 239–243.

Robertson, W. D., and Cherry, J. A. (1995). In situ denitrification of septic-system nitrate using reactive porous media barriers: Field trials. *Ground Water* **33** (1), 99–111.

Van Wazer, J. E. (1973). "The Compounds of Phosphorus, Environmental Phosphorus Handbook." Wiley, New York.

Waite, T. D., Davis, J. A., Payne, T. W., Waychunas, G. A., and Xu, N. (1994). Uranium(VI) adsorption to ferrihydrite: Application of a surface complexation model. *Geochem. Cosmochim. Acta* **58**, 5465–5478.

Chapter 8

Treatability Study of Reactive Materials to Remediate Groundwater Contaminated with Radionuclides, Metals, and Nitrates in a Four-Component Permeable Reactive Barrier

James Conca,* Elizabeth Strietelmeier,† Ningping Lu,† Stuart D. Ware,† Tammy P. Taylor,† John Kaszuba,† and Judith Wright‡
*Los Alamos National Laboratory, Carlsbad, New Mexico 88220
†Los Alamos National Laboratory, Los Alamos, New Mexico 87545
‡UFA Ventures, Inc., Richland, WA 99352

The treatment of a shallow multicontaminant plume of $^{239,240}Pu$, ^{241}Am, ^{90}Sr, nitrate, and perchlorate in Mortandad Canyon, Los Alamos, New Mexico, was investigated in the laboratory using a multiple permeable reactive barrier consisting of four sequential layers. These layers include a polyelectrolyte-impregnated porous gravel for flocculating colloids, an Apatite II layer for plutonium, americium, and strontium immobilization, a layer of pecan shells as a biobarrier to nitrate and perchlorate; and a limestone gravel layer for any anionic species that may slip through the other layers, especially those of americium–carbonate. These layers can perform multiple functions, e.g., the pecan shells also sorb strontium very well and the Apatite II also remediates nitrate and perchlorate very well. Nitrate, perchlorate, plutonium,

Handbook of Groundwater Remediation Using Permeable Reactive Barriers

americium, and ^{90}Sr concentrations were reduced to below their maximum concentration limits (MCL) and usually to below detection limits in laboratory studies. The materials for this particular multiple barrier are inexpensive and readily available.

I. INTRODUCTION

The purpose of this study was to investigate the performance of alternative reactive barrier materials for possible use in a multiple permeable reactive barrier (PRB) at Mortandad Canyon. A barrier refers to anything that functions to prevent, bar, or retard the passage or movement of anything else, e.g., water, chemicals, trade goods, or ideas. In hydrology, traditional hydraulic barriers include impermeable walls, liners or caps made of bentonite, grout, asphalt, other clay minerals, precipitates of calcium carbonate or other chemicals, geotextiles, gels, or any material that can be made relatively impermeable. Containment of contaminant plumes can be accomplished by preventing the flow of the entire system, fluid and contaminant, through these conventional hydraulic barriers. More selective barriers have been developed that are barriers to one or a few components of the system, but not a barrier to the bulk of the system. Under unsaturated conditions, a capillary barrier consisting of a sharp boundary between two materials with greatly dissimilar pore sizes will act as an advective barrier to water but will not be a barrier to vapor flow. Likewise, a saturated, compacted bentonite clay liner will act as an advective barrier to water and vapor but will not be a barrier at all to aqueous diffusion of molecular species for which sorption to bentonite is not significant, e.g., most anionic species. Permeable reactive barriers to particular chemical contaminants consist of a water permeable material with specific chemical reactivities toward one or more chemical constituents via mechanisms such as adsorption, exchange, oxidation–reduction, or precipitation. These barriers can have varying ranges of specificity, e.g., adsorption of specific species by a modified zeolite (Sullivan *et al.*, 1994), precipitation of metals and radionuclides by apatite (Conca *et al.*, 2000; Bostick *et al.*, 1999), or overall reduction of the system by zero valent iron, other iron phases, or microbial activity (Blowes *et al.*, 1997; Tratnyek *et al.*; 1997; Puls *et al.*, 1999).

PRBs can be emplaced physically by (1) trenching and replacing the natural substrate with reactive media to form a wall of new material or (2) mixing or injecting reactive components into the substrate. Either way, the migrating contaminant plume encounters the reactive material in the barrier, the contaminant is sequestered, altered, or degraded, and the water moves on through the barrier and exits the system successfully treated. PRBs require precise knowledge of both the hydraulic and the chemical nature of the system. Impermeable barriers and PRBs can be used in combination to control both the hydrological and the chemical nature of the treatment zone. Funnel and gate systems, or other systems that use impermeable barriers to channel groundwater flow, are well suited for these applications (Starr and Cherry, 1994).

Mixed waste plumes will often require PRBs consisting of multiple layers of components configured specifically for a particular site and contaminants. Materials used in the field must be effective, inexpensive, and readily available in multiple-ton quantities to be able to treat large volumes of water, or soil. Possible materials include zero valent iron, compost, apatites, MgO, carbonates, pecan shells, peat moss, lime, and other materials that have been investigated by researchers to varying degrees. In order to be cost effective and produced in multiple ton quantities, most materials will come from mining production, agricultural waste production, or industrial waste production, with few modifying steps. The materials used in this study fit these criteria.

Radionuclides, including strontium-90 (^{90}Sr), plutonium-238,239,240 (^{238,239,240}Pu), and americium-241 (^{241}Am), have been detected since 1963 (LANL, 1997) in the shallow alluvial groundwater system that exists in Mortandad Canyon at Los Alamos National Laboratory (LANL) in Los Alamos, New Mexico. These radionuclides are stable as dissolved species in the groundwater and in the suspended phase where they adsorb onto colloids consisting of calcium carbonate, silica, ferric hydroxide, and solid organic carbon (LANL, 1997). Nitrogen species (nitrate and total Kjeldahl nitrogen) also occur in the alluvial groundwater in Mortandad Canyon. Perchlorates have been discharged at concentrations above 1 mg/liter (ppm) and occur in the Mortandad Canyon alluvial groundwater at levels above 300 μg/liter (ppb). Perchlorates were not monitored until recently because there were no regulatory limits. However, because the state of California instituted regulatory limits of 18 μg/liter, and the United States Environmental Protection Agency (EPA) is considering an MCL of 32 μg/liter for perchlorate, the concentrations in Mortandad Canyon may become a concern.

As a group, the contaminants in Mortandad Canyon can occur at concentrations up to several times the public dose limit (DCG) for the radionuclides or MCL for the nitrate and perhaps the perchlorate. Discharges into the canyon containing radionuclides, perchlorate, and nitrate have varied dramatically since 1963, generally dropping off with time. Most contaminants are assumed to come from the TA-50 water treatment facility near the head of the canyon, which presently discharges approximately 20,000 gallons per day over a short time period. Table I gives both discharge water chemistry from TA-50 during 1997 and some alluvial groundwater chemistry at one of the Mortandad Canyon observation wells (MCO-3) in 1997 (LANL, 1997) just upstream of the proposed emplacement site of a multiple PRB. The chemistry

Table I

Discharge Water and Alluvial Groundwater Chemistry in Mortandad Canyon during 1997

Constituent	Concentration in discharge water (mg/liter or pCi/liter)	DCG (pCi/liter) or MCL (mg/liter)	Concentration in groundwater from well MCO-3 (mg/liter or pCi/liter)
Perchlorate	>> 1 mg/liter	0.032 mg/liter[a]	0.36 mg/liter
^{238}Pu	11.39 pCi/liter	1.6 pCi/liter	16.2 pCi/liter ($4.0 \times 10^{-15} M$)
$^{239, 240}$Pu	4.58 pCi/liter	1.2 pCi/liter	12.5 pCi/liter ($8.4 \times 10^{-13} M$)
^{241}Am	58.0 pCi/liter	1.2 pCi/liter	14.1 pCi/liter ($1.7 \times 10^{-14} M$)
^{90}Sr	28.5 pCi/liter	8.0 pCi/liter	20.2 pCi/liter ($1.6 \times 10^{-15} M$)
Nitrate	68. mg/liter	10 mg/liter	3.86
Ammonium	2.15		—
Calcium	102		22.1
Magnesium	0.5		2.8
Potassium	11.7		7.9
Sodium	140		40
Silica	50.7		46
Iron	<0.01		—
Bicarbonate	251		105
Carbonate	24.4		—
Sulfate	28		12.0
Chloride	24.4		8.0
Fluoride	1.01		1.08
Total dissolved solids	686		274
pH	8.47		7.5
Specific conductance (μS/cm)	1,372		288
Ionic strength	0.0137 *M*		—

[a]Possible regulatory limit being proposed.

of discharges and groundwater varies greatly over time, e.g., nitrate varies from a few milligrams per liter to over 50 mg/liter.

The stratigraphy of Mortandad Canyon at the proposed PRB site consists of about 7 ft of alluvium composed of sand and gravel in a matrix of silt and clay overlying about 10 to 12 ft of clay and silt sediments that have resulted from weathering of tuff in place. A multiple PRB is proposed to be emplaced across the canyon near this point and down to the tuff bedrock to ensure maximum contact with groundwater. The hydraulic conductivity of the sediments ranges from less than 10^{-5} cm/s locally in the silts and clays to 10^{-2} cm/s in the sands. Therefore, to avoid hydrologic complications in the overall flow in the canyon, the barrier materials need to be chosen with conductivities well above 0.1 cm/s to prevent plugging of the PRB. As shown later, the materials chosen are gravel size and have conductivities greater than 1 cm/s, total porosities greater than 40%, and average pore sizes greater than 0.5 cm. Therefore, the flux of groundwater through the barrier will be determined by the hydraulic conductivity of the undisturbed sediments on either side of the barrier, plus any boundary properties that develop as a result of the barrier.

Numerous studies have been performed for zero valent iron, iron in other forms, polysulfides, granulated activated charcoal (GAC), and compost (Benner *et al.*, 1999; Blowes *et al.*, 1997; Fruchter, 1996; Tratnyek *et al.*; 1997; Puls *et al.*, 1999; Williamson *et al.*, 2000). Results of these studies are well known, having resulted in many field deployments (Goldstein *et al.*, 2000; Naftz *et al.*, 2000; Hocking *et al.*, 2000; Wickramanayake *et al.*, 2000). Sequential treatment has been investigated by researchers, including Morkin *et al.* (2000), who delineated the challenges facing the treatment of multiple contaminants with multiple materials.

The multiple PRB chosen for this study and this site consists of four sequential layers: a polyelectrolyte-impregnated porous volcanic gravel for flocculating colloids, an Apatite II layer for plutonium, americium, and strontium stabilization, a layer of pecan shells as a biobarrier substrate to nitrate and perchlorate; and a limestone gravel layer for anionic species that may not be trapped by other layers, especially those of americium–carbonate. These layers can perform multiple functions, e.g., the pecan shells also sorb strontium very well and the Apatite II also remediates nitrate and perchlorate very well. The two types of gravels are inexpensive and easily available in ton quantities. Pecan shells are an agricultural waste that is easy to obtain in ton quantities and are fairly durable in the subsurface given its high lignin content. Apatite II is an inexpensive, primarily amorphous form of a carbonated hydroxy-apatite that has random nanocrystals of apatite embedded in it, resulting in the efficient and rapid precipitation of various phosphate phases of metals and radionuclides. Apatite II is an efficient nonspecific surface sorber and is readily available in multiple-ton quantities.

II. MATERIALS AND METHODS

Past investigations and feasibility test results were used to focus on a design for the multiple PRB in Mortandad Canyon. Results are described in the following sections. Batch tests and one-dimensional (1–D) column tests conducted using Mortandad Canyon groundwaters and reactive media are also described later. On the basis of these results, large-scale columns and aquifer-box tests were conducted as final feasibility tests prior to emplacement, and the results were used for the final emplacement design. Figure 1 gives a schematic of how batch and column tests were used to reflect field conditions. Analytical techniques included ion chromatography, inductively couple plasma (ICP) emission and mass spectroscopy, liquid scintillation counting, and transmission electron microscopy, plus standard methods for measuring pH and dissolved oxygen.

Groundwater collected directly from Mortandad Canyon was used as the stock solution in all new feasibility studies (LANL, 1997). For contaminated studies, plutonium, americium, strontium, nitrate (NO_3), or perchlorate (ClO_4) was added in sufficient quantities for particular experiments. A large volume of uncontaminated groundwater was collected from Mortandad Canyon well MCO-5 for subsequent laboratory studies.

Batch sorption studies were performed as initial screening tools for evaluating solid media. Batch studies generally consisted of a small amount of substrate combined with 10 times the amount of contaminated water by mass. The combination is shaken for 24 h or longer depending on the estimate of kinetic reaction times. The concentration of contaminant in the water was measured to determine how much contaminant had been removed from solution by interaction with the substrate.

Column studies were performed to reflect more relevant field conditions. These studies monitor the contaminant concentrations and other solution properties in the effluent exiting the column following contact with reactive media. Results are often given as a breakthrough curve that shows the normalized concentration of the contaminant as a function of the number of pore volumes exiting the column (Fig. 1). This normalizes the tests to any size system. The 1-D column studies were small-scale columns 4.8 cm in inside diameter and 6 to 10 cm in length. These were loaded with small grain-size splits of the various materials and run at a range of flow rates to provide a scientific basis and design criteria for larger columns. Flow rates were up to five times the natural velocity of groundwater in Mortandad Canyon. Data from these types of column studies were also used to evaluate rate-limited immobilization of contaminants. Feasibility studies with larger columns, in

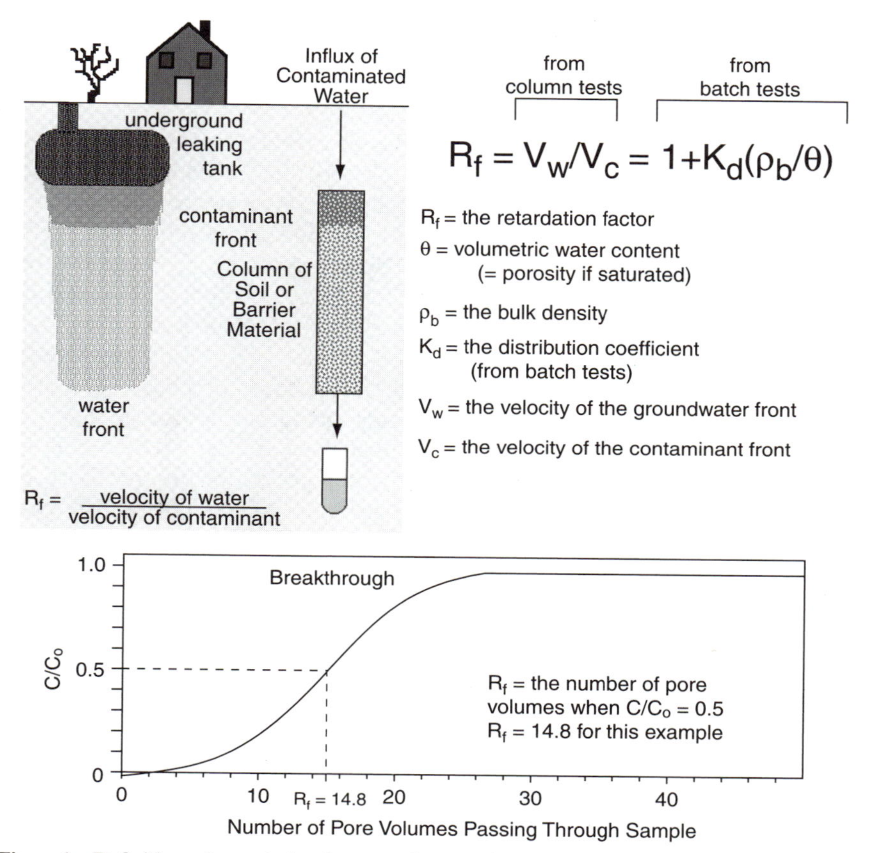

Figure 1 Definition of retardation factor and contaminant breakthrough behavior. Batch and column studies are related to behavior in the field by the aforementioned relationship. The S shape of the breakthrough curve reflects hydrodynamic dispersion. The number of inflection points, the shape of the curve, and the absolute position along the axis reflect the number, distribution, and type of sorption sites. C/C_o is the concentration of the specific chemical species in the effluent solution divided by the concentration in the starting solution.

which flow rates, material sizes, and juxtaposition of materials were scaled closer to actual field conditions, were performed as a test of the multiple PRB. In addition, two-dimensional (2-D) aquifer cell studies were performed to approach a realistic remediation scenario with the media contacting each other *in situ*.

A. Barrier Materials

1. Porous Gravel

A porous gravel coated with a material to capture colloids was chosen as the first layer in the PRB to retard or prevent the movement of colloids in the Mortandad Canyon groundwater. Significant evidence indicates that enhanced transport of actinides occurs in subsurface geologic media through their association with colloid-sized particles in groundwater, resulting in colloid-facilitated transport (Kersting *et al.*, 1999). The colloids have a high potential for adsorbing these contaminants because of their large surface area and generally negative charge. The resulting pseudo-colloids, i.e., contaminants attached to the surface of silica, clay, iron oxides, and other noncontaminant colloidal materials, move greater distances faster than dissolved species in subsurface aquifers. These colloid-sized contaminants include radionuclides, metals, pesticides, organics, viruses, bacteria, and synthetic materials such as asbestos fibers. The transport of radionuclides as pseudo-colloids has been described by many authors at multiple DOE sites, including LANL, the Nevada Test Site, and others (Triay *et al.*, 1997). The rapid movement of colloidal contaminants through the subsurface is also the subject of numerous modeling studies as cited by Kale (1993).

Kale (1993) developed a concept of a colloid barrier using glass beads and a cationic, highly soluble polyelectrolyte with the ability to capture and agglomerate colloids. The polyelectrolyte used was a commercially available material, polydialylldimethylammonium chloride, named Catfloc. This material is used in the wastewater treatment industry to flocculate small particles and was selected for its ability to adsorb onto silica substrates, e.g., silica colloids, rock, or glass beads. Catfloc was also chosen for this study because of the low concentrations necessary for good performance, because of its the relatively low cost, and because it is not easily biodegraded. The laboratory-scale studies used glass beads and centimeter-sized rock chips of the gravel layer and Mortandad Canyon groundwater with and without colloids, at concentrations of 2×10^8 particles/ml. The biodegradability of Catfloc was evaluated in studies where Catfloc acted as the sole carbon source in a nutrient culture medium inoculated with Mortandad Canyon groundwater. The culture medium also contained potassium nitrate and sodium phosphate as a supply of nitrogen and phosphorous nutrients to ensure that nutrient deficiency was not a growth-limiting factor. Aliquots of the culture medium were taken and stained for total microbial counts.

2. Apatite II

A special form of the mineral apatite, Apatite II, was chosen as the second layer in the PRB to treat metals, particularly Pu. Feasibility studies were performed to determine the metal stabilization potential of various materials with respect to acid mine drainage in northern Idaho and actinide-contaminated waters at DOE sites (Wright *et al.*, 1995; Conca, 1997; Bostick *et al.*, 1999; Conca *et al.*, 2000; Runde, 2000). Materials investigated included different sources of apatite (various mineral sources, synthetic, commercial, cowbone charred or uncharred, and Apatite II), zeolites (clinoptilolite and chabazite), various polymers, including Dowex II, C-sorb, Cabsorb, Mersorb, peat moss, zero valent iron, hematite, goethite, and activated charcoal. The Apatite II performed the best with respect to all metals studied, i.e., lead, cadmium, zinc, uranium, strontium, and plutonium, but with different mechanisms depending on the metal and the conditions. Because of this generally good performance for many metals under various conditions, Apatite II was chosen for this Mortandad Canyon study.

Apatite II can be obtained in almost any grain size from powdered to gravel size. The gravel size has hydraulic conductivities exceeding 10 cm/s and a bulk density between 0.5 and 1.0 g/cm^3. Apatite is ideal for stabilizing many metals because it instigates the heterogeneous nucleation of metal-apatite phases under environmental conditions (Lower *et al.*, 1998). The excellent stabilization efficiency comes from the extremely low solubility products of the resultant metal-apatites, e.g., for uranium–phosphate (autunite) $K_{sp} = 10^{-49}$ and for lead–apatite (pyromorphite) $K_{sp} = 10^{-76}$. Table II lists some solubility products as examples [the reactions are given in MINTEQ-A2; Geochem Software Inc. (1994)]. There are several apatite sources with widely varying reactivities and properties, but most are not appropriate for metal remediation. For best results, the apatite should (1) have as much carbonate ion substituted as possible, (2) have no fluorine substitution,

Table II
Stabilities of Some Relevant Phases[a]

Mineral phase	Solubility product (log K_{sp})	Mineral phase	Solubility product (log K_{sp})
$Pb_5(PO_4)_3(OH, Cl)$	−76.5	$Am(PO_4)$	−24.8
$Ca(UO_2)_2(PO_4)_2 \bullet 10H_2O$	−49.0	$Pu(PO_4)$	−24.4
$Sr_5(PO_4)_3(OH)$	−51.3	UO_2HPO_4	−10.7
$Zn_3(PO_4)_2$	−35.3	$Cd_3(PO_4)_2$	−32.6

[a]From MINTEQ-A2; Geochem Software Inc. (1994).

(3) have no trace metals initially in the structure, (4) be poorly crystalline or even amorphous, and (5) have high internal porosity (Conca *et al.*, 2000). For these reasons, traditional phosphate ores and cow bone, charred or not, are not as effective as Apatite II.

The nominal composition of Apatite II is $Ca_{5-x}Na_x(PO_4)_3(OH)_{1-y}(CO_3)_y$, where $x < 0.2$ and $y < 0.05$. The high reactivity of Apatite II relative to other apatites comes from the unusual structure of the solid. Most of the material is amorphous hydroxy calcium phosphate with respect to both X-Ray Diffraction (XRD) and Transmission Electron Microscopy (TEM). However, there are random nanocrystals of well-crystallized apatite embedded throughout the amorphous phase (Fig. 2). Also, there is substantial carbonate substitution, which makes the structure less stable and more reactive (Wright *et al.*, 1990). However, once metals are in the apatite

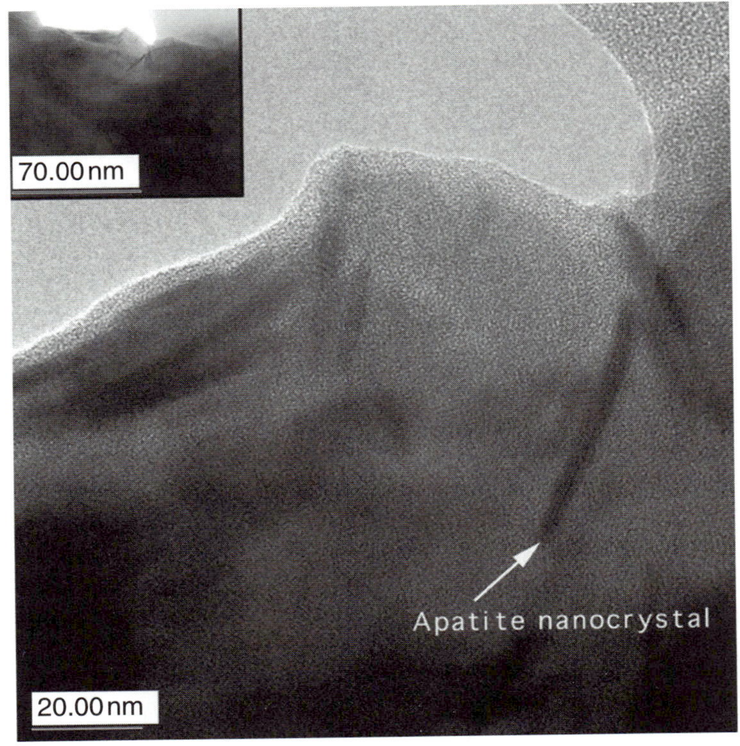

Figure 2 HR-TEM image of Apatite II showing general amorphous nature with random nanocrystal inclusions of crystalline apatite (low–resolution image is inserted in upper left corner). Copyright 2000, A. A. BALKemg, Reprinted with permission.

structure, it is much more stable than the initial Apatite II. Because of this arrangement, the amorphous structure is relatively reactive and provides a sufficient excess of phosphate ion to solution, locally exceeding the solubility limit of many metal-phosphate phases and resulting in precipitation, particularly for lead, uranium, and plutonium. At the same time, the nanocrystals provide the apatite structure for nucleating the precipitates. Other apatites, such as phosphate rock, bone char, or reagent-grade tricalcium phosphate, are crystalline (Fig. 3 and 4), have little or no carbonate substitution, have substantial fluorine substitution, have significant trace metal substitution (strontium, barium, uranium, and others), or have less microporosity, all of which reduce the reactivity and make them less optimal for metal stabilization. The concentration of phosphate ion in solution in equilibrium with Apatite II is in the 10- to 100-mg/liter (ppb) range and depends on solution

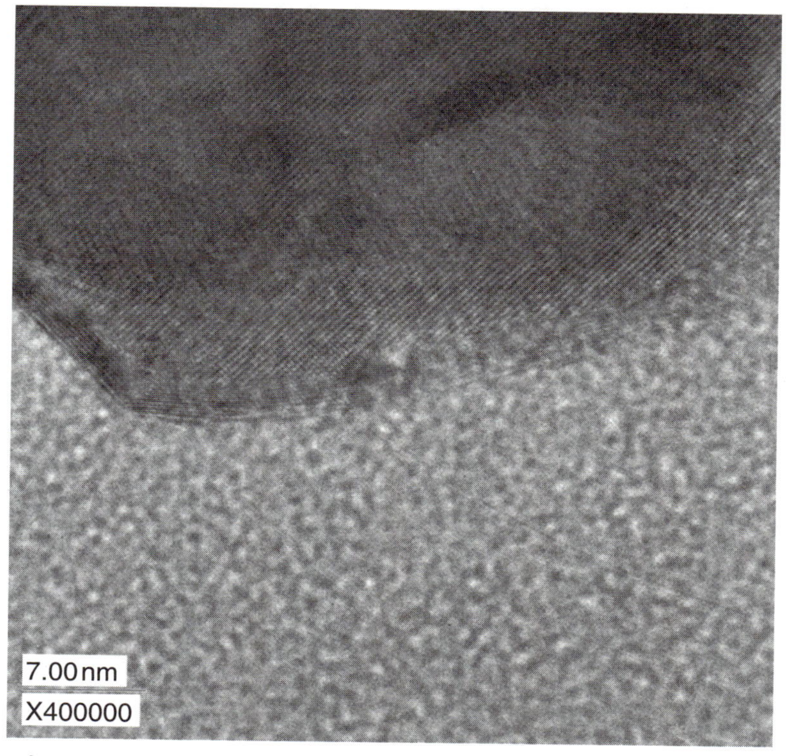

Figure 3 HR-TEM image of mineral apatite (NC phosphate rock) showing the crystalline nature of the apatite. Copyright 2000, A. A. BALKemg, Reprinted with permission.

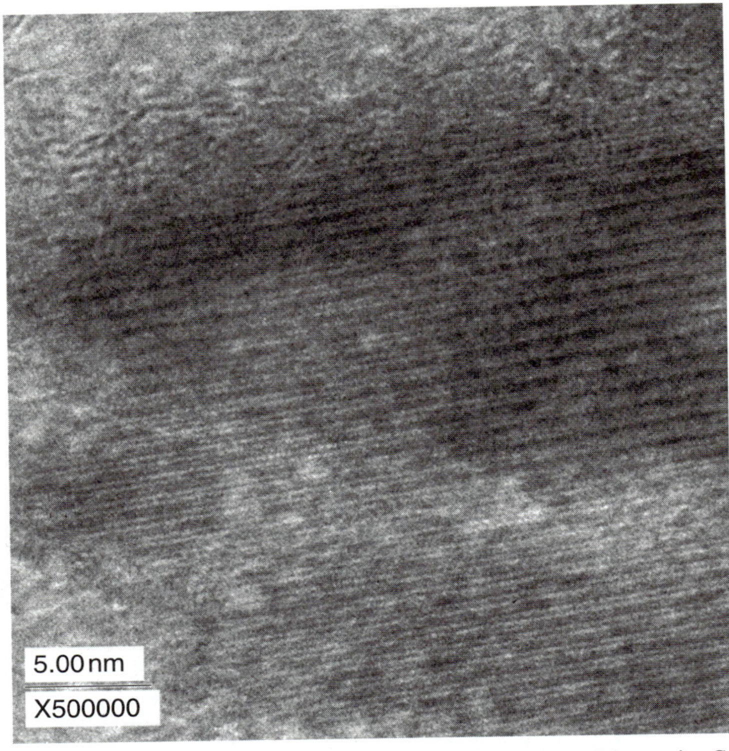

Figure 4 HR-TEM image of bone char showing the crystalline nature of the apatite. Copyright 2000, A. A. BALKemg, Reprinted with permission.

chemistry, particularly pH. It is important, however, because these low values avoid the phosphate-loading issues common to organophosphates and soluble inorganic phosphates such as most fertilizers.

The Apatite II was found to leach ^{87}Sr at a constant concentration, regardless of the strontium concentration in the influent water. For batch tests with radioactive ^{85}Sr, this was not a problem, as the two isotopes are easily distinguished. However, during the column studies that could use only ^{87}Sr because of laboratory activity limits, there was no way to distinguish the difference between the ^{87}Sr leached from, or sorbed onto, the Apatite II. It was found that ^{87}Sr leached from the Apatite II in columns at a constant concentration of 0.85 mg/liter, regardless of influent concentration, e.g., columns with influent ^{87}Sr concentrations of 2 or 0 mg/liter both showed effluent concentrations of about 0.85 mg/liter. The column study described in Section III,B,3 used ^{85}Sr and so avoided this problem. Further studies will use

by first cleaning the gravel in weak (~0.15 *M*) hydrochloric acid by overnight soaking. Next, the gravel was rinsed in MCO-5 water and immersed in a colloidal coagulant polymer solution (Catfloc ~2 g/liter) overnight. Finally, the gravel was given a final rinse in MCO-5 water and kept immersed in MCO-5 water to guard against unknown effects of allowing the coagulant-treated gravel to dry.

Apatite II samples treated with phosphoric acid were prepared by overnight soaking in 0.5 *M* phosphoric acid followed by rinsing in MCO-5 water and air drying. As shown later, this treatment did not increase performance sufficiently to warrant its use in the field. Any field application will use the Apatite II as received in the specified grain size, and not treated in any way.

Three radionuclides were utilized in these batch tests: $^{85}Sr^{2+}$, $^{239}Pu^{5+}$, and $^{241}Am^{3+}$. ^{85}Sr concentrations were quantified using a Packard Cobra Gamma Counter, whereas both ^{239}Pu and ^{241}Am were analyzed on a Packard Model 2500 liquid scintillation counter (LSC). Separate batch sorption studies were repeated for each radionuclide. The ^{85}Sr batch sorption study used a liquid-to-solid ratio of approximately 10:1 (150 ml solution and 15 g solid). Studies with ^{239}Pu and ^{241}Am used a liquid-to-solid ratio of approximately 15:1 (225 ml solution and 15 g solid). Apatite/limestone mixtures were formed by adding another 2 of limestone, an amount that modified the liquid–solid ratios only slightly. The starting concentration of each radionuclide solution was $7.6 \times 10^{-9} M\,^{85}Sr$, $2.4 \times 10^{-8} M\,^{239}Pu$, and $8.5 \times 10^{-12} M\,^{241}Am$.

Aggregate materials were equilibrated with water from well MCO-5 before batch sorption procedures began. Dry aggregates were weighed out and placed in 250-ml round Teflon bottles with sterilized water from well MCO-5 and agitated for 2 weeks prior to beginning batch studies. Agitation was accomplished by placing the bottles horizontally onto laboratory shakers. At the completion of the 2-week equilibration phase, the solids were drained of water from well MCO-5, water from well MCO-5 containing a spike of known radionuclide concentration was added, and agitation was continued.

Water from well MCO-5 was sampled and analyzed over time to determine sorption behavior and kinetics. Sampling of the container solutions was performed using pipettes. To remove particulates, each sample was centrifuged at 16,500 rotations per minute for an hour in Teflon vials. Samples were then split for radionuclide and water chemistry analysis. Because the ratio of liquid to solid changed for each sample interval, solution volumes were carefully measured and accounted for in calculating the sorption coefficient isotherms.

For the biobarrier studies, pecan shells (2 g) or pecan shells and dog food (2 g/0.2 g) were loaded into Oak Ridge tubes, and 20 ml of water from well MCO-5 was added for a solution/solid ratio of 10:1 (and a pecan shell/dog food ratio of 10:1). Reaction tubes were incubated at room temperature on a shaker. Successive samples were taken on the day after the tubes were loaded, day 1, and on day 2, day 7, day 14, and day 21 in most cases. Nitrate, nitrite, and ammonia concentrations were determined using EPA method 300.0, an ion chromatographic method. MPN analysis involved the use of a nitrate-reducing MPN method developed from methods found in Collins *et al.* (1995).

C. Column Test Methods

1. Single Column Tests with ^{87}Sr, Nitrate, and Colloids

A single column study to evaluate pecan shells with respect to nitrate, strontium and colloids in water from well MCO-5 used single 6-cm-long, 4.8-cm-inside-diameter columns. The column was packed with pecan shells with no food source and was infused with water at a flow rate of 0.13 ml/min. The groundwater was spiked with 2 mg/liter ^{87}Sr, 50 mg/liter sodium nitrate ($NaNO_3$) (for a total nitrate concentration of 100 mg/liter), and 2×10^8 fluorescent colloids per liter of water. Fluorescent polymer (polystyrene) microsphere colloids (1 μm) were obtained from Duke Scientific Corp. (Palo Alto, CA). Sieve analysis of the pecan shells indicated that 75.9% of the shells were 2.00–4.75 mm in size, whereas 24.1% were 1.00–2.00 mm in size. Approximately 6.5-ml aliquots were retrieved automatically by an Isco Foxy Junior fraction collector. ICP analysis of ^{87}Sr and nitrate-ion chromatography were performed on the effluent following collection from the column, whereas fluorescent colloids were counted using a microscopic technique (Abdel-Fattah and Reimus, 2000).

2. Sequential Column Tests with ^{87}Sr and Nitrate

Four 1–D columns were arranged in sequence to evaluate the ability of various reactive media to remove contaminants from flowing groundwater. These columns were packed with materials selected based on the results of batch sorption treatability studies and were run at a range of flow rates up to five times the natural velocity of groundwater in Mortandad Canyon. Material included gravel coated with Catfloc, volcanic rock (basalt) coated with Catfloc, Apatite II, pecan shells, pecan shells mixed with dog food (10:1, respectively), and limestone.

MCO-5 water was spiked with various contaminants to desired concentrations. Columns were packed with various reactive media through which contaminated groundwater was flushed at a constant flow rate. The columns were constructed of borosilicate glass with a 4.8-cm inside diameter and are equipped with an adjustable end plate for bed length variation from 1 to 13 cm.

All media used in the four sequential columns were sieved to less than 4.75 mm size fraction but greater than 1.00 mm. Catfloc-coated gravel or basalt was prepared by contacting 2 g/liter Catfloc solution with the gravel or basalt overnight on a rotator (60 rpm; approximately 2:1, liquid–solid ratio). Apatite II was used as obtained or was soaked in 90°C hot water to further remove residual organics. Pecan shells were used as received or crushed to the desired size fractions. In the first column study only, dog food was added to the pecan shells (1:10, dog food:pecan shells) to serve as a carbon source stimulant for microbial growth. All column studies were designed to run until breakthrough, however, the Los Alamos Cerro Grande fire resulted in complete shut down of the study before breakthrough was achieved.

3. Sequential Large-Scale Column Tests with ^{239}Pu, ^{241}Am, ^{85}Sr, Nitrate, and Perchlorate

Four columns were manufactured and machined from transparent acrylic sheets and tubing. The columns measured 5.5 in. (~14 cm) in internal diameter and 12 in. (~30 cm) in length. Two insteps were milled into each end of the acrylic tubing to allow the emplacement of perforated Teflon rings. The perforated rings created an area of dead volume and slight backpressure to allow dispersion of the tracer before contacting the column media. Flow ports were established on each end cap of the columns using 1/4 X 28 thread peek flow fittings.

Four aggregates were chosen for use in the columns. The selection of materials was based on batch sorption studies. The columns were packed with these materials and filled with MCO-5 water. The columns were filled with deaerated water and purged of large air pockets. No attempts were made to eliminate small air bubbles. The columns were connected in line to each other via 1/16 in. inside diameter Teflon tubing. Deaired MCO-5 water was pumped through the columns so that flow introduced into the first column flowed through the second, third, and out the fourth, respectively. Column media were equilibrated with MCO-5 water by flushing the water through the columns at approximately 20 ml/h for 2 weeks prior to injection of contaminants. Prior to injecting the column network with spiked MCO-5 water, the columns were equilibrated to the injection flow rate by adjusting the flow rate of the pump to the planned injection rate 72 h in advance of the injection. The pore volume of the column network was slightly greater then 11 liters, distributed approximately equally within the four columns.

A 100-ml pulse of MCO-5 water spiked with $7 \times 10^{-9} M\ ^{85}Sr$, $5 \times 10^{-7} M\ ^{239}Pu$ (750,000 pCi), and $1 \times 10^{-9} M\ ^{241}Am$ was injected into the columns after the 2-week equilibration period. This plutonium concentration represents about 600,000 times the field concentrations, about 60,000 liters at the field concentrations, or about 60 years of flow at present plutonium concentrations. This strontium concentration represents about 440,000 times the field concentrations, about 44,000 liters at the field concentrations, or about 44 years of flow at present ^{90}Sr concentrations. This americium concentration represents about 100,000 times the field concentrations, about 10,000 liters at the field concentrations, or about 10 years of flow at present ^{241}Am concentrations. Nitrate and perchlorate were at the field concentrations of 3.9 and 0.4 mg/liter, respectively. No colloids were injected as part of this particular column study. The purpose of the pulse was to see if rapid contaminant loading would be stable against future flow or if sorbed radionuclides would be remobilized. The spiked water was injected via a KD Scientific syringe pump. After a 1-h injection, a Scilog Chemtec piston pump then maintained the approximate 100 ml/h flow of uncontaminated MCO-5 water. Initially, an Isco Foxy funnel fraction collector was used to collect the elutants hourly into Teflon collection bottles. After the first 3 days, this was changed to a large collection bottle sampled approximately every 72 h. The elutants were analyzed for radionuclide breakthrough and solution chemistry. Radionuclide breakthrough was analyzed using the same instruments as described in the batch sorption studies. However, because plutonium and americium have similar radioactive energy emissions, differentiating between these two radionuclides was not done. Only an analysis of gross α decay was performed. Direct analysis of ^{85}Sr was possible because ^{85}Sr is the only γ emitter in the study.

4. Two-Dimensional Aquifer Box Tests with ^{87}Sr, Nitrate, and Perchlorate

A 2-D aquifer cell was manufactured from aluminum. The inside dimensions of the rectangular box were 32 cm (height) $\times$ 62 cm (length) $\times$ 5 cm (width). Rectangular end chambers, constructed of aluminum alloy, were screened over the entire vertical height of the box. The bottom and back of the box were aluminum, and a glass plate served as the front (see Fig. 12 in Section IIIB4). The pore volume of the cell was approximately 2 liter. ^{87}Sr and perchlorate in water from well MCO-5 were at concentrations of 2 mg/liter and 300 μg/liter, respectively. During the first 60 days, influent nitrate concentrations were held constant at approximately 125 mg/liter. Thereafter, nitrate concentrations were increased to 250 mg/liter in order to evaluate the ability of microorganisms in the PRB materials to reduce irregular nitrate concentrations. The cell was constructed to allow discrete sampling between each layer interface.

III. RESULTS

A. Batch Tests

1. Colloid Tests in Mortandad Canyon Groundwater

The laboratory-scale studies used glass beads and centimeter-sized rock chips of the gravel layer and Mortandad Canyon groundwater with and without colloids at concentrations of 2×10^8 particles/ml. These studies were performed to determine the effect of a number of different parameters on agglomeration of microspheres on the centimeter-sized rock surfaces. There was a strong dependence on Catfloc concentration and colloid concentration, but the presence of Catfloc generally increased colloid removal from solution by at least three orders of magnitude. The Catfloc was not significantly degraded after 1 year.

2. Batch Tests for ^{239}Pu, ^{241}Am, and ^{90}Sr in Mortandad Canyon Groundwater

Results of the batch sorption studies are presented as K_d versus time for ^{239}Pu, ^{241}Am, and ^{85}Sr in Figs. 5, 6, and 7, respectively. Results indicate that untreated Apatite II has a large affinity for all three radionuclides. All combinations of Apatite II displayed large affinities for plutonium. Limestone/untreated apatite mixtures and treated Apatite II also display large affinities for americium. Apatite II has greater affinities for all three radionuclides compared to mineral apatite. Materials that will function as components of the multiple PRB, such as pecan shells in the biobarrier, were evaluated in batch sorption studies to determine radionuclide capture properties. These materials displayed much less affinity for radionuclides than apatite.

Although the amounts of plutonium used were too small for phase identification, the extremely high K_d values are consistent with the extremely low solubilities of plutonium-phosphates and are consistent with studies of apatites showing precipitation with respect to lead and uranium (Ma *et al.*, 1993; Moody and Wright, 1995; Bostick *et al.*, 1999; Runde, 2000). For plutonium, the various samples of apatite exhibited the highest sorption values. The mineral apatite was slower to react than the Apatite II, but eventually reached the same levels. The other materials still showed reasonably high K_d values

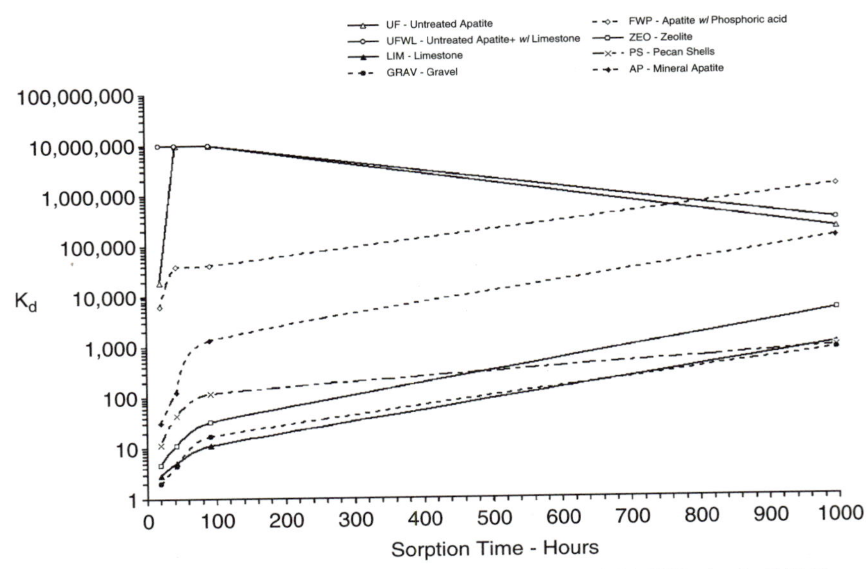

Figure 5 Sorption of Pu-239 over a 42-day time span. Copyright 2000, A. A. BALKemg, Reprinted with permission.

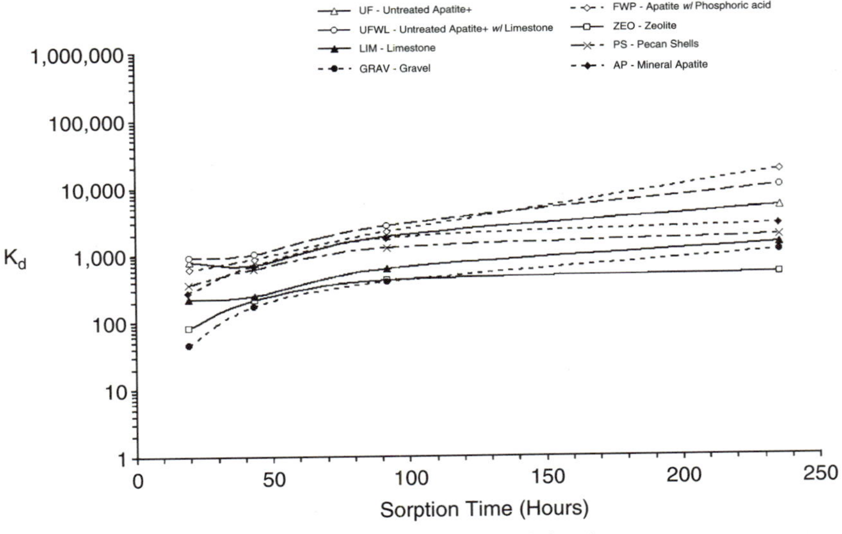

Figure 6 Sorption of Am-241 over a 10-day time span.

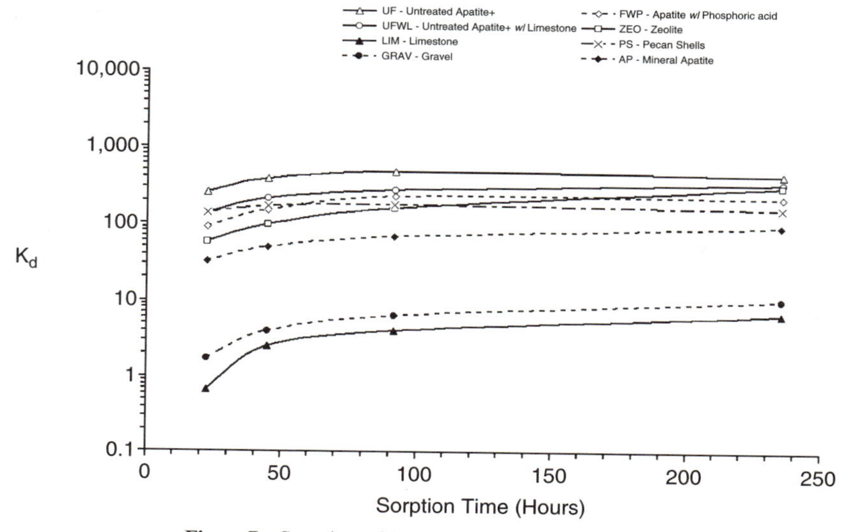

Figure 7 Sorption of Sr-85 over a 10-day time span.

for plutonium, up to 1000. For americium, the apatites exhibited K_d values between 1000 and 10,000, and the other materials had K_d values below 1000. ^{85}Sr showed the least sorption affinity for any of these materials. While Apatite II performed the best with K_d values of about 500, all were less than 1000. The large amount of ^{87}Sr that leached from the Apatite II during column tests makes it difficult to evaluate these batch tests with ^{85}Sr. There was probably significant isotopic exchange, which may account for most of the sorption. Geochemically, strontium has a strong affinity for apatite phases and precipitates readily in phosphate phases in nature, including coprecipitation with calcium in bone materials (Wright, 1990). The poor showing in laboratory tests may also reflect kinetic effects.

3. Biobarrier Batch Tests for Nitrate

The biobarrier support material selected was pecan shells. Batch tests were performed to determine the effects of initial sterilization of materials, effects of readily available food sources, the efficiency of the medium, and the time necessary for inducing microbial population growth to sufficient levels to reduce nitrate concentrations in the groundwater. The effect of a food source on the microbial activity was investigated using dry dog food as a readily accessible food source. Two studies were used to determine the amount of bacterial growth and contaminant degradation in nitrate-contaminated water. Each study contained five different sets of samples: (1) the water and the pecan shells were both sterilized (sterile control), (2) the pecan shells were sterilized and the water was nonsterile, thereby providing bacteria (sterile PS, unsterile H_2O), (3) the water was sterilized and the pecan shells were not (unsterile PS, sterile H_2O), (4) both the water and the pecan shells were nonsterile (unsterile control), and (5) filtered nitrate-contaminated water without any substrate. Duplicate samples were run under each set of conditions in each study. Another set of experiments was run with dog food as an additive along with pecan shells. Experiments were run both with field concentrations of nitrate (~20 mg/liter) and with high nitrate (~600 mg/liter).

Results for the pecan shells alone with the 600-mg/liter nitrate-contaminated groundwater are shown in Fig. 8. These results are similar for all runs. All filtered water samples with no substrate showed no reduction in nitrate concentrations. Sterilizing the substrates (pecan shells or the pecan shells and the dog food) diminished the degree of nitrate reduction more than sterilizing the groundwater. Experiments with nonsterilized media resulted in a greater reduction of nitrate and shorter induction periods needed to form a sufficiently active microbial population to significantly reduce nitrate concentrations, especially with both components providing bacteria. Nonsterile substrates did contain denitrifying bacteria. In two out of three experiments,

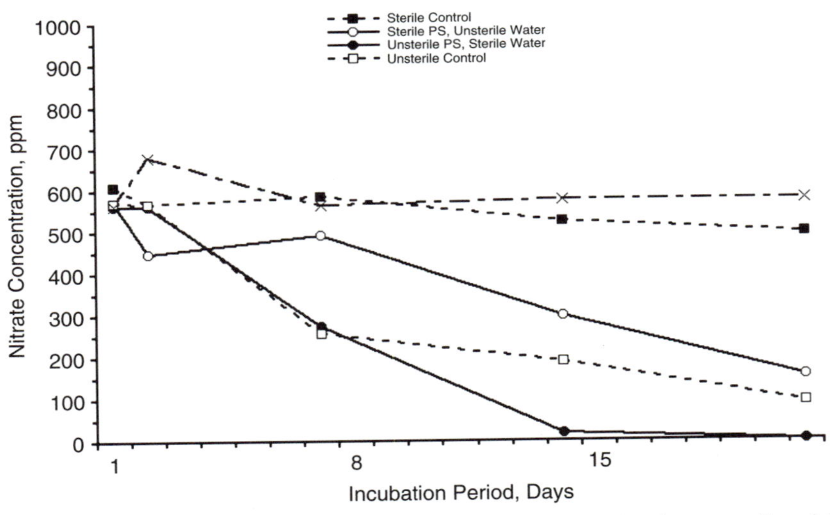

Figure 8 Nitrate removal over a 21-day period of Sr-85 over a 10-day time span. Copyright 2000, A. A. BALKemg, Reprinted with permission.

the presence of dog food reduced the induction periods needed to form an active microbial population by about half, particularly at high nitrate concentrations. With pecan shells alone, the experiments with high nitrate concentrations required about twice as long as those with low nitrate concentrations to reduce nitrate levels to half the initial concentrations.

The results demonstrate that a healthy microbial population with denitrifying activity was established that can reduce even high nitrate concentrations to undetectable levels within about 1 week with the pecan shell/dog food combination and in about twice that time without the dog food. As shown later, Apatite II has sufficient residual organics to serve as a readily accessible food source to the adjacent biobarrier so that no additional food source is needed. The rate of denitrification will be adequate to reduce nitrate in the proposed multiple PRB for Mortandad Canyon given the flow rates and proposed dimensions.

B. Column Studies

1. Pecan Shells with ^{87}Sr, NO_3^-, and Polystyrene Colloids

A single column study was undertaken to determine if pecan shells are a feasible material for the biobarrier part of the multiple PRB for degradation of nitrate. ^{87}Sr and colloids were also included to observe their behavior. The groundwater was spiked with 2 mg/liter ^{87}Sr, 50 mg/liter $NaNO_3$ (for a total nitrate concentration of 100 mg/liter), and 2×10^8 fluorescent colloids per liter of water. ^{87}Sr was removed to below detection limits throughout the course of the experiment, an unexpected advantageous result. Nitrate concentrations initially appeared to be reduced, however, they reached a maximum concentration of 60 mg/liter and decreased with time to approximately 30 mg/liter. This incomplete removal of nitrate probably resulted from insufficient time being provided to establish an active microbial community, and no food source being available. Only eight pore volumes were put through the column, which at the 0.13-ml/min flow rate took just over 2 days. As was seen later in column tests with Apatite II upstream of the pecan shells, the presence of an available food source from the Apatite II led to rapid degradation of nitrate in the Apatite II. The pecan shells did not result in complete removal of colloids from the influent solution, however, they did reduce the colloid concentration by at least an order of magnitude, providing evidence that pecan shells could serve as a secondary trap for the retention of colloids that may escape the primary colloid barrier material. The average pH of collected effluents was 6.92 ± 0.10, and the average dissolved oxygen (DO) concentrations were 4.91 ± 0.12 mg/liter.

It is difficult to compare the results for the pecan shells between batch and column tests because the conditions are so different. The water-to-substrate ratio is much greater in batch tests, and the residence time in batch tests was much longer (21-day incubation times in the batch tests and only 2 days in the column tests). In the subsequent sequential column tests with all layers, the nitrate was unexpectedly removed by the Apatite II layer and so the performance of the pecan shells with respect to the nitrate could not be evaluated in a long-term column study.

2. Sequential Column Tests with ^{87}Sr and Nitrate

Mortandad Canyon groundwater retrieved from well MCO-5 was spiked with 2 mg/liter ^{87}Sr and was introduced to the columns at a constant flow rate of 0.2 ml/min. Nitrate concentrations in the water from well MCO-5 were measured to be approximately 50 mg/liter; therefore, no additional nitrate was added to the groundwater at first, but nitrate concentrations were increased to 205 mg/liter in the influent on 4/14/00 (see Fig. 10). Four columns, approximately 10 cm long with a 4.8-cm inside diameter and loaded with the reactive media, were arranged in series as depicted in Fig. 9. Column 1 was loaded with Catfloc-coated gravel (0.5 liter of 2 g/liter Catfloc solution contacted with 280 g of gravel for 12 h), column 2 with Apatite II prewashed with hot water, column 3 with crushed pecan shells mixed with dog food to enhance microbial proliferation, and column 4 with limestone. Pore volumes were about 70 ml for each column, and water residence times were about 6 h for each of the columns. Three-way valves were placed between columns so that samples could be collected following contact with each reactive medium. Approximately 5.5-ml aliquots were retrieved manually at regular time periods between the columns while an Isco Foxy Junior fraction collector collected effluent from column 4 following complete treatment through the column sequence. ICP analysis of ^{87}Sr and nitrate ion chromatography ensued following the collection of effluent from the columns.

Figure 10 illustrates ^{87}Sr and nitrate concentrations detected in the effluent following treatment through all four columns. Nitrate was removed to below detection limits throughout the course of the study except on the day that the concentration was increased to 205 mg/liter, after which it dropped to below detection limits for the rest of the experiment. ^{87}Sr concentrations rarely exceeded the 0.85-mg/liter leached baseline value of Apatite II. The pH and dissolved oxygen (DO) concentration were monitored daily for each of the four columns. A comparison of average pH and DO concentration values, along with the influent water from well MCO-5, is presented in Table IV.

It appears from the DO concentrations that Apatite II and its inherent residual organics were supporting a sufficient denitrifying microbial

Figure 9 Sequential column tests of multiple PRB for Mortandad Canyon. Copyright 2000, A. A. BALKemg, Reprinted with permission.

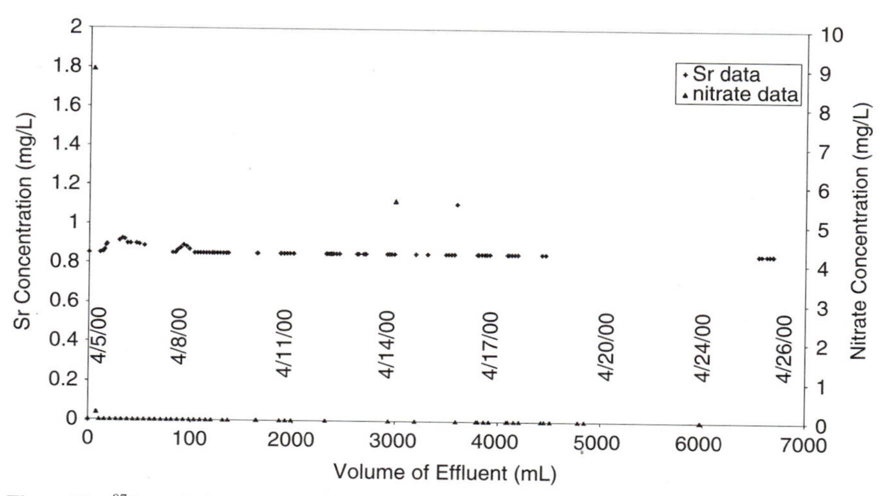

Figure 10 ^{87}Sr and nitrate concentrations in effluent from the sequential column study. Copyright 2000, A. A. BALKemg, Reprinted with permission.

Table IV

MCO-5 Water pH and DO Concentration Values, and Daily Average pH and DO Concentration Values for Individual Columns

Source description	pH	DO (mg/liter)
MCO-5 well water	8.03 ± 0.27	6.34 ± 0.20
Column 1 (Catfloc-coated gravel)	7.82 ± 0.36	5.34 ± 0.15
Column 2 (Apatite II)	6.96 ± 0.11	1.89 ± 0.17
Column 3 (pecan shells + dog food)	6.47 ± 0.30	2.35 ± 0.20
Column 4 (limestone)	6.94 ± 0.19	2.00 ± 0.13

population to explain the removal of nitrate from solutions exiting Apatite II. The pH results in Table IV are consistent with the observation that apatite minerals buffer the pH to about neutral. The pecan shells and limestone do not have much effect in this column study because of the removal of nitrate by Apatite II and the buffering effect of Apatite II. There was concern that the limestone would raise the pH back toward 8, but the water residence time in the coarse limestone gravel is probably not long enough. Again the experiment was interrupted before breakthrough occurred.

3. Sequential Column Tests with ^{239}Pu, ^{241}Am, ^{85}Sr, Nitrate, and Perchlorate

Four aggregates were chosen for use in the four sequential columns in this test (Fig. 11, Table V). A 100-ml pulse of MCO-5 water spiked with $7 \times 10^{-9} M\, ^{85}$Sr, $5 \times 10^{-7} M\, ^{239}$Pu (750,000 pCi), and $1 \times 10^{-9} M\, ^{241}$Am was injected into the columns after the 2-week equilibration period. The use of ^{85}Sr avoided the leaching problem of ^{87}Sr in the determination of total strontium sorption in this column test. As stated earlier, these radionuclide concentrations represent about 60 years of flow at present plutonium concentrations, 44 years of flow at present ^{90}Sr concentrations, and 10 years of flow at present ^{241}Am concentrations. Nitrate and perchlorate were at the field concentrations of 3.9 and 0.4 mg/liter, respectively. No colloids were injected as part of this test. Relating a high-concentration pulse to long-term flow at lower concentrations is valid only if there is no degradation of removal efficiency of the solid over this time period.

Analysis of column elutants was conducted for 3 months until the experiment was interrupted. After injection of the contaminant pulse, approxi

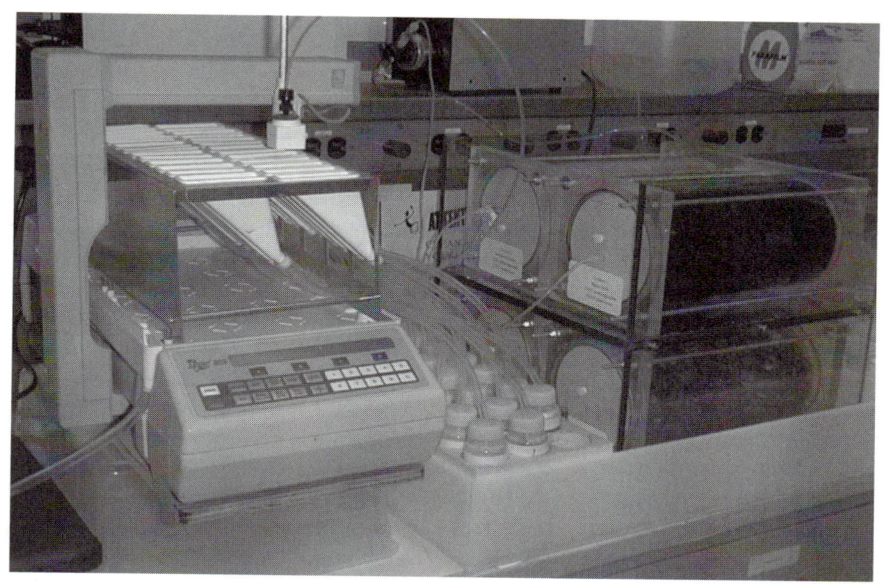

Figure 11 Feasibility column network.

Table V
Column Volumes and Masses

Media	Column 1 (Catfloc-treated volcanic rock)	Column 2 (Apatite II)	Column 3 (pecan shells)	Column 4 (limestone)
Aggregate dry mass (g)	1910	2722	1510	8663
Pore volume (ml)	3217	2705	3125	2304

mately 300 pore volumes exited the columns with no detectable nitrate, perchlorate, plutonium, americium, or strontium (neither ^{85}Sr nor ^{87}Sr; no ^{85}Sr exited Apatite II and no ^{87}Sr leached from Apatite II exited the pecan shells) in the effluent. The immobilized plutonium, americium, and strontium, therefore, appear stable against future subsequent flow, at least for this number of pore volumes, a condition that needs to be addressed if the spent barrier materials need to be left in place after the plume is treated.

4.2 Two-Dimensional Aquifer Box Tests with ^{87}Sr, Nitrate, and Perchlorate

A 2-D aquifer cell was assembled and loaded with basalt coated with Catfloc, Apatite II, pecan shells, and limestone arranged in sequential vertical layers between sand similar to that in the field design (Fig. 12). ^{87}Sr and perchlorate in water from well MCO-5 were injected in a constant concentration of 2 mg/liter and 300 μg/liter, respectively. Nitrate was injected at a constant concentration of 125 mg/liter until day 60 when it was increased to 250 mg/liter to evaluate the ability of microorganisms in PRB materials to reduce irregular nitrate concentrations. With the flow rate of about 85 ml/h and a total cell pore volume of about 2 liter, the water residence time was about 1 day, the same as that observed in Mortandad Canyon shallow groundwater (1 pore volume per day).

Results were obtained for 100 days of continuous contaminant injection. Figure 13 shows results exiting the aquifer box test after passing through all layers. The cell was constructed to allow discrete sampling between each layer interface. Samples taken between the first sand layer and the volcanic gravel (layer 1) and the volcanic rock and Apatite II (layer 2) showed no measurable change in a concentration for any of the contaminants. Effluents from all other layers showed the same results as in Fig. 13, i.e., Apatite II determined the contaminant concentration for all effluent. As was seen in previous tests,

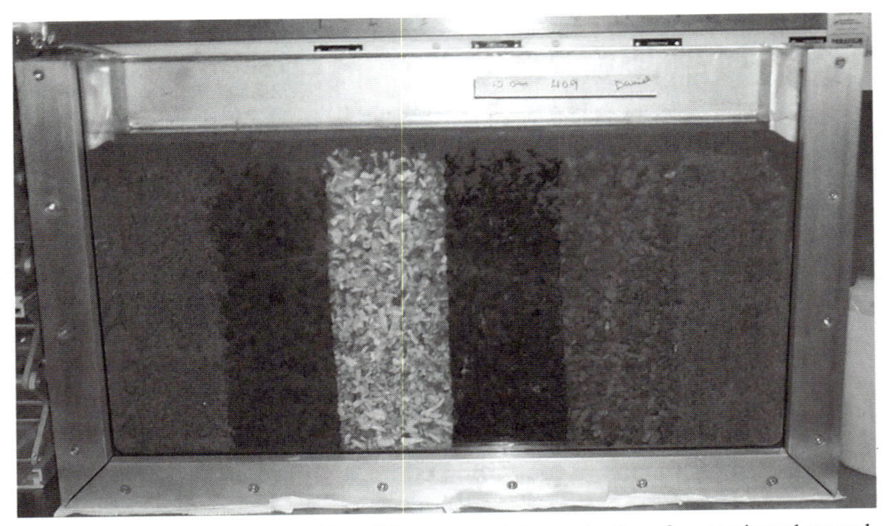

Figure 12 Two-dimensional aquifer cell for treatment demonstration of contaminated groundwater. Copyright 2000, A. A. BALKemg, Reprinted with permission.

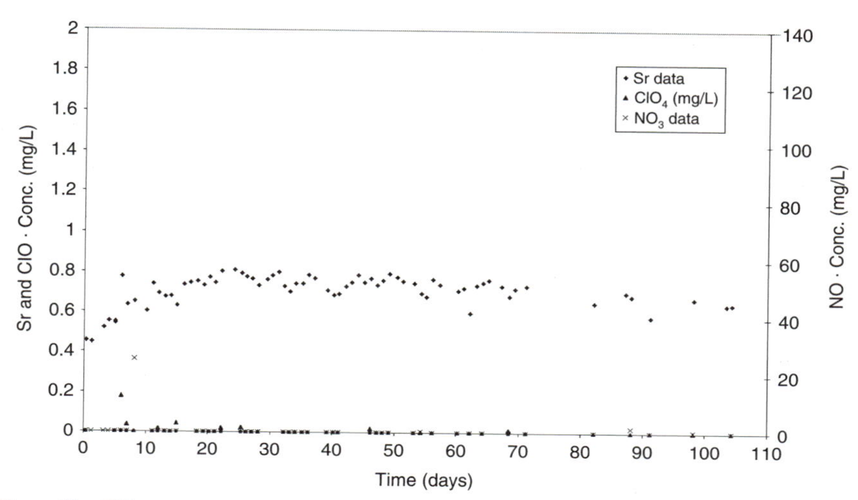

Figure 13 Effluent concentrations for strontium, nitrate, and perchlorate exiting the multiple PRB in the aquifer test. Copyright 2000, A. A. BALKemg, Reprinted with permission.

^{87}Sr was leached from Apatite II at a constant concentration of 0.85 mg/liter and was not significantly reduced by passage through the remaining layers after the first week. All effluents following Apatite II showed a strontium concentration of about 0.85 mg/liter. The Apatite II layer reduced the nitrate and the perchlorate before entering the pecan shells to below detection limits in almost all samples. Detection limits for perchlorate were 3 μg/liter (ppb).

IV. CONCLUSION

The feasibility studies performed, thus far, indicate that a multiple PRB composed of four layers in sequence, Catfloc-coated volcanic gravel, Apatite II, pecan shells, and limestone gravel, could remediate shallow groundwater at Mortandad Canyon, contaminated with $^{238,\ 239,\ 240}$Pu, ^{241}Am, ^{90}Sr, nitrate, and perchlorate, sufficiently to meet regulatory concerns. However, the performance period cannot be determined from the existing results because the feasibility studies were not able to result in breakthrough over the time periods of the study. Feasibility column tests for all contaminants, using ^{85}Sr, should be run to completion, or breakthrough, prior to installation in the field. Because the materials used are inexpensive, a large amount can be used to ensure conservative performance. If the barrier performs as well as desired in the field, the total expected inventory of radionuclides in the shallow aquifer is low enough that complete removal by so many tons of barrier material will not result in activities high enough for the barrier to become a hazardous or radioactive waste. The nitrate and perchlorate are reduced to harmless components and also will not cause the barrier to become a hazardous waste. Therefore, the materials could be left in place as long as regulatory approval could be obtained.

ACKNOWLEDGMENTS

We gratefully acknowledge Patrick Longmire for help of all kinds, scientific and spiritual, and Tom Baca for having the vision to recognize a better future.

REFERENCES

Abdel-Fattah, A. I., and Reimus, P. W. (2000). "Colloid Attachment and Detachment Kinetics from Direct Visualization Experiments," Report to the U.S. Department of Energy, Nevada Operation Office/Underground Test Area (UGTA) Program, April 2000.

Abdelouas, A., Lu, Y. M., Lutze, W., and Nuttall, H. E. (1998a). Reduction of U(VI) to U(IV) by indigenous bacteria in contaminated groundwater. *J. Contamin. Hydrol.* **35**, 217–233.

Abdelouas, A., Lutze, W., and Nuttall, H. E. (1998b). Reduction of nitrate and uranium by indigenous bacteria. *J. French Acad. Sci.* (*Cs R. Acad. Sci, Ser. II Fascicule A Sci Terre Planet.*) **327**, 25–29.

Attaway, H., and Smith, M. (1993). Reduction of perchlorate by an anaerobic enrichment culture. *J. Indust. Microbiol.* **12**, 408–412.

Benner, S. G., Blowes, D. W., Gould, W. D., Herbert, R. B., Jr., and Ptacek, C. J. (1999). Geochemistry of a permeable reactive barrier for metals and acid mine drainage. *Environm. Scie. Technol* **33**, 2793–2799.

Benner, S. G., Blowes, D. W., and Ptacek, C. J. (1997). A full-scale porous reactive wall for the prevention of acid mine drainage. *Groundwater Monitor. Rev.* Fall 1997, 99–107.

Bowman, R. S., and Sullivan, E. J. (1995). Surfactant-modified zeolites as permeable barriers to organic and inorganic groundwater contaminants. *In* "Proceedings, Environmental Technology Development through Industry Partnership Conference," Vol. 2, pp. 392–397.

Blowes, D. W., Ptacek, C. J., and Jambor, J. L. (1997). In-situ remediation of Cr(VI)-contaminated groundwater using permeable reactive walls: Laboratory studies. *Environm. Sci. Technol.* **31**, 3348–3357.

Bostick, W. D., Jarabek, R. J., Bostick, D. A., and Conca, J. (1999). Phosphate-induced metal stabilization: Use of apatite and bone char for the removal of soluble radionuclides in authentic and simulated DOE groundwaters. *Adva. Environ. Res.* **3**, 488–498.

Chen, X.-B., Wright, J. V., Conca, J. L., and Peurrung, L. M. (1997). Evaluation of heavy metal remediation using mineral apatite. *Water Air Soil Pollut.* **98**, 57–78.

Collins, C. H., Lyne, P. M., and Grange, J. M. (1995). "Microbiological Methods", 7th Ed. Butterworth-Heinemann Ltd., Oxford.

Conca, J. L. (1997). Phosphate-Induced Metal Stabilization (PIMS), Final Report to the U. S. Environmental Protection Agency #68D60023, Res. Triangle Park, NC.

Conca, J. L., Lu, N., Parker, G., Moore, B., Adams, A., Wright, J. V., and Heller, P. (2000). PIMS—Remediation of metal contaminated waters and soils, *In* "Remediation of Chlorinated and Recalcitrant Compounds" (G. B. Wickramanayake, A. R. Gavaskar, and A. Chen, eds.), Vol. 7, pp. 319–326. Battelle Memorial Institute, Columbus, OH.

Fruchter, J. S. (1996). In Situ Redox Manipulation Field Injection Test Report – Hanford 100 H Area, Pacific Northwest National Laboratory Technical Report PNNL-11372, Richland, WA.

Geochem Software Inc. (1994). Mac MINTEQ-A2: Aqueous Geochemistry for the MacIntosh, Published by Geochem Software, Inc., Reston, VA.

Goldstein, K. J., O'Hannesin, S., McDonald, S., Gaule, C., Anderson, G. A., Marsh, R., and Senick, M. (2000). Dual permeable reactive barrier walls remediate chlorinated hydrocarbon contamination, *In* "Remediation of Chlorinated and Recalcitrant Compounds" (Wickramanayake, Gavaskar, Gibbs, and Means, eds.), Vol. 6, pp. 273–280. Battelle Memorial Institute, Columbus, OH.

Herman, D. C., and Frankenberger, W. T., Jr. (1998). Microbial-mediated reduction of perchlorate in groundwater. *J. Environ. Qual.* **27**, 750–754.

Hocking, G., Wells, S. L., and Ospina, R. I. (2000). Deep reactive barriers for remediation of VOCs and heavy metals. *In* "Remediation of Chlorinated and Recalcitrant Compounds" (G. B. Wickramanayake, A. R. Gavaskar, and A. Chen, eds.), Vol. 6, pp. 307–314. Battelle Memorial Institute, Columbus, OH.

Kale, R. P. (1993). "Groundwater Remediation Using Polyelectrolytes". M. S. thesis, Univ. of New Mexico.

Kersting, A. B., Efurd, D. W., Finnegan, D. L., Rokop, D. J., Smith, D. K., and Thompson, J. L. (1999). Migration of plutonium in groundwater at the Nevada Test Site. *Nature* **397**, 56–59.

LANL. (1997). Work Plan for Mortandad Canyon, LA-UR-97–3291, Los Alamos National Laboratory, Los Alamos, NM.

Lower, S. K., Maurice, P. A., Traina, S. J., and Carlson, E. H. (1998). Aqueous Pb sorption by hydroxylapatite: Applications of atomic force microscopy to dissolution, nucleation and growth studies. *Am. Mineral.* **83**, 147–158.

Ma, Q. Y., Traina, S. J., and Logan, T. J. (1993). In situ lead immobilization by apatite. *Environ. Sci. Technol.* **27**, 1803–1810.

Moody, T. E., and Wright, J. (1995). Adsorption Isotherms: North Carolina Apatite Induced Precipitation of Lead, Zinc, Manganese and Cadmium from Bunker Hill 4000 Soil, Technical Report BHI-00197, Bechtel Hanford, Richland, WA.

Moody, T. E., and Wright, J. V. (1996). Permeable Barrier Materials for Strontium Immobilization, Bechtel Hanford Technical Report BHI-00857, Bechtel Hanford, Inc., Richland, WA.

Morkin, M, Devlin, J. F., Barker J. F., and Butler B. J. (2000). In situ sequential treatment of a mixed contaminant plume. *J. Contamin. Hydrol.* **45**, 283–302.

Naftz, D. L., Fuller, C. C., Davis, J. A., Piana, M. J., Morrison, S. J., Freethey, G. W., and Rowland, R. C. (2000). Field demonstration of PRBs to control uranium contamination in groundwater, *In* "Remediation of Chlorinated and Recalcitrant Compounds" (G. B. Wickramanayake, A. R. Gavaskar, and A. Chen, eds.), Vol. 6, pp. 281–290. Battelle Memorial Institute, Columbus, OH.

Puls, R. W., Blowes, D. W., and Gillham, R. W. (1999). Long-term performance monitoring for a permeable reactive barrier at the U. S. Coast Guard Support Center, Elizabeth City, North Carolina. *J. Hazard. Mater.* **68**, 109–124.

Runde, W. (1992). The chemical interactions of actinides in the environment. *Los Alamos Sci.* **26**, 338–357.

Runde, W., Meinrath, G., and Kim, J. I. (1992). A study of solid-liquid phase equilibria of trivalent lanthanide and actinide ions in carbonate systems. *Radiochim. Acta* **58/59**, 93–100.

Starr, R. C., and Cherry, J. A. (1994). In situ remediation of contaminated groundwater: The funnel-and-gate system. *Groundwater* **32**, 456–476.

Sullivan, E. J., Bowman, R. S., and Haggerty, G. M. (1994). Sorption of inorganic oxyanions by surfactant-modified zeolites. *In* "Spectrum 94, Proceedings of the Nuclear and Hazardous Waste Management International Topical Meeting," Vol. 2, pp. 940–945.

Tratnyek, P. G., Johnson, T. L., Scherer, M. M., and Eykholt, G. R. (1997). Remediating groundwater with zero-valent metals: Kinetic considerations in barrier design. *Groundwater Monitor. Remed.* **17**, 108–114.

Triay, I. R., Meijer, A., Cisneros, M. R., Miller, G. G., Mitchell, A. J., Ott, M. A., Hobart, D. E., Palmer, P. D., Perrin, R. E., and Aguilar, R. D. (1991). Sorption of americium in tuff and pure minerals using synthetic and natural groundwaters. *Radiochim. Acta* **52/53**, 141–145.

Triay, I. R, Meijer, A., Conca, J. L., Kung, K. S., Rundberg, R. S., and Strietelmeier, B. A. (1997). "Summary and Synthesis Report on Radionuclide Retardation for the Yucca Mountain Site Characterization Project" (R. C. Eckhardt, ed.), Los Alamos National Laboratory Technical Report LA-13262-MS, Los Alamos, NM.

Wallace, W., Beshear, S., Willims, D., Hospadar, S., and Owens, M. (1998). Perchlorate reduction by a mixed culture in an up-flow anaerobic fixed bed reactor. *J. Indust. Microbiol. Biotechnol.* **20**, 126–131.

Waybrant, K. R., Blowes, D. W., and Ptacek, C. J. (1998). Selection of reactive mixtures for use in permeable reactive walls for treatment of mine drainage. *Environ. Sci. Technol.* **32**, 1972–1979.

Wickramanayake, G. B., Gavaskar, A. R. and Chen, S. C. (2000). "Remediation of Chlorinated and Recalcitrant Compounds", Vol. 6. Battelle Memorial Institute, Columbus, OH.

Williamson, D., Hoenke, K., Wyatt, J., Davis, A., and Anderson, J. (2000). Construction of a funnel-and-gate treatment system for pesticide-contaminated groundwater, *In* "Remediation of Chlorinated and Recalcitrant Compounds" (G. B. Wickramanayake, A. R. Gavaskar, and A. Chen, eds.), Vol. 6, pp. 257–264. Battelle Memorial Institute, Columbus, OH.

Wright, J. (1990). Conodont apatite: Structure and geochemistry, *In* "Metazoan Biomineralization: Patterns, Processes and Evolutionary Trends" (J. Carter, ed.), pp. 445–459. Van Nostrand Reinhold, New York.

Wright, J., Conca, J. L., Repetski, J., and Clark, J. (1990). Microgeochemistry of some lower ordovician cordylodans from Jilin, China, *In* "1st International Senckenberg Conference and 5th European Conodont Symposium (ECOS V) Contributions III" (W. Ziegler, ed.), Vol. 118, pp. 307–332. Courier Forschungsinstitut Senckenberg, Frankfurt, Germany.

Wright, J. V., Peurrung, L. M., Moody, T. E., Conca, J. L., Chen, X., Didzerekis, P. P., and Wyse, E. (1995). In Situ Immobilization of Heavy metals in Apatite Mineral Formulations, Technical Report to the Strategic Environmental Research and Development Program, Department of Defense, PNL, Richland, WA.

Wright, J., Schrader, H., and Holser, W. T. (1987). Paleoredox variations in ancient oceans recorded by rare earth elements in fossil apatite. *Geochim. Cosmochim. Acta* **51**, 631–644.

Zumft, W. G. (1997). Cell biology and molecular basis of denitrification. *Microbiol. Mol. Biol. Rev.*, **61**, 533–616.

Part III

Evaluations of Chemical and Biological Processes

Chapter 9

Evaluation of Apatite Materials for Use in Permeable Reactive Barriers for the Remediation of Uranium-Contaminated Groundwater

Christopher C. Fuller*, Michael J. Piana†, John R. Bargar‡, James A. Davis*, and Matthias Kohler§

*U.S. Geological Survey, Menlo Park, California 94025
†NEC Electronics, Process Engineering Division, Roseville, California 95747
‡Stanford Synchrotron Radiation Laboratory, Stanford, California 94309
§Colorado School of Mines, Environmental Science and Engineering Department/Division, Golden, Colorado 80401

Phosphate rock, bone meal, bone charcoal, and pelletized bone charcoal apatites were evaluated for use in a demonstration of permeable reactive barriers (PRB) for uranium [U(VI)] removal from groundwater at Fry Canyon, Utah. Apatites were studied because phosphate supplied by apatite dissolution may cause precipitation of uranyl phosphate. Bone meal and bone charcoal sorbed 1.5 to 2 orders of magnitude more U(VI) than phosphate rock in batch and column experiments. Phosphate rock was found unsuitable for PRB use because of rapid breakthrough and low sorption capacity. In contrast, U(VI) breakthrough of 12 mg/liter U(VI) occurred at 216 column pore volumes (*cv*) for bone meal. Although the bone meal and bone charcoal were highly effective at sequestering U(VI), the low permeability of these fine-grained materials and/or potential for clogging from bacterial growth would require 10-fold or greater dilution with coarse grain materials to maintain

Handbook of Groundwater Remediation Using Permeable Reactive Barriers

adequate PRB permeability. Instead, a porous, pelletized bone charcoal was chosen for the PRB demonstration because of its high permeability and intermediate U(VI) sorption capacity compared to the other materials tested, with 50% breakthrough occurring after 100 *cv*. U(VI) sorption was largely reversible for all materials with up to 80% of total sorbed U(VI) released during elution with U(VI)-free groundwater. Bone char pellets recovered from the upgradient face of the PRB 18 months after installation had sorbed U(VI) equal to the maximum sorbed in column tests. X-ray absorption spectroscopy and X-ray diffraction of U(VI)-reacted apatites in laboratory tests and recovered from the PRB indicated uranyl phosphate precipitation did not occur for dissolved U(VI) concentrations characteristic of Fry Canyon. Instead, sorption occurred by formation of a uranyl surface complex (adsorption) as was observed with synthetic hydroxyapatite. Because surface complexation likely dominates sorption, an apatite-based PRB will require close monitoring and removal of the solid phase as breakthrough occurs when adsorption equilibrium is attained.

I. INTRODUCTION

A. Background

Use of permeable reactive barriers (PRB) for the removal of dissolved contaminants has increased in recent years as a low-cost alternative to pump and treat methods for groundwater remediation (Blowes *et al.*, 1997; Shoemaker *et al.*, 1995; Morrison and Spangler, 1993). PRBs have the advantage of lower operational and maintenance costs, as water flow through the PRB is driven by the natural hydraulic gradient. *In situ* treatment by a PRB results from either degradation of contaminants to nontoxic forms or transformation to an immobile phase (Blowes *et al.*, 1997). Ideally, the removal process should be largely irreversible and can be initiated with a low-cost, high-permeability material. Most PRB applications to date have been deployed for the removal of organic solvents from groundwater systems, although a few have been used for the removal of radionuclides and trace elements (Benner *et al.*, 1999; Blowes *et al.*, 2000; Morrison *et al.*, 2001).

Development and evaluation of reactive materials are critical steps in the design of PRBs for groundwater remediation. This chapter focuses on the evaluation and choice of commercially available, inexpensive apatites for use in the Fry Canyon, Utah, field demonstration of PRBs for the remediation of uranium [U(VI)] contamination in groundwater (Naftz *et al.*, 2000). The mechanism of U(VI) sorption by these materials was determined from X-ray absorption spectroscopic (XAS) and X-ray diffraction (XRD) measurements of U(VI) apatites (Fuller *et al.*, 2002a,b) and the results are discussed here with regard to their implications for use of apatite in PRBs.

B. Previous Studies of Metal Immobilization by Apatite

Hydroxyapatite [$Ca_5(PO_4)_3OH$], HA, has been found very effective in sorbing many divalent cations, including lead (Pb), copper, zinc (Zn), cadmium, and strontium (Ma *et al.*, 1993; Ruby *et al.*, 1994, Xu *et al.*, 1994; Valsami-Jones *et al.*, 1998; Middleburg and Comans, 1991; Collin, 1960). Proposed removal processes for metal sorption include cation exchange, incorporation into the apatite structure, and precipitation of metal phosphates. Most investigations on metal interaction with HA have focused on Pb because the effectiveness of HA in transforming Pb to low solubility phases suggests good potential for use of HA in remediation strategies (Laperche and Traina, 1998). Immobilization of Pb by HA in contaminated soils and groundwater has been shown to occur by precipitation of lead phosphates, notably hydroxypyromorphite, by reaction of Pb with phosphate released from HA (Xu and Schwartz, 1997; Ma *et al.*, 1994, 1995; Zhang *et al.*, 1997). Jeanjean *et al.* (1995) observed extensive removal of U(VI) upon addition of synthetic HA to uranyl solutions, although uranyl phosphates could not be detected by XRD.

The motivation for testing apatite in a PRB application for U(VI) is based on the low solubility of uranyl phosphates (Langmuir, 1978; Sandino and Bruno, 1992), which suggests that natural apatites could be effective in lowering dissolved U(VI) in groundwater. The hypothesis for uranium immobilization is that HA dissolution would provide a source of dissolved phosphate ([P]) with which aqueous U(VI) may precipitate to form a uranyl phosphate phase, such as hydrogen, calcium, magnesium, potassium, or sodium autunite (e.g., [$Ca(UO_2)_2(PO_4)_2 \bullet 10H_2O$]) (Sowder *et al.*, 1996; Sandino and Bruno, 1992; Arey *et al.*, 1999). The addition of HA to batch experiments of U-contaminated acidic sediment and soils resulted in a decrease of U(VI) release to solution (Arey *et al.*, 1999). In these systems, dissolved U(VI) was sequestered by Al and Mn phosphate phases that precipitated in response to the dissolution of Al and Mn from the acidic sediments and elevated dissolved phosphate from HA dissolution. Formation of uranyl phosphate phase such as autunite was observed upon exposing HA to high concentration uranyl nitrate solutions (Bostick *et al.*, 1999; Ordonez-Regil *et al.*, 1999). In those experiments, dissolution of HA by the acidic uranyl nitrate solution and subsequent neutralization likely resulted in greatly exceeding the solubility of uranyl phosphate phases.

Fuller *et al.* (2002a) reported quantitative sorption of dissolved uranium by synthetic HA over a wide range of total U(VI) concentrations up to an equivalent to the total P in the experimental system. U(VI) sorption was determined to occur by surface complexation on HA at dissolved U(VI) concentrations relevant to the contaminated groundwater at Fry Canyon. At higher U(VI) concentrations, formation of uranyl phosphate phases resulted.

C. Focus of Present Study

Because the cost of synthetic HA is prohibitive for use in PRB applications, lower cost, commercially available natural apatite materials were evaluated. Bone meal has been proposed as an inexpensive source of apatite for soil remediation because of its ability to immobilize Pb and Zn (Hodson *et al.*, 2000). Analysis of the solid phase recovered from column-leaching experiments indicated association of Pb and Zn with P phases. Phosphate rock also has been proposed as an inexpensive P source for the remediation of Pb-contaminated soils and water (Ma *et al.*, 1995). In both cases, the natural apatites provide a source of phosphate by dissolution and subsequent immobilization of the contaminant by the precipitation of sparingly soluble metal phosphates.

This chapter presents results for laboratory batch uptake and column experiments to evaluate U(VI) sorption by phosphate rock, bone meal, and bone charcoal apatites for potential use as PRB-reactive materials. A highly porous, pelletized bone charcoal was also evaluated. Other critical properties of materials for PRB use that were tested include permeability, the extent of reversibility of U(VI) removal, and the release of solutes detrimental to water quality. Uranium removal by apatite recovered from the Fry Canyon PRB demonstration is compared to laboratory results.

II. LABORATORY EVALUATION OF APATITES FOR URANIUM SORPTION

A. Apatite Materials

Three phosphate rock samples were obtained from mining companies in Florida, North Carolina, and Utah (Table I). The mined rock had been separated from accessory minerals and crushed to fine sand or silt grain size prior to receipt. Fertilizer-grade bone meals from two processing methods (Dale Alley Corp., St. Joseph, MO) and a powdered bone charcoal (EM Scientific) were also tested. Porous, pelletized bone charcoal prototypes (CP) were manufactured by Cercona of America, Inc. (Dayton, OH) for evaluation. Pellet formulation varied by temperature (>1000°C), presence or absence of air, sources of bone meal, and binder composition. All samples were used as received.

B. Physical Characteristics

XRD analysis indicated that P-bearing minerals in the phosphate rocks were fluoroapatite and carbonated fluoroapatite. The bone meal and bone charcoal samples contained hydroxyapatite. The pelletized bone charcoal contained whitlockite, a dehydrated form of apatite. Physical characteristics of the three types of materials are presented in Table I. Specific surface area determined by single point N_2 gas adsorption typically was higher for the finer grain-sized materials with the exception of steamed bone meal (BB2), which had the smallest surface area ($0.3\,m^2/g$). This bone meal had the highest P release (17 mg P/liter) in batch experiments.

The bone char pellets had very high permeability and high internal porosity (Table II). The permeability of the bone char pellets was determined from the dependence of flow rate on hydraulic head through packed columns using a modified Marriot bottle apparatus. Permeability, expressed as hydraulic conductivity (*K*, m/day), was calculated using Darcy's law and corrected for conductivity of the apparatus. Measured permeability greatly exceeded the aquifer permeability at Fry Canyon (17 to 26 m/day, see Chapter 14). Internal pellet volume was determined from weight loss upon drying of water-saturated pellets. Analysis of breakthrough curves of tritiated water indicated that essentially the entire intraparticle porosity of bone char pellets was accessible to flow. Total column pore volume (*cv*) was calculated from tritiated water breakthrough curves and agreed within 5% of a gravimetric measure of total porosity. This comparison indicates that there was rapid exchange of water between interparticle porosity and essentially all internal porosity.

C. Batch Uptake Experiments

1. Methodology

Batch uptake experiments provide a test of the relative efficiency of each apatite material for U(VI) sorption under similar solution conditions and in

Table I
Physical Characteristics of Apatite Materials Used in Laboratory Evaluation

Material	ID	Source	Surface area (m^2/g)	Grainsize	Solid concentration[a] (g/liter)	Phosphate concentration[b] (mg P/liter)	pH[c]
Phosphate rock	SF	SF Phosphates, LTD	4.2	75% 125–1000 μm	100	0.9	7.2–7.5
		Vernal, UT		25% <125 μm			
Phosphate rock	PCS	PCS Phosphates	14.4	95% 125–1000 μm	100	14	7.3–7.5
		Raleigh, NC		5% <125 μm			
Phosphate rock	CF	CF Industries	12.2	51% 125–1000 μm	10	0.8	7.2–7.3
		Plant City, FL		47% <125 μm			
Bone meal, cooked	BB1	Fertilizer company	6.6	55%>500 μm	10	1.0	6.5–6.6
		Dale Alley		45% <500 μm			
		Company St Joseph, MO					
Bone meal, steamed	BB2	Fertilizer company	0.3	100% < 63 μm	100	17	6.4–6.6
		Dale Alley Company St Joseph, MO					
Bone charcoal	BK	EM Scientific	64	3–15 μm	10	5.3	7.5–7.5
Bone char pellet	CP3	Cercona of America	33	7.5% > 4 mm	10	0.9	7.2–7.4
				65.3 % > 2–4 mm			
				23.7 % > 1–2 mm			
				3.5% < 1 mm			
Bone char pellet	CP5	Cercona of America	44	13.8 % > 4 mm	10	7.8	6.9–7.1
				42.9% > 2–4 mm			
				39.4% > 1–2 mm			
				3.8% < 1 mm			

[a] Solid concentration is the mass of apatite per liter of artificial groundwater in batch experiments.
[b] Phosphate concentration is the amount of dissolved phosphorus released in the absence of dissolved uranium.
[c] pH is the range of pH after uranium uptake.

Table II
Physical Properties of Bone Char Phosphate Pellet Formulations CP3 and CP5[a]

Property	CP3	CP5	Units
Grain size			
% >4 mm	7.5	13.8	Percent
% 2–4 mm	65.3	42.9	Percent
% 1–2 mm	23.7	39.4	Percent
% <1 mm	3.5	3.8	Percent
Hydraulic conductivity	734	431	m/day
Specific surface area	33	44	m^2/g
Interparticle porosity	20	19	Percent
Intraparticle volume	0.83	0.59	cm^3/g pellets
Intraparticle porosity	55	51	Percent
Phosphate concentration[b]	0.9	7.8	mg P/liter

[a]CP5 used in PRB field demonstration.
[b]Dissolved phosphate concentration in batch experiment after 24-h equilibration with artificial groundwater in the absence of uranium.

the absence of transport limitation to uptake. Uranium sorption was measured in an artificial groundwater (AGW) of similar major ion, pH (7.0), and alkalinity (480 mg/liter as calcium carbonate) to Fry Canyon groundwater (Naftz *et al.*, 2000). Under these conditions, dissolved uranium exists predominantly in the +6 oxidation state as carbonate complexes of the uranyl dioxocation (UO_2^{2+}) and is denoted as U(VI). The apatite solid (10 or 100 g solid per liter depending on grain size, Table I) was equilibrated with the AGW for 24 h prior to the addition of U(VI). The U(VI) was added in stepwise increments ($\leq$ 2400 μg/liter) to yield total U(VI) concentrations of 2400 to 48,000 μg/liter. Dissolved U(VI) ([U]) ranges from 1400 to over 20,000 μg/liter at Fry Canyon (Naftz *et al.*, 2000). The pH was adjusted to 7.0 with dilute acid or base. Because no measurable change in U(VI) uptake was observed after 24 h of reaction in preliminary experiments, samples were reacted for 48 h on a shaker table or end-over-end rotator. The final pH was then measured, the sample centrifuged (10 min at 16,000 *g*), and the supernatant retained for [U], dissolved cations and phosphorous ([P]), analyses. Release of [P] from apatite solids was measured in batch experiments in the absence of U(VI).

Initial batch experiments used a ^{233}U tracer with U(VI) concentration determined by liquid scintillation counting of the ^{233}U alpha decay [precision

±2%; method detection limit (mdl) 5 μg/liter for ^{233}U-labeled 2400 μg/liter total U]. Dissolved U(VI) was determined in subsequent batch and all column experiments by kinetic phosphorescence analysis (mdl 0.06 μg/liter, precision ±2%). Uptake was determined by the change in [U]. Dissolved calcium ([Ca]) and [P] were measured by inductively coupled plasma optical emission spectrometry (ICP-OES; mdl 0.01 and 0.1 mg/liter, respectively). Although the ICP-OES measures all forms of [P], it is assumed that [P] is comprised largely of orthophosphate species from here on.

2. U(VI) Sorption in Batch Experiments

Uranium sorption in batch experiments was compared on a mass basis instead of equal surface area because the mass of reactive material that can be used in a PRB application is limited by volumetric dimensions. The extent of U(VI) sorbed by the various apatite materials varied considerably (Fig. 1). Bone meal and bone charcoal sorbed 1.5 to 2 orders of magnitude more U(VI) per gram than phosphate rock. More than 98% of a total dissolved U(VI) of 4800 μg/liter was sorbed by bone meal and bone charcoal, compared to 54 to 80% by phosphate rock. Uranium sorption by bone char pellets (CP3 and CP5) was about a factor 10 lower than the bone meal and bone charcoal (Fig. 1). U(VI) sorption was independent of the amount of [P] released from the solid prior to the addition of U(VI) (Table I). The final pH of batch solutions increased between 0.2 and 0.5 units for phosphate rock, bone charcoal, and bone char pellets and decreased 0.4 to 0.6

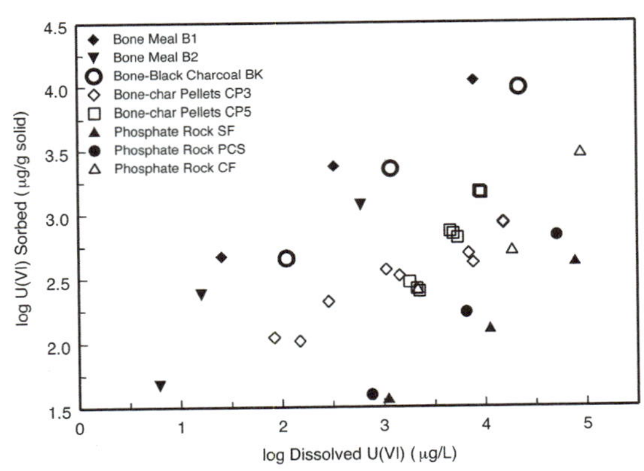

Figure 1 Uranium sorption by apatite materials in batch experiments with artificial groundwater. The log of sorbed uranium [μg U(VI) sorbed per gram of solid] versus the log of equilibrium dissolved uranium (μg/liter).

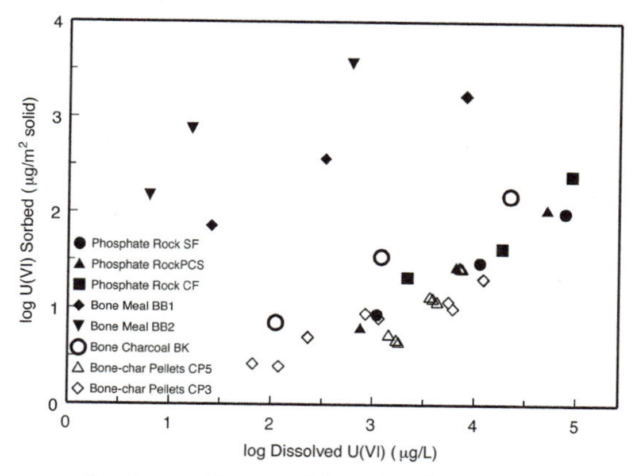

Figure 2 Uranium sorption by apatite materials in batch experiments normalized to specific surface area. The log of sorbed uranium [μg U(VI) sorbed per m^2 of solid] versus the log of equilibrium dissolved uranium (μg/liter).

units for bone meal from the initial pH of 7.0 (Table I). The amount U(VI) sorption among the solids did not vary with pH for a given total U(VI) concentration.

The reactivity of apatites was compared by normalizing U(VI) removal to specific surface area, assuming that the N_2 surface area represents reactive surface site density (Davis and Kent, 1990). A narrower range of U(VI) sorption was observed among materials when compared this way, with the exception of the bone meals, which had significantly greater U(VI) uptake per m^2 (Fig. 2).

D. Column Experiments

Column tests were used to determine the volume of U(VI)-contaminated groundwater that can be treated prior to U(VI) breakthrough and the U(VI) sorption capacity of potential apatite materials for conditions characteristic of the Fry Canyon aquifer.

1. Methodology

Glass columns (15 or 30 cm length, 1-cm inside diameter) fitted with 20-μm end-cap screens were used for the bone meal and phosphate rock experiments. Column outlets were fitted with 0.45-μm syringe filters. Initial tests of columns packed with bone meal failed because of clogging either from bacterial growth or from loss of integrity. Because of the clogging and/or small grain size, bone meal and phosphate rock were diluted 10-fold by weight with Ottawa sand (20 to 30 mesh) to maintain adequate permeability. The sand had negligible U(VI) sorption in a batch experiment. A column test with the bone charcoal was not feasible due to elution of this very fine-grained (3 to 15 μm) material out of the column. The large particle size of the bone char pellets enabled packing into a column (2.5 cm diameter) without dilution with sand.

Packed columns were flushed with carbon dioxide (CO_2) and then flushed with deionized water to dissolve the CO_2. This method was effective in eliminating gas pockets. Upward flow of AGW with 12 mg/liter U(VI) was started immediately using gravity feed (10 ml/h). The AGW feed reservoir was equilibrated continuously with 2% CO_2 in air to maintain a constant pH of 7±0.1 pH units. Effluent pH of the bone char pellet columns was measured daily during the first week, then weekly, using a flow cell electrode. Effluent pH was measured at the end of the column experiment for phosphate rock and bone meal apatites. Reversibility of U(VI) sorption was tested by changing the influent to U-free AGW after U(VI) breakthrough.

Total *cv* was determined from dry solid and water-saturated weights. The ratio of effluent to influent concentration (C/C_o) typically is plotted versus *cv*. Because of dilution with sand and likely differences in the accessible intra-particle porosity, C/C_o was plotted versus the milliliters of column effluent passed per gram reactive material (ml/g) to allow direct comparison of U(VI) breakthrough among the various materials. Total *cv* of the bone char pellet columns was determined from ^3H breakthrough and included the accessible internal porosity of bone char pellets. Cumulative U(VI) sorption was calculated by integrating the difference between total U(VI) input and outflow concentration [U(VI) uptake], divided by the mass of apatite material in the column. Total sorption capacity is defined as the maximum cumulative sorption of U(VI) at complete breakthrough (100%).

2. Column Results

Column breakthrough curves are presented in Fig. 3, 4, and 5 for phosphate rock, bone meal, and CP3 bone char pellets, respectively. Effluent pH was 7.2±.1 and 7.1±.1 for the duration of the CP3 and CP5 bone char pellet column tests. The pH at the end of column test ranged from 7.2 to 7.5 for the other materials (Table III). Consistent with batch experiments, phosphate rock had the lowest U(VI) removal. Breakthrough to $C/C_o = 0.5$ was rapid (within 14–68 ml/g or 7 to 18 nominal column volumes) for phosphate rock (Fig. 3, Table III). The sorption capacity for phosphate rock ranged from 160 to 410 ppm. In contrast, the greatest U(VI) sorption was observed with bone

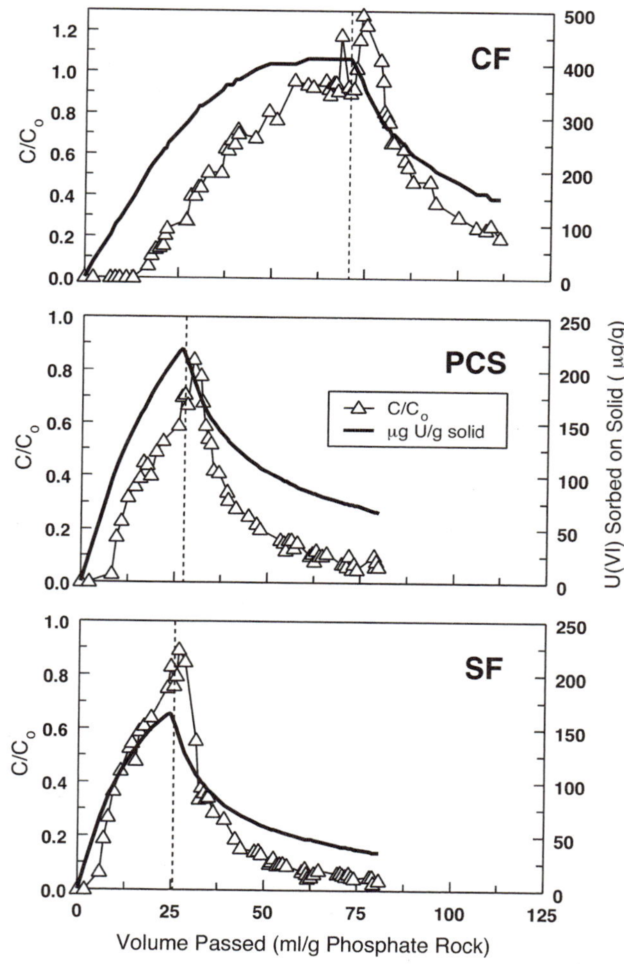

Figure 3 Column breakthrough and reversibility for phosphate rock samples (CF, PCS, SF) plotted as the ratio of effluent-to-influent dissolved uranium (C/C_o, Δ) versus, the volume of solution passed in units of milliliter per gram of phosphate rock. Cumulative uranium sorbed per gram of phosphate rock (μg/g) is depicted by the solid line. The vertical dashed line indicates change of influent to uranium-free artificial groundwater to test for the reversibility of uranium sorption. Concentration of uranium is effluent during the reversibility test is the ratio of measured effluent dissolved uranium to the U(VI) concentration of influent during the uptake limb of the experiment.

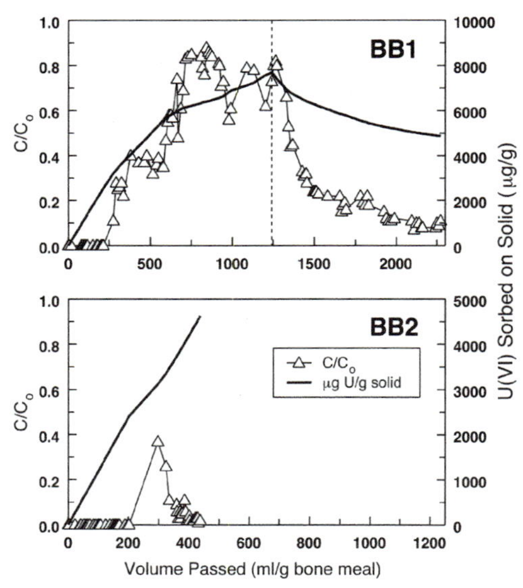

Figure 4 Column breakthrough and reversibility for fertilizer-grade bone meals (BB1 and BB2) plotted as the ratio of effluent-to-influent dissolved uranium (C/C_o, Δ) versus the volume of solution passed in units of milliliter per gram of bone meal. Cumulative uranium sorbed per gram of bone meal (μg/g) is depicted by the solid line. The vertical dashed line indicates change of influent to uranium-free artificial groundwater to test for the reversibility of uranium sorption. Concentration of uranium is effluent during the reversibility test is the ratio of measured effluent dissolved uranium to the U(VI) concentration of influent during the uptake limb of the experiment.

meal. Breakthrough ($C/C_{o=0.5}$) required nearly 600 ml/g (216 *cv*) of groundwater for the BB1 bone meal (Fig. 4a). The sorption capacity of this bone meal was 7700 ppm. The column with steamed bone meal (BB2) was terminated after 437 ml/g (155 *cv*) due to clogging (Fig. 4b), likely because of bacterial growth. No significant U(VI) breakthrough was measured to this point, with a cumulative U(VI) sorption of 4600 ppm.

The column test with bone char pellets with iron binder (CP1) clogged within 5.5 ml/g (18 *cv*) due to oxidation of iron, which cemented pellets at the inlet portion of the column. In addition, effluent pH rose to 9, characteristic of zero valent iron material (Gu *et al.*, 1998). Pellets calcined in air (CP2) had complete U(VI) breakthrough within 13 ml/g (4 *cv*). The bone char pellet formulation (CP3) that performed best was fired in the absence of air and was made with a phosphate binder and retained carbon. Breakthrough to $C/C_o = 0.5$ occurred at 122 ml/g (104 *cv*) with complete breakthrough (100%) occurring at 181 ml/g (618 *cv*) (Fig. 5a). The sorption capacity of the CP3 bone char pellets was 1370 ppm U(VI).

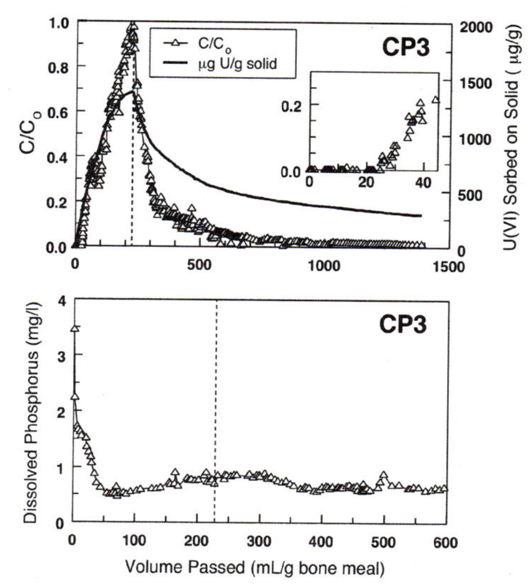

Figure 5 (a) Column breakthrough and reversibility test for Cercona bone char pellets (CP3). The ratio of effluent-to-influent dissolved uranium (C/C_o, Δ) versus the volume of solution passed in units of milliliter per gram of bone char pellets. Cumulative uranium sorbed per gram of bone char pellets (μg/g) is depicted by the solid line. The vertical dashed line indicates change of influent to U(VI)-free artificial groundwater to test for the reversibility of uranium sorption. Concentration of uranium is effluent during the reversibility test is the ratio of measured effluent dissolved uranium to the U(VI) concentration of influent during the uptake limb of the experiment. (b) Dissolved phosphorous (mg/liter) in column effluent versus volume of solution passed in units of milliliter per gram of bone char pellets. The vertical dashed line depicting the change to U(VI)-free artificial groundwater is shown for reference because of the different x-axis scale.

E. Reversibility of U(VI) Sorption

An important criterion for PRB application is the potential for re-Release of the contaminant from the PRB. Conditions favoring contaminant release from a PRB are thought to occur after the contaminant plume has passed and uncontaminated groundwater enters the PRB. The potential for release of U(VI) from apatite was evaluated in column tests after 100% breakthrough of U(VI) by passing U(VI)-free AGW through the columns. In all three types of materials tested, most of the sorbed U(VI) was released from the solid. Desorption was characterized by an initial faster release until the effluent concentration reached a C/C_o of about 0.2. A slower release followed that continued over many column pore volumes (Figs. 3–5). In some cases, an increase in effluent concentration to $C/C_o > 1$ occurred over the first several *cv* of elution, indicating rapid desorption. About 65% of the U(VI) uptake

Table III

Summary of U(VI) Breakthrough and Sorption Capacity for Column Tests of apatite Materials with Artificial Groundwater with 12,000 μg/liter Dissolved U(VI)[a]

Material ID	Volume to 50% breakthrough (ml/g[b])	cv^c	Total U(VI) sorption capacity[d]	Column effluent pH[e]
Phosphate rock				
SF	14	7	160	7.3
PCS	20	11	220	7.4
CF	68	18	410	7.5
Bone meal				
BB1	594	216	7700	7.2
BB2[f]	437	155	4600	—
Bone char pellets[g]				
CP3	122	104	1370	7.2±.1
CP5[h]	206	252	650	7.1±.1
CP5[h,i]	222	278	580	7.1±.1

[a]See Table I for explanation of material ID.
[b]Volume passed in milliliter per gram of apatite
[c]Column pore volumes (*cv*) passed. *cv* estimated for phosphate rock and bone meal. *cv* measured by tritium breakthrough for bone char pellets.
[d]Uranium uptake at 100% breakthrough [mg U(VI) per gram solid].
[e]pH is the column effluent pH measured at the end of experiment for phosphate rock and bone meal and the average pH for the duration of CP3 and CP5 experiments.
[f]Column terminated at <20% breakthrough due to clogging.
[g]Bone char pellets packed undiluted.
[h]2400 μg/liter U(VI) influent.
[i]Column eluted with 80 *cv* prior to introduction of U(VI). Volume passed reported does not include pretreatment volume.

was released from phosphate rock within 30 to 50 ml/g of U-free AGW passed, at which point the effluent concentration was less than 0.1 C/C_o and the experiments were terminated (Fig. 3). The initial release of sorbed U(VI) was about 20 and 40% of total U(VI) sorbed by bone meal and bone char pellets, respectively (Figs. 4a and 5b). Subsequently, the rate of release decreased and release continued through hundreds of pore volumes. The BB2 bone meal column reversibility test was terminated after 1030 ml/g of elution because of clogging. At that point, about 35% of the sorbed U(VI) had been released. After about 1160 ml/g of elution (990 pore volumes), 80% of the total U(VI) removed by the bone char had been released back to groundwater. Reversibility tests indicate that groundwater with low U(VI) entering an apatite PRB could result in release of most of the sorbed U(VI) back to the groundwater unless the material was removed. However, about half of the U(VI) released from bone meal or bone charcoal apatites would occur over a long period.

F. Effects of Apatite on Water Quality

An important consideration in PRB reactive material choice is the potential impact on water quality down gradient of the PRB. Release of dissolved phosphate was observed in column tests for all materials and was characterized by an initial pulse concentration that decreased to a plateau concentration. For example, phosphate release from CP3 bone char pellets was initially 3.5 mg/liter [P] and decreased to about 0.7 mg/liter after 44 ml/g (37 *cv*, Fig. 5b). Dissolved phosphate remained relatively constant (0.68 ± 0.11 mg/liter) throughout the U(VI) breakthrough and elution. This sustained release of [P] would be detrimental to aquatic systems down gradient of a PRB installation if no subsequent processes removed [P] from groundwater. At the Fry Canyon study site, sorption of [P] to iron oxide coatings on aquifer sediments is expected to remove most [P] prior to discharge into Fry Creek. No significant changes in the concentration of other solutes or pH were observed during the duration of column tests.

G. Apatite Material Chosen for the PRB Field Demonstration

1. Selection of Material

Rapid breakthrough and low U(VI) sorption capacity indicated that phosphate rock would be unsuitable as a PRB material for U(VI) remediation. Fertilizer-grade bone meal had the highest U(VI) sorption capacity on a mass basis. However, potential clogging due to bacterial growth would require 10-fold or greater dilution with coarse grain materials to maintain permeability, thereby decreasing the sorption capacity of a PRB. In addition, the introduction of organic material from a bone meal PRB to the aquifer could adversely affect water quality at a remediation site. The bone charcoal powder also would require dilution to maintain permeability. In addition, the small particles (3–15 μm) could be transported out of the PRB. The high permeability of the CP3 bone char pellets allows direct use in a PRB, offsetting its 5 to 10 times lower U(VI) sorption capacity compared to bone meal and bone charcoal powder. In addition, no clogging of the bone char column was experienced during the 1-year testing period. Based on these results, pelletized bone char was chosen for the field demonstration.

Due to the large quantity ($3\,m^3$) required for the PRB and a different supplier of bone meal, modifications to the firing process were required that resulted in a different formulation of bone char pellets than the CP3 material used for laboratory testing. Material (CP5) used for the PRB demonstration at Fry Canyon, Utah, was composed of 12–15 weight percent aluminum phosphate (as Al_2O_3: P_2O_5, from binder), 10–12% carbon (from bone meal), and 74–76% calcium phosphate (from bone meal). Prior to PRB installation, batch U(VI) sorption and hydraulic conductivity of the CP5 formulation were measured. Little difference in U(VI) sorption in batch experiments was observed compared to the CP3 material (Fig. 1). Permeability was about 35% lower than the CP3 (Table II) but still greatly exceeded the aquifer permeability.

2. Column Tests of Bone Char Pellet Formulation Used in Fry Canyon PRB

Column tests with the CP5 pellets were conducted after barrier emplacement. A lower dissolved U(VI) concentration (2400 μg/liter) was used to better simulate the actual groundwater entering the bone char PRB at Fry Canyon (1400 to 7100 μg/liter; see Chapter 14). Breakthrough ($C/C_o = 0.5$) of U(VI) occurred at 206 ml/g or 253 *cv* (Fig. 6a). Both U(VI) sorption at breakthrough and total sorption capacity for the CP5 material were about half that determined for the CP3 material (Table III). In addition, a significantly higher dissolved [P] was measured for both the initial release and the plateau concentration for the CP5 material (Fig. 6c). Variations in the pellet binder formulation and fertilizer-grade bone meal source, as well as the manufacturing conditions, may account for these differences in U(VI) uptake and phosphate release between these materials. Considerably more fluctuation was observed in the CP5 effluent U(VI) concentration than with CP3 pellets. In contrast, no significant differences in variability of flow rate, pH, or other constituents were observed. The cause of variability in the effluent U(VI) concentration is unknown.

A second column test was conducted to test U(VI) removal after the initial release of [P]. Uranium-free AGW was passed through the bone char pellet column for 80 *cv* (64 ml/g) prior to introduction of U(VI). During this pretreatment, effluent [P] decreased from an initial maximum of 360 to 10 mg/liter within 20 *cv* (16 ml/g, Fig. 6c). The large initial release likely resulted from dissolution of excess binder. A relatively constant [P] (5.4 ± 0.4 mg/liter) was observed during the last 30 *cv* of pretreatment. Breakthrough of U(VI) occurred after about the same volume of influent as the

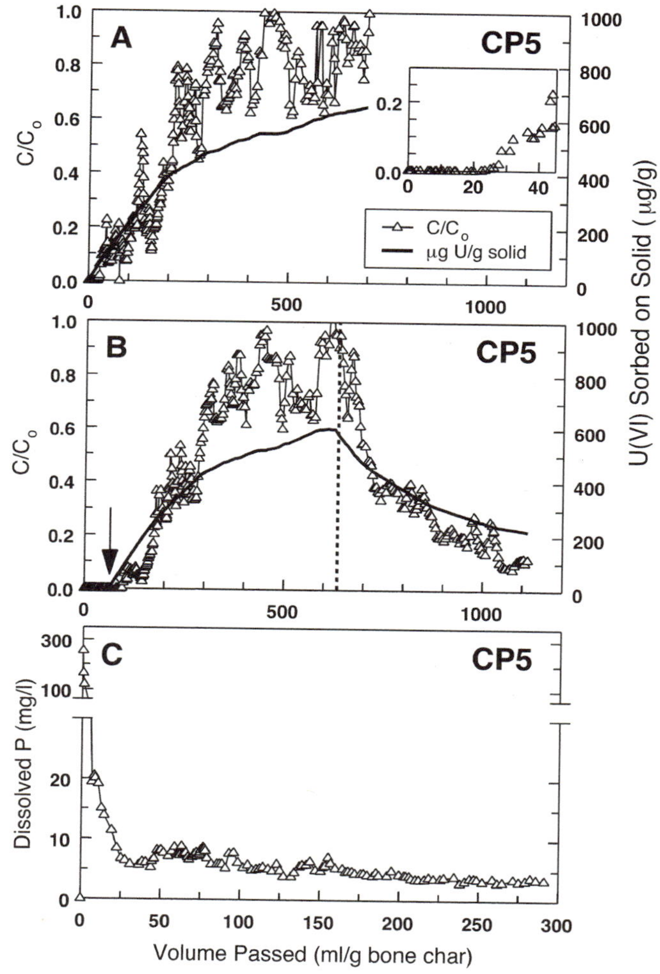

Figure 6 (a) Column breakthrough and reversibility tests for Cercona bone char pellet formulation (CP5) used in the Fry Canyon PRB demonstration. The ratio of effluent-to-influent dissolved uranium (C/C_o, Δ) versus the volume of solution passed in units of milliliter per gram of bone char pellets. Cumulative uranium sorbed per gram of bone char pellets (μg/g) is depicted by the solid line. (b) The effect of pretreatment of bone char pellets with 80 *cv* (64 ml/g) of uranium-free artificial groundwater prior to U(VI) inflow with the arrow denoting the start of U(VI) inflow. The vertical dashed line indicates change of influent to uranium-free artificial ground water to test for the reversibility of uranium sorption. Concentration of uranium is effluent during the reversibility test is the ratio of measured effluent dissolved uranium to the U(VI) concentration of influent during the uptake limb of the experiment. (c) Dissolved phosphorous (mg/liter) in column effluent versus volume of solution passed in units of milliliter per gram of bone char pellets. Data are for column experiment depicted in a.

untreated solid, with comparable total U(VI) sorption capacity (Fig. 6b, Table III). This comparison is consistent with the removal process occurring by sorption on the bone char solid rather than by precipitation with the dissolved phosphate. The extent and rate of U(VI) desorption were similar to that observed for the CP3 pellets (Figs. 5 and 6).

Because U(VI) sorption was largely reversible, a third column test was conducted to determine if the material could be reused after elution of sorbed U(VI). At the end of the reversibility test for CP5, about 220 ppm or about 30% of total U(VI) sorbed remained after 600 *cv* of elution. A portion of the material was repacked into a new column, and AGW with 12 mg/liter U(VI) was passed. An additional 500 ppm U(VI) was sorbed at complete breakthrough, indicating little change in the total U(VI) capacity due to prior sorption and desorption steps (data not shown). These results suggest potential for reuse of the bone char pellets used in a PRB application, which may lower operation and decommissioning costs if the reactive material can be rejuvenated *in situ*.

III. PROCESSES OF U(VI) SORPTION

Knowledge of the contaminant removal process is critical for evaluating the effectiveness of materials used in PRB-based remediation applications. The long-term stability of sorbed contaminants is dependent on the contaminant immobilization process. Ideally, U(VI) sorbed by apatite should remain immobilized with respect to changes in water chemistry to minimize U(VI) rerelease after a contaminant plume has passed through a PRB.

The mechanisms of U(VI) sorption by apatites have been investigated using XAS and synchrotron-source XRD (Fuller *et al.*, 2002a,b). XAS is an element-specific, bulk method that provides quantitative information on the local coordination environment of the element of interest. The energy of X-ray absorption and X-ray absorption near edge structure (XANES) is characteristic of the oxidation state of the element of interest. At energies above the critical absorption energy, photons are ejected from the absorbing atom and scattered by surrounding atoms both constructively and destructively, giving rise to extended X-ray absorption fine structure (EXAFS) absorption spectra. EXAFS spectra are used to derive interatomic distance, coordination numbers, and identity of atoms surrounding the element of interest. The spectral contribution arising from first and second shell coordination of U by low atomic number atoms such as oxygen and phosphorus can be successfully fit from EXAFS spectra of uranyl phosphate and carbonate phases (Thompson *et al.*, 1997). The mechanisms of uranium sorption are then derived from the bonding environment of uranium determined by fitting

EXAFS spectra. Details of data collection, analysis, and interpretation of spectra are presented in Fuller *et al.* (2002a).

The mechanisms of U(VI) sorption by a synthetic HA were evaluated by EXAFS and XRD data for U(VI)-reacted HA samples prepared in batch experiments (Fuller *et al.*, 2001a). In the absence of HA solid, a uranyl phosphate precipitated from solution that was equilibrated with HA and filtered prior to the addition of uranium (pH 6.9, initial [P] 6.5 mg/liter, [Ca] 7.6 mg/liter). The precipitate was identified as chernikovite $H_2(UO_2)_2(PO_4)_2 \cdot 10H_2O$ by XRD. EXAFS spectra and XRD of uranium-reacted HA with > 7000 ppm sorbed U(VI) indicated chernikovite precipitation. In contrast, analyses of EXAFS and XRD data indicated that U(VI)–phosphate, – hydroxide, and – carbonate solids were not present in U-reacted HA at sorbed concentrations ≤4700 ppm U(VI). Instead, fits to EXAFS spectra were consistent with the formation of an innersphere complex U(VI) with either Ca or P sites on the HA surface.

Although solubility constants for chernikovite are subject to uncertainty (Grenthe *et al.*, 1992), the log ion activity product for chernikovite formation ($\log IAP_{CHN} = 2\log\{H^+\} + 2\log\{UO_2^{2+}\} + 2\log\{PO_4^{-3}\}$, where {} are free ion activities) provides an indicator of precipitation of this solubility-limiting phase. The log IAP_{CHN} calculated using aqueous stability constants summarized in Table IV ranged from −54.6 to −55 for the homogeneous precipitation of chernikovite in the absence of HA solid and averaged −56 for

Table IV

Stability Constants for Uranyl Solution Complexes[a]

Reaction	$\log \beta^*(I = 0)$[b]
$UO_2^{2+} + H_2O \Leftrightarrow UO_2OH^+ + H^+$	−5.20
$UO_2^{2+} + 2H_2O \Leftrightarrow UO_2(OH)_{2,\ aq} + 2H^+$	−11.50[c]
$UO_2^{2+} + 3H_2O \Leftrightarrow UO_2(OH)_3^- + 3H^+$	−20.00[d]
$UO_2^{2+} + 4H_2O \Leftrightarrow UO_2(OH)_4^{2-} + 4H^+$	−33.0
$2UO_2^{2+} + H_2O \Leftrightarrow (UO_2)_2OH^{3+} + H^+$	−2.70
$2UO_2^{2+} + 2H_2O \Leftrightarrow (UO_2)_2(OH)_2^{2+} + 2H^+$	−5.62
$3UO_2^{2+} + 4H_2O \Leftrightarrow (UO_2)_3(OH)_4^{2+} + 4H^+$	−11.90
$3UO_2^{2+} + 5H_2O \Leftrightarrow (UO_2)_3(OH)_5^+ + 5H^+$	−15.55
$3UO_2^{2+} + 7H_2O \Leftrightarrow (UO_2)_3(OH)_7^- + 7H^+$	−31.00
$4UO_2^{2+} + 7H_2O \Leftrightarrow (UO_2)_4(OH)_7^+ + 7H^+$	−21.9
$UO_2^{2+} + CO_3^{2-} \Leftrightarrow UO_2CO_{3,\ aq}$	9.67[e]
$UO_2^{2+} + 2CO_3^{2-} \Leftrightarrow UO_2(CO_3)_2^{2-}$	16.94

(*continues*)

Table IV *(continued)*

Reaction	$\log \beta^*(I = 0)^b$
$UO_2^{2+} + 3CO_3^{2-} \Leftrightarrow UO_2(CO_3)_3^{4-}$	21.60
$3UO_2^{2+} + 6CO_3^{2-} \Leftrightarrow (UO_2(_3(CO_3)_6^{6-}$	54.0
$2UO_2^{2+} + CO_3^{2-} + 3H_2O \Leftrightarrow (UO_2)_2CO_3(OH)_3^- + 3H^+$	−0.86
$3UO_2^{2+} + CO_3^{2-} + 3H_2O \Leftrightarrow (UO_2)_3CO_3(OH)_3^+ + 3H^+$	0.66
$11UO_2^{2+} + 6CO_3^{2-} + 12H_2O \Leftrightarrow (UO_2)_{11}(CO_3)_6(OH)_{12}^{2-} + 12H^+$	36.43
$H^+ + CO_3^{2-} \Leftrightarrow HCO_3^-$	10.329
$2H^+ + CO_3^{2-} \Leftrightarrow H_2CO_3^*$	16.683
$CO_{2(g)} + H_2O \Leftrightarrow H_2CO_3^*$	−1.472
$UO_2^{2+} + NO_3^- \Leftrightarrow UO_2NO_3^+$	0.3
$UO_2^{2+} + Cl^- \Leftrightarrow UO_2Cl^+$	0.17
$UO_2^{2+} + 2Cl^- \Leftrightarrow UO_2Cl_{2,\ aq}$	−1.1
$UO_2^{2+} + SO_4^{2-} \Leftrightarrow UO_2SO_{4,\ aq}$	3.15
$UO_2^{2+} + 2SO_4^{2-} \Leftrightarrow UO_2(SO_4)_2^{2-}$	4.14
$SO_4^{2-} + H^+ \Leftrightarrow HSO_4^-$	1.98[e]
$UO_2^{2+} + PO_4^{3-} \Leftrightarrow UO_2PO_4^-$	13.23
$UO_2^{2+} + PO_4^{3-} + H^+ \Leftrightarrow UO_2HPO_{4,\ aq}$	19.59
$UO_2^{2+} + PO_4^{3-} + 2H^+ \Leftrightarrow UO_2H_2PO_4^+$	22.82
$UO_2^{2+} + PO_4^{3-} + 3H^+ \Leftrightarrow UO_2H_3PO_4^{2+}$	22.46
$UO_2^{2+} + 2PO_4^{3-} + 4H^+ \Leftrightarrow UO_2(H_2PO_4)_{2,\ aq}$	44.04
$UO_2^{2+} + 2PO_4^{3-} + 5H^+ \Leftrightarrow UO_2(H_2PO_4)(H_3PO_4)^+$	45.05
$PO_4^{3-} + H^+ \Leftrightarrow HPO_4^{2-}$	12.35[e]
$PO_4^{3-} + 2H^+ \Leftrightarrow H_2PO_4^-$	19.562[e]
$PO_4^{3-} + 3H^+ \Leftrightarrow H_3PO_{4,\ aq}$	21.702[e]

[a]Values from Grenthe *et al.* (1992) unless indicated otherwise.
[b]Values are for zero ionic strength.
[c]Values from Silva (1992).
[d]Values from Sandino and Bruno (1992).
[e]Values from Silva *et al.* (1995).

HA samples with >10,000 ppm sorbed U(VI) with chernikovite identified (Fuller *et al.*, 2002a). In batch sorption samples with sorbed concentrations ≤4700 ppm U(VI), the log IAP_{CHN} was ≤ -58, consistent with the conclusion from the fits to EXAFS spectra that surface complexation effectively outcompeted the precipitation of chernikovite. In these samples, the initial dissolved uranium concentrations (≥ 7600 μg/liter) resulted in log $IAP_{CHN} > -52$, indicating that the initial conditions greatly exceeded chernikovite solubility, but precipitation was not observed.

The U(VI) sorption processes by bone meal and bone charcoal were also determined using XAS and XRD data (Fuller *et al.*, 2002b). The U-L_{III} X-ray absorption edge positions and XANES spectra of U(VI)-reacted bone meal, bone charcoal, and bone char pellets closely matched those for U(VI) model compounds. These results indicate that sorbed uranium remained primarily in the +6 oxidation state with no significant reduction, despite the high organic C content and potential for bacterial growth of bone meal (Fuller *et al.*, 2001b).

Qualitatively, EXAFS spectra of U(VI)-reacted bone meal, bone charcoal, and bone char pellets did not match uranyl phases such as phosphates (e.g., autunite), carbonates, or oxides for sorbed concentrations of <4000 ppm. Fits to these data indicated a split or a highly disordered oxygen equatorial shell common to all samples and consistent with the formation of a uranyl surface complex, as observed previously for synthetic hydroxyapatite (Fuller *et al.*, 2001a) and other absorbents (Bargar *et al.*, 2000; Sylwester *et al.*, 2000). The elevated carbonate concentration of AGW resulted in ternary complexes of carbonate with the uranyl surface complex, as observed for U(VI) adsorbed on hematite (Bargar *et al.*, 2000). At higher sorbed U(VI) concentrations (>4000 ppm), analysis of EXAFS spectra for bone meal and bone char indicates that sorption includes a component of uranyl phosphate precipitation, likely chernikovite, as observed for the synthetic HA system. Chernikovite precipitation was detected in batch samples of CP3 bone char pellets with sorbed concentrations in excess of the sorption capacity determined in column experiments. This may be because the CP3 material used for these XAS measurements was ground prior to U(VI) sorption.

Because the solid to water ratio is much greater in a PRB (380 g/liter for the Fry Canyon apatite PRB) than in batch sorption samples prepared for XAS (1 to 10 g/liter) and dissolved U(VI) is lower, the surface complexation process likely would dominate U(VI) sorption in a PRB because of the much greater surface site concentration. This can be illustrated by comparing the log IAP_{CHN} for batch and column experiments to the log IAP_{CHN} for chernikovite precipitation in the presence of synthetic HA (−56, Fuller *et al.*, 2001a). For example, a log IAP_{CHN} of −56 was calculated for the CP3 batch sample (12 mg/liter initial dissolved uranium) after U(VI) sorption. Precipitation of chernikovite in this sample was evident from the EXAFS spectrum. In contrast, the log IAP_{CHN} for the effluent from the CP3 bone char pellet column (12 mg/liter influent dissolved uranium, Fig. 5) ranged from −58.4 at 10% breakthrough to −57 at 100% breakthrough. These values are consistent with U(VI) sorption in the column by surface complexation instead of chernikovite precipitation observed in the batch system. This difference may result from the higher surface site concentration in the column compared to the batch system, despite initial [P] during the first 20 ml/g of flow (Fig. 5b) that results in exceeding chernikovite solubility for the influent dissolved U(VI). Subsequently, [P] decreases to concentrations where the log IAP_{CHN} is below the solubility of chernikovite even at 100% U(VI) breakthrough.

IV. PERFORMANCE OF THE FRY CANYON BONE CHAR APATITE PRB

A. Lifetime Estimate

The column test with the CP5 material can be used to estimate the potential lifetime of the PRB at Fry Canyon for comparison with the measured performance in the field demonstration. A flow rate of 324 liter/day into the bone char PRB was calculated from ionic tracer tests and is equivalent to 173 PRB pore volumes over the 1130 days of operation assuming a total porosity of 0.7 (see Chapter 14). At this flow rate, 50% breakthrough would occur after 1646 days (4.5 years) for a 1.4-mg/liter dissolved U(VI) inflow assuming breakthrough occurs at 252 pore volumes as observed in the column test (Table III). The lifetime estimate is an upper limit because the U(VI) concentration used (1400 μg/liter) is at the low end of the measured range entering the PRB. Based on the measured decrease in U(VI) within the PRB, total U(VI) removal for the bone char PRB was estimated at 270 g of U(VI) over 1130 days of operation (see Chapter 14). This calculation assumed that only the right half of the PRB had flow-through conditions because of the oblique angle of the PRB to the average groundwater flow. The estimated U(VI) removal equals about 350 ppm, assuming sorption over the entire mass of bone char in the right half of the PRB. This sorbed concentration is about half of the total sorption capacity of the solid measured in column tests (650 ppm). Both the estimated lifetime to 50% breakthrough and the sorbed concentration indicate that the PRB is approaching failure. In contrast, dissolved U(VI) within the PRB at the end of this period was less than 5% of influent concentration. Although the observed barrier performance appears much better than these estimates, complexities in flow paths suggest a large amount of uncertainty in the estimate of volumetric flow through the PRB (see Chapter 14).

B. Analysis of Bone Char Apatite From the PRB

Two core samples of the bone char pellets were recovered from the PRB in May 1999, after 18 months of operation, using a Geoprobe corer fitted with a hydraulic hammer. Because the corer entered the PRB at a 40° angle relative to land surface, the location of the cores within the PRB could not be determined reliably and do not represent profiles into the PRB or along flow paths. The plastic liner containing the sample was capped immediately upon recovery and was stored at 4°C until analysis. No attempt was made to isolate PRB from oxygen. Dissolved oxygen ranged from 3.9 to 7.5 mg/liter in water samples collected 2 weeks prior to coring from samplers located 15 cm from the upgradient face of the PRB. The presence of pea gravel adjacent to bone char indicates cores sampled within the first 10 cm of the upgradient face of the PRB. Pea gravel was placed on the upgradient side of the PRB during installation to facilitate uniform flow into the PRB (see Chapter 14). Cores were sectioned for bulk chemical and XAS analyses within 3 weeks of collection. A split of each core interval containing the bone char pellets was air dried, ground, and extracted with 1 *N* HNO_3 for 24 h. Samples were centrifuged and supernatants analyzed for U(VI) by KPA. XAS spectra were collected on ground, wet samples following methods described in Fuller *et al.* (2001a).

The total U(VI) concentration ranged from 180 to 690 ppm for the recovered bone char pellets. The highest concentration sample was from within 5 cm of the upgradient face and had a U(VI) concentration (690 ppm) approximately equal to the maximum U(VI) sorbed in column tests for the CP5 pellets (650 ppm). Because of heterogeneity in flow into the PRB and incomplete coring of the PRB, total U(VI) removal calculated from the volume treated (see Chapter 14) cannot be compared to the U(VI) concentration of the bone char pellets recovered from the PRB.

XAS measurements of the PRB core samples indicate that sorbed uranium was in the +6 oxidation state at the time of analysis. Fits to EXAFS spectra were consistent with the surface complexation of uranium by the apatite with no evidence of precipitation of chernikovite or other uranyl phosphate solid phases (Fuller *et al.*, 2002b). Because uranyl phosphates were not detected in the PRB sample that had total sorbed U(VI) equal to the capacity measured in column experiments, surface complexation likely was the dominant removal process at the dissolved U(VI) concentration entering the PRB. In contrast to this result, the log IAP for chernikovite formation calculated using the measured dissolved U(VI) in the most upgradient PRB sampler closest to the core locations ranged from −56.1 to −56.2 during the first 4 months of operation when measured [P] was 10 mg/liter. These log IAP_{CHN} values are approaching the log IAP_{CHN} observed for chernikovite precipitation in the presence of HA (−56; Fuller *et al.*, 2001a), suggesting that chernikovite precipitation may have occurred, but was not detected in the samples recovered from the PRB. However, the IAP_{CHN} calculated for dissolved U(VI) and [P] measured at this sampler (−58.7) 1 week prior to coring was much lower than the log IAP for chernikovite precipitation, consistent with the conclusion from EXAFS data that U(VI) removal in the PRB occurred largely by surface complexation.

V. CONCLUDING REMARKS

Bone meal and bone charcoal were much more effective than phosphate rock for the removal of U(VI) from groundwater. Pelletized bone charcoal was chosen for the PRB demonstration because this material could be used directly without dilution to obtain the required permeability. U(VI) sorption by this material was intermediate to other materials tested. The release of phosphate from the apatite material may be detrimental to down gradient water quality if sorption by aquifer material does not sequester phosphate. U(VI) sorption by apatite was dominated by surface complexation at concentrations relevant to contaminated groundwaters. Because EXAFS measurements of bone char pellets recovered from the Fry Canyon PRB indicated that U(VI) removal had occurred by surface complexation, U(VI) breakthrough likely will occur when adsorption equilibrium is reached over the entire flow path through the PRB. As a result, an apatite-based PRB will require close monitoring and removal of the solid phase because of breakthrough when adsorption equilibrium is attained.

VI. REFERENCES

Arey, J. S., Seaman, J. C., and Bertsch, P. M. (1999). Immobilization of uranium in contaminated sediments by hydroxyapatite addition. *Environ. Sci. Technol.* **33**, 337–342.

Bargar, J. R., Reitmeyer, R., Lenhart, J. J., and Davis, J. A. (2000). Characterization of U(VI)-carbonato ternary complexes on hematite: EXAFS and electrophoretic mobility measurements. *Geochim. Cosmochim. Acta* **64**, 2737–2749.

Benner, S. G., Herbert, R. B., Blowes, D. W., Ptacek, C. J., and Gould, D. (1999). Geochemistry and microbiology of a permeable reactive barrier for acid mine drainage. *Environ. Sci. Technol.* **33**, 2793–2799.

Blowes, D. W., Ptacek, C. J., Benner, S. G., McRae, C. W. T., Bennett, T. A., and Puls, R. W. (2000). Treatment of inorganic contaminants using permeable reactive barriers. *J. Contamin. Hydrol.* **45**, 123–137.

Blowes, D. W., Ptacek, C. J., and Jambor, J. L. (1997). In-situ remediation of Cr(VI)-contaminated groundwater using permeable reactive walls: Laboratory studies. *Environ. Sci. Technol.* **31**, 3348–3357.

Bostick, W. D., Stevenson, R. J., Jarabek, R. J., and Conca, J. L. (1999). Use of apatite and bone char for the removal of soluble radionuclides in authentic and simulated DOE groundwater. *Adv. Environ. Res.* **3**, 488–498.

Collin, R. L. (1960). Strontium-calcium hydroxyapatite solid solutions precipitated from basic, aqueous solutions. *J. Am. Chem. Soc.* **82**, 5067–5069.

Davis, J. A., and Kent, D. B. (1990). Surface complexation modeling in aqueous geochemistry. *In* "Mineral–Water Interface Geochemistry" (M. F. Hochella, and A. F. White, eds.), pp. 177–248. Mineralogical Society of America, Washington, DC.

Fuller, C. C., Bargar, J. R., Davis, J. A., and Piana, M. J. (2002a). Mechanisms of uranium interactions with apatite: Implications for ground-water remediation. *Environ. Sci. Technol.* **36**, 158–165.

Fuller, C. C., Bargar, J. R., Davis, J. A., Piana, M. J., and Kohler, M. (2001b). Mechanisms of uranium interactions with apatite: Evaluation of sorption by bone meal and bone charcoal. Submitted for Publication.

Grenthe, I., Fuger, J., Konings, R. J. M., Lemire, R. J., Muller, A. B., Nguyen-Trung, C., and Wanner, H. (1992). "Chemical Thermodynamics 1: Chemical Thermodynamics of Uranium." North-Holland, Elsevier, Amsterdam.

Gu, B., Liang, L., Dickey, M. J., Yin, X. and Dai, S. (1998). Reductive precipitation of uranium(VI) by zero-valent iron. *Environ. Sci. Technol.* **32**, 3366–3373.

Hodson, M. E., Valsami-Jones, E., and Cotter-Howells, J. D., (2000). Bone meal additions as remediation treatment for contaminated soil. *Environ. Sci. Technol.* **34**, 3501–3507.

Jeanjean, J., Rouchard, J. C., and Fedoroff, T. M. (1995). Sorption of uranium and other heavy metals on hydroxyapatite. *J. Radioanal. Nuc.Chem. Lett.* **201**, 529–539.

Langmuir, D. (1978). Uranium solution-mineral equilibria at low temperatures with applications to sedimentary ore deposits. *Geochim. Cosmochim. Acta* **42**, 547–569.

Laperche, V., and Traina, S. J. (1998). Immobilization of Pb by hydroxyapatite. *In* "Adsorption by Geomedia" (E. A. Jenne, ed.), pp. 255–276. Academic Press, San Diego.

Ma, Q. Y., Traina, S. J., and Logan, T. J. (1993). In situ lead immobilization by apatite. *Environ. Sci. Technol.* **27**, 1803–1810.

Ma, Q. Y., Logan, T. J., and Traina, S. J. (1995). Lead immobilization from aqueous solutions and contaminated soils using phosphate rocks. *Environ. Sci. Technol.* **29**, 1118–1126.

Ma, Q. Y., Logan, T. J., Traina, S. J., and Ryan, J. A. (1994). The effects of NO_3^-, Cl^-, SO_4^{2-} and CO_3^{2-} on Pb^{2+} immobilization by hydroxyapatite. *Environ. Sci. Technol.* **28**, 408–418.

Middleburg, J. J., and Comans, R. N. J. (1991). Sorption of cadmium on hydroxyapatite. *Chem. Geol.* **90**, 45–53.

Morrison, S. J., Metzler, D. R., and Carpenter, C. E. (2001). Uranium precipitation in a permeable reactive barrier by progressive irreversible dissolution of zerovalent iron. *Environ. Sci. Technol.* **35**, 385–390.

Morrison, S. J., and Spangler, R. R. (1993). Chemical barriers for controlling groundwater contamination. *Environ. Prog.* **12**, 175–181.

Naftz, D. L., Morrison, S. J., Feltcorn, E. M., Freethey, G. W., Fuller, C. C., Piana, M. J., Wilhelm, R. G., Rowland, R. C., Davis, J. A., and Blue, J. E. (2000). Field demonstration of permeable reactive barriers to remove dissolved uranium from groundwater, Fry Canyon, Utah, September 1997 through September 1998, Interim report. U.S. Environmental Protection Agency Report EPA 402-C-00-001.

Ordonez-Regil, E., Romero Guzman, E. T., and Ordonez-Regil, E. (1999). Surface modification in natural fluoroapatite after uranyl solution treatment. *J. Radioanal. Nucl. Chem.* **240**, 541–545.

Ruby. M. V., Davis, A., and Nicholson, A. (1994). In situ formation of lead phosphates in soils as a method to immobilize lead. *Environ. Sci. Technol.* **28**, 646–654.

Sandino, A., and Bruno, J. (1992). The solubility of $(UO_2)_3(PO_4)_2 \cdot 4H_2O(s)$ and the formation of U(VI) phosphate complexes: Their influence in uranium speciation in natural waters. *Geochim. Cosmochim. Acta* **56**, 4135–4145.

Shoemaker, S. H., Greiner, J. F., and Gillham, R. W. (1995). Permeable reactive barriers. *In* "Assessment of Barrier Containment Technologies" (R. R., Rumer, and J. K. Mitchell, eds.), pp. 310–348. Proceedings of the International Containment Technology Workshop, Baltimore, MD.

Silva, R. J., (1992). Mechanisms for the retardation of uranium(VI) migration. *Mat. Res. Soc. Symp. Proc.* **257**, 323–330.

Silva, R. J. Bidoglio, G., Rand, M. H., Robouch, P. B., Wanner, H., and Puigdomenech, I. (1995). "Chemical Thermodynamics 2: Chemical Thermodynamics of Americium." North-Holland, Elsevier, Amsterdam.

Sowder, A. G., Clark, S. B., and Fjeld, R. A. (1996). The effect of silica and phosphate on the transformation of schoepite to becquerelite and other uranyl phases. *Radiochim. Acta* **74**, 45–49.

Sylwester, E. R., Hudson, E. A., and Allen, P. G. (2000). The structure of U(VI) complexes on silica, alumina, and montmorillonite. *Geochim. Cosmochim. Acta* **64**, 2431–2438.

Thompson, H. A., Brown, G. E., and Parks, G. A. (1997). XAFS spectroscopic study of uranyl coordination in solids and aqueous solution. *Am. Min.* **82**, 483–496.

Valsami-Jones, E., Ragnarsdottir, K. V., Putnis, A., Bosbach, D., Kemp, A. J., and Cressey, G. (1998). The dissolution of apatite in the presence of aqueous metal cations at pH 2–7. *Chem. Geol.* **151**, 215–233.

Xu, Y., and Schwartz, F. W. (1994). Lead immobilization by hydroxyapatite in aqueous solutions. *J. Contamin. Hydrol.* **15**, 187–206.

Xu, Y., Schwartz, F. W., and Traina, S. J. (1994). Sorption of Zn^{2+} and Cd^{2+} on hydroxyapatite surfaces. *Environ. Sci. Technol.* **28**, 1472–1480.

Zhang, P, Ryan, J. A., and Bryndzia, L. T. (1997). Pyromorphite formation from goethite adsorbed lead. *Environ. Sci. Technol.* **31**, 2673–2678.

Chapter 10

Sulfate-Reducing Bacteria in the Zero Valent Iron Permeable Reactive Barrier at Fry Canyon, Utah

Ryan C. Rowland
U.S. Geological Survey, Salt Lake City, Utah 84119

A zero valent iron (ZVI) permeable reactive barrier (PRB) was installed in a shallow, colluvial aquifer contaminated with uranium in Fry Canyon, Utah, in September 1997. Aerobic and anaerobic iron corrosion reactions in the ZVI PRB have created a highly reducing, oxygen-depleted, and hydrogen gas-enriched geochemical environment in the PRB that is favorable for sulfate-reducing bacteria (SRB). Stable sulfur isotope, microbiologic, and geochemical evidence indicates that SRB are active in the ZVI PRB. The value of $\delta^{34}S$ in sulfate increased from −3.6 to −1.4 ± 0.4 permil (‰) in water from deep wells along an assumed flow path and from 0.7 to 3.3 ± 0.4 ‰ in water from shallow wells along the same flow path. $\delta^{34}S$ data for sulfate and sulfide in water from a well in the center of the PRB indicate that sulfate is enriched in $\delta^{34}S$ 23.6 ‰ relative to sulfide. Stable sulfur isotope and sulfate concentration data indicate that Rayleigh-type distillation is occurring in the ZVI PRB with a resulting isotopic enrichment factor of −9.0 ‰. Enumeration of SRB in groundwater by using direct plate counts indicates that the population of SRB is two orders of magnitude larger in the ZVI PRB relative to an upgradient well. Inverse geochemical modeling coupled with groundwater velocity measurements indicates that 32.6 cm^3 of iron sulfide precipitates in the first 0.15 m of the modeled flow path in the ZVI PRB each year, corresponding to 12% of its void space. Despite isotopic, microbiologic, and

Handbook of Groundwater Remediation Using Permeable Reactive Barriers

geochemical evidence, sulfide phases were not detected on material cored from the ZVI PRB in May 1999.

I. INTRODUCTION

The geochemical environment in zero valent iron (ZVI) permeable reactive barriers (PRBs) is strongly reducing, oxygen depleted, alkaline, and enriched in hydrogen gas (Naftz *et al.*, 2000; Puls *et al.*, 1999; Sass *et al.*, 1998; Reardon, 1995). These conditions are a result of the following aerobic and anaerobic iron corrosion reactions (U.S. Environmental Protection Agency, 1998; Reardon, 1995):

$$Fe^{0}{}_{(s)} + H_2O_{(l)} + 1/2\,O_{2(aq)} = Fe^{2+}{}_{(aq)} + 2\,OH^{-}{}_{(aq)} \tag{I}$$

$$Fe^{0}{}_{(s)} + 2\,H_2O_{(l)} = Fe^{2+}{}_{(aq)} + H_{2(g)} + 2\,OH^{-}{}_{(aq)}. \tag{II}$$

During aerobic iron corrosion (reaction I), one mole of ZVI is oxidized by dissolved oxygen (DO), producing 1 mol of ferrous iron (Fe^{2+}) and 2 mol of hydroxide (OH^-). In the absence of DO, 1 mol of ZVI reduces water (reaction II) forming 1 mol of Fe^{2+}, 1 mol of hydrogen gas (H_2), and 2 mol of OH^-. Both reactions are driven in part by the high standard oxidation potential (E^{o}_{ox}) of the following iron oxidation half-cell reaction:

$$Fe^0 = Fe^{2+} + 2\,e^-, \tag{III}$$

where E^{o}_{ox} of reaction III equals 440 mV (Sherman *et al.*, 2000). In other words, ZVI is a strong reductant whose outer shell electrons will transfer readily to species susceptible to reduction, including many groundwater contaminants. ZVI has been shown to reduce redox-sensitive metal contaminants and radionuclides such as chromium (Cr^{6+}) (Puls *et al.*, 1999), and uranium (U^{6+}) (Gu *et al.*, 1998), respectively, to less soluble forms. Several classes of organic contaminants, including many halogenated organic (Orth and Gillham, 1996; Sayles *et al.*, 1997; Eykholt and Davenport, 1998; Kiilerich *et al.*, 2000) and nitroaromatic (Singh *et al.*, 1999; Devlin *et al.*, 1998) compounds, are reduced to less toxic or nontoxic forms by ZVI.

The contaminant reduction reactions are believed to be surface mediated (Weber, 1996; Gu *et al.*, 1998); therefore, the rate of contaminant removal and total mass of contaminant removed by ZVI PRBs strongly depends on the amount of reactive ZVI surface area in contact with contaminated groundwater (Sherman *et al.*, 2000). It follows that the long-term performance of a ZVI PRB can be shortened by chemical reactions that reduce the reactive surface area. The alkaline conditions created in ZVI PRBs favor the precipitation of carbonate minerals such as aragonite ($CaCO_3$) and siderite

($FeCO_3$), as shown in reactions IV and V, respectively (Vogan *et al.*, 1998; McMahon *et al.*, 1999; Naftz *et al.*, 2000; see Chapter 15).

$$Ca^{2+}{}_{(aq)} + HCO^{-}{}_{3(aq)} \rightarrow CaCO_{3(s)} + H^{+}{}_{(aq)} \quad (IV)$$

$$Fe^{2+}{}_{(aq)} + HCO^{-}{}_{3(aq)} \rightarrow FeCO_{3(s)} + H^{+}{}_{(aq)} \quad (V)$$

In addition, iron corrosion by-products such as $Fe(OH)_2$, akaganeite (β-FeOOH), goethite (α-FeOOH), and green rust II [$4\,Fe(OH)_2 \bullet 2\,FeOOH \bullet FeSO_4 \bullet 4\,H_2O$] are observed in ZVI PRB cores and/or laboratory columns of ZVI and may inhibit ZVI reactivity (Phillips *et al.*, 2000; Puls *et al.*, 1999; Gu *et al.*, 1999). However, little is known about biogeochemical mechanisms that could reduce reactive surface area in ZVI PRBs. This could be a concern because the geochemical environment in ZVI PRBs is ideal for a class of bacteria known collectively as sulfate-reducing bacteria (SRB).

SRB are heterotrophic, obligate anaerobic bacteria that favor strongly reducing conditions [< -200 mV, relative to the standard hydrogen reference electrode (SHE)] (Postgate, 1984). They have been detected in numerous aquatic environments, including deep (Jones *et al.*, 1989; Olson *et al.*, 1981) and shallow aquifers (Martino *et al.*, 1998; Jakobsen *et al.*, 1999). In oxygenated water, SRB will grow in anaerobic microniches or in complex bacterial consortia consisting of aerobic bacteria overlying anaerobic bacteria (Postgate, 1984; Chapelle, 1993). SRB use short-chained organic carbon, such as lactate or H_2 as electron donors, and sulfate anions ($SO_4{}^{2-}$) as electron acceptors in their metabolic pathway (Postgate, 1984). The overall reactions for respiration with lactic acid and H_2, respectively, are

$$\begin{aligned}&2\,CH_3CHOHCOOH_{(aq)} + SO_4{}^{2-}{}_{(aq)} \rightarrow \\ &CH_3COOH_{(aq)} + H_2S_{(g)} + 2\,HCO_3{}^{-}{}_{(aq)}, \text{ and}\end{aligned} \quad (VI)$$

$$4\,H_{2(g)} + SO_4{}^{2-}{}_{(aq)} \rightarrow S^{2-}{}_{(aq)} + 4\,H_2O_{(l)}. \quad (VII)$$

Reactions VI and VII are examples of dissimilatory sulfate reduction (DSR), where $SO_4{}^{2-}$ is completely reduced to sulfide (S^{2-}) by the bacteria, which is released back into the environment. This should not be confused with assimilatory sulfate reduction, where SRB reduce $SO_4{}^{2-}$ for use in proteins that are needed for various functions by the organisms. Regardless of the electron donor used in DSR, 1 mol of S^{2-} is generated for every mol of $SO_4{}^{2-}$ that is reduced. The S^{2-} produced by SRB will react readily with metal cations such as Fe^{2+}, cadmium (Cd^{2+}), copper (Cu^{2+}), nickel (Ni^{2+}), lead (Pb^{2+}), and zinc (Zn^{2+}) to produce sparingly soluble sulfide minerals, a process that has been used in PRBs made of organic matter to immobilize metal contaminants in acid mine drainage (Benner *et al.*, 1999; see Chapter 17). Aerobic iron corrosion in ZVI PRBs creates the strongly reducing and oxygen-depleted environment required by SRB. Anaerobic iron corrosion provides a suitable electron donor for SRB, H_2. If SO_4^{2-} and SRB are present in the aquifer containing the ZVI PRB, it is likely that a population of SRB will establish itself in the PRB and begin DSR with the concomitant precipitation of iron sulfide (FeS):

$$Fe^{2+}{}_{(aq)} + SO_4{}^{2-}{}_{(aq)} + 4\,H_{2(g)} \overset{SRB}{\rightarrow} FeS_{(s)} + 4\,H_2O_{(aq)}. \qquad (VIII)$$

The FeS created in reaction VIII may form on the surface of ZVI particles in a ZVI PRB; therefore, this biogeochemical reaction could represent an additional mechanism for the loss of reactive surface area and porosity in ZVI PRBs.

The biogeochemical mechanism shown in reaction VIII is receiving more attention in the PRB community because mineralogic and microbiologic data from several pilot and field-scale ZVI PRBs indicate the presence and activity of SRB. For example, Phillips *et al.* (2000) found iron sulfides as coatings and infilings on ZVI material from a PRB in the uranium plume at the Y-12 site in Oak Ridge, Tennessee. Analysis of phospholipid fatty acids on ZVI core samples revealed that SRB were present in the ZVI PRB and were probably responsible for formation of the sulfides. FeS was found on ZVI filings from a funnel-and-gate PRB at Moffett Airfield in Mountain View, California, and was attributed to enhanced anaerobic microbial activity in the PRB (Sass *et al.*, 1998). Liang *et al.* (1997) detected iron sulfides on Fe filings deployed in canisters used to treat groundwater contaminated with trichloroethylene in Piketon, Ohio. Gu *et al.* (1999) showed that SRB will colonize ZVI filings and actively reduce SO_4^{2-} in laboratory columns. Solid-phase analysis at the end of the experiment showed that FeS had formed on the ZVI filings, and most probable number enumeration of SRB documented elevated populations in the ZVI filings.

In February 1999, SRB were detected in water samples from the ZVI PRB in Fry Canyon, Utah. The ZVI PRB is part of a demonstration project carried out by the U.S. Geological Survey (USGS), Water Resources Division, with funding from the U.S. Environmental Protection Agency, Office of Air and Radiation (Naftz *et al.*, 2000). The purpose of the demonstration project is to compare the U removal efficiencies of three PRB materials installed in a shallow (< 6 m) colluvial aquifer. To assess the presence, activity, and potential impacts of SRB in the ZVI PRB, a four-step study was undertaken. The objective of this chapter is to present the results of this study and to discuss their implications for the long-term performance of the ZVI PRB in Fry Canyon.

II. SITE DESCRIPTION

The Fry Canyon demonstration site is located in southeastern Utah at an abandoned uranium upgrader facility (Fig. 1). About 36,000 metric tons of uranium tailings containing 0.02% U_2O_3 were left at the site after upgrader operations ceased in 1960 (see Chapter 14). Chemical weathering of the tailings followed by infiltration have resulted in elevated U concentrations in the shallow colluvial aquifer beneath the site. Pre-PRB installation monitoring done in 1996 and early 1997 showed dissolved U concentrations to be between 110 and 16,300 μg/liter in wells completed in the tailings area; dissolved U concentrations ranged between 60 and 80 μg/liter in a well upgradient of the tailings area. Groundwater quality data collected from well FC-3 (Fig. 1) during this same monitoring phase showed that the aquifer is rich in calcium (Ca^{2+}) and SO_4^{2-} (Fig. 2), oxidizing [average Eh = 349 mV (SHE)], near neutral pH, and aerobic (average DO = 1 mg/liter). Select groundwater and surface water quality data collected during the duration of the demonstration project can be found in Wilkowske *et al.* (2001). Aquifer properties such as hydraulic conductivity, transmissivity, and porosity can be found in Chapter 14.

The PRB was installed adjacent to Fry Creek, an ephemeral stream on the north end of the tailings area (Fig. 1). A funnel-and-gate PRB design was used for the demonstration project. This design consists of three permeable gates, each filled with a different PRB material, impermeable walls between the permeable gates, and impermeable wing walls at each end of the structure to channel groundwater through the permeable gates. ZVI was placed in the center gate, and for simplicity the center gate will be referred to as the ZVI PRB in the remainder of this report. The other two PRB materials are amorphous ferric oxyhydyroxide (AFO) and bone charcoal phosphate (PO_4) (see Chapter 9). The funnel-and-gate PRB was keyed into the Permian-age Cedar Mesa Sandstone underlying the colluvial aquifer. A 0.46-m-wide pea gravel layer was placed on the upgradient side of the PRB to facilitate uniform flow of U-contaminated groundwater through the gates (Fig. 3). The ZVI PRB is 2.1 m wide by 0.91 m long by 1.1 m deep. The ZVI PRB contains an array of 0.5-cm-diameter polyvinyl chloride (PVC) monitoring wells screened at deep (0 to 0.15 m above the Cedar Mesa Sandstone) and shallow (0.41 to 0.56 m above the Cedar Mesa Sandstone) depths. Two 5-cm-diameter PVC monitoring wells were placed in the center of each gate and screened from 0.1 to 1.52 m above the Cedar Mesa Sandstone. Two 5-cm-diameter PVC wells, screened from 0.1 to 1.52 m above the Cedar Mesa Sandstone, were installed in the gravel pack in front of the ZVI PRB and about 1.5 m downgradient of the ZVI PRB.

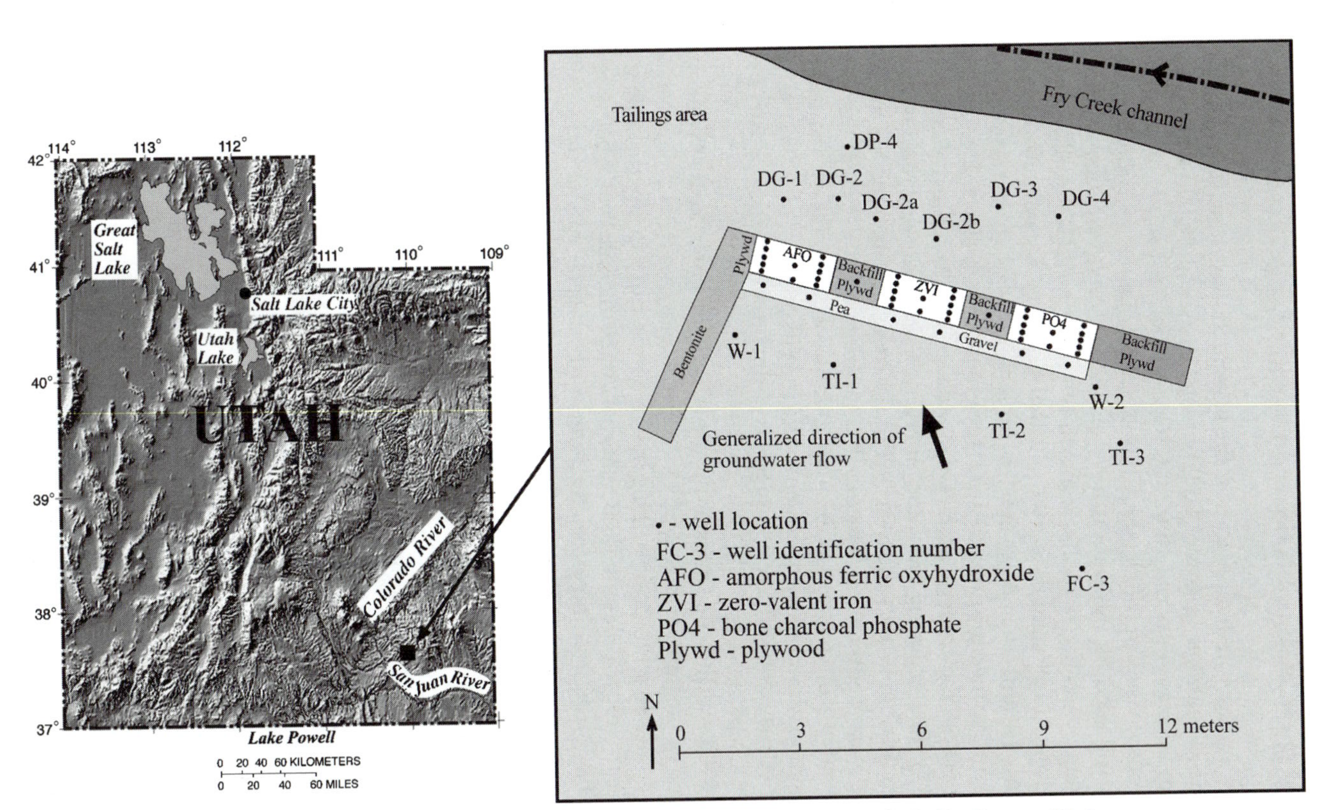

Figure 1 Location and sketch map of the Fry Canyon demonstration site in Fry Canyon, Utah.

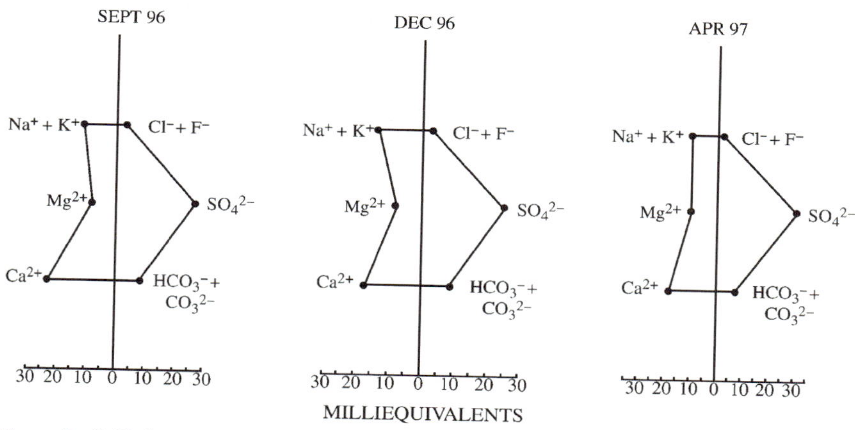

Figure 2 Stiff diagrams showing the relative amounts of major ions in water from well FC-3 during the prepermeable reactive barrier installation monitoring phase at Fry Canyon, Utah.

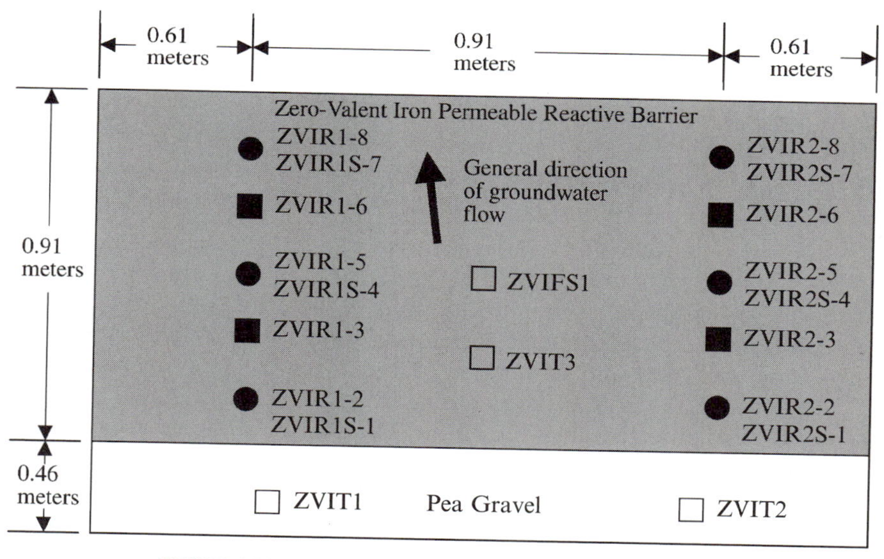

Figure 3 Schematic diagram showing monitoring well location and designation in the ZVI permeable reactive barrier, Fry Canyon, Utah. The drawing is not to scale.

The ZVI used in the ZVI PRB was in the form of porous, well-rounded pellets with diameters ranging from 0.3 to 2.0 mm. The hydraulic conductivity of the ZVI was about 10 times that of the native aquifer at the time of installation (see Chapter 14). Groundwater velocity and direction were measured periodically in the ZVI PRB with a heat-pulse flow meter. Groundwater velocities ranged from less than 1 m/day to greater than 5 m/day, and flow directions were north to northwest (see Chapter 14). Bromide (Br^-) tracer tests conducted in April 1999 yielded an average groundwater velocity of 0.7 m/day in the ZVI PRB (see Chapter 14).

III. METHODOLOGY

A four-step study was undertaken to assess the presence, activity, and impacts of SRB in the ZVI PRB. This section describes each step. The use of trade, product, industry, or firm names is for descriptive purposes only and does not imply endorsement by the U.S. Government.

A. Stable Sulfur Isotopes

Water samples for SO_4^{2-} sulfur isotopic analysis were collected in June 2000 from select wells in and near the ZVI PRB. One liter of water was purged from the 0.5-cm-diameter monitoring wells in the ZVI PRB, 3.8 liters was purged from the 5-cm-diameter monitoring wells in the ZVI PRB and the adjacent gravel pack, and 10 liters was purged from the 5-cm-diameter wells located outside the ZVI PRB and adjacent gravel pack. Limited volumes of water were purged from wells in the ZVI PRB and the adjacent gravel pack because of their close proximity to one another. One liter of unfiltered water was collected in field-rinsed polyethylene bottles after each well was purged. Samples were acidified to about pH 3 with ultrapure hydrochloric acid (HCl). After acidification, samples were bubbled vigorously with ultrapure nitrogen gas (N_2) for 40 min to strip hydrogen sulfide gas (H_2S). One S^{2-} sulfur isotope sample was collected from well ZVIFS1 in a silver nitrate ($AgNO_3$) trap according to the protocol of Carmody *et al.* (1997). Silver sulfide (Ag_2S) formed in the trap was separated from the residual $AgNO_3$ solution in a centrifuge and rinsed with deionized water. The stable sulfur isotope samples were analyzed by the USGS stable isotope laboratory in Reston, Virginia.

Relative amounts of the heavy stable sulfur isotope, ^{34}S, and light stable sulfur isotope, ^{32}S, in SO_4^{2-} and S^{2-} are expressed as the difference between the measured ratios of ^{34}S to ^{32}S in the sample and in the Canyon Diablo Troilite (CDT) or the Vienna Canyon Diablo Troilite (VCDT) standards. This difference is reported as δ values in permil units (‰) and is computed as

$$\delta^{34}S(‰) = \frac{R_{sample} - R_{std}}{R_{std}} \times 1000, \tag{1}$$

where $R_{sample} = {}^{34}S/{}^{32}S$ in the sample and $R_{std} = {}^{34}S/{}^{32}S$ in the standard. Reactions involving SO_4^{2-} or S^{2-} often cause changes in $\delta^{34}S$ in these species, a process called isotopic fractionation. SRB preferentially reduce isotopically light SO_4^{2-} ($^{32}SO_4^{2-}$), resulting in the enrichment of ^{34}S in residual SO_4^{2-} and ^{32}S in the S^{2-} produced during the reaction (Nakai and Jensen, 1964; Faure, 1986; Clark and Fritz, 1997). Laboratory investigations at near-surface temperatures (10–30 °C) have shown that SRB can produce enrichments of $\delta^{34}S$ in SO_4^{2-} relative to S^{2-} of 10 to 46 ‰, with an average enrichment of 25 ‰ (Kaplan and Rittenberg, 1964; Kemp and Thode, 1968; Nakai and Jensen, 1964; Clark and Fritz, 1997; Canfield, 2001). Chemical or abiotic reduction of sulfate at temperatures ranging from 18 to 50 °C is possible thermodynamically and yields an average enrichment of $\delta^{34}S$ in sulfate relative to S^{2-} of 22 ‰ (Harrison and Thode, 1957); however, the kinetics of abiotic sulfate reduction at near-surface temperatures are extremely slow (Machel *et al.*, 1995). There is no direct evidence for abiotic reduction of SO_4^{2-} at near-surface temperatures by ZVI (Gu *et al.*, 1999); therefore, this process should be negligible in ZVI PRBs.

Several investigators have used stable sulfur isotope signals to verify or reject DSR in groundwater systems (Dockins *et al.*, 1980; Strebel *et al.*, 1990; Robertson and Schiff, 1994). Stable sulfur isotope signals, such as increasing $\delta^{34}S$ in SO_4^{2-} along a flow path and/or an enrichment of $\delta^{34}S$ in SO_4^{2-} relative to S^{2-} on the order of 25 ‰, can be used to verify the presence of DSR in a ZVI PRB.

B. Microbiological Assessment

Biological activity reaction test kits (BARTs) manufactured by HACH were inoculated with water from wells in and near the ZVI PRB concurrent with the stable sulfur isotope sampling in June 2000. A BART consists of a sterile sample bottle containing a buoyant media ball that has a diameter slightly smaller than the inside diameter of the sample bottle. The theory behind its operation is that aerobic bacteria present in a water sample will consume any DO and form a biological film at the air–water interface between the media ball and the sample container. The film effectively terminates the diffusion of atmospheric oxygen trapped in the headspace of the sample container, allowing DSR to proceed (if SRB are present in the sample) (Cullimore, 1993). The BART samples were incubated in the dark at room temperature and observed daily for the presence of black precipitates (iron sulfides). If black precipitates were visible at the end of the incubation period in a sample, it was scored as positive for the presence of SRB. To avoid cross-contamination, pump tubing was sterilized with isopropyl alcohol and rinsed with deionized water between sample sites.

Direct plate counts were completed on unfiltered groundwater samples collected from a well in the ZVI PRB and an upgradient well in January 2001. Groundwater samples for plate count enumeration were collected in sterilized polyethylene culture tubes after the appropriate volume of water was purged from each well (see Section III,A). Pump tubing was sterilized between sample sites as described earlier. The sample bottles were filled completely to minimize oxygenated headspace and packed on ice. Postgate's medium E (Postgate, 1984), adjusted to pH 8.6, was used to grow the bacteria. Plates were incubated in the dark under anaerobic conditions at room temperature for 14 days. At the end of the incubation period, the black colonies were counted. The samples were processed at the Center for Bioremediation at Weber State University in Ogden, Utah.

C. Mineralogic Investigation

Cores collected from the ZVI PRB in May 1999 were analyzed for the presence of sulfide phases by X-ray diffraction (XRD) and scanning electron microscopy with energy dispersive X-ray analytical capability (SEM-EDX). The cores were collected with a Geoprobe fitted with a hydraulic impact hammer. Samples were collected in plastic sleeves that lined the core barrel. Immediately after coring, ZVI samples were bottled and sealed under an argon atmosphere and stored at 4 °C. Two methods were used to prepare the ZVI samples for XRD analysis: (1) alcohol glass slide smears of finely ground ZVI pellets and (2) alcohol glass slide smears of finely ground surface precipitates that were shaken from the ZVI pellets in a sonicating bath. Both sample types were ground in an acetone solution to minimize oxidation. Samples for XRD analysis were analyzed within minutes of air drying. Whole ZVI pellets were selected randomly for SEM-EDX analysis. Samples were air dried, attached to sample studs with carbon tape, and analyzed within minutes of drying.

D. Geochemical Modeling

Mass balance geochemical modeling with NETPATH (Plummer *et al.*, 1994) was combined with results from Br^- tracer tests conducted in April 1999 to help quantify the rate of sulfide mineral precipitation in the ZVI PRB. NETPATH is an inverse geochemical computer modeling program; it computes the masses of plausible minerals and/or gases that must enter or exit a solution between an arbitrarily chosen start point and end point in a flow path to account for the observed changes in water quality between the two points (Plummer *et al.*, 1983, 1994; Plummer and Back, 1980). Assumptions inherit in mass balance modeling are (1) the change in water composition between arbitrary starting and end points in a groundwater system can be attributed only to effects of reactions and (2) the chemical composition of the water remains constant with time at a given point in the flow path, i.e., steady-state conditions prevail in the system (Plummer *et al.*, 1994). The net chemical reaction for NETPATH modeling is

$$\begin{aligned}&\text{initial solution} + \text{``reactant phases''} \rightarrow \\ &\text{final solution} + \text{``product phases,''}\end{aligned} \qquad \text{(IX)}$$

where "reactant phases" and "product phases" are the minerals and/or gases consumed and produced, respectively, in the reaction and initial solution and final solution are the chemical composition of water at the start point and end point, respectively, along the flow path. Inputs to NETPATH include chemical analyses of water from two points along a flow path; plausible phases that enter or exit the water between two points along the flow path, i.e., the "reactant" and "product" phases in reaction IX; and the minimum number of chemical constraints (elements) necessary to define the plausible phases.

Water quality data used in the NETPATH model were collected in June 2000 in conjunction with the stable sulfur isotope samples. A Hydrolab Mini-Sonde Water Quality Multiprobe fitted with a gas tight flow-through chamber was used to measure temperature, pH, DO, and Eh. Eh values are relative to the SHE. Field alkalinity values were determined in filtered samples with a HACH digital titrator containing a 1.6 *N* sulfuric acid cartridge. Samples were filterd with $0.45-\mu m$ pore size filters. Fe^{2+} and soluble sulfide ($HS^- + S^{2-}$) were measured in unfiltered samples with a Chemetrics brand spectrophotometer. Water samples for anion analysis were filtered in the field and collected in 125-ml field-rinsed polyethylene bottles. Water samples for cation analysis were also filtered on site and collected in 125-ml acid-rinsed bottles. After collection, cation samples were preserved with 1 ml of ultrapure concentrated nitric acid. One duplicate and one process blank sample were also collected and analyzed for quality assurance of data. The concentration of major anions in each groundwater sample was measured by ion chromatography. Major cations were measured by inductively coupled plasma optical emission spectrometry. Water analyses were conducted at the USGS Research Laboratories in Menlo Park, California.

IV. RESULTS

A. Stable Sulfur Isotope Evidence for Dissimilatory Sulfate Reduction

Stable sulfur isotope samples were collected from seven wells in the ZVI PRB (ZVIR1S-1, ZVIR1-2, ZVIR1S-4, ZVIR1-5, ZVIR1S-7, ZVIR1-8, and ZVIFS1), one upgradient well (FC-3), and one downgradient well (DG-2a) (Figs. 1 and 3). SO_4^{2-} concentrations are plotted against the $\delta^{34}S$ in SO_4^{2-}($\delta^{34}S$-SO_4^{2-}) in Fig. 4. $\delta^{34}S$-SO_4^{2-} increases from −3.6 to −1.4 ‰ in water from the deep wells along an assumed flow path in the PRB (i.e., wells ZVIR1-2, ZVIR1-5, and ZVIR1-8) and from 0.7 to 3.3 ‰ in water from the shallow wells along the same flow path (i.e., ZVIR1S-1, ZVIR1S-4,

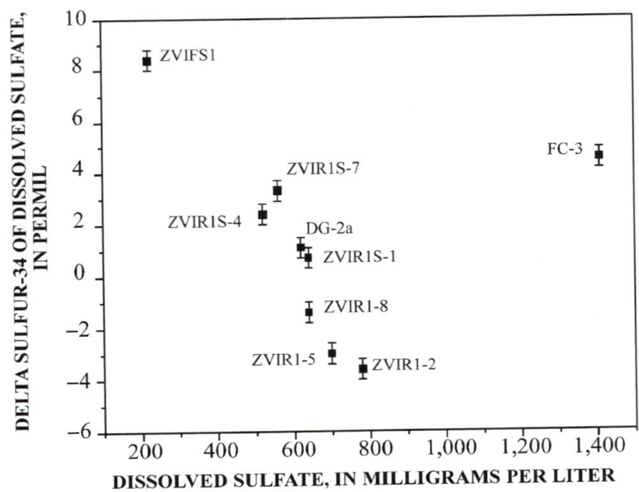

Figure 4 Relation between $\delta^{34}S$-SO_4^{2-} (relative to the Vienna Canyon Diablo Troilite standard) and SO_4^{2-} concentrations for water from wells in and near the zero valent iron permeable reactive barrier, Fry Canyon, Utah.

and ZVIR1S-7). Water from well ZVIFS1, located in the center of the ZVI PRB, had the lowest SO_4^{2-} concentration (220 mg/liter) and the highest $\delta^{34}S\text{-}SO_4^{2-}$ value (8.4 ‰). Water from well FC3, upgradient of the gravel pack, had the highest SO_4^{2-} concentration (1460 mg/liter) and a relatively heavy $\delta^{34}S\text{-}SO_4^{2-}$ value (4.5 ‰). Water from the downgradient well, DG-2a, had a $\delta^{34}S\text{-}SO_4^{2-}$ value of 1.1 ‰ and a SO_4^{2-} concentration of 620 mg/liter. The $\delta^{34}S$ in S^{2-} ($\delta^{34}S\text{-}S^{2-}$) in water from well ZVIFS1 was −15.2 ‰.

Overall, water from wells in the ZVI PRB shows a strong trend for increasing $\delta^{34}S\text{-}SO_4^{2-}$ with decreasing SO_4^{2-} concentration, indicating that DSR is occurring. In addition, $\delta^{34}S\text{-}SO_4^{2-}$ was enriched 23.6 ‰ relative to $\delta^{34}S\text{-}S^{2-}$ in well ZVIFS1. This value is very close to the average fractionation measured in the laboratory studies of DSR discussed in Section III, A. The value also compares well with the results of Dockins *et al.* (1980), which measured an average enrichment of $\delta^{34}S$ in SO_4^{2-} relative to S^{2-} of 32.3 ‰ in Montana groundwaters undergoing DSR.

The values for $\delta^{34}S\text{-}SO_4^{2-}$ and SO_4^{2-} concentration data from water in wells in and downgradient of the ZVI PRB closely follow a Rayleigh-type distillation process. Rayleigh-type distillation applies in systems in which the isotopic composition of the product is controlled by equilibrium or kinetic isotope reactions but is removed instantaneously after its formation (Fritz and Fontes, 1980). Rayleigh distillation can be described as

$$\delta_R = \delta_I + \varepsilon \ln f, \tag{2}$$

where δ_R and δ_I are the residual and initial $\delta^{34}S\text{-}SO_4^{2-}$ values, ε is the isotopic enrichment factor, and f is the fraction of initial SO_4^{2-} remaining. A plot of ln f vs $\delta^{34}S\text{-}SO_4^{2-}$ is shown in Fig. 5 and includes data from all wells in the ZVI PRB and the downgradient well, DG-2a. The SO_4^{2-} concentration in water from well ZVIT1 was used as the initial SO_4^{2-} concentration to compute ln$f$. The linearity of the plot ($R^2 = 0.92$) indicates Rayleigh distillation is occurring in and downgradient of the ZVI PRB. The slope of the best-fit regression line gives $\varepsilon = -9.0$ ‰. This value compares well to the results of Strebel *et al.* (1990) and Robertson and Schiff (1994), who computed ε values of −9.7 and −15.5 ‰, respectively, in studies of two different shallow aquifers subject to DSR.

The presence of Rayleigh-type distillation with an ε value documented in other studies of DSR is strong evidence that DSR is the major mechanism for SO_4^{2-} declines in the ZVI PRB. Formation of green rust II would probably result in a small depletion of $\delta^{34}S$ in residual SO_4^{2-} (i.e., < 2.0 ‰), as is observed during the formation of gypsum ($CaSO_4 \bullet 2H_2O$) (Thode and Monster, 1965). If the formation of green rust II were a significant sink for SO_4^{2-} in the ZVI PRB, ε would be much less negative. Also of significance is the location of data from water in well DG-2a on the Rayleigh plot, which

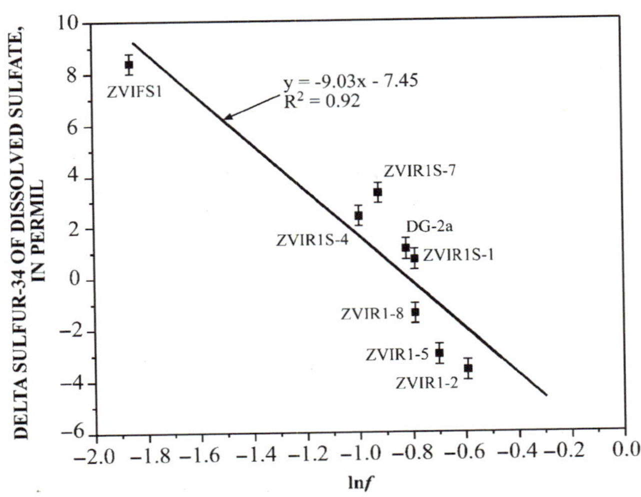

Figure 5 Rayleigh plot of $\delta^{34}S$-SO_4^{2-} (relative to the Vienna Canyon Diablo Troilite standard) and SO_4^{2-} data collected from wells in and downgradient of the zero valent iron permeable reactive barrier, Fry Canyon, Utah.

indicates that a zone of sulfate reduction has developed downgradient of the PRB. This may be a result of migration of anaerobic and hydrogen gas-enriched water from the ZVI PRB to well DG-2a.

B. Microbiological Evidence for the Presence of Sulfate Reducing Bacteria

BART kits were inoculated with water from wells FC3, ZVIR1-2, ZVIT1, ZVIR1S-1, ZVIR1-2, ZVIR1S-4, ZVIR1-5, ZVIR1S-7, ZVIR1-8, ZVIFS1, and DG-2a. Results indicate that SRB were present in unfiltered water from all wells sampled; however, SRB were not present in filtered water from well ZVIR1-5, confirming that SRB must be present in order for iron sulfides to form in inoculated BART kits. The presence of SRB in water from the upgradient well, FC-3, was not surprising, as it is hypothesized that SRB in the native aquifer were transported via groundwater flow into the ZVI PRB.

The direct plate counts completed in January 2001 indicate that the concentration of SRB is two orders of magnitude higher in water from the ZVI PRB relative to water from an upgradient well. The average concentration of SRB was 10 cfu/ml ($n = 3$, SD $= 2$) in water from well FC-3 and 1483 cfu/ml ($n = 3$, SD $= 275$) in water from well ZVIFS1. The combined results of

BART and direct plate counts show that a population of SRB has established itself in the ZVI PRB.

C. Mineralogic Data

Core samples collected from the ZVI PRB in May 1999 were analyzed with XRD and SEM-EDX to investigate the mineralogy of the PRB. The exact PRB intervals cored by the Geoprobe were difficult to determine due to uncertainty regarding the distance of the Geoprobe from the PRB and the angle of penetration. Therefore, no attempt is made to give the exact position of the samples in the ZVI PRB. However, all samples were collected from the saturated zone in the ZVI PRB and were probably within 0.6 m of the gravel pack–ZVI interface. XRD analysis of the finely ground ZVI pellets revealed only two phases: ZVI and plustite (FeO) (Fig. 6). These were the only two phases identified in an unreacted ZVI pellet by XRD. It is likely that the large amount of these phases relative to precipitates that formed on the surface of PRB samples overwhelmed the XRD patterns of the whole, crushed ZVI pellets. Somewhat better resolution of surface precipitates was achieved with the sonicated samples. In addition to ZVI and FeO, XRD analysis of sonicated samples revealed the presence of β-FeOOH and aragonite ($CaCO_3$) (Fig. 6). SEM-EDX analysis indicated the presence of $FeCO_3$. The elemental weight percentages in the iron carbonate phase determined by EDX spectra compare well with pure $FeCO_3$ (Table I). Large surface area bulk scans of ZVI pellets revealed the presence of sulfur. However, sulfur phases could not be identified directly with SEM-EDX analysis.

Table I

Elemental Weight Percentages, Determined with Energy-Dispersive X-Ray Analysis, in a Phase on the Surface of a ZVI Pellet from the PRB in Fry Canyon, Utah[a]

Element	Weight percentage in phase on ZVI sample	Weight percentage in pure $FeCO_3$
C	14.95	10.4
O	40.33	41.4
Mg	2.18	0.0
Si	1.8	0.0
S	0.64	0.0
Fe	40.1	48.2
Total	100.0	100.0

[a]Weight percentages of each element in pure $FeCO_3$ are shown for comparison.

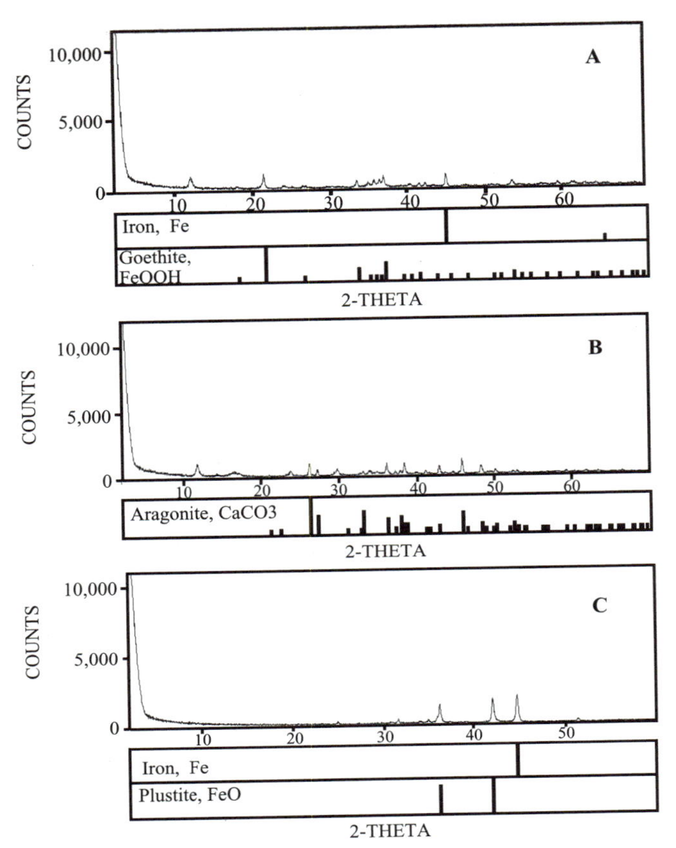

Figure 6 X-ray diffraction patterns of select samples from the zero valent iron permeable reactive barrier in Fry Canyon, Utah. Phases such as goethite and aragonite are discernible in sonicated samples (A and B); ZVI and FeO dominated XRD patterns from unsonicated, whole samples of ZVI (C).

In light of isotopic and microbiological evidence, it is very likely that sulfide phases are forming in the ZVI PRB. There are several reasons why XRD and SEM-EDX analyses did not find sulfide phases. First, the volume of ZVI cored from the PRB and subsequently analyzed was very small compared to the total volume of the PRB. It is possible that sulfate reduction is not distributed uniformly throughout the PRB and that the cores missed the zones of sulfate reduction. Second, sulfide phases formed as a result of DSR may be in the form of amorphous FeS (Neal *et al.*, 2001) and would be invisible to XRD and difficult to discern with SEM-EDX. Third, the core samples may have been exposed to oxygen and rapid oxidation of iron sulfide phases would have occurred (Simpson *et al.*, 2000). Fourth, sulfides could be precipitating as colloidal-sized particles and transported out of the PRB. It is recommended that a wet chemical analysis for sulfides be included in any mineralogical study of ZVI PRB core material.

E. Geochemistry and Geochemical Modeling

Chemical analysis of water from wells along an assumed flow path (ZVIT1 $\rightarrow$ ZVIR1-2 $\rightarrow$ ZVIR1-5 $\rightarrow$ ZVIR1-8 $\rightarrow$ DG2a) shows large decreases in the concentrations of DO, alkalinity, Ca^{2+}, Mg^{2+}, and SO_4^{2-} and large increases in the concentration of Fe^{2+} and pH between wells ZVIT1 and ZVIR1-2 (Fig. 7). DO concentrations in the ZVI PRB were below the detection limit of 0.10 mg/liter. The changes in these parameters are consistent with anaerobic and aerobic iron corrosion, carbonate mineral precipitation, and DSR in the ZVI PRB.

The trend in alkalinity and Ca^{2+} appears to reverse as groundwater travels between wells ZVIR1-5 and ZVIR1-8. This may indicate an incorrect flow path assumption between these wells. The concentrations of Ca^{2+}, Mg^{2+}, alkalinity, and SO_4^{2-} decline between the ZVI PRB and well DG-2a. The decline in the cation concentrations is attributed to continued carbonate mineral precipitation, and the decline in SO_4^{2-} is attributed to DSR. Field measurements of soluble sulfide concentrations along the flow path were below detection limit (0.06 mg/liter) in all wells except DG-2a, where 0.15 mg/liter was measured. The absence of soluble sulfide in wells along the flow path in the ZVI PRB can be explained by the rapid formation of iron sulfide phases, such as troilite (FeS). The concentration of soluble sulfide was greater than 1 mg/liter in well ZVIFS1, which is not on the assumed flow path. Water from this well had a relatively small amount of Fe^{2+} (0.90 μg/liter), which may limit the formation of iron sulfide phases. Charge balance errors computed with chemical data were less than 5%, and duplicate and process blank samples verified the quality of chemical data.

To better understand chemical data collected along the assumed flow path, a mass balance geochemical model was constructed with NETPATH (Plummer *et al.*, 1994). The reactions hypothesized to account for the changes in water quality along each increment of the flow path are based on reactions II, IV, and VII (see Section I). The plausible phases used in NETPATH were $CaCO_3$, $FeCO_3$, magnesite ($MgCO_3$), ZVI, $Fe(OH)_2$, and FeS. $MgCO_3$ was included as a plausible phase to quantify the mass of Mg^{2+} that coprecipitates during the formation of aragonite. $Fe(OH)_2$ was included as a sink for Fe^{2+}. The constraints were carbon (C), S, Fe, Ca, Mg, and redox state (RS). RS provides a means to account for electron transfer and must be included as a

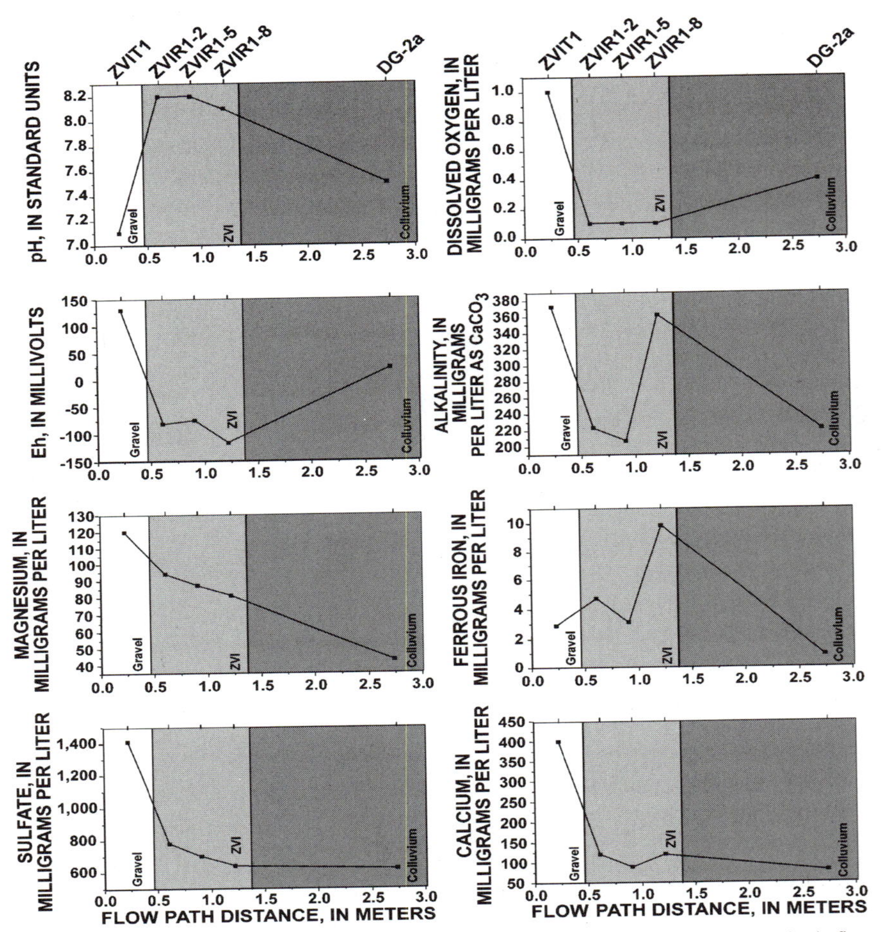

Figure 7 Results of chemical analyses for select parameters along an assumed hydrologic flow path in and near the zero valent iron permeable reactive barrier, Fry Canyon, Utah.

constraint for any oxidation–reduction reaction. The conventions defining RS can be found in Plummer and Back (1980) and Plummer *et al.* (1994).

A complexity to the model arises because the measured decrease in inorganic C (computed from the alkalinity concentrations) among the flow path segments ZVIT1 to ZVIR1-2, ZVIR1-2 to ZVIR1-5, and ZVIR1-8 to DG-2a cannot precipitate out the measured loss of Ca^{2+} and Mg^{2+} (as carbonate phases) in these segments. To provide an additional sink for Ca^{2+} and Mg^{2+} in the model, the minerals portlandite [$Ca(OH)_2$] and brucite [$Mg(OH)_2$] were included as plausible phases. In reality, the formation of these two phases in the ZVI PRB is unlikely. They are included in the model to achieve mass balance. Additional mineralogic work is needed to identify the apparent noncarbonate Ca^{2+} and Mg^{2+} sinks in the ZVI PRB.

The model indicates that most of the mass transfer occurs between wells ZVIT1 and ZVIR1-2 and is probably confined to the first 0.15 m of the ZVI PRB (Fig. 8). A total of 4.18 mmol of Ca^{2+} and Mg^{2+} carbonates and 6.58 mmol of FeS precipitated in this flow path increment. The model indicates that a significant amount of $Ca(OH)_2$ (3.90 mmol) precipitates to balance the observed decrease in Ca^{2+} concentration. According to the model, 26.4 mmol of ZVI is required to corrode under anaerobic conditions to produce enough H_2 to sustain the level of SO_4^{2-} reduction implied by chemical data.

The level of mineral precipitation drops off substantially in the succeeding flow path increments. A total of 2.78 mmol of Ca^{2+} and Mg^{2+} carbonate and 1.67 mmol of FeS precipitate along the remainder of the flow path. The mass of H_2 required for SO_4^{2-} reduction in the remainder of the flow path is only

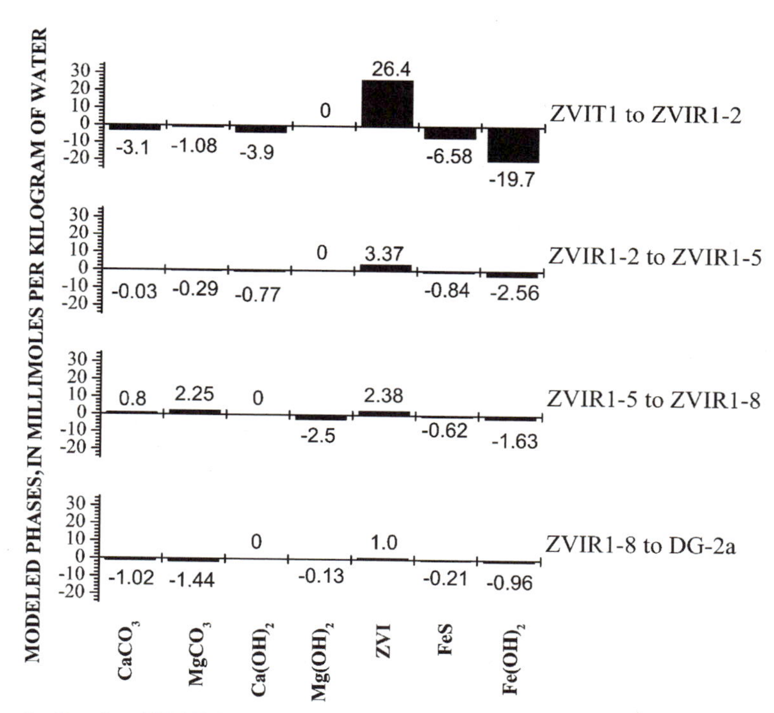

Figure 8 Results of NETPATH modeling along an assumed hydrologic flow path in and near the zero valent iron permeable reactive barrier, Fry Canyon, Utah. A positive value indicates dissolution of a solid phase, and a negative value indicates precipitation of a solid phase.

25% of that required between wells ZVIT1 and ZVIR1-2. In addition, the model shows that a small amount of $Ca(OH)_2$ (0.77 mmol) precipitates between wells ZVIR1-2 and ZVIR1-5 and that a small amount of $Mg(OH)_2$ (2.63 mmol) precipitates between wells ZVIR1-5 to DG-2a to balance the observed loss of Ca^{2+} and Mg^{2+} with the loss of inorganic C.

The model is supported by thermodynamic calculations, mineralogic data, and stable sulfur isotope data. The thermodynamic speciation program WATEQ4F (Ball and Nordstrom, 1991) was used to compute the saturation index (SI) of minerals contained in its thermodynamic database. The SI is quantified as

$$\mathrm{SI} = \log_{10}(Q/K_{\mathrm{eq}}), \tag{3}$$

where Q is the ion activity product and K_{eq} is the equilibrium constant for the mineral dissolution reaction of interest. For a given mineral, a positive SI indicates that it is supersaturated in the solution (it should precipitate from solution), a negative SI indicates that it is undersaturated in the solution (it should dissolve into solution), and a SI at or near zero indicates that it is in equilibrium with the solution (it should not dissolve or precipitate in solution). SIs for select minerals are shown in Table II. Water from each well in the ZVI PRB was supersaturated with respect to aragonite. XRD and SEM-EDX analyses confirmed the presence of aragonite on ZVI particles. The SI of FeS could not be calculated in wells finished upgradient and in the ZVI PRB because field measurements for soluble sulfide were below the detection limit. However, the lack of soluble sulfide concentrations greater than the detection limit does not rule out DSR and formation of FeS. The abundant Fe^{2+} in the ZVI PRB and the low solubilities of iron sulfide phases could rapidly remove

Table II

Summary of Select Saturation Indices for Water from Wells along an Assumed Flow Path in and near the ZVI PRB in Fry Canyon, Utah[a]

Well	Saturation index for aragonite	Saturation index for gypsum	Saturation index for FeS
ZVIT1	0.285	−0.228	—
ZVIR1-2	0.645	−0.828	—
ZVIR1-5	0.546	−0.978	—
ZVIR1-8	0.845	−0.898	—
DG-2a	−0.104	−1.029	0.626

[a]Saturation indices were computed with WATEQ4F. The concentration of sulfide was below the detection limit in water from wells ZVIT1, ZVIR1-2, ZVIR1-5, and ZVIR1-8; therefore, saturation indices for FeS are not given for those wells.

soluble sulfide via precipitation reactions. Soluble sulfide and a small amount of Fe^{2+} were detected in water from well DG-2a, resulting in a large positive SI for FeS. Sulfide mineral formation along the modeled flow path is strongly supported by the Rayleigh-type distillation removal of SO_4^{2-} and the enrichment of ^{34}S in SO_4^{2-} relative to S^{2-} in well ZVIFS1. The absence of an additional sulfur sink along the modeled flow path is supported by the relatively large negative SI values for gypsum ($CaSO_4 \bullet 2H_2O$) (Table II). More mineralogic work is needed to identify the presence of noncarbonate Ca^{2+} and Mg^{2+} sinks in the ZVI PRB. H_2 should also be measured in the ZVI PRB to confirm or reject the mechanism of DSR used in the NETPATH model.

V. CONCLUSIONS

The anaerobic, highly reducing, and H_2-enriched geochemical environment in the ZVI PRB in Fry Canyon, Utah, is sustaining an active population of SRB. The stable sulfur isotope and SO_4^{2-} data from wells in the ZVI PRB and a downgradient well show that sulfur is removed by DSR through a Rayliegh-type distillation process. The enrichment factor computed from the Rayleigh plot ($\varepsilon = -9.0$ ‰) is close to values measured in other field investigations of DSR in groundwater systems. The enrichment of $\delta^{34}S$ in SO_4^{2-} relative to S^{2-} of 23.6 ‰ in water from well ZVIFS1 is close to values that have been measured in independent laboratory and field investigations of DSR. Furthermore, thermodynamic speciation calculations and stable sulfur isotope data indicate that sulfide precipitation is the only sink for sulfur in the PRB. The distribution of SO_4^{2-} concentrations indicates that most of the sulfide precipitation is occurring in the first 0.15 m of the PRB.

SO_4^{2-} concentrations in water from wells ZVIT1 and ZVIR1-2 in June 2000 can be used in conjunction with Br^- tracer tests done in April 1999 to estimate the loss of porosity in the first 0.15 m of the ZVI PRB due only to DSR. Br^- tracer tests indicate that the average groundwater velocity is 0.7 m/day in the ZVI PRB (see Chapter 14). Assuming a flow path width of 1.07 m, a depth of 1.10 m, and an effective porosity of 0.55, 0.450 m^3/day of water enters the ZVI PRB. Computing with the density of troilite (FeS, 4.61g/cm^3), 56.8 cm^3 of FeS precipitates per day in the first 0.15 m of the ZVI PRB. The volume of void space in the first 0.15 m of the flow path in the ZVI PRB is 97, 103 cm^3 (assuming flow path dimensions of 1.07 m wide by 1.1 m deep by 0.15 m long, and an effective porosity of 0.55). Each year, the production of FeS via DSR consumes about 21% of the void space in the first 0.15 m of the ZVI PRB. A correspondingly large amount of reactive surface is probably consumed by sulfide production. SO_4^{2-} concentration data in water from wells ZVIT1 and ZVIR1-2 collected between December 1998 and June 2000 indicate that an average of 12% of the porosity in the first 0.15 m of the ZVI PRB is consumed each year via DSR. These estimates assume steady-state conditions with respect to groundwater velocity, saturated thickness, water chemistry, and DSR. The estimates also assume that the gravel pack upgradient of the ZVI PRB and the ZVI PRB are homogeneous with respect to hydraulic properties and water quality.

The presence and activity of SRB in the ZVI PRB are probably beneficial on short time scales because the S^{2-} they produce can react with contaminants such as Cd^{2+}, Cu^{2+}, Ni^{2+}, Pb^{2+}, and Zn^{2+} to make sparingly soluble sulfide phases. In addition, SRB can use U^{6+} as a terminal electron acceptor, which results in the formation of insoluble U^{4+} phases (Lovley *et al.*, 1993). However, the volume of FeS precipitation in the ZVI PRB because of DSR could significantly shorten the effective life span of the barrier by consuming porosity and reactive surface area; therefore, the presence of SRB in the ZVI PRB in Fry Canyon is not desirable.

REFERENCES

Ball, J. W., and Nordstrom, D. K. (1991). User's manual for WATEQ4F, with revised thermodyanamic data base and test cases for calculating speciation of major, trace, and redox elements in natural waters. U.S. Geological Survey Open-File Report 91–9183.

Benner, S. G., Blowes, D. W., Gould, W. D., Herbert, R. B., and Ptacek, C. J. (1999). Geochemistry of a permeable reactive barrier for metals and acid mine drainage. *Environ. Sci. Technol.* **33**, 2793–2799.

Canfield, D. E. (2001). Isotope fractionation by natural populations of sulfate-reducing bacteria. *Geochim. Cosmochim. Acta.* **65**, 1117–1124.

Carmody, R. W., Plummer, N. L., Busenberg, E., and Coplen, T. B. (1997). Methods for collection of dissolved sulfate and sulfide and analysis of their sulfur isotopic composition. U.S. Geological Survey Open-File Report 97–234.

Chapelle, F. H. (1993). "Ground-water Microbiology and Geochemistry". Wiley, New York.

Clark, I., and Fritz, P. (1997). "Environmental Isotopes in Hydrogeology". Lewis Publishers, New York.

Cullimore, D. R. (1993). "Practical Manual of Groundwater Microbiology". Lewis Publishers, Michigan.

Devlin, J. F., Klausen, J., and Schwarzenbach, R. P. (1998). Kinetics of nitroaromatic reduction on granular iron in recirculating batch experiments. *Environ. Sci. Technol.* **32**, 1941–1947.

Dockins, W. S., Olson, G. J., McFeters, G. A., and Turbak, S. C. (1980). Dissimilatory bacterial sulfate reduction in Montana groundwaters. *Geomicrobiol. J.* **2**, 83–98.

Eykholt, G. R., and Davenport D. T. (1998). Dechlorination of the chloroacetanilide herbicides, alachlor and metolachlor by iron metal. *Environ. Sci. Technol.* **32**, 1482–1487.

Faure, G. (1986). "Principles of Isotope geology". Wiley, New York.

Fritz, P., and Fontes, J. Ch. (1980). "Handbook of Environmental Isotope Geochemistry". Elsevier, New York.

Gu, B., Liang, L., Dickey, M. J., Yin, X., and Dai, S. (1998). Reductive precipitation of uranium(VI) by zero-valent iron. *Environ. Sci. Technol.* **32**, 3366–3373.

Gu, B., Phelps, T. J., Liang, L., Dickey, M. J., Roh, R., Kinsall, B. L., Palumbo, A. V., and Jacobs, G. K. (1999). Biogeochemical dynamics in zero-valent iron columns: Implications for permeable reactive barriers. *Environ. Sci. Technol.* **33**, 2170–2177.

Harrison, A. G., and Thode, H. G. (1957). The kinetic isotope effect in the chemical reduction of sulphate. *Trans. Faraday Soc.* **53**, 1648–1651.

Jakobsen, R., and Postma, D. (1999). Redox zoning, rates of sulfate reduction and interactions with Fe-reduction and methanogenesis in a shallow sandy aquifer, Romo, Denmark. *Geochim. Cosmochim. Acta.* **63**, 137–151.

Jones, R. E., Beeman, R. E., and Suflita, J. M. (1989). Anaerobic processes in the deep terrestrial subsurface. *Geomicrobiol. J.* **7**, 117–130.

Kaplan, I. R., and Rittenberg, S. C. (1964). Microbiological fractionation of sulphur isotopes. *J. Gen. Microbiol.* **34**, 195–212.

Kemp, A. L., and Thode, H. G. (1968). The mechanism of the bacterial reduction of sulphate and of sulphite from isotope fractionation studies. *Geochim. Cosmochim. Acta* **32**, 71–91.

Kiilerich, O., Wosdchow, L., and Nielsen, C. (2000). Field results from the use of a permeable reactive wall. *In* "Chemical Oxidation and Reactive Barriers: Remediation of Chlorinated and Recalcitrant Compounds" (G. B. Wickramanayake, A. R. Gavaskar, and A. Chen, eds.), pp. 377–384. Battelle Press, Columbus, OH.

Liang, L., Korte, N. E., Goodlaxson, J. D., Pickering, D. A., Zutman, J. L., West, O. R., Anderson, F. J., Welch, C. A., Pelfrey, M. J., and Dickey, M. J. (1997). A field-scale test of trichloroethylene dechlorination using iron filings for the x- 120/x-749 groundwater plume. ORNL/TM-13217. Oak Ridge National Laboratory, Oak Ridge, TN.

Lovley, D. R., Roden, E. E., Phillips, E. J. P., and Woodward, J. C. (1993). Enzymatic iron and uranium reduction by sulfate-reducing bacteria. *Mar. Geol.* **113**, 41–53.

Machel, H. G., Krouse, H. R., and Sassen, R. (1995). Products and distinguishing criteria of bacterial and thermochemical sulfate reduction. *Appl. Geochem.* **10**, 373–389.

Martino, D. P., Grossman, E. L., Ulrich, G. A., Burger, K. C., Schlichenmeyer, J. L., Suflita, J. M., and Ammerman, J. W. (1998). Microbial abundance and activity in a low-conductivity aquifer system in east-central Texas. *Microbial Ecol.* **35**, 224–234.

McMahon, P. B., Dennehy, K. F., and Sandstrom, M. W. (1999). Hydraulic and geochemical performance of a permeable reactive barrier containing zero-valent iron, Denver Federal Center. *Ground Water* **37**, 396–404.

Naftz, D. L., Fuller, C. C., Davis, J. A., Piana, M. J., Morrison, S. J., Freethey, G. W., and Rowland, R. C. (2000). Field demonstration of permeable reactive barriers to control uranium contamination in ground water. *In* "Chemical Oxidation and Reactive Barriers: Remediation of Chlorinated and Recalcitrant Compounds" (G. B. Wickramanayake, A. R. Gavaskar, and A. Chen, eds.), pp. 282–289. Battelle Press, Columbus, OH.

Nakai, N., and Jensen, M. L. (1964). The kinetic isotope effect in the bacterial reduction and oxidation of sulfur. *Geochim. Cosmochim. Acta* **28**, 1893–1912.

Neal, A. L., Techkarnjanaruk, S., Dohnalkova, A., McCready, D., Peyton, B. M., and Geesey, G. G. (2001). Iron sulfides and sulfur species produced at hematite surfaces in the presence of sulfate- reducing bacteria. *Geochim. Cosmochim. Acta* **65**, 223–235.

Olson, G. J., Dockins, W. S., McFeters, G. A., and Iverson, W. P. (1981). Sulfate-reducing and methanogenic bacteria from deep aquifers in Montana. *Geomicrobiol. J.* **2**, 327–340.

Orth, S. W., and Gillham, R. W. (1996). Dechlorination of trichloroethene in aqueous solution using ZVI. *Environ. Sci. Technol.* **30**, 66–71.

Phillips, D. H., Gu. B., Watson, D. B., Roh, Y., Liang, L., and Lee, S. Y. (2000). Performance evaluation of a zerovalent iron reactive barrier; mineralogical characteristics. *Environ. Sci. Technol.* **34**, 4169–4176.

Plummer, N. L., and Back, W. (1980). The mass balance approach: Application to interpreting the chemical evolution of hydrologic systems. *Am. J. Sci.* **280**, 130–142.

Plummer, N. L., Parkhurst, D. L., and Thorstenson, D. C. (1983). Development of reaction models for ground-water systems. *Geochim. Cosmochim. Acta* **47**, 665–668.

Plummer, N. L., Prestemon, E. C., and Parkhurst, D. L. (1994). An interactive code (NETPATH) for modeling net geochemical reactions along a flow path, version 2.0. U.S. Geological Survey Water-Resources Investigation Report 94–4169.

Postgate, J. R. (1984). "The Sulphate-Reducing Bacteria", 2nd Ed. Cambridge Univ. Press, Cambridge, MA.

Puls, R. W., Paul, C. J., and Powell, R. M. (1999). The application of in situ permeable reactive (zero-valent iron) barrier technology for the remediation of chromate-contaminated groundwater: A field test. *Appl. Geochem.* **14**, 989–1000.

Reardon, E. J. (1995). Anaerobic corrosion of granular iron: Measurement and interpretation of hydrogen evolution rates. *Environ. Sci. Technol.* **29**, 2936–2945.

Robertson, W. D., and Schiff, S. L. (1994). Fractionation of sulphur isotopes during biogenic sulphate reduction below a sandy forested recharge area in south-central Canada. *J. Hydrol.* **158**, 123–134.

Sass, B. M., Gavaskar, A. R., Gupta, N., Yoon, W., Hicks, J. E., O'Dwyer, D., and Reeter, C. (1998). Evaluating the Moffet Field permeable barrier using groundwater monitoring and geochemical modeling. *In* "Designing and Applying Treatment Technologies: Remediation of Chlorinated and Recalcitrant Compounds" (G. B. Wickramanayake, B. Godage, and R. E. Hinchee, eds.), pp. 169–175. Battelle Press, Columbus, OH.

Sayles, G. D., You, G., Wang, M., and Kupferle, M. J. (1997). DDT, DDD, and DDE dechlorination by zero-valent iron. *Environ. Sci. Technol.* **31**, 3448–3454.

Sherman, M. P., Darab, J. G., and Mallouk, T. E. (2000). Remediation of Cr(VI) and Pb(II) aqueous solutions using supported, nanoscale zero-valent iron. *Environ. Sci. Technol.* **34**, 2564–2569.

Simpson, S. L., Apte, S. C., and Batley, G. E. (2000). Effect of short-term resuspension events on the oxidation of cadimium, lead, and zinc sulfide phases in anoxic estuarine sediments. *Environ. Sci. Technol.* **34**, 4533–4537.

Singh, J., Comfort, S. D., and Shea, P. J. (1999). Iron-mediated remediation of RDX-contaminated water and soil under controlled Eh/pH. *Environ. Sci. Technol.* **33**, 1488–1494.

Strebel, O., Bottcher, J., and Fritz, P. (1990). Use of isotopic fractionation of sulfate-sulfur and sulfate-oxygen to assess bacterial desulfurication in a sandy aquifer. *J. Hydrol.* **121**, 155–172.

Thode, H. G., and Monster, J. (1965). Sulfur-isotope geochemistry of petroleum, evaporites, and ancient seas. Memoir of the American Association of Petroleum Geologists, 367–377.

U.S. Environmental Protection Agency (1998). Permeable reactive barrier technologies for contaminant remediation. EPA/600/R- 98/125.

Vogan, J. L., Butler, B. J., Odziemkowski, M. S., Friday, G., and Gillham, R. W. (1998). Inorganic and biological evaluation of cores from permeable iron reactive barriers. *In* "Designing and Applying Treatment Technologies: Remediation of Chlorinated and Recalcitrant Compounds" (G. B. Wickramanayake, B. Godage, and R. E. Hinchee, eds.), pp. 163–168. Battelle Press, Columbus, OH.

Weber, E. J. (1996). Iron-mediated reductive transformations: Investigation of reaction mechanism. *Environ. Sci. Technol.* **30**, 716–719.

Wilkowske, C. D., Rowland, R. C., and Naftz, D. L. (2001). Selected hydrologic data for the field demonstration of three permeable reactive barriers near Fry Canyon, Utah, 1996–2000. U.S. Geological Open-File Report 01–361.

Chapter 11

Biogeochemical, Mineralogical, and Hydrological Characteristics of an Iron Reactive Barrier Used for Treatment of Uranium and Nitrate

Baohua Gu, David B. Watson, Debra H. Phillips, and Liyuan Liang
Environmental Sciences Division, Oak Ridge National Laboratory, Oak Ridge, Tennessee 37831

A permeable iron reactive barrier was installed in late November 1997 at the U.S. Department of Energy's Y-12 National Security Complex in Oak Ridge, Tennessee. The overall goal of this research was to determine the effectiveness of the use of zero valent iron (Fe^0) to retain or remove uranium and other contaminants such as technetium and nitrate in groundwater. The long-term performance issues were investigated by studying the biogeochemical interactions between Fe^0 and groundwater constituents and the mineralogical and biological characteristics over an extended field operation. Results from nearly 3 years of monitoring indicated that the Fe^0 barrier was performing effectively in removing contaminant radionuclides such as uranium and technetium. In addition, a number of groundwater constituents such as

This manuscript has been authored by a contractor of the U.S. Government under contract No. DE-AC05-00OR22725. Accordingly, the U.S. Government retains a nonexclusive, royalty-free license to publish or reproduce the published form of this contribution, or allow others to do so, for U.S. Government purposes.

bicarbonates, nitrate, and sulfate were found to react with the Fe^0. Both nitrate and sulfate were reduced within or in the influence zone of the Fe^0 with a low redox potential (i.e., low Eh). An increased anaerobic microbial population was also observed within and in the vicinity of the Fe^0 barrier, and these microorganisms were at least partially responsible for the reduction of nitrate and sulfate in groundwater. Decreased concentrations of Ca^{2+} and bicarbonate in groundwater occurred as a result of the formation of minerals such as aragonite ($CaCO_3$) and siderite ($FeCO_3$), which coincided with the Fe^0 corrosion and an increased groundwater pH. A suite of mineral precipitates was identified in the Fe^0 barrier system, including amorphous iron oxyhydroxides, goethite, ferrous carbonates and sulfides, aragonite, and green rusts. These minerals were found to be responsible for the cementation and possibly clogging of Fe^0 filings observed in a number of core samples from the barrier. Significant increases in cementation of the Fe^0 occurred between two coring events conducted at $\sim$ 1 year apart and appeared to correspond to the changes in an apparent decrease in hydraulic gradient and connectivity. The present study concludes that while Fe^0 may be used as an effective reactive medium for the retention or degradation of many redox-sensitive contaminants, its long-term reactivity and performance could be severely hindered by its reactions with other groundwater constituents, and groundwater flow may be restricted because of the buildup of mineral precipitates at the soil/Fe^0 interface. Depending on site biogeochemical conditions, the rate of Fe^0 corrosion may increase; therefore, the life span of the Fe^0 barrier could be shorter than predicted in previous studies ($\sim$ 15–30 years).

I. INTRODUCTION

The use of zero valent iron (Fe^0) filings to remediate groundwater contaminated with chlorinated organic compounds, heavy metals, and radionuclides has received considerable attention in recent years (Gillham *et al.*, 1994; Gu *et al.*, 1999; Liang *et al.*, 2000; O'Hannesin and Gillham, 1998; Puls *et al.*, 1999; Scherer *et al.*, 2000; Sivavec *et al.*, 1997; Tratnyek *et al.*, 1997). Although the mechanisms for degrading or immobilizing these contaminants with Fe^0 are not completely understood (Roberts *et al.*, 1996; Sivavec *et al.*, 1997), it has been shown that Fe^0 is a promising reactive medium because of its efficiency in degrading or retaining contaminants and its relatively low cost. Consequently, the Fe^0-based reactive barrier treatment has been widely emplaced for passive, long-term applications for groundwater remediation.

While the focus of many applications was on utilizing an Fe^0-based barrier treatment system to remove or retain environmental contaminants, relatively few studies have paid particular attention to other biogeochemical reactions that may occur simultaneously as Fe^0 corrodes in groundwater. It is recognized that groundwater geochemistry plays a significant role in determining rates of Fe^0 corrosion, its surface reactivity, mineral precipitation and/or barrier clogging, microbial activity, and consequently the long-term performance of the Fe^0 treatment system (Gu *et al.*, 1999; Liang *et al.*, 2000; Scherer *et al.*, 2000). Although the roles of dissolved O_2 and pH in determining Fe^0 reactivity and precipitation chemistry are well established, interactions between Fe^0 and other groundwater constituents, such as HCO_3^-, NO_3^-, SO_4^{2-}, and some metal cations, such as Ca^{2+}, are less well studied and defined. In particular, because SO_4^{2-} , NO_3^-, and HCO_3^- are all corrosive to Fe^0 (Agrawal *et al.*, 1995; Gui and Devine, 1994; Lipczynska-Kochany *et al.*, 1994; Odziemkowski *et al.*, 1998) and are found commonly in groundwater at contaminated sites, these groundwater constituents are of great significance in influencing both geochemical and biological interactions and the barrier-clogging processes.

A potential limitation of Fe^0 technology is the deterioration of the Fe^0 reactive media by corrosion and the subsequent precipitation of minerals that may cause cementation and decreased permeability of the Fe^0 barrier or the surrounding soil. Few studies are available concerning the mineralogical and long-term performance characteristics of Fe^0-based barriers (Gu *et al.*, 1999; Mackenzie *et al.*, 1997; O'Hannesin and Gillham, 1998). However, data indicate that a suite of mineral precipitates can occur rapidly in the Fe^0 barrier system, and flow restriction could occur under certain biogeochemical conditions (Gu *et al.*, 1999; Liang *et al.*, 1997; Mackenzie *et al.*, 1997; Phillips *et al.*, 2000). Minerals such as goethite, magnetite, ferrous carbonates, sulfides, green rusts, and calcite have been reported in both laboratory and field investigations (Gu *et al.*, 1999; Phillips *et al.*, 2000; Pratt *et al.*, 1997). For example, a substantially decreased flow rate was observed over a 6-month period in a series of Fe^0-filled canisters used for treating trichloroethylene-contaminated groundwater at the Portsmouth Gaseous Diffusion Plant (Piketon, Ohio) (Liang *et al.* 1997). Post treatment analysis of Fe^0 filings showed cementation of the iron grains, possibly as a result of precipitation of iron sulfides, oxyhydroxides, and carbonates. Clogging has also been reported in laboratory and pilot-scale studies with Fe^0 filings as reactive media (Johnson and Tratnyek, 1994; Scherer *et al.*, 1998). At the Lowry Air Force base (AFB) in Denver, Colorado, and at Elizabeth City, North Carolina, sites, green rusts (i.e., a mixture of partially reduced/oxidized iron oxyhydroxides and sulfate) were observed in barrier materials (Edwards *et al.*, 1996; Puls *et al.*, 1999). At the Hill AFB, Utah, site, precipitation of iron and calcium carbonates was concluded to be responsible for a 14% porosity reduction within a few months of operation (Shoemaker *et al.*, 1995). In contrast, mineral precipitation was not observed after 1 year of operation in a reactive barrier at the Borden, Ontario, site.

This chapter presents results obtained from ~3 years of groundwater monitoring to evaluate the performance of an Fe^0 reactive barrier used primarily for the retention (or degradation) of uranium (U) and other contaminants, such as technetium (Tc) and nitrate (NO_3^-), at the U.S. Department of Energy's Y-12 National Security Complex in Oak Ridge, Tennessee. Emphasis was given to biogeochemical interactions between Fe^0 and groundwater constituents, mineralogical and hydrological characteristics, and related long-term performance issues of the reactive barrier system.

II. BARRIER SITE DESCRIPTION

A. Site Hydrogeology

Past waste disposal activities at the Oak Ridge Y-12 S-3 ponds have created a mixed waste plume of contamination in the underlying unconsolidated residuum and competent shale bedrock. The plume is more than 400 ft deep directly beneath the ponds and extends ~4000 ft along geologic strike both east and west of the ponds. S-3 ponds consisted of four unlined ponds constructed in 1951 on the west end of the Y-12 plant. The ponds had a storage capacity of 40 million liters (or ~10 million gallons). Liquid wastes, composed primarily of nitric acid plating wastes, containing various metals and radionuclides (e.g., Ni, Cr, U, and Tc) were disposed of in the ponds until 1983. Volatile organic compounds such as tetrachloroethylene and acetone were also disposed in the ponds, although only low levels of chlorinated organic contaminants and acetone were detected in the groundwater at the barrier site. Pond wastes that remained were neutralized and denitrified in 1984, and the site was capped and paved thereafter (Cook *et al.*, 1996; SAIC, 1996, 1997).

The geology that underlies the site is primarily the Nolichucky shale bedrock that dips approximately 45° to the southeast and has a strike of N55E (parallel to Bear Creek Valley). Overlying the bedrock is unconsolidated material that consists of weathered bedrock (referred to as residuum or saprolite), alluvium, colluvium, and man-made fill materials. Silty and clayey residuum comprises most of the unconsolidated material in this area. The residuum overlying the Nolichucky shale is typically between 5 and 10 m (~20 and 30 ft) thick. Between the unconsolidated residuum and the competent bedrock is a transition zone of weathered fractured bedrock. Remnant fracturing in the residuum and transition zone increases the permeability relative to the silt and clay matrix. The shallow groundwater flow direction west of the S-3 ponds is generally to the southeast with a horizontal gradient of approximately 0.016 ft/ft (SAIC, 1996). Additionally, upward vertical hydraulic gradients were identified at the site and are as high as 0.25 ft/ft between the competent bedrock and the transition zone, and ~0.12 ft/ft between the transition zone and the shallow unconsolidated zone (Watson *et al.*, 1999).

In general, the groundwater plume near the S-3 ponds is composed primarily of nitrate, bicarbonate, uranium, technetium, and other metal ions. These inorganic metal contaminants include low levels of heavy metals such as Ni, Cr, Co, Cd, Zn, and Pb. The plume is stratified, with the distribution of contaminants dependent on geochemical characteristics of the contaminants and groundwater. For example, nitrate and technetium (as pertechnetate, TcO_4^-), which are not particularly reactive with soil minerals, have the most extensive distribution in groundwater. Uranium and heavy metals that are more reactive are not as deep and have not migrated as extensively away from the ponds. Three major groundwater migration pathways to Bear Creek and its tributaries have been identified during the Bear Creek Valley treatability study (SAIC, 1997).

- Pathway 1 is located just to the south of and adjacent to the S-3 ponds and is a shallow pathway to the upper reach of Bear Creek. Contaminants in this pathway include uranium, technetium, nitrate, metals, and high total dissolved solids (TDS).
- Pathway 2 is another shallow pathway to the upper reaches of Bear Creek and is located approximately 600 ft downstream west from pathway 1. This pathway is thought to be associated with an old burial Bear Creek stream channel. The iron/gravel trench barrier is located at this site and is targeted to treat primarily low levels of uranium and nitrate (described in detail in Section II,B).
- Pathway 3 consists of contaminated groundwater in bedrock that is migrating to the west along strike in the Nolichucky Shale. Contaminants in this deeper pathway discharge to tributaries of Bear Creek and include technetium, nitrate, metals, and high TDS. Pathway 3 contains much less uranium than the other pathways.

B. Installation of the Iron–Gravel Trench Barrier

A permeable iron–gravel trench barrier was constructed at the pathway 2 site in late November 1997 as part of the technology demonstration using Fe^0 to retain or remove uranium and other contaminants as groundwater passively passes through the Fe^0 treatment medium. The pathway 2 site is predominantly a shallow pathway for the migration of uranium-contaminated groundwater (~1 mg/liter) to the upper reach of Bear Creek. The nitrate concentrations are generally lower (~20–150 mg/liter) at pathway 2 than other areas of the groundwater plume, but they have been detected at levels above 1000 mg/liter in some of the deeper piezometers because of the upward vertical hydraulic gradient. Technetium is generally detected at levels below 600 pCi/liter, and TDS concentrations (~1000 mg/liter) are generally lower than in the shallow plume at pathway 1 and deeper parts of the S-3 plume. Uranium-contaminated groundwater is discharging to the creek near pathway 2 through seeps adjacent to the headwaters of Bear Creek.

The trench dimensions are ~225 ft in length, 2 ft in width, and ~30 ft in depth (to bedrock) with an Fe^0-filled midsection of ~26 ft in length between two ~100-ft sections of granite/quartz-pea gravel (Fig. 1). Guar-gum biopolymer slurry was used during trench excavation to prevent the walls from collapsing, and Peerless Fe^0 filings (about −1/2 to 25 mesh size, from Peerless Metal Powders and Abrasives, Detroit, MI) were used as the reactive medium in the midsection of the barrier. A total of ~80 tons of Peerless iron filings was placed in the middle section of the trench to a depth of ~18–20 ft above

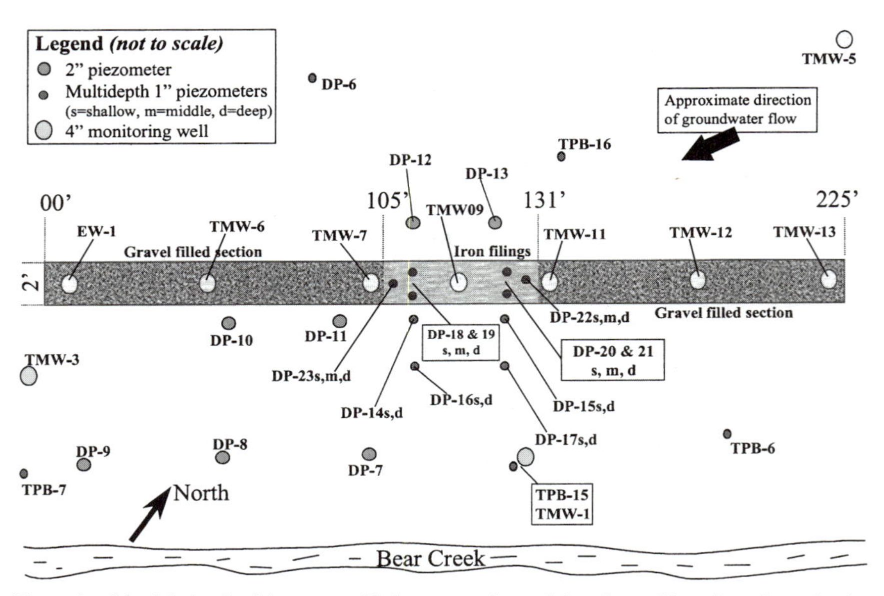

Figure 1 "Aerial view" of the permeable iron–gravel trench barrier and location of monitoring wells and piezometers at the Y-12 site, Oak Ridge, Tennessee. Trench dimensions are approximately 225 ft long, 30 ft deep, and 2 ft wide with the Fe^0-filled midsection of ~26 ft.

the bottom of the trench, or 10 to 12 ft below the original ground surface. Sediments around the barrier are heterogeneous mixtures of fill materials, native soil, saprolite, and rock fragments (Watson *et al.*, 1999). Undisturbed soil and saprolite from the Nolichucky Shale formation are present near the bottom of the barrier. The barrier trench is oriented nearly parallel (or in a small angle) to the direction of groundwater flow and was designed to direct groundwater flow through the iron treatment zone using both the natural groundwater gradient and the permeability contrast between the iron/gravel in the trench and the native silt/clay outside the trench. Hydraulic monitoring at the site indicated that the hydraulic gradient is ~0.025 ft/ft across the site but this gradient flattens to 0.01 ft/ft in the vicinity of the barrier trench. Approximately 48 piezometers, including six multiport wells in the Fe^0 barrier, were installed at the site (Fig. 1). Additional information regarding the site and its hydrogeology has been reported elsewhere (Phillips *et al.*, 2000; Watson *et al.*, 1999).

After completion of the trench installation, the biopolymer Guar gum was broken down with an enzyme (LED-4 jell breaker, GeoCon, Inc.) that was circulated through the trench for 2 days. That process was followed by a 5-day step-drawdown pumping test and a 7-day pumping test in the trench. The primary purpose of these initial pumping tests was to determine the hydraulic conductivity and groundwater flow conditions for a remediation scenario at the site.

C. Groundwater and Core Sampling and Analysis

Periodic groundwater samples were collected for the determination of contaminant metals/radionuclides (e.g., U, Tc, Cr), geochemical parameters [e.g., major cations and anions, pH, Eh, dissolved oxygen (DO), and total organic carbon (TOC)], and microbial characteristics from some selected monitoring wells. Groundwater pH, Eh, DO, and temperature were measured in line (without exposing to the air) by means of a YSI XL600M multiparameter probe (Yellow Springs Instruments, CO), which had been precalibrated. Ferrous ion (Fe^{2+}) and sulfide (S^{2-}) were determined immediately after taking samples in the field by the colorimetric technique using HACH kits equipped with a DR/2000 spectrophotometer (HACH, Loveland, CO).

The filtered groundwater samples (using 0.45-μm in-line filters) were also collected in two separate containers, either acidified or unacidified with concentrated nitric acid. The acidified samples were used for elemental analysis such as Ca, Mg, Fe, Al, Mn, Ni, Cd, Cr, Cu, Pb, Zn, Co, Sr, Na, K, and Si by inductively coupled plasma – atomic emission spectroscopy (Thermo

Jarrell Ash PolyScan Iris Spectrometer). Unacidified samples were used for the analysis of major anions, including nitrate (NO_3^-), sulfate (SO_4^{2-}), chloride, and phosphate, by means of an ion chromatograph equipped with a conductivity detector (Dionex, Sunnyvale, CA). Aliquots of the unacidified samples were also used for the analyses of TOC and total inorganic carbon (TIC) by means of a total organic carbon analyzer (TOC-5000A, Shimadzu, Tokyo). The TIC was then converted to bicarbonate (HCO_3^-) or carbonate concentrations in groundwater.

An aliquot of the acidified groundwater sample was also used for the analysis of total U(VI) by means of a laser-induced kinetic phosphorescence analyzer (ChemChek, KPA-11). However, because of a relatively low U(VI) concentration and the interference (or quenching effects) of other groundwater constituents such as Cl^- and Ca^{2+} in groundwater, the samples were purified and preconcentrated using UTEVA resin extraction columns (Eichrom, IL). Experimentally, an aliquot of groundwater sample (10–20 ml) was acidified with nitric acid to a minimum concentration of 3 *M*, in which uranyl (UO_2^{2+}) forms complexes with NO_3^- (Horwitz *et al.*, 1992). The uranyl–nitrate complexes were subsequently sorbed by the UTEVA resin, leaving other groundwater constituents unsorbed (or in the leachate). The column was then leached with dilute HNO_3 (1 m*M*), in which uranyl–nitrate complexes are unstable (or dissociate) and therefore desorbed from the resin column. The dilute HNO_3 leachate (with uranium) was collected and analyzed for U(VI) content as described earlier, and the detection limit was better than 0.01 μg/liter.

About 1.2 and 2.5 years after the Fe^0 barrier was installed, soil and iron core samples were collected in polyurethane tubes from the barrier and its adjacent fill materials and used for analyses of surface morphology, mineral precipitation, uranium content, and microbial characteristics. The mineralogical characterization of the core materials was performed using X-ray diffraction (XRD) operated at 45 kV and 40 mA (Scientag XDS-2000 diffractometer, Sunnyvale, CA). To determine mineral deposition, morphology, and elemental composition, Fe^0 core samples were also carbon coated with a Bio-Rad carbon sputter coater and examined immediately by means of a JEOL ISM-35CF scanning electron microscope (SEM) equipped with an energy-dispersive X-ray analyzer (EDX) (Tokyo, Japan). Note that all core samples were stored in Ar-purged airtight PVC tubes before use so as to minimize oxidation. Additionally, it is pointed out that the sample preparation and timing are critical because such minerals as green rusts are particularly sensitive to oxidation and drying methodologies, and details regarding sample preparation and analytical procedures can be found elsewhere (Gu *et al.*, 2001; Phillips *et al.*, 2000, 2001).

III. BIOGEOCHEMICAL REACTIONS IN THE IRON REACTIVE BARRIER

Although the ultimate goal of constructing the Fe^0 reactive barrier is to retain or degrade target contaminants of concern in groundwater, it must be realized that a range of biogeochemical reactions occur simultaneously because Fe^0 reacts not only with contaminant chemicals but also with a number of natural groundwater constituents. As illustrated in Fig. 2, corrosion of Fe^0 in groundwater generates soluble Fe^{2+} ions, dissolved H_2, and an increased groundwater pH. More important, corrosion of Fe^0 results in a low redox potential (or Eh) and provides excess electrons for the reduction of a number of redox-sensitive contaminant metal species (e.g., uranyl, pertechnetate, and chromate), chlorinated organic compounds, and other groundwater constituents (e.g., nitrate and sulfate). Depending on the site biogeochemical conditions, these reactions are of great importance because they not only affect the reactivity and long-term performance of Fe^0 to retain or remove contaminants, but also determine the rate of Fe^0 corrosion and, hence, the life span of the permeable Fe^0 reactive barrier.

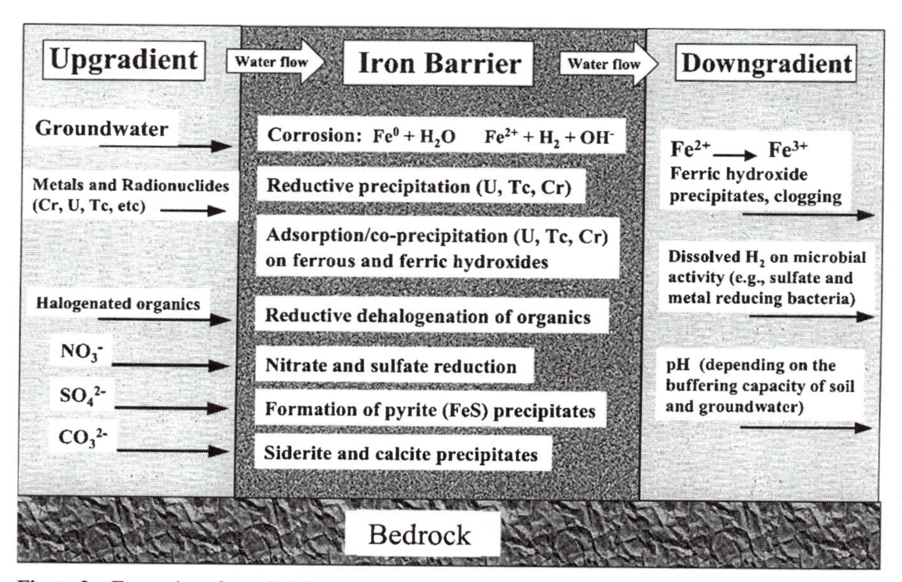

Figure 2 Examples of possible biogeochemical reactions occurring within and in the vicinity of the iron reactive barrier.

A. Groundwater pH, Eh, and Ferrous Ion

Corrosion of Fe^0 in groundwater follows two general pathways. In the presence of dissolved oxygen, Fe^0 corrodes according to the reaction

$$Fe^0 + H_2O + \frac{1}{2}O_2 \rightarrow Fe^{2+} + 2\,OH^- \quad (1)$$

However, under anaerobic conditions (e.g., oxygen consumed by the above reaction or by anaerobic microorganisms), Fe^0 can react with water according to

$$Fe^0 + 2\,H_2O \rightarrow Fe^{2+} + H_2 + 2\,OH^-. \quad (2)$$

Both of these reactions result in a decreased redox potential but an increased solution pH as 2 mol of OH^- are formed per mole of Fe^0 oxidized. As shown in Fig. 3, the site groundwater pH (upgradient) is generally stabilized at ~6.5. However, upon reaction with Fe^0 in the barrier, groundwater pH within the Fe^0 increased and stabilized at from ~7.5 up to ~10 under field conditions. However, groundwater pH remained at ~6.5 in downgradient wells (except those in the Fe^0 barrier, DP-23s,m), and corrosion of Fe^0 in the barrier appeared to have little impact on the downgradient soil based on observed pH values (Fig. 3). This observation may be attributed to the relatively high pH-buffering capacity of clay minerals and organic matter in downgradient soil. The groundwater pH in monitoring well from TMW-7 appeared to be somewhat high (up to ~9.5) because this well is situated in the gravel trench downgradient of the Fe^0 barrier, which is low in pH-buffering capacity.

Groundwater redox potential (Eh), however, decreased dramatically in those monitoring wells both within and downgradient of the Fe^0 barrier (Fig. 4). Although the site groundwater is generally aerobic, with Eh values mostly positive (over 200 mV) (Fig. 4 upgradient), a generally low Eh was observed within the Fe^0 barrier (< -200 mV). This decrease in Eh may directly result from the consumption of dissolved O_2 and the production of dissolved H_2 as groundwater reacted with Fe^0 in the reactive zone. Eh values were also in the negative range in most of the downgradient monitoring wells. Note that a low Eh was also observed in the upgradient TMW-11 and DP-22s monitoring wells because the DP-22 well is also located within the Fe^0 barrier and the TMW-11 is located adjacent to the Fe^0 barrier. The use of Guar gum could also have resulted in an increased anaerobic microbial activity (Gu *et al.*, 2001) and thus may have contributed to a low redox potential in the TMW-11 monitoring well.

Ferrous iron (Fe^{2+}), one of the major by-products of Fe^0 corrosion in groundwater, may have a significant impact on water quality and cause

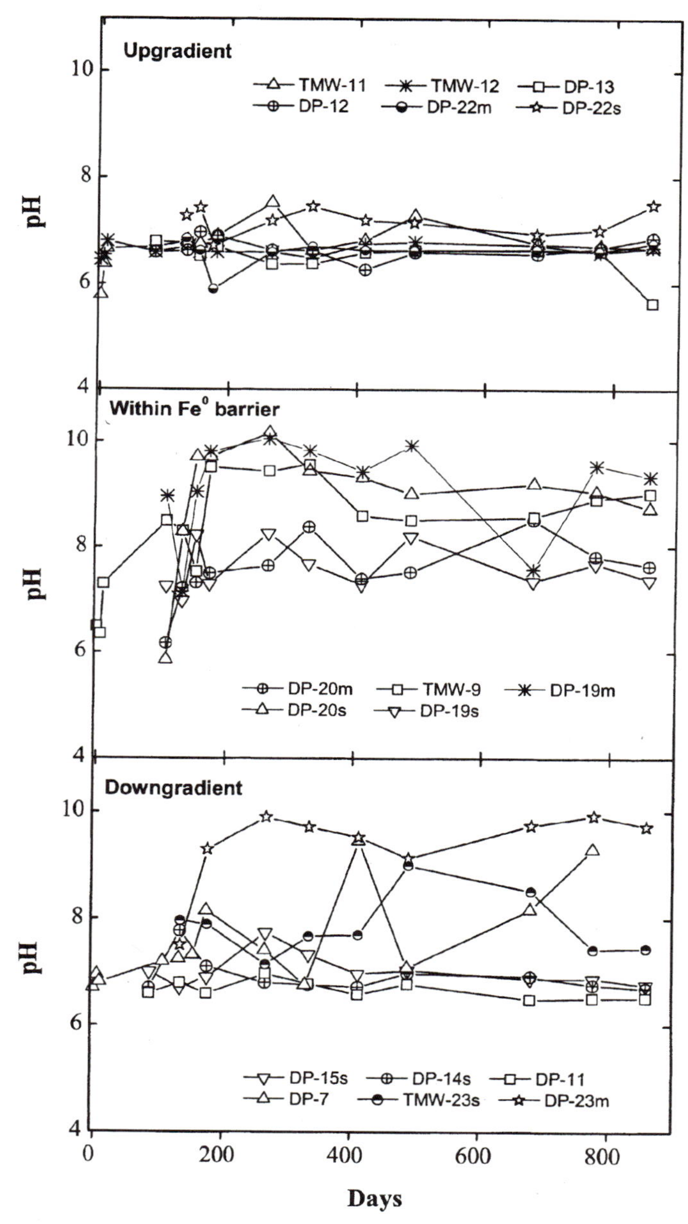

Figure 3 Groundwater pH in some selected monitoring wells within and in the vicinity of the Fe^0 reactive barrier at the Y-12 site, Oak Ridge, Tennessee. The barrier was installed on November 30, 1997 (time = 0).

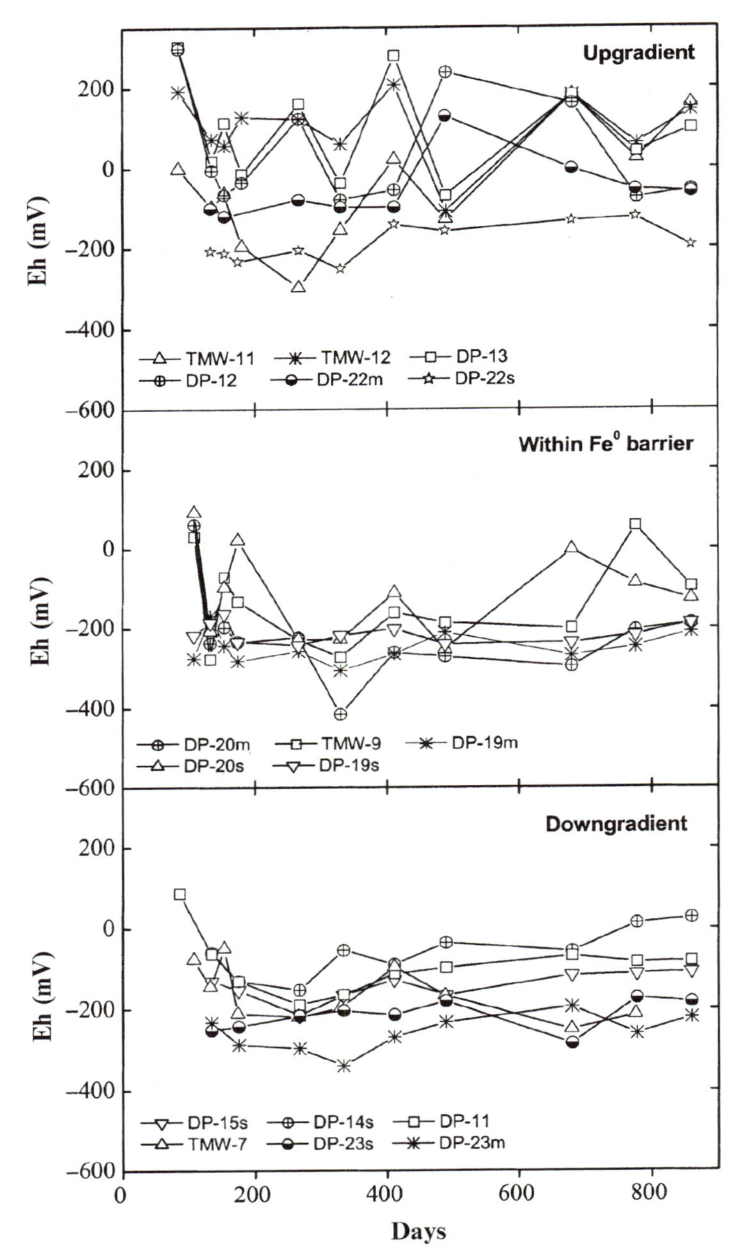

Figure 4 Groundwater redox potential (Eh) in some selected monitoring wells within and in the vicinity of the Fe^0 reactive barrier at the Y-12 site, Oak Ridge, Tennessee.

clogging of soil porous structure as it oxidizes and precipitates out in the downgradient barrier or the soil. In fact, discharge of Fe^{2+} to Bear Creek had been one of the major concerns initially regarding the implementability of Fe^0 barriers at the site. However, we found that Fe^{2+} concentrations in groundwater were relatively low after a few months of operation of the Fe^0 barrier in the field (Fig. 5). Initially, the Fe^{2+} concentration was found to be extremely high (up to ~ 150 mg/liter) in the center well (TMW-9) of the Fe^0 barrier, and total iron concentration reached levels as high as ~700 mg/liter (data not shown). Relatively high Fe^{2+} concentrations were also observed in the pea gravel section adjacent to the Fe^0 barrier for TMW-7 and TMW-11 (Fig. 6). However, Fe^{2+} concentrations decreased rapidly over the first few months after the Fe^0 barrier was installed (such as in the TMW-9 and TMW-7 wells). Only a few monitoring wells within the Fe^0 barrier showed a slightly high Fe^{2+} concentration but was <15 mg/liter in general (e.g., DP-22s,m; DP-20s,m; DP-23s) (Fig. 5). As discussed later, Fe^{2+} ions may be precipitated as FeS, $FeCO_3$, or be further oxidized as Fe^{3+}, which forms relatively insoluble iron oxyhydroxides in the Fe^0 barrier. These results suggest that Fe^{2+} discharge as a result of Fe^0 corrosion is not always a significant concern at this site. However, the rate of Fe^0 corrosion and the production of Fe^{2+} depend on both groundwater pH and constituent concentrations, such as nitrate, sulfate, and bicarbonate, which accelerate the corrosion process (Agrawal and Tratnyek, 1996; Agrawal *et al.*, 1995; Gu *et al.*, 1999). Depending on removal rates via precipitation and sorption, high concentrations of ferrous ion may persist under certain conditions.

The high initial Fe^{2+} and total iron concentrations observed in groundwater (Fig. 5) may be attributed in part to the following factors: (1) a rapid initial oxidation of Fe^0 filings (particularly some fine iron particles) when they were emplaced into the groundwater; (2) the use of Guar gum and, subsequently, the addition of enzyme (used to break up the Guar gum), which resulted in a decreased groundwater pH in a short time period; and (3) increased microbial activity (or respiration), which may also contribute to an increased corrosion rate of Fe^0.

B. Reactions between Nitrate and Fe^0

Groundwater at the barrier site is contaminated with relatively high levels of nitrate (NO_3^-) at ~20–150 mg/liter at the pathway 2 site; in some deep monitoring wells or piezometers, levels >1000 mg/liter NO_3^- were observed as a result of the migration of deep contaminated groundwater and the upward vertical hydraulic gradients at the site. Within and in the vicinity of the Fe^0 barrier, however, the NO_3^- concentration were low to nondetectable (Fig. 7),

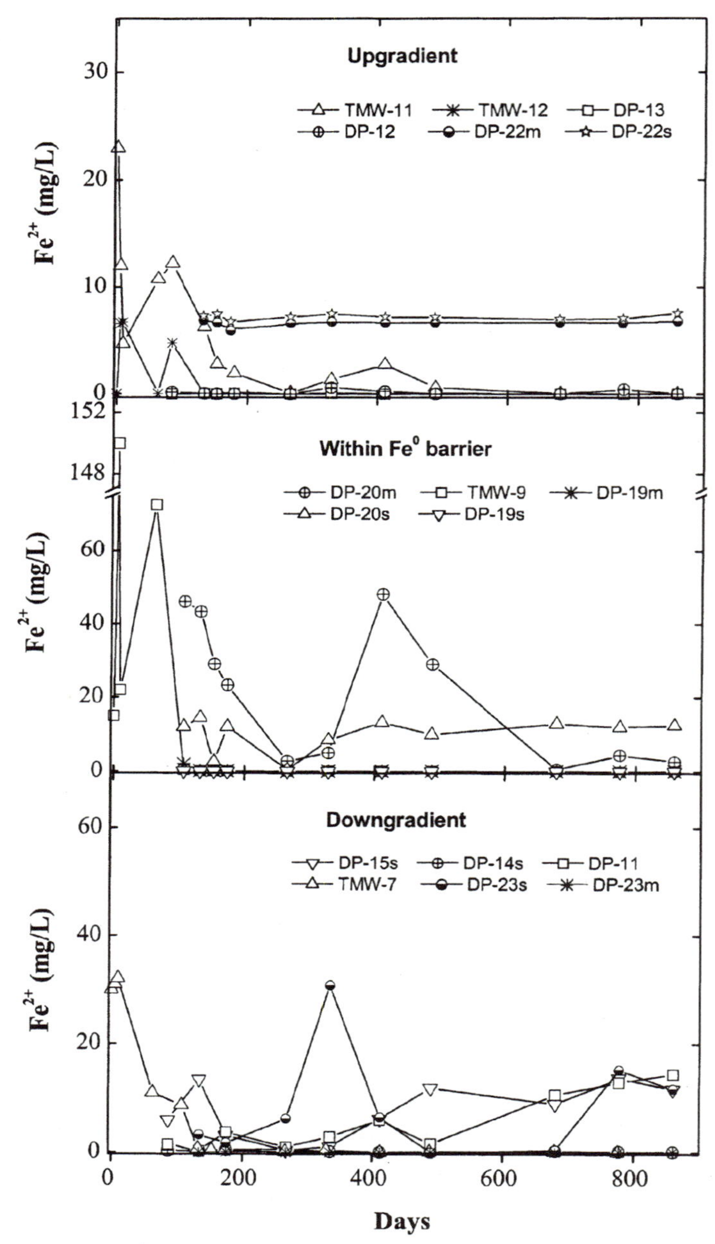

Figure 5 Ferrous (Fe^{2+}) concentration in some selected monitoring wells within and in the vicinity of the Fe^0 reactive barrier at the Y-12 site, Oak Ridge, Tennessee.

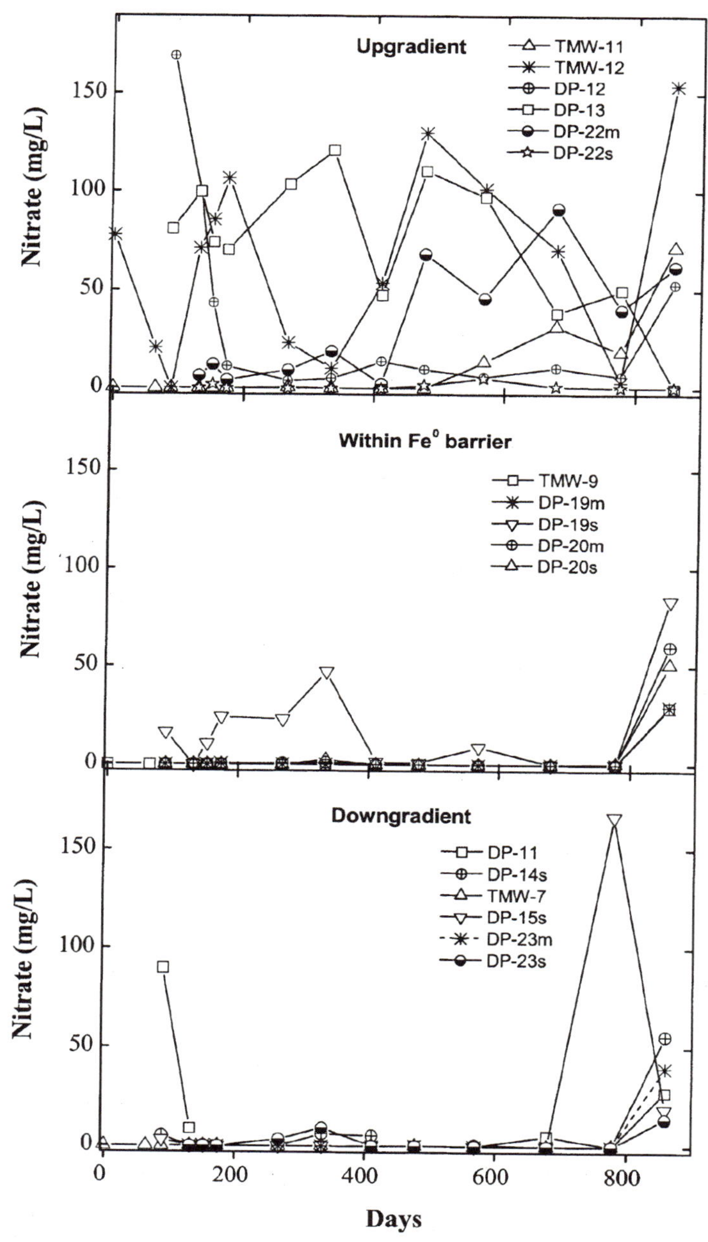

Figure 6 Nitrate concentration profiles in some selected monitoring wells within and in the vicinity of the Fe^0 reactive barrier at the Y-12 site, Oak Ridge, Tennessee.

suggesting that NO_3^- was effectively degraded as the groundwater passed through the Fe^0 barrier. The nitrate concentrations were also found to be low or nondetectable downgradient of the Fe^0 barrier, including monitoring wells or piezometers TMW-7, DP-11, DP-14s, DP-15s, and many others that are not shown in Fig. 6. Even in some upgradient monitoring wells or piezometers (adjacent to the Fe^0 barrier), a low NO_3^- concentration was observed in monitoring wells such as TMW-11 and DP-12. These observations suggest that NO_3^- is degraded readily in the reducing zone of influence by Fe^0 corrosion.

The reduction of NO_3^- observed in the downgradient and some upgradient monitoring wells may be partially attributed to denitrification by microorganisms. As reported previously (Gu *et al.*, 2001), an enhanced anaerobic microbial population was observed in soils both downgradient and upgradient of the Fe^0 barrier. This was presumably related to a low Eh and an increased level of dissolved H_2 in groundwater (a by-product of Fe corrosion), which served as electron donors for the microbial reduction of NO_3^-. However, direct abiotic reduction of NO_3^- by Fe^0 filings should not be ruled out within the Fe^0 barrier (Gu *et al.*, 1997; Huang *et al.*, 1998).

In the laboratory, the reduction of NO_3^- and its associated by-products by Fe^0 filings was evaluated in the presence or absence of peat materials and/or denitrifying bacteria. Results indicated that nitrate was reduced effectively by Fe^0, despite a relatively high initial NO_3^- concentration (6000 mg/liter) used in these laboratory batch experiments (Fig. 7a). The degradation half-life by Fe^0 alone was found to be on the order of ~1–2 weeks by assuming a pseudo-first-order reaction kinetics, and more than 60% of NO_3^- was degraded after about 2 weeks of reaction. The addition of peat materials (from Wards Scientific) was found to enhance the reduction rate of NO_3^-, with a decreased reaction half-life on the order of ~2 days. More than 95% of the NO_3^- was degraded in a 1-week period. However, note that reduction rates did not increase significantly with the addition of a toluene-degrading denitrifying bacterium, *Azoarcus tolulyticus* Tol-4, into the Fe^0 and peat mixture (Chee-Sanford *et al.*, 1996). In fact, addition of this denitrifying bacterium directly into Fe^0 filings did not increase denitrification either, probably because of a high NO_3^- concentration and a relatively high pH condition (up to ~10) in the reactant solutions or an unfavorable environment for the microbial reduction of NO_3^-. The presence of peat (with indigenous microbes in the peat), however, buffered the pH of the reactant solution (pH <8.5) so that a substantially enhanced NO_3^- reduction rate was observed under these conditions. Nevertheless, results of these laboratory experiments are consistent with the field monitoring results and demonstrate that Fe^0 is an effective reactive medium for removing NO_3^-, in addition to degrading chlorinated organics and sequestering some redox-sensitive metals or radionuclides, as reported previously

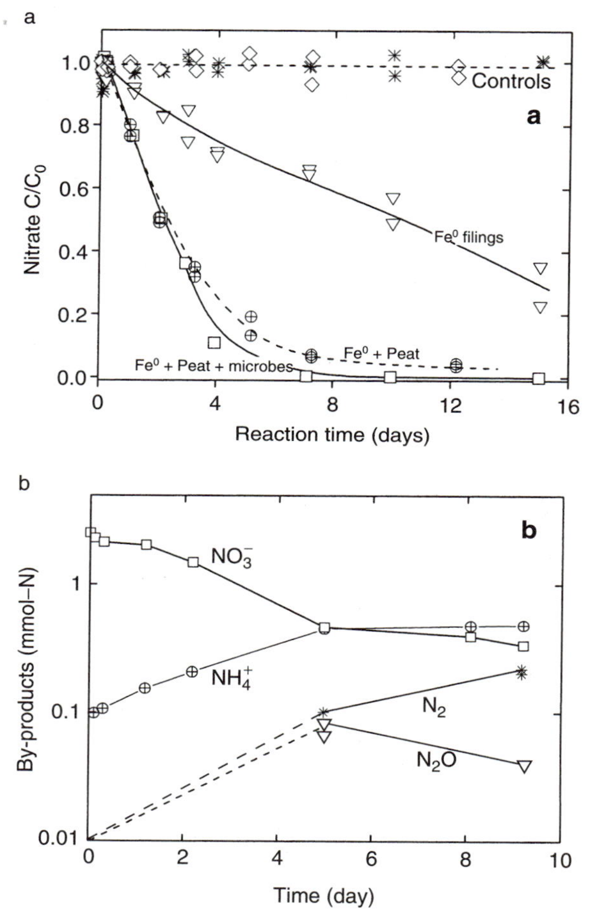

Figure 7 (a) Nitrate degradation kinetics with Fe^0 filings, or a mixture of Fe^0 filings with Wards peat and/or an *Azoarcus tolulyticus* Tol-4 denitrifying bacterium. Initial NO_3^- concentration was 6000 mg/liter in these laboratory studies, and controls were added with Wards peat or Tol-4 bacterium only (without addition of Fe^0 filings). (b) Identification of NO_3^- degradation by-products with Fe^0 filings and Wards peat.

(Blowes *et al.*, 1997; Cantrell *et al.*, 1995; Gillham *et al.*, 1994; Gu *et al.*, 1998; McMahon *et al.*, 1999).

The reaction by-products between NO_3^- and Fe^0 and peat mixtures were also examined. Results (Fig. 7b) indicate that a portion of NO_3^- (~25%) was converted to ammonia (NH_4^+) in the aqueous solution, and a large percentage of NO_3^- may have been degraded as N_2 or N_2O gases. A good mass balance was not obtained in these batch kinetic experiments, largely because of the loss of N_2 and N_2O gases to the headspace or atmosphere.

C. Sulfate Reduction

A decreased concentration of sulfate (SO_4^{2-}) was also observed within the Fe^0 barrier. As illustrated in Fig. 8, sulfate was the highest in the upgradient soil and pea gravel portion of the barrier. Upon entering the Fe^0 portion of the trench, sulfate was found to be reduced substantially at all levels. For example, at the multilevel monitoring wells of DP-19 and DP-20, SO_4^{2-} concentrations were significantly lower than those in the upgradient wells. In particular, sulfate was largely removed or degraded in some of the downgradient monitoring wells in soil (e.g., DP-14s), in iron (DP-23s,m), and in gravel (TMW-7). These observations provide evidence of sulfate reduction in the zone of Fe^0 influence, although the mechanisms of sulfate reduction are still a subject of investigation. These observations are also consistent with previous studies that show that groundwater SO_4^{2-} concentrations decreased through the Fe^0 barriers at the Moffett Field and Lowry AFB sites, at the Elizabeth City, U.S. Coast Guard site (Puls *et al.*, 1999), and in the laboratory-simulated column studies with a continuous input of SO_4^{2-} and HCO_3^- solutions (Gu *et al.*, 1999).

The reduction of SO_4^{2-} resulted in the formation of sulfide (S^{2-}), although much of the sulfide produced may have been precipitated rapidly as FeS because of its low solubility (K_{sp} on the order of 10^{-18}). This explains a relatively low S^{2-} concentration observed in most of the monitoring wells (data not shown). Nevertheless, sulfide concentrations were found to be somewhat higher in those monitoring wells adjacent to the iron (TMW-11 and TMW-7) and within the iron barrier (TMW-9) than in those monitoring wells upgradient of the Fe^0 barrier (TMW-12 and DP-12). The exact mechanism of SO_4^{2-} reduction to S^{2-} is not yet clear because there is no direct evidence showing an abiotic reduction of SO_4^{2-} by Fe^0, although reduction of sulfonic acid to S^{2-} by Fe^0 has been reported (Lipczynska-Kochany *et al.*, 1994). However, a decreased SO_4^{2-} concentration in the barrier could be at least partially attributed to reduction by sulfate-reducing microorganisms (Gu *et al.*, 2001).

Using phospholipid fatty acids (PLFA) and DNA analyses (Dowling *et al.*, 1986; Guckert *et al.*, 1986; Tunlid and White, 1991; Zhou *et al.*, 1996), an increased microbial population was observed within and in the vicinity of the Fe^0 barrier. The microbial population was found to be on the order of 10^5 to 10^6 cells/ml groundwater, which is substantially higher than that found in the background soil, located ~50 ft upgradient of the Fe^0 barrier (Gu *et al.*,

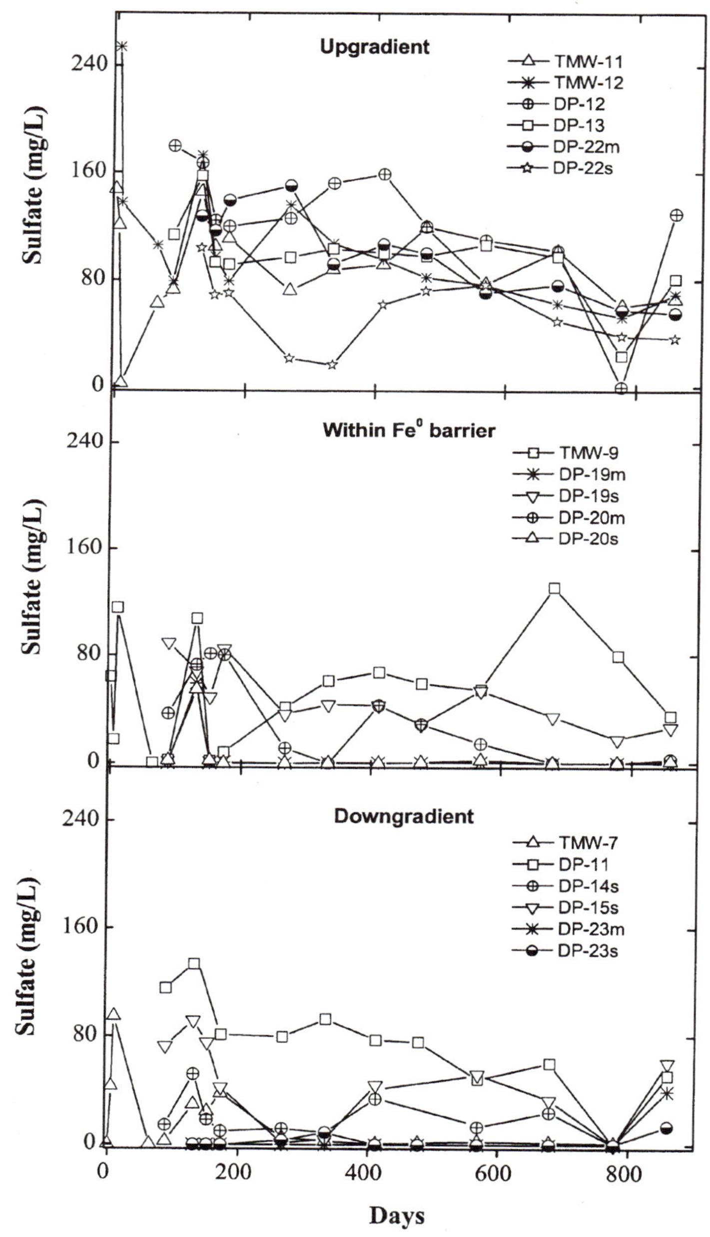

Figure 8 Sulfate concentration profiles in some selected monitoring wells within and in the vicinity of the Fe^0 reactive barrier at the Y-12 site, Oak Ridge, Tennessee.

2001). More importantly, perhaps, diversified microbial communities were also detected in groundwater by examining the characteristic fatty acid profiles or lipid biomarkers, although PLFA analysis is unable to identify the specific functional groups of microorganisms. Many microbial species may have similar PLFA patterns. Therefore, DNA analysis based on polymerase chain reactions (or PCR analysis) was used to further identify different functional groups of microorganisms. As reported previously, sulfate-reducing bacteria appeared to be one of the most abundant microorganisms identified in both groundwater and core samples obtained within and in the vicinity of the Fe^0 barrier. Both sulfate-reducing and denitrifying bacteria were found to be the highest in TMW-11 (upgradient adjacent to the Fe^0 barrier) and DP-11 (~3 ft downgradient of the Fe^0 barrier). These observations provide additional evidence that a decreased SO_4^{2-} concentration within the Fe^0 portion of the trench could be a result of microbial reduction of SO_4^{2-} to S^{2-} under anaerobic conditions. Hydrogen generated by the corrosion of Fe^0 (and the initial use of Guar gum for trench excavation) could have played a significant role in stimulating the growth of these anaerobic microorganisms (Gu *et al.*, 1999).

D. Interactions between Fe^0 and Contaminant Metals

1. Uranium Removal

At the Y-12 site, uranium is the main driver for this groundwater remediation using the permeable Fe^0 reactive barrier because uranium poses the major potential health and environmental risks within Bear Creek Valley. Its concentration trends within and in the vicinity of the Fe^0 barrier are plotted in Fig. 9 for some monitoring wells and piezometers. Results indicate that uranium is removed effectively within the Fe^0 barrier and that uranium concentrations in the Fe^0 barrier (e.g., TMW-09, DP-19s,m, and DP-20s,m) were generally very low (<0.01 mg/liter) in comparison with uranium concentrations in the upgradient monitoring wells. With the exception of DP-11, low amounts of uranium were also found in the downgradient monitoring wells (e.g., DP-23s,m, and DP-14s, TMW-7), suggesting that the Fe^0 barrier is performing well in removing or retaining uranium from the contaminated groundwater. Similarly, uranium concentrations in other monitoring wells and piezometers (such as DP-10, DP-18s,m,d; DP-21s,m; DP-22s) were found to be low or below the detection limit in the Fe^0 barrier (data not shown). These observations are generally consistent with previous laboratory studies that show uranium can be reduced effectively and rapidly by Fe^0 filings (Gu *et al.*, 1998). However, a consistently high uranium was detected in the

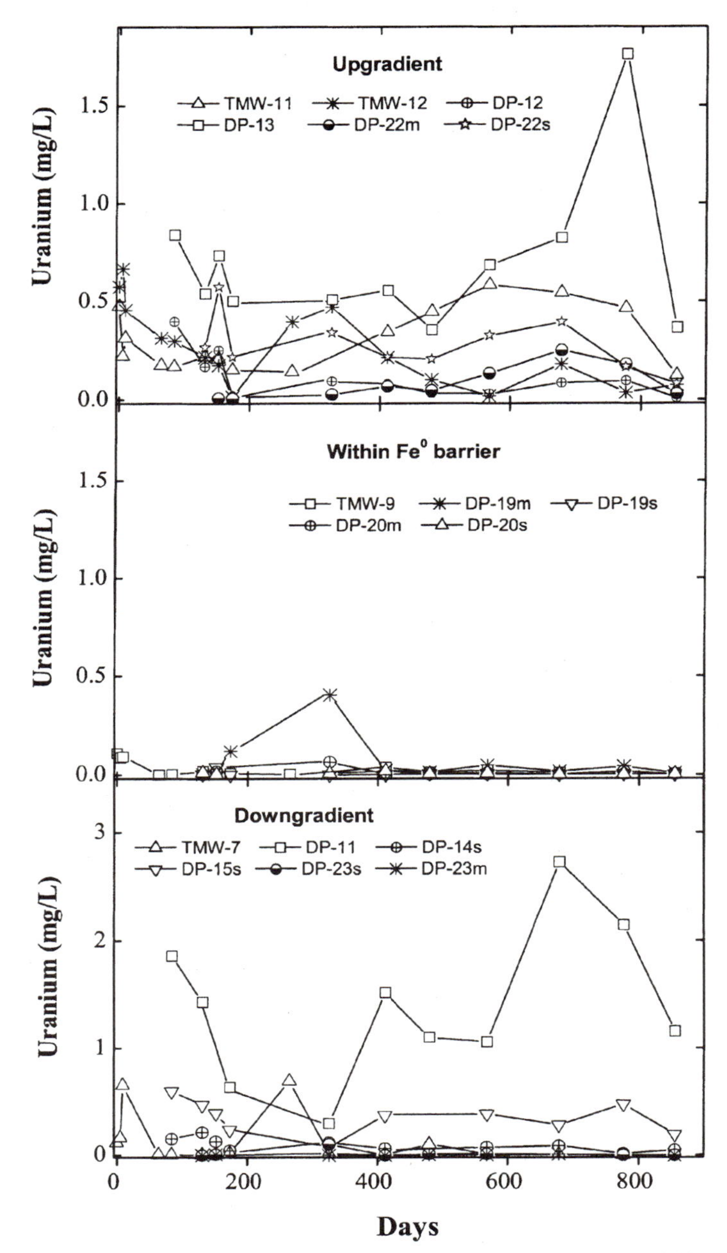

Figure 9 Uranium concentration profiles in some selected monitoring wells within and in the vicinity of the Fe^0 reactive barrier at the Y-12 site, Oak Ridge, Tennessee.

downgradient DP-11 well, and its concentration was even higher than those found in the upgradient monitoring wells. This observation was attributed primarily to the fact that some uranium hot spots existed in such heterogeneous filled materials as evidenced by our recent soil core sampling and analysis near DP-11 (conducted in September 2001). Another possible explanation was due to an upward flow of the contaminated groundwater as a result of the vertical hydraulic gradient (described in Section II).

Uranium retention by Fe^0 filings was also evidenced by the analysis of uranium content in Fe^0 core samples (Fig. 10). These angled cores were collected ~15 months after installation of the Fe^0 barrier, and detailed sample preparation and analysis are given elsewhere (Phillips *et al.*, 2000). Although only trace quantities of uranium are present in the contaminated groundwater, an elevated amount of uranium was detected in Fe^0 core materials, particularly in those samples near the interface (between the soil and Fe^0 filings) where groundwater enters the Fe^0 barrier. The greatest concentration of uranium in the Fe^0 barrier occurred at the shallow, upgradient interface, but uranium concentration decreased dramatically over a short distance. These observations suggest that once uranium enters the Fe^0 reactive barrier, it is sequestered rapidly *in situ* as a result of either reductive precipitation of

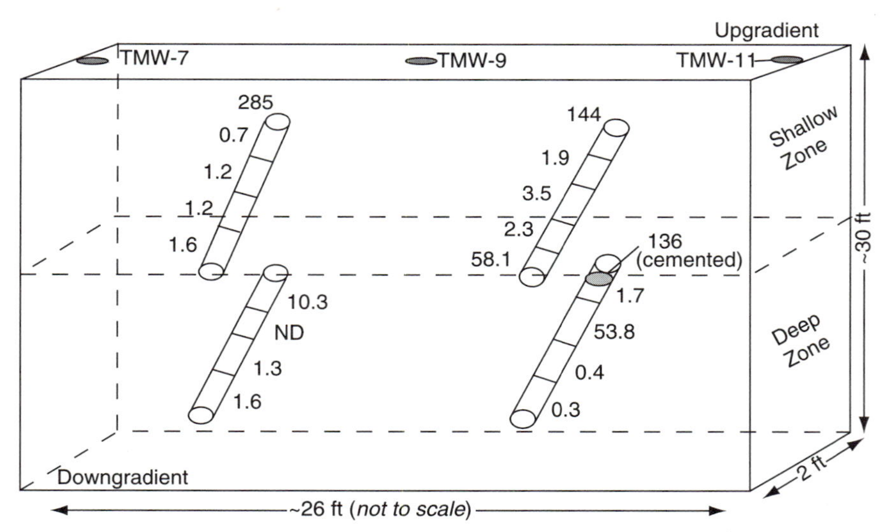

Figure 10 Distribution of uranium (mg/kg) in core samples within the Fe^0 portion of the iron/gravel trench barrier. Angled core samples (60°) were collected from upgradient to downgradient of the Fe^0 barrier in February 1999 (Reprinted with permission from Phillips *et al.*, 2000. Copyright 2000, American Chemical Society).

relatively insoluble U(IV) species or surface adsorption of U(VI) species on the Fe^0 corrosion products (e.g., iron oxyhydroxides) (Cantrell *et al.*, 1995; Gu *et al.*, 1998).

Reductive precipitation of U(VI) to U(IV) species by Fe^0 is believed to be one of the dominant mechanisms for uranium removal and is thermodynamically favorable, according to the following stoichiometric reactions (Baes and Mesmer, 1976; Gu *et al.*, 1998; Morse and Choppin, 1991):

$$Fe^{2+} + 2e^- \rightarrow Fe(0) \qquad \varepsilon^0 = -0.440V \tag{3}$$

$$UO_2{}^{2+} + 4H^+ + 2e^- \rightarrow U(IV) + 2H_2O \qquad \varepsilon^0 = 0.327V \quad \textit{or} \quad \varepsilon = -0.07\text{ V at pH8.} \tag{4}$$

The reduced U(IV) readily forms oxyhydroxide precipitates in aqueous solution. Such a reductive reaction mechanism has been evaluated by laboratory batch equilibrium studies and by sensitive fluorescence spectroscopic analysis (Gu *et al.*, 1998). Fluorescence spectra gave direct evidence of U(VI) reduction by Fe^0 because the reduced U(IV) species do not fluorescence. The batch equilibrium studies provided additional evidence because, regardless of the initial added uranium concentration in solution (up to 20,000 mg/liter), no detectable amounts of uranium were found in equilibrium solutions after reaction with Fe^0. These results are indicative of a reductive precipitation process rather than a simple sorption process in which uranyl distributes or partitions between the solution and solid phases depending on the adsorption affinity and capacity on the adsorbent surfaces. However, sorption by iron oxides and other adsorbent materials was found to be much less effective than Fe^0 filings in removing uranium from the solution. A much higher equilibrium U(VI) concentration was observed in samples treated with these materials than in those treated with Fe^0 filings. Using the X-ray photoelectron spectroscopy (XPS) technique, Fiedor *et al.* (1998) also reported that U(VI) was readily reduced to U(IV) species ($\sim$75%) by reacting with an Fe^0 coupon under anaerobic conditions, although a large solution to Fe^0 ratio was used in these laboratory studies.

However, as the corrosion products of Fe^0, such as iron oxyhydroxides, accumulate on Fe^0 surfaces, uranium removal through sorption or coprecipitation could not be ruled out (Fiedor *et al.*, 1998; Gu *et al.*, 1998; Hsi and Langmuir, 1985; Morrison *et al.*, 1995). Unfortunately, because of a relatively low amount of uranium retained by the Fe^0 filings and a possible reoxidation of reduced U(IV) species during sample preparation and extraction, no attempts were made to distinguish whether uranium was reductively precipitated or sorbed by iron oxyhydroxides in these Fe^0 barrier materials. On the basis of previous laboratory studies (Cantrell *et al.*, 1995; Fiedor *et al.*, 1998; Gu *et al.*, 1998), a speculation is that a majority of uranium could have been retained by the reductive precipitation process because of a strong reducing environment within the Fe^0 barrier (with a high solid to solution ratio). Understanding of U(VI) removal mechanisms through either reductive precipitation or sorption/coprecipitation has important environmental implications because the reduced U(IV) species on Fe^0 surfaces could be potentially reoxidized when it is exposed to the air or dissolved O_2 in a matter of a few hours or days (Gu *et al.*, 1998). Similarly, the sorbed U(VI) species could be desorbed and therefore remobilized as groundwater geochemistry changes.

2. Pertechnetate and Chromate Reduction

The effectiveness of the Fe^0 barrier in removing other contaminant metals, such as pertechnetate (TcO_4^-) and chromate (CrO_4^{2-}), was also monitored, although they are not major contaminants of concern because of their low concentration in the site groundwater. Technetium (as TcO_4^-) is a radioactive β emitter with an extremely long half-life (2.1×10^5 years), and chromate is a hazardous heavy metal. Results indicate that TcO_4^- was also retained effectively by the Fe^0 barrier and that generally less than 40 pCi/liter (or $\sim$2 ng/liter) of ^{99}Tc was detected in those monitoring wells within or adjacent downgradient of the Fe^0 barrier. The influent groundwater (in upgradient monitoring wells) contained $\sim$600 pCi/liter of TcO_4^-. The chromate concentrations in the upgradient and downgradient groundwater were mostly below the detection limit. As with the U(VI) species, the mechanisms of TcO_4^- and CrO_4^{2-} removal by Fe^0 filings are attributed to a reductive precipitation process, which was confirmed in laboratory batch kinetic studies (Fig. 11). As Fe^0 corrodes in water, CrO_4^{2-} can be reduced to Cr^{3+}, which is easily hydrolyzed and precipitated as $Cr(OH)_3$. Similarly, TcO_4^- is reduced to TcO_2 or $Tc(OH)_4$ as precipitates. The reduction kinetics of these contaminant metals or radionuclides appears to be extremely fast, and nearly 100% of UO_2^{2+}, TcO_4^-, and CrO_4^{2-} was removed after they contacted Fe^0 in water in a short time period (Fig. 11). The reductive precipitation or removal of CrO_4^{2-} has also been reported previously in the Fe^0 reactive barrier used to remediate groundwater contaminated with CrO_4^{2-} and trichloroethylene at the U.S. Coast Guard site at Elizabeth City, North Carolina (Blowes *et al.*, 1997; Puls *et al.*, 1999).

E. Interactions with Calcium and Carbonates

The groundwater at the barrier site contains high concentrations of both Ca^{2+} and bicarbonate because of the presence of calcium-rich bedrock, the

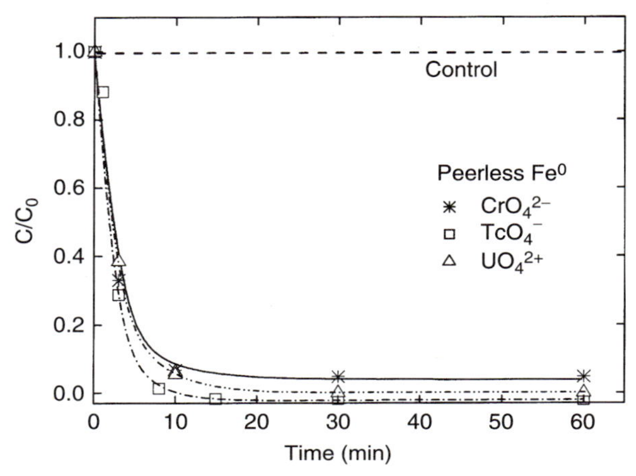

Figure 11 Reaction kinetics of contaminant metals (UO_2^{2+}, CrO_4^{2-}, and TcO_4^-) by Peerless Fe^0 filings in water. Initial concentrations (C_0) were 1000 mg U/liter, 10 mg Cr/liter, and 6 μg Tc/liter, respectively.

calcareous Nolichucky shale, and strong nitric acid leachate from the S-3 ponds and because of the neutralization of acid wastes by limestone in 1984. An analysis of groundwater carbonate/bicarbonate and Ca^{2+} indicates that these groundwater constituents were partially retained or precipitated within the Fe^0 barrier (Figs. 12 and 13). As shown in Fig. 12, Ca^{2+} concentrations in the upgradient side of the Fe^0 barrier (e.g., TMW-11, TMW-12, DP-12, and DP-13) appeared to be relatively constant (between ~120 and 200 mg/liter). However, the Ca^{2+} concentrations within the Fe^0 barrier from TMW-9, DP-19, and DP-20 monitoring wells were about an order of magnitude lower than those found in the upgradient monitoring wells, suggesting that Ca^{2+} was retained by the Fe^0 barrier. A relatively low Ca^{2+} concentration was also observed in many of the downgradient monitoring wells (e.g., DP-23s,m, TMW-7, and DP-14s).

Examination of carbonate/bicarbonate concentrations in groundwater revealed that these constituents were also partially removed (Fig. 13), suggesting that calcium carbonate precipitation is probably one of the dominant mechanisms responsible for decreased concentrations of Ca^{2+} and bicarbonate in groundwater. These results can be expected because of an increased groundwater pH as Fe^0 corrodes in the barrier and a resulting shift from bicarbonate to carbonate species in the groundwater. As has been reported previously, an increased pH and relatively high concentrations of Ca^{2+} and bicarbonate in the groundwater could have induced the chemical precipitation of Ca–carbonate and/or of a mixture of Fe– and Ca–carbonate and

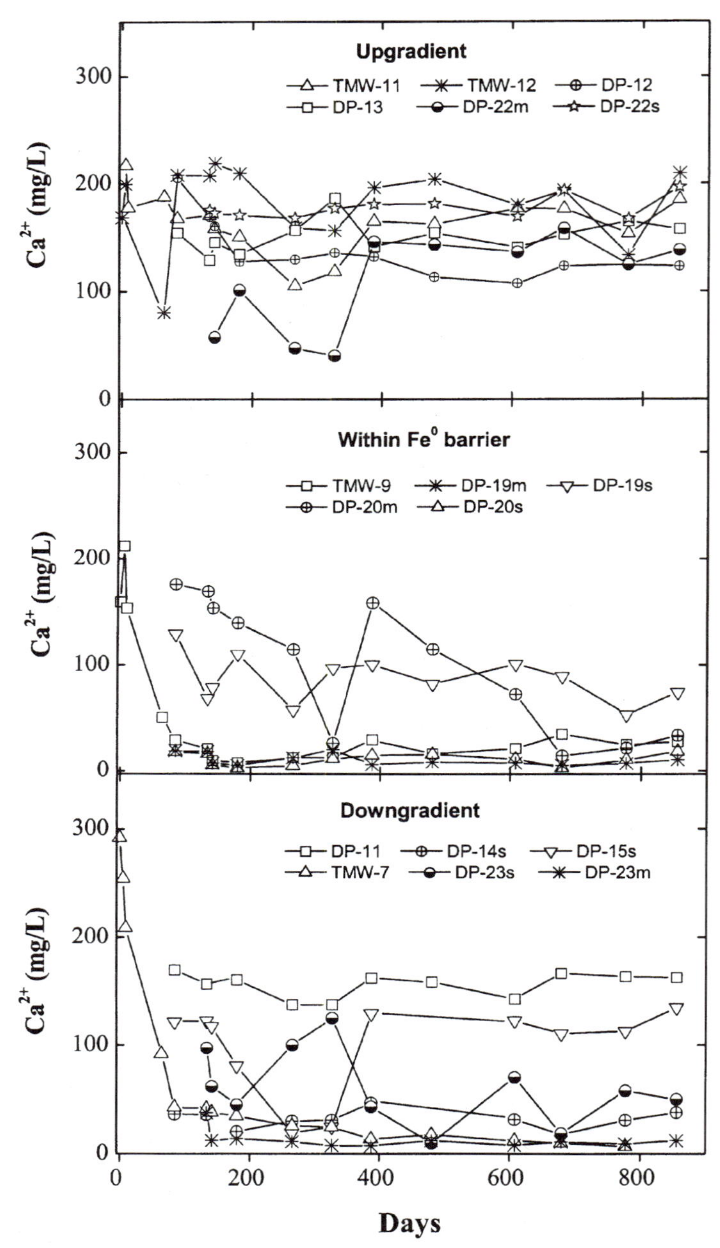

Figure 12 Calcium concentration profiles in some selected monitoring wells within and in the vicinity of the Fe^0 reactive barrier at the Y-12 site, Oak Ridge, Tennessee.

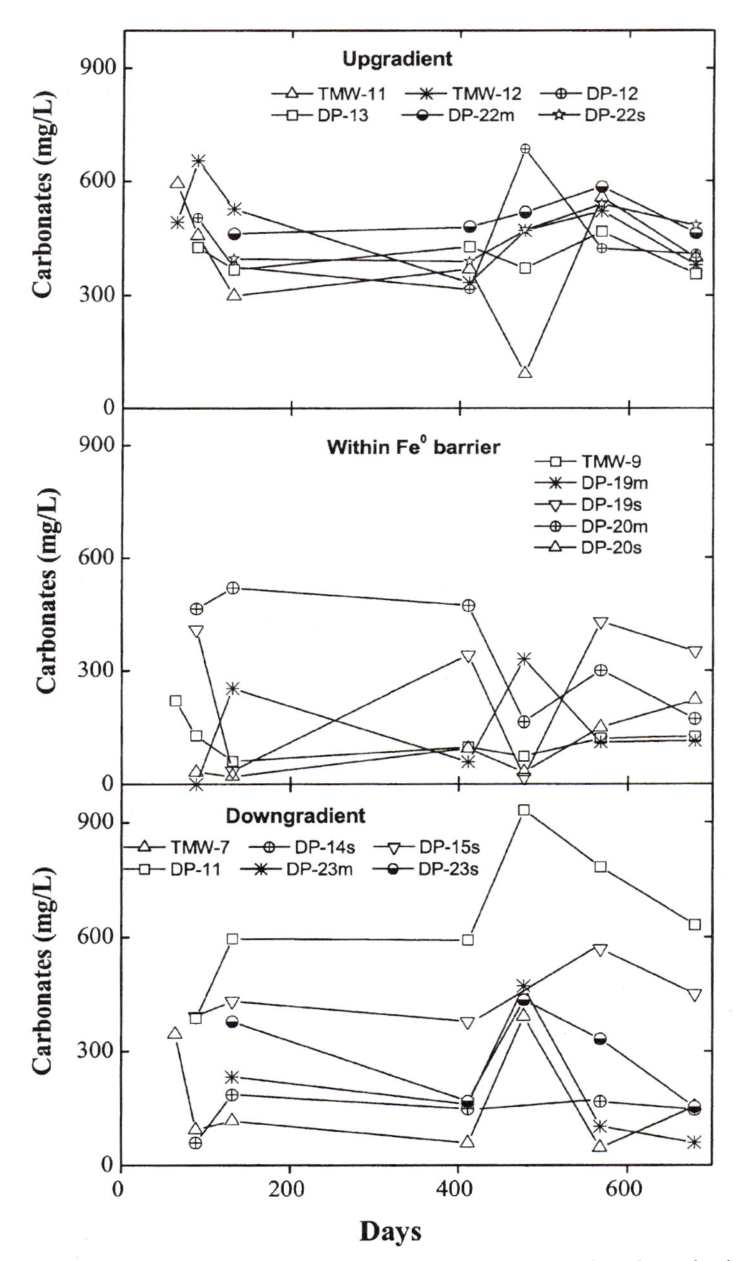

Figure 13 Carbonate/bicarbonate concentration profiles in some selected monitoring wells within and in the vicinity of the Fe^0 reactive barrier at the Y-12 site, Oak Ridge, Tennessee.

oxyhydroxide coprecipitates (Gu *et al.*, 1999; Phillips *et al.*, 2000). Similarly, by examining the concentration profiles of Mg^{2+} (data not shown), the Mg^{2+} concentration within the Fe^0 barrier also decreased over time because it also could form carbonate precipitates or coprecipitates with iron oxyhydroxides (Phillips *et al.*, 2000).

IV. HYDROLOGICAL AND MINERALOGICAL CHARACTERISTICS

A. Hydraulic Properties and Connectivity in the Fe^0 Barrier

Hydraulic gradients across the pathway 2 barrier site have remained relatively stable and consistent from east to west (magnitude of approximately 0.02). Figure 14 shows a comparison of groundwater levels and flow directions on May 6, 1998 and on May 17, 2000, respectively. Results indicate that the general flow patterns at the site have not changed since the start of the installation of the trench. Groundwater monitoring results over the past 3 years also suggest that increases and decreases in the gradients across the reactive barrier site appear to be related primarily to recharge during precipitation events and seasonal fluctuations. The hydraulic gradient in the trench across the iron has also remained consistently from east to west (Fig. 14) with an average gradient of approximately 0.008. The magnitude of the gradient changes during recharge events, but the direction of groundwater flow has been consistently toward the west. However, closer inspection of gradient fluctuations within the trench and the Fe^0 barrier seems to indicate that cementation within the iron may be starting to impact groundwater flow through the iron. Figure 15 shows groundwater elevations and gradients among three monitoring wells located in the pathway 2 trench. Since the spring of 1999 (or ~500 days after the barrier was installed), recharge events appear to have a more pronounced impact on hydraulic gradients (or fluctuations) observed among wells located upgradient (TMW-11), within (TMW-9), and downgradient (TMW-7) of the Fe^0 barrier. These observations suggest that the connectivity of the iron and gravel in the upgradient portion of the trench to the iron and gravel in the downgradient portion of the trench may be decreasing over time due to mineral precipitation and/or cementation in the Fe^0 barrier, as discussed in Sections IV,B. Coincidentally, the nitrate concentrations in upgradient well TMW-11 and DP-22m (located the furthest upgradient but still in the Fe^0 barrier) have an increasing trend starting in the spring of 1999 (Fig. 6). These data suggest

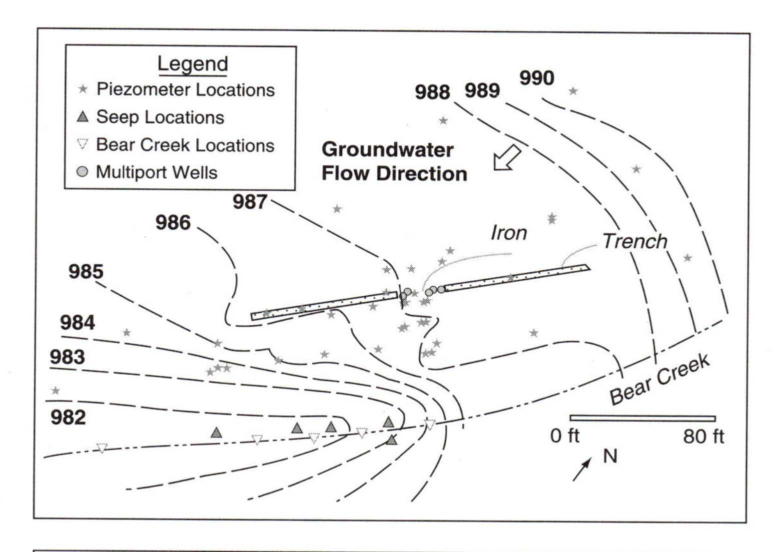

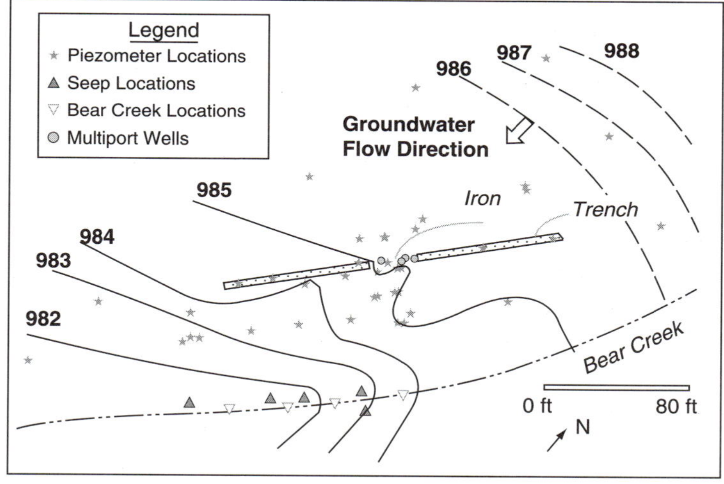

Figure 14 Comparison of groundwater levels (ft) measured on May 6, 1998 (a) and May 17, 2000 (b) at the Y-12 reactive barrier site, Oak Ridge, Tennessee.

that cementation of the iron in the upgradient portion of the trench may be causing a decrease in iron reactivity, hydraulic connectivity, and the beginning stages of system clogging.

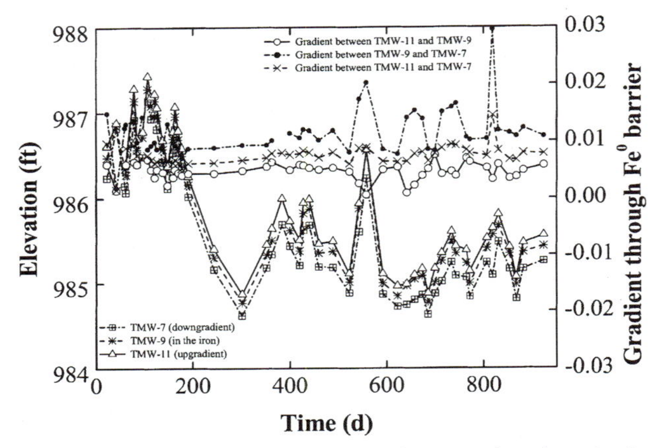

Figure 15 Groundwater elevations and gradients (ft) in some selected monitoring wells upgradient or downgradient of the iron reactive barrier at the Y-12 site, Oak Ridge, Tennessee.

B. Mineral Precipitates and their Occurrence

Iron corrosion in groundwater results in the formation of ferrous or ferric ions (when dissolved O_2 is present), which ultimately form iron oxyhydroxide mineral precipitates because of their low solubility (Fig. 2). It is not surprising, therefore, that many investigators observed iron oxyhydroxides to be the predominant minerals found in iron reactive barriers (Gu *et al.*, 1999; Phillips *et al.*, 2000; Pratt *et al.*, 1997; Roh *et al.*, 2000). Core samples were taken ~1.2 and 2.5 years after the Fe^0 barrier was installed, and X-ray diffraction (XRD) analysis revealed akaganeite (β-FeOOH) as the major iron mineral precipitant throughout the cores, whereas goethite (α-FeOOH) was present to a lesser extent (Phillips *et al.*, 2000). Although they were not detected by XRD analysis, amorphous iron oxyhydroxide deposits were also observed throughout the iron core materials by means of scanning electron microscope (SEM) and energy-dispersive X-ray (EDX) spectroscopic analyses. Presumably, these amorphous iron oxyhydroxides gradually transform to crystalline akaganeite and goethite within the barrier. The formation of akaganeite may be related to a relatively high concentration of chloride in groundwater entering the trench because, in laboratory studies, akaganeite is commonly observed as the dominant mineral phase from the precipitation of ferric chloride (Schwertmann and Cornell, 1991). The presence of goethite within the Fe^0 barrier instead of lepidocrocite, which has been reported in laboratory column studies (Gu *et al.*, 1999), could result from relatively high dissolved O_2 and bicarbonate contents of the groundwater. It has been reported that the formation of goethite is favored over the formation of lepidocrocite when carbonates or CO_2 is present in the system (Schwertmann and Taylor, 1977).

Although to a lesser extent, green rusts were also observed as corrosion products in the Fe^0 barrier, and similar observations have been reported previously (Gu *et al.*, 1999; Phillips *et al.*, 2000; Roh *et al.*, 2000). However, green rusts are not stable and can transform into crystalline iron minerals quickly when exposed to the air or subjected to oven drying. Therefore, care must be taken in sample preservation and preparation in order to observe green rusts in the iron barrier material (Phillips *et al.*, 2001).

In addition to iron oxyhydroxide minerals, analysis of Fe^0 core materials indicated the presence of abundant calcium carbonates such as aragonite ($CaCO_3$) and siderite ($FeCO_3$) (Fig. 16). These results are consistent with decreased concentrations of calcium and carbonates in groundwater within and downgradient of the Fe^0 barrier, as shown in Figs. 12 and 13. Crystalline aragonite was observed throughout the core materials of the Fe^0 barrier, and its structure and forms were identified by both SEM and EDX analyses (Fig. 16a). As indicated previously, relatively high concentrations of Ca^{2+} and carbonates, coupled with a relatively high pH within the Fe^0 barrier, may be largely responsible for the precipitation of $CaCO_3$ minerals (Phillips *et al.*, 2000; Roh *et al.*, 2000).

The formation of ferrous carbonate (i.e., siderite) offers another mechanism for a decreased carbonate or bicarbonate concentration in the Fe^0 barrier. Ferrous iron is one of the major by-products of Fe^0 corrosion in groundwater; it is thus conceivable that the formation of siderite can be a favorable reaction when high amounts of carbonate are present in groundwater, particularly at a relatively high pH condition (Mackenzie *et al.*, 1999). Figure 16b shows the SEM image of cubic-shaped siderite minerals in the Fe^0

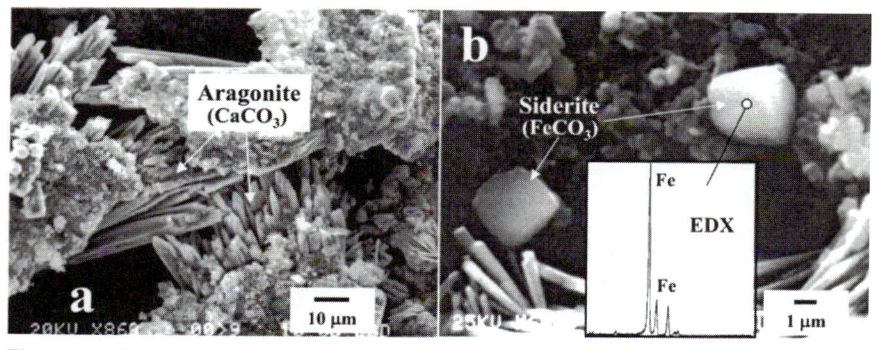

Figure 16 Carbonate minerals in the Fe^0 barrier: (a) characteristic aragonite ($CaCO_3$) crystals and (b) cubed-shaped siderite ($FeCO_3$) and its composition identified by energy-dispersive X-ray (EDX) analysis.

barrier. As shown in Fig. 13, bicarbonate contents from monitoring wells such as TMW-9 were particularly low and could be attributed largely to its precipitation with both Ca^{2+} and Fe^{2+} to form carbonate minerals. However, siderite precipitation was much less extensive than aragonite precipitation. The presence of siderite was detected only in patches in some of the iron core samples. Several factors may contribute to these observations. A relatively high pH and high carbonate but low Ca^{2+} concentrations favor the formation of siderite (Phillips *et al.*, 2000). However, a relatively low pH (about neutral) and a low carbonate concentration shift the chemical equilibrium in favor of $Fe(OH)_2$ precipitation. High Ca^{2+} concentrations in groundwater may compete with Fe^{2+} for carbonate and form $CaCO_3$ minerals as described earlier.

Precipitation of amorphous ferrous sulfide (FeS) was detected by SEM–EDX in most of the core samples from the Fe^0 barrier (Fig. 17). The morphology of FeS appeared to be rounded or bytrodial, and it was commonly observed as coatings on Fe^0 filings or other mineral deposits on iron surfaces. Many Fe^0 particles were completely encrusted in FeS, and these coated Fe^0 filings remained black (rather than rusty) after drying (by vacuum rinsing with acetone), especially those core materials obtained near the interface where groundwater enters the Fe^0 barrier (Phillips *et al.*, 2000). Note that although there appears to be a high occurrence of FeS, crystalline pyrite was not detected by XRD, perhaps because of the noncrystallinity of the FeS structure. As stated previously, FeS formation in the Fe^0 barrier may be largely attributed to the reduction of SO_4^{2-} to S^{2-} by sulfate-reducing bacteria under a highly reducing environment in the Fe^0 barrier. Similar observations have also been reported in both laboratory-simulated iron columns and field experiments with a relatively high influent sulfate concentration

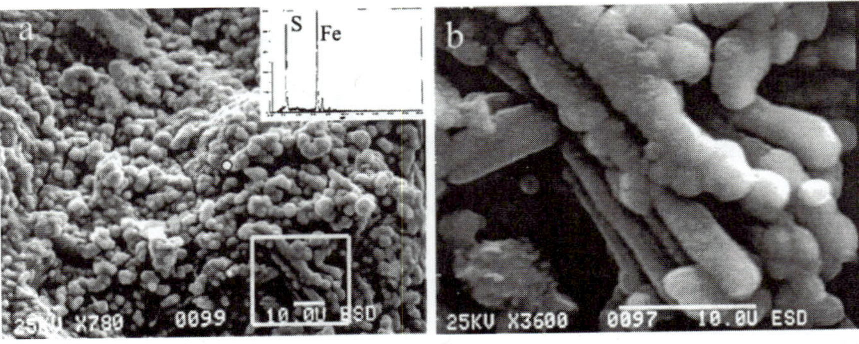

Figure 17 Photomicrographs of amorphous FeS in the Fe^0 barrier: (a) rounded or bytrodial amorphous FeS coating materials in a cemented zone of the Fe^0 barrier and (b) enlargement of an area of (a) showing FeS coatings on aragonite crystals.

(Gu *et al.*, 1999; Phillips *et al.*, 2000). However, although reduction of SO_4^{2-} was observed in Fe^0 barriers at the U.S. Coast Guard Support Center at Elizabeth City, North Carolina, and at an industrial facility in upstate New York (Puls *et al.*, 1999; Vogan *et al.*, 1999), no appreciable amounts of FeS precipitates were observed. These observations may be related to a relatively low SO_4^{2-} concentration (<20 mg/liter) present in the groundwaters at these sites.

C. Implications for Long-Term Performance

The occurrence of a suite of mineral precipitates could have serious implications for the long-term performance of Fe^0 reactive barriers. Specifically, these mineral precipitates commonly exist as coating and cementing materials on Fe^0 surfaces. They not only reduce the reactivity of Fe^0 and thus its capacity to degrade or retain target contaminants of concern, but also cause cementation and clogging of the reactive Fe^0 filings. Ultimately, they may result in reduced hydraulic conductivity or the diversion of groundwater through the barrier. However, site groundwater geochemistry and contaminant concentration may determine the rate and forms of mineral precipitant formation and thus the life span of Fe^0 permeable reactive barriers. For groundwater of relatively low ionic strength, McMahon *et al.* (1999) estimated a 0.35% yearly loss of total porosity in iron-reactive media at the Denver Federal Center site. However, a fast corrosion rate and mineral precipitant formation have been observed at the Oak Ridge Y-12 barrier site, where groundwater contains relatively high concentrations of bicarbonate, sulfate, nitrate, calcium, and magnesium. Isolated spots of clogged and/or cemented Fe^0 media were observed only ~1.2 years after the installation of the Fe^0 barrier at the Oak Ridge site (Phillips *et al.*, 2000).

More extensive cementation and clogging of iron reactive media were found in iron core materials taken ~2.5 years after the installation of the Fe^0 barrier, particularly at the soil/barrier interfaces where groundwater enters the Fe^0 barrier. The cemented iron cores appeared to be hard to break (Fig. 18a), and close examination (by SEM) revealed an extensive iron corrosion and subsequent mineral precipitation on Fe^0 surfaces (Fig. 18b). SEM–EDX analysis of a polished cross section of cemented iron filings (Fig. 19) indicated that iron oxyhydroxides were the primary mineral precipitates accumulated on or between individual iron particles as a thick rind (Phillips *et al.*, 2000). These iron oxyhydroxides may therefore be largely responsible for the cementation of Fe^0 particles in the barrier. Similarly, Mackenzie *et al.* (1999) reported the portion of an iron column clogged with iron oxyhydroxides to be a hardened solid mass that decreased hydraulic

Figure 18 Photomicrographs showing (a) cemented Fe^0 filings and (b) enlarged SEM image of (a).

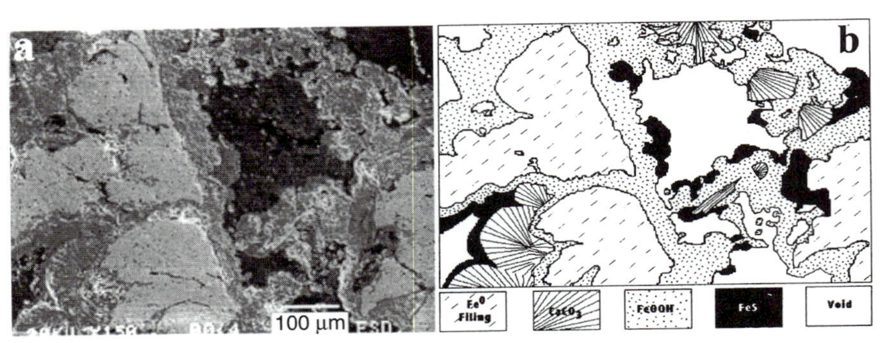

Figure 19 (a) Polished cross section of cemented Fe^0 filings as shown in Fig. 17 and (b) the outline of (a) showing Fe0 filings bound together by iron oxyhydroxides and aragonite coated with amorphous FeS (Reprinted with permission from Phillips *et al.*, 2000. Copyright 2000, American Chemical Society).

conductivity greatly. Additionally, precipitation and formation of aragonite and FeS minerals were also at least partially responsible for binding iron particles in cemented iron core samples (Fig. 19). In particular, FeS precipitates exist primarily in the form of coatings on iron surfaces and thus may act as active binding agents as well. It is interesting to note the sequence of mineral precipitation on iron surfaces as viewed by SEM of the polished sections (Fig. 19). Iron oxyhydroxides appeared to precipitate first as a corrosion product on Fe^0 surfaces. The formation of aragonite followed in response to an increased groundwater pH as a result of iron corrosion, and FeS precipitates then followed as coatings or void fillers on aragonite and iron oxyhydroxide surfaces. The initial delay in the formation of FeS compared with the other precipitates could perhaps be due to the greater length of time needed for accumulation of a microbial population to facilitate sulfate reduction in the barrier (Gu *et al.*, 1999; Phillips *et al.*, 2000).

Based on an average Fe^0 filing thickness of ~0.5 to 1.25 mm, Phillips *et al.* (2000) estimated that these Fe^0 filings could be completely corroded within ~5 to <10 years under the specific site geochemical conditions. This estimated life span of an iron reactive barrier is substantially shorter than the life spans that have been estimated previously, ~15 to 30 years (Gillham *et al.*, 1994; Liang *et al.*, 2000; McMahon *et al.*, 1999), and may be explained by the fact that the site groundwater contains relatively high levels of NO_3^- and HCO_3^-, both of which are known to accelerate the corrosion of Fe^0 (Davies and Burstein, 1980; Gu *et al.*, 1999; Huang *et al.*, 1998). It is also important to note that mineral precipitation and iron cementation appeared to occur progressively with time. Within ~1.5 years, spotted cementations of Fe^0 filings were observed mostly at the interface where groundwater enters the Fe^0 barrier. Cementation extended and further developed downgradient of the Fe^0 barrier, as observed in the second coring event (~2.5 years after the barrier was installed). An important implication of these observations is that such an uneven distribution of iron corrosion and mineral precipitation could potentially result in early system clogging at the interface regions and therefore shorten the functional lifetime of *in situ* Fe^0 barriers. Therefore, close attention should be given to areas in the barrier that seem more vulnerable to corrosion, mineral precipitation, and subsequent cementation (e.g., where groundwater first enters the barriers). Particular attention should also be given to the geochemical composition and concentration in groundwater, which may largely determine the corrosion rate and thus the life span of Fe^0 reactive barriers.

ACKNOWLEDGMENTS

We thank M. A. Bogle and N. D. Farrow for their technical assistance in field groundwater and core sampling activities and MSE Technology Applications (W.C. Goldberg) for the first round core sampling. Funding for this research was supported by the Contaminants Focus Area of the Office of Environmental Management (EM-50), U.S. Department of Energy, under contract DE-AC05–00OR22725 with Oak Ridge National Laboratory, which is managed by UT-Battelle, LLC.

REFERENCES

Agrawal, A., and Tratnyek, P. G. (1996). Reduction of nitro aromatic compounds by zero-valent iron metal. *Environ. Sci. Technol.* **30**, 153–160.

Agrawal, A., Tratnyek, P. G., Stoffyn-Egli, P., and Liang, L. (1995). Processes affecting nitro reduction by iron metal: Mineralogical consequences of precipitation in aqueous carbonate environments. *209th ACS Natl. Meet. Am. Chem. Soc.* **35**, 720–723.

Baes, C. F., and Mesmer, R. E. (1976). "The Hydrolysis of Cations." Wiley, New York.

Blowes, D. W., Ptacek, C. J., and Jambor, J. L. (1997). In-situ remediation of Cr(VI)-contaminated groundwater using permeable reactive walls: Laboratory studies. *Environ. Sci. Technol.* **31**, 3348–3357.

Cantrell, K. J., Kaplan, D. I., and Wietsma, T. W. (1995). Zero-valent iron for the in situ remediation of selected metals in groundwater. *J. Hazard. Mater.* **42**, 201–212.

Chee-Sanford, J. C., Frost, J. W., Fries, M. R., Zhou, J., and Tiedje, J. M. (1996). Evidence for acetyl coenzyme A and cinnamoyl coenzyme A in the anaerobic toluene mineralization pathway in Azoarcus tolulyticus Tol-4. *Appl. Environ. Microbiol.* **62**, 964–973.

Cook, P. G., Solomon, D. K., Sanford, W. E., Busenberg, E., Plummer, L. N., and Poreda, R. J. (1996). Inferring shallow groundwater flow in saprolite and fractured rock using environmental tracers. *Water Res.* **32**, 1501–1509.

Davies, D. H., and Burstein, G. T. (1980). The effects of bicarbonate on the corrosion and passivation of iron. *Corr.* **36**, 416–422.

Dowling, J. E., Widdel, F., and White, D. C. (1986). Phospholipid ester-linked fatty acid biomarkers of acetate-oxidizing sulfate reducers and other sulfide-forming bacteria. *J. Gen. Microbiol.* **132**, 1815–1825.

Edwards, R. W., Duster, D., Faile, M., Gallant, W., Gibeau, E., Myller, B., Nevling, K., and O'Brady, B. (1996). "RTDF Permeable Reactive Barriers Action Team Meeting," August 15–16, San Francisco, CA.

Fiedor, J. N., Bostick, W. D., Jarabek, R. J., and Farrell, J. (1998). Understanding the mechanism of uranium removal from groundwater by zero-valent iron using X-ray photoelectron spectroscopy. *Environ. Sci. Technol.* **32**, 1466–1473.

Gillham, R. W., Blowes, D. W., Ptacek, C. J., and O'Hannesin, S. F. (1994). Use zero-valent metals in in-situ remediation of contaminated groundwater. *In* "In-Situ Remediation: Scientific Basis for Current and Future Technologies" (G. W. Gee and N. R. Wing, eds.), pp. 913–929. Battelle Press, Columbus, Richland.

Gu, B., Liang, L., Dickey, M. J., Yin, X., and Dai, S. (1998). Reductive precipitation of uranium(VI) by zero-valence iron. *Environ. Sci. Technol.* **32**, 3366–3373.

Gu, B., Liang, L., Dickey, M. J., Yin, X., and Watson, D. (1997). In situ reactive barriers for simulteneous treatment of radionuclides and chlorinated organic contaminants. *In* "RTDF Permeable Reactive Barriers Action Team Meeting," pp. 46–47. September 18–19, Virginia Beach, VA.

Gu, B., Phelps, T. J., Liang, L., Dickey, M. J., Roh, Y., Kinsall, B. L., Palumbo, A. V., and Jacobs, G. K. (1999). Biogeochemical dynamics in zero-valent iron columns: Implications for permeable reactive barriers. *Environ. Sci. Technol.* **33**, 2170–2177.

Gu, B., Zhou, J. Z., Watson, D. B., Phillips, D. H., and Wu, L. (2001). Microbiological characteristics in a zero-valent iron reactive barrier. *Environ. Monit. Ass.*

Guckert, J. B., Hood, M. A., and White, D. C. (1986). Phospholipid, ester-linked fatty acid profile changes during nutrient deprivation of *Vibrio cholerae*: Increases in the *trans/cis* ratio and productions of cyclopropyl fatty acids. *Appl. Environ. Microbiol.* **52**, 794–801.

Gui, J., and Devine, T. M. (1994). The influence of sulfate ions on the surface enhanced Raman spectra of passive films formed on iron. *Corr. Sci.* **36**, 441–462.

Horwitz, E. P., Dietz, M. L., Chiarizia, R., and Diamond, H. (1992). Separation and preconcentration of uranium from acidic media by extraction chromatography. *Anal. Chim. Acta* **266**, 25–37.

Hsi, C. K. D., and Langmuir, D. (1985). Adsorption of uranyl onto ferric oxyhydroxides: Application of the surface complexation site-binding model. *Geochim. Cosmochim. Acta* **49**, 1931–1941.

Huang, C. P., Wang, H. W., and Chiu, P. C. (1998). Nitrate reduction by metallic iron. *Water Res.* **32**, 2257–2262.

Johnson, T. L., and Tratnyek, P. G. (1994). A column study of geochemical factors affecting reductive dechlorination of chlorinated solvents by zero-valent iron. *In* "In-Situ Remediation: Scientific Basis for Current and Future Technologies" (G. W. Gee and N. R. Wing, eds.), Vol. 2, pp. 931–947. Battelle Pacific Northwest Laboratories, Pasco, WA.

Liang, L., Korte, N., Goodlaxson, J. D., Clausen, J., Fernando, Q., and Muftikian, R. (1997). Byproduct formation during the reduction of TCE by zero-valence iron and palladized iron. *GWMR Winter* 122–127.

Liang, L., Korte, N., Gu, B., Puls, R., and Reeter, C. (2000). Geochemical and microbiological reactions affecting the long-term performance of in situ iron barriers. *Adv. Environ. Res.* **4**, 273–286.

Lipczynska-Kochany, E., Harms, S., and Nadarajah, N. (1994). Degradation of carbon tetrachloride in the presence of iron and sulphur containing compounds. *Chemosphere* **29**, 1477–1489.

Mackenzie, P. D., Horney, D. P., and Sivavec, T. M. (1999). Mineral precipitation and porosity losses in granular iron columns. *J. Hazard. Mater.* **68**, 1–17.

Mackenzie, P. D., Sivavec, T. M., and Horney, D. P. (1997). Extending hydraulic lifetime of iron walls. *In* "International Containment Technology Conference and Exhibition," pp. 781–786. St. Petersburg, FL.

McMahon, P. B., Dennehy, K. F., and Sandstrom, M. W. (1999). Hydraulic and geochemical performance of a permeable reactive barrier containing zero-valent iron, Denver Federal Center. *Ground Water* **37**, 396–404.

Morrison, S. J., Spangler, R. R., and Tripathi, V. S. (1995). Adsorption fo uranium(VI) on amorphous ferric oxyhydroxide at high concentrations of dissolved carbon(IV) and sulfur(VI). *J. Contamin. Hydrol.* **17**, 333.

Morse, J. W., and Choppin, G. R. (1991). The chemistry of transuranic elements in natural waters. *Rev. Aquat. Sci.* **4**, 1–22.

Odziemkowski, M. S., Schumacher, T. T., and Reardon, E. J. (1998). Mechanism of oxide film formation on iron in simulating groundwater solutions: Raman spectroscopic studies. *Corr. Sci.* **40**, 371–389.

O'Hannesin, S. F., and Gillham, R. W. (1998). Long-term performance of an in situ "Iron Wall" for remediation of VOCs. *Groundwater* **36**, 164–172.

Phillips, D. H., Gu, B., Watson, D. B., and Roh, Y. (2001). Impact of sample preparation on mineralogical analysis of iron reactive barrier materials. *Environ. Sci. Technol.*

Phillips, D. H., Gu, B., Watson, D. B., Roh, Y., Liang, L., and Lee, S. Y. (2000). Performance evaluation of a zero-valent iron reactive barrier: Mineralogical characteristics. *Environ. Sci. Technol.* **34**, 4169–4176.

Pratt, A. R., Blowes, D. W., and Ptacek, C. J. (1997). Products of chromate reduction on proposed subsurface remediation material. *Environ. Sci. Technol.* **31**, 2492–2498.

Puls, R. W., Paul, C. J., and Powell, R. M. (1999). The application of in situ permeable reactive (zero-valent iron) barrier technology for the remediation of chromate-contaminated groundwater: A field test. *Appl. Geochem.* **14**, 989–1000.

Roberts, A. L., Totten, L. A., Burris, D. R., and Campbell, T. J. (1996). Reductive elimination of chlorinated ethylenes by zero-valent metals. *Environ. Sci. Technol.* **30**, 2654–2659.

Roh, Y., Lee, S. Y., and Elless, M. P. (2000). Characterization of corrosion products in the permeable reactive barriers. *Environ. Geol.* **40**, 184–194.

SAIC (1996). Bear Creek Valley characterization area technology demonstration action plan. SY/EN-5479. Science Applications International Corporation, Oak Ridge, TN 37831.

SAIC (1997). Phase I report on the Bear Creek Valley Treatability study, Oak Ridge Y-12 Plant, Oak Ridge, Tennessee. Y/ER-285. Science Applications International Corporation, Oak Ridge, TN 37831.

Scherer, M. M., Balko, B. A., and Tratnyek, P. G. (1998). The role of oxides in reduction reactions at the metal-water interface. *In* "Kinetics and Mechanisms of Reactions at the Mineral-Water Interface" (D. Sparks and T. Grundl, eds.). American Chemical Society, Washington, DC.

Scherer, M. M., Richter, S., Valentine, R. L., and Alvarez, P. J. (2000). Chemistry and microbiology of permeable reactive barriers for in situ groundwaer cleanup. *Crit. Rev. Environ. Sci. Technol.* **30**, 363–411.

Schwertmann, U., and Cornell, R. M. (1991). "Iron Oxide in the Laboratory: Preparation and Characterization," VCH Pub., New York.

Schwertmann, U., and Taylor, R. M. (1977). *In* "Minerals in Soil Environments" (J. Dixon and S. B. Weed, eds.), pp. 145–180. Soil Sci. Soc. Amer.

Shoemaker, S. H., Greiner, J. F., and Gillham, R. W. (1995). Permeable reactive barriers. *In* "Assessment of Barrier Containment Technologies" (R. R. Rumer and J. K. Mitchell, eds.), Section 11, pp. 301–353.

Sivavec, T. M., Mackenzie, P. D., Horney, D. P., and Baghel, S. S. (1997). Redox-active media for permeable reactive barriers. *In* "2nd International Containment Technology Conference," pp. 753–759. St. Petersburg, FL.

Tratnyek, P. G., Johnson, T. L., Scherer, M. M., and Eykholt, G. R. (1997). Remediating ground water with zero-valent metals: Chemical considerations in barrier design. *GWMR* **17**, 108–114.

Tunlid, A., and White, D. C. (1991). Biochemical analysis of biomass, community structure, nutritional status and metabolic activity of the microbial communities in soil. *In* "Soil Biochemistry" (J. M. Bollag and G. Stotzky, eds.), Vol. 7, pp. 229–262.

Vogan, J. L., Focht, R. M., Clark, D. K., and Graham, S. L. (1999). Performance evaluation of a permeable reactive barrier for remediation of dissolved chlorinated solvents in groundwater. *J. Hazard Mater.* **68**, 97–108.

Watson, D., Gu, B., Phillips, D. H., and Lee, S. Y. (1999). Evaluation of permeable reactive barriers for removal of uranium and other inorganics at the Department of Energy Y-12 Plant, S-3 disposal ponds. ORNL/TM-1999/143, Oak Ridge National Laboratory, Oak Ridge, TN.

Zhou, J. Z., Bruns, M. A., and Tiedje, J. M. (1996). DNA recovery from soils of diverse composition. *Appl. Environ. Microbiol.* **62**, 461–468.

Chapter 12

Analysis of Uranium-Contaminated Zero Valent Iron Media Sampled from Permeable Reactive Barriers Installed at U.S. Department of Energy Sites in Oak Ridge, Tennessee, and Durango, Colorado

Leah J. Matheson and Will C. Goldberg
MSE Technology Applications, Inc.
Butte, Montana 59701
W. D. Bostick and Larry Harris
Materials & Chemistry Laboratory, Inc. Oak Ridge, Tennessee 37830

Permeable reactive barriers, composed of zero valent iron (ZVI), have been installed and operated at U.S. Department of Energy sites located at the Y-12 Plant in Oak Ridge, Tennessee, and at the uranium mill tails repository near Durango, Colorado. The ZVI medium is intended to remove select toxic solutes, especially uranium (U), from contaminated groundwater. Core samples from these barrier installations were sampled after prolonged exposure to contaminated groundwater ($\sim$ 1.2 year of operation at the Y-12 site and up to 3 years of operation at the Durango site). Core samples were protected from exposure to the ambient atmosphere by packaging them in argon-purged containers for shipment to an off-site laboratory. The concern was to protect media from exposure to oxygen, which could alter the valence state of the treatment medium itself and of the metal deposits therein. Samples were subjected to a battery of analytical techniques, including X-ray photoelectron spectroscopy (XPS), a surface-sensitive technique that can be used to determine the average valence state of elements. Major findings in this study were that (1) ZVI media had extensive surface deposition of various mineral phases (e.g., calcite and amorphous iron sulfides), (2) U was present within the media at somewhat modest levels (e.g., ≤ 0.2 wt%), and (3) U on the ZVI surface was at least partially oxidized.

I. INTRODUCTION

Corroding iron [Fe^0, or zero valent iron (ZVI)] has been shown to provide reducing equivalents to a variety of oxidized compounds, including chlorinated alkanes and alkenes, nitroaromatics, and inorganic species such as metals. Some recent reviews for the mechanisms of these types of reactions and their application to remediation of contaminated environments include Matheson and Tratnyek (1994), U.S. EPA (1998), Scherer *et al.* (1999), and Tratnyek *et al.* (2001). Fe provides electrons via

$$Fe^0 \rightarrow Fe^{2+} + 2\,e^-, \tag{1}$$

which can reduce compounds as shown in Eq. (2) for halogenated or nitroaromatic compounds or as shown in Eq. (3) for metals/inorganic species:

$$RX + 2\,e^- + H^+ \rightarrow RH + H^- \tag{2}$$

$$M^{n+} + 2\,e^- \rightarrow M^{(n-2)+}. \tag{3}$$

Water can effectively compete for the electrons produced in Eq. (1). In anaerobic aqueous systems, Fe corrodes as shown in

$$2\,H_2O + Fe^0 \rightarrow Fe^{2+} + H_2 + 2\,OH^-. \tag{4}$$

Some of the ferrous ion (Fe^{2+}) reaction product may remain surface bound, where it is postulated to facilitate some surface-mediated reductive reactions (Charlet *et al.*, 1998). Oxygen is also an effective electron acceptor; if present in the aqueous system, it is removed rapidly by reactions with the Fe^0 [Eq. (5)] and/or associated Fe^{2+} [Eq. (6)].

$$2\,Fe^0 + O_2 + 2\,H_2O \rightarrow 2\,Fe^{2+} + 4\,OH^-, \tag{5}$$

$$4\,Fe^{2+} + O_2 + 10\,H_2O \rightarrow 4\,Fe(OH)_3(s) + 8\,H^+. \tag{6}$$

In Eq. (6), $Fe(OH)_3$ (*s*) is solid-phase hydrolyzed ferric oxide, or common "rust," initially an amorphous precipitate that may be transformed in time to crystalline iron oxide phases (e.g., goethite, α-FeOOH).

Cantrell *et al.* (1995) demonstrated that ZVI can be an effective medium for the removal of soluble U (i.e., U^{6+} as the uranyl ion, UO_2^{2+}), a toxic and radiological contaminant of concern at many U.S. Department of Energy (DOE) sites and at other sites supporting the nuclear fuel cycle (Morrison *et al.*, 2001). However, the operative mechanism for such removal is the subject of considerable debate (see, *e.g.*, Wersin *et al.* 1994; Cantrell *et al.*, 1995; Grambow *et al.*, 1996; Fiedor *et al.*, 1998; Gu *et al.*, 1998; Abdelouas *et al.*, 1999 a, b; Farrell *et al.*, 1999; Qui *et al.*, 2000; Morrison *et al.*, 2001). The probable dominant mechanisms include sorption of U^{6+} complex ions onto solid-phase oxidized Fe corrosion products (e.g., "rust") that are normally present at or near the surface of the Fe, and/or reductive precipitation [represented generically in Eq. (3)].

It is well known that many oxides and other mineral phases of Fe (including many of those normally present in soil minerals) are potent sorbents of U^{6+} (see, e.g., Ho and Miller, 1986; Tricknor, 1994; Grambow *et al.*, 1996; Farrell *et al.*, 1999; Moyes *et al.*, 2000). Thus, U^{6+} may be attracted and held near the oxidized surface of ZVI, where surface-mediated electron transport may be facilitated.

Farrell *et al.*, (1999) estimated the thermodynamic equilibrium outcome for the reduction of U^{6+} by ZVI, as described in Eqs. (7) and (8):

$$UO_2^{2+} + Fe^0 \rightarrow UO_{2(s)} + Fe^{2+}, \tag{7}$$

$$E(\text{volt}) = 0.661 + 0.295 \log [UO_2^{2+}] - 0.295 \log [Fe^{2+}]. \tag{8}$$

The predictions cited earlier are based on the precipitation of reduced U (as U^{4+} in the form of crystalline uraninite, UO_2), and the value of the computed cell potential (E) given in Eq. (8) is positive, indicating an energetically favorable outcome. However, an energetically favorable predictive outcome at equilibrium does not address the possible effect of electron transfer kinetics. For example, the reduction of As^{5+} to As^{3+} by ZVI is similarly predicted to be energetically favorable (Farrell *et al.*, 2001), but long-term exposure of arsenate solution to ZVI results in As being bound to the Fe surface as the oxidized species (As^{5+}) (see Farrell *et al.*, 2001; Lackovic *et al.*, 2000). Under environmentally relevant conditions of relatively low temperature and near-neutral pH values, many redox reactions involving actinides are notoriously sluggish (Rao and Sagi, 1962; Murphy and Shock, 1999).

Reductive precipitation is a preferred contaminant removal mechanism in the sense that U^{4+} species are generally much less soluble than their U^{6+}

counterparts, and thus reduced U should be less mobile in the environment. The solubility of UO_2 solid phase can vary considerably as a function of its degree of crystallinity and the aqueous phase composition (Kertes and Guillaumont, 1985), but approaches a limiting value near 1 μg/liter in strongly reductive near-neutral solution.

The aqueous solubility of U^{6+} (UO_2^{2+}) may be increased significantly through complex formation with common groundwater ligands, especially the carbonate ion (Langmuir, 1978; Scanlan, 1977; Meinrath *et al.*, 1996). In contrast, U^{4+} is much less prone to complexation, although it may hydrolyze to yield polynuclear species (and often mobile colloids) of the form $U(OH)_n^{4-n}$ (Kertes and Guillaumont, 1985; von Gunten and Benes, 1995). The net result of oxidation state and complexation effects is that, under favorable conditions, the aqueous solubility of U^{4+} species may be many orders of magnitude lower than those of U^{6+} (Miyahara, 1992). However, uranium that was precipitated initially as U^{4+} or mixed valent oxide (e.g., by chemical-, electrochemical-, or biological-mediated reduction) may be susceptible to rapid oxidation to more mobile U^{6+} when it is exposed to air or to dissolved oxygen in the aqueous phase (Katz and Rabinowitch, 1951; Duff *et al.*, 1999).

It is apparent that the surface condition of the ZVI medium and the site-specific water chemistry (e.g., dissolved oxygen, influent solution temperature and pH values, solution ionic strength, metal complexing agents, and a myriad of reactions to compete for the electrons released by corroding iron) all influence the probable success of ZVI for the treatment of contaminated groundwaters. Permeable reactive barriers, composed of ZVI, have been installed and operated at U.S. DOE sites located at the Y-12 plant in Oak Ridge, Tennessee, and at the uranium mill tails repository near Durango, Colorado. The ZVI medium is intended to remove select toxic solutes, especially soluble uranium, from contaminated groundwater. Core samples from these barrier installations were sampled after prolonged exposure to contaminated groundwater ($\sim$ 1.2 year of operation at the Y-12 site and up to 3 years of intermittent operation at the Durango site). Core samples were protected from exposure to the ambient atmosphere by packaging them in argon-purged containers for shipment to an off-site laboratory. The concern was to protect media from exposure to oxygen, which could alter the valence state of the treatment medium itself and of the metal contaminant deposits therein.

X-ray photoelectron spectroscopy (XPS) is a surface-selective technique that provides information on the local chemical environment of near-surface elements. This technique has been used in previous investigations at our laboratory (Fiedor *et al.*, 1998) and elsewhere (Wersin *et al.*, 1994; Qiu *et al.*, 2000; Halada *et al.*, 2001) to monitor the valence state of surface-bound uranium. Samples for XPS were transferred from the protective inert atmosphere within the shipping containers to a nitrogen-purged radiochemical glovebox and hence to the high-vacuum analytical instrument without intermediate exposure to air. Iron on the media was found to be oxidized (i.e., no truly "zero valent" material remained at the near-surface due to corrosion by the groundwater; see Eq. (4)]. Sulfur was present in both reduced (sulfide) and oxidized (sulfate) forms, probably as amorphous ferrous sulfide and as calcium sulfate (gypsum). Uranium was present at the surface of the Y-12 ZVI medium in a mixed valent state (i.e., it was at least partially oxidized). In contrast, uranium from the Durango cores was found to be essentially completely oxidized at the near-surface of the media. These findings suggest that the reaction depicted in Eq. (7) does not attain equilibrium under field phase contact conditions.

In field applications of ZVI to treat contaminated groundwater, the buildup of certain mineral phases (e.g., due to precipitation of sparingly soluble metal sulfate, sulfide, and/or carbonate minerals) within the ZVI medium may cause plugging of solution flow or may physically block access to the reactive surface. Significant *crystalline* phases within groundwater-exposed ZVI media were identified by X-ray diffraction (XRD) as synthetic iron, quartz (SiO_2), calcite ($CaCO_3$), and magnetite (Fe_3O_4). The bulk phase U concentration within ZVI media was relatively low (typically ≤ 0.2 wt%, as estimated by radioactivity measurements and X-ray fluorescence), and no specific uranium phases were identified.

II. FIELD SITE DESCRIPTIONS

A. Bear Creek Valley at the Y-12 Plant in Oak Ridge, Tennessee

The Bear Creek Valley (BCV) watershed is located within the U.S. DOE Oak Ridge Reservation and encompasses multiple disposal units containing hazardous and radiological wastes that originated from former operations at the Y-12 plant. At the S-3 pond area, operated from 1951 to 1984, liquid wastes, including depleted uranium in nitric acid solution, were disposed into unlined surface impoundments. The ponds were filled and capped in 1988; however, liquid wastes formerly disposed at the ponds have percolated into the surrounding subsurface, contaminating the groundwater. The groundwater contaminant of principal concern for the protection of human health is soluble uranium [present as the uranyl ion [UO_2^{2+})]. A number of treatability studies were conducted to identify media with the potential to remove soluble uranium from representative site water compositions. Results indicated that zero valent iron, as metal scrap or filings, was the most promising candidate among those media evaluated (SAIC, Phase 1 Report, 1997).

A permeable reactive barrier (PRB), composed of ZVI filings poured into a subterrarium trench, was installed at the BCV migration pathway 2 site to intercept and treat contaminated groundwater. Water in this pathway, near the Y-12 S-3 site, is characterized by elevated uranium concentrations, but does not have the elevated levels of nitrate ion that occur in other site-contaminated groundwater. This PRB was sampled by MSE Technology Applications staff in February 1999 after ~14 months of exposure to relatively low levels of soluble U (generally < 1 mg/liter) in the influent groundwater. As described in Section III, sample cores were prepared and transported in a manner to minimize possible exposure to oxygen in ambient air.

B. Bodo Canyon Site at Durango, Colorado

The Uranium Mill Tailings Remedial Action (UMTRA) Ground Water Program has been operating a PRB demonstration facility at the Bodo Canyon uranium mill tailing disposal site near Durango, Colorado, since October 1995. Chapter 15 describes the site and the configuration of PRB demonstration units. One unit, designated as PRB C, contained ZVI foam plates, operated intermittently for about 3 years and treated an estimated 300,000 gallons of contaminated seep water before becoming plugged. PBR C was opened in July 1999, and treatment media were sampled extensively. Chapter 15 describes some analytical data obtained for the media. Duplicate samples, prepared and shipped in a manner to minimize possible exposure to oxygen in ambient air, were sent to the Materials and Chemistry Laboratory in Oak Ridge, Tennessee, for additional analysis, including determination of the valence state for near-surface deposits containing uranium.

III. SAMPLING AND ANALYSIS METHODS

A. Sample Collection

Core samples from the Oak Ridge Y-12 site were collected from angle core holes oriented approximately perpendicular to the ZVI-filled trench and at a dip of approximately 60° from horizontal (Figs. 1 and 2). A GeoProbe

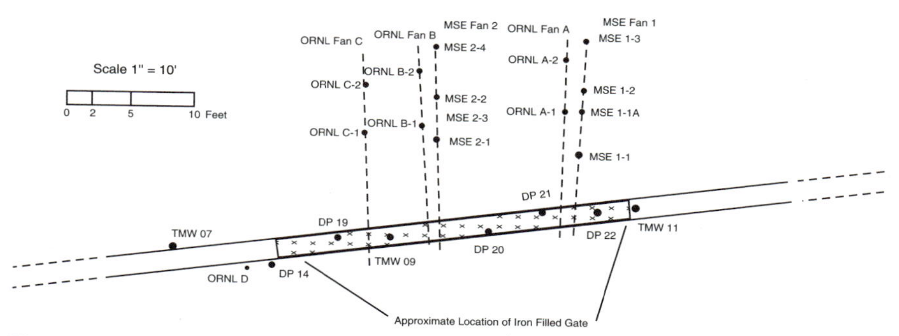

Figure 1 Plan view of locations of core holes drilled at the Oak Ridge, Tennessee, Y-12 pathway 2 permeable reactive barrier. With reference to this figure, the direction of general water flow is diagonal from upper right to lower left. Analysis of MSE Fans 1 & 2 are reported in this chapter. Adjacent area fans were collected for examination at the Oak Ridge National Laboratory (ORNL).

MacroCore sampler system was used, with 3.8-cm ($1\frac{1}{2}$ in.) outer diameter (OD) acrylic liners (section lengths $\sim$ 1.2 m). Core samples were collected to encompass the upstream and downstream edges of the ZVI zone. The ends of each core sample liner were stuffed with plastic wrap and fitted with end caps promptly after they were removed from the trench. Liner caps were slit to allow the contents of the tubes to be purged with inert gas (argon) before resealing them with vinyl tape. The tape was color coded to identify "top" and "bottom" of the strike. The sealed liners containing the core samples and an inert blanket of argon gas were placed within argon-sparged 10-cm (4-in.) OD polyvinyl chloride (PVC) shipping tubes that were fitted with gas flow valves. Tubes containing the core samples were flushed with an additional three nominal bed volumes of argon before the valves were sealed.

Samples from the PRB-C at Durango, Colorado, were collected in an analogous manner (see Chapter 15) in order to minimize exposure of media to air. The cores were removed from the ZVI strata with use of a handyman jack and subsequently placed in an argon-sparged shipping tube. Two sets of cores were collected: one from the approximately 1.3-m depth of the inlet baffle and one from the entire depth of the outlet baffle. The cores were shipped in $\sim$ 0.6-m sections under an inert atmosphere.

Upon receipt at the Materials and Chemistry Laboratory in Oak Ridge, Tennessee, the PVC shipping tubes were opened and the sealed core sleeves

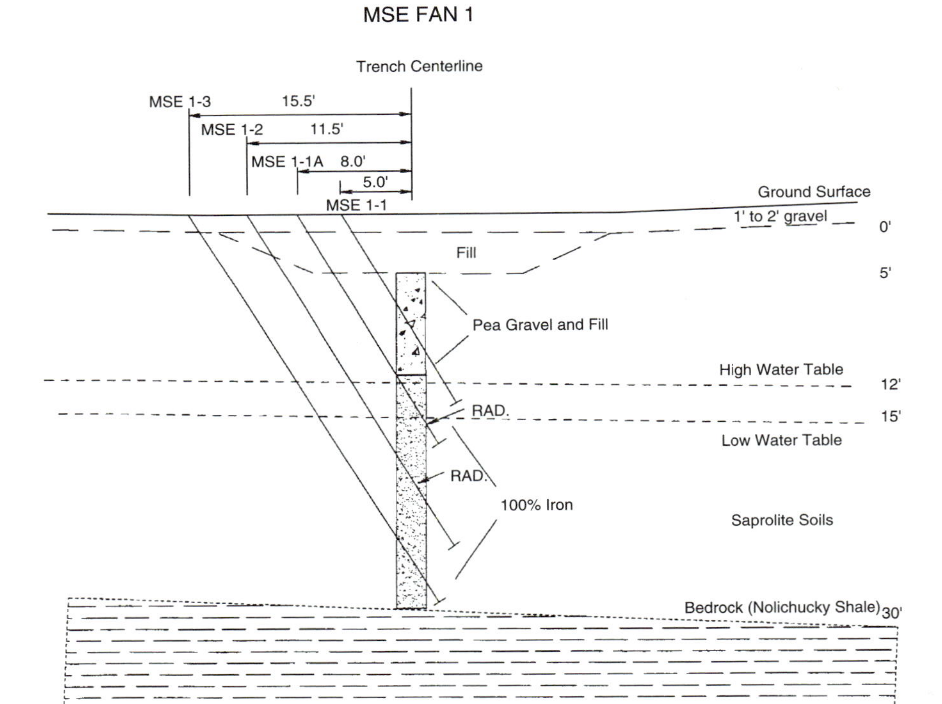

Figure 2 Cross section of the first fan of core holes drilled at the Oak Ridge, Tennessee, Y-12 pathway 2 permeable reactive barrier. The direction of general groundwater flow is toward the viewer. Within this fan, elevated levels of gross radioactivity (labeled "RAD") were only found in cores MSE 1-1A and 1-2, near the flow inlet interface with the iron medium. The second fan of core holes is similar (MSE Fan 2 not show).

were removed promptly to a nitrogen-purged radiological glove box. The cores were opened within the glovebox, where samples of material were taken from along the length of the core. One set of samples remained within the protective (anoxic) atmosphere of the glovebox, and a second set of subsamples was removed via the interlock to air dry. The protected samples were subsequently examined by XPS to determine the valence state, whereas the unprotected samples were examined by other analytical techniques that are not sensitive to the valence state of the constituent elements.

B. Radiological Survey

Air-dried subsamples of ZVI (1–3 g) were counted with use of a Tennelec Series 5 E gas proportional counter to profile the spatial distribution of radioactivity (α and β particles) along a core sample. These types of low-energy radioactive particles are subject to self-shielding, and the measured activities were compared to reference materials prepared in-house with use of known amounts of uranyl nitrate to allow an approximate estimate of uranium content.

C. Scanning Electron Microscopy (SEM) With Energy-Dispersive X-Ray Spectroscopy (EDS)

SEM rasters a finely focused electron beam over the area to be imaged. The image (photomicrograph) is formed by either the collection of (inelastically scattered) secondary electrons (SE) or backscattered electrons (BSE). SE emission is determined to a large extent by the surface topography, whereas backscattered electron emission is determined largely by the compositional differences within the specimen.

The image contrast of BSE is proportional to the atomic number of the elements present; deposits containing heavier elements, such as uranium, are identified readily by the relatively bright images produced (in phase-contrast to lighter element substrates). The images can be collected digitally to allow computer-assisted image analysis (e.g., morphology, feature analysis, particle size distributions). When the electron beam (typically 10–50 keV accelerating voltage) impinges on the sample, X rays representative of the elements present are also generated. The X-ray spectrum for the area rastered is analyzed by EDS. This elemental analysis can also be used to obtain elemental analysis of a small, discrete portion of the samples or to create X-ray dot maps (the image collected is an X-ray intensity distribution with the spatial resolution of the sample image).

Two SEM–EDS systems were used in this investigation. The Hitachi S-4500 system, with an attached PGT Prism detector, was used for morphology imaging. Due to image-forming electronics, the S-4500 system gives superior image resolution for magnetic substrates, such as ZVI filings. The Hitachi S-5000 SEM, with associated Noran X-ray microanalysis systems, has superior sensitivity for some of the lighter elements of interest (e.g., C, N, O). Both SEM systems are capable of magnification up to approximately 1,000,000 $\times$ (i.e., to a spatial resolution down to $\sim$ 6 Å), although the magnetic character of the ZVI limited the resolution for morphology imaging with use of the

Hitachi S-5000 system. Conductive samples were analyzed in an uncoated condition, mounted on carbon tape.

C. X-Ray Diffraction

XRD analyzes samples to determine the crystalline phases present in samples. X-ray diffraction data were obtained on a Phillips 3100 high-angle diffractometer using Cu Kα ($\lambda = 1.5418$ Å) radiation, typically operated at 45 kV and 40 mA. Data were collected at a step rate of 0.02° 2 theta per minute.

D. X-Ray Fluorescence (XRF)

XRF is a simple, nondestructive method for qualitative and quantitative analysis of the elemental composition in a material. Accurate quantitative analysis requires the preparation of standards in a matrix composition and morphology very similar to the sample being analyzed. In this investigation, a standardless matrix correction data reduction algorithm was used to provide a quantitative estimate for the composition of a sample. Our instrumentation is a Phillips PW-1480 X-ray spectrophotometer with sequential wavelength dispersive spectral analysis.

E. X-Ray Photoelectron Spectroscopy

X-ray photoelectron spectroscopy, or electron spectroscopy for chemical analysis (ESCA), is a surface-sensitive spectroscopic tool that provides information about the chemical state of elements located at the near-surface of a solid (effective sampling depth generally 10–100 nm). XPS data reflect the average surface composition over a relatively large area.

Samples in the form of relatively coarse powders or particles were mounted on doubly sticky carbon tape in a glovebox that was maintained under a nitrogen pressure. Once mounted, the samples were placed in a transfer chamber that was placed in an N_2-filled bag and subsequently transferred to the XPS sample chamber without exposure to air. XPS data were obtained using a Perkin Elmer physical electronics XPS system. For purposes of chemical state identification and for peak deconvolution and other mathematical manipulations of XPS data, detail scans must be obtained for precise peak location and for accurate registration of peak shapes (Wagner *et al.*,

1979). The as-recorded spectrum was deconvoluted, smoothed, and shifted to compensate for charging effects; the energy shift was based on the adventitious hydrocarbon C 1*s* photoelecton peak at a binding energy of 284.6 eV. Subsequently, the spectral contour was curve-fitted using the XPS system computer program.

For low-concentration uranium samples (e.g., those received from the Y-12 site), it was necessary to record the spectrum over a period of 3–4 h to improve photoelectron counting statistics. Long-term exposure of the samples to the X-ray beam at ultra high vacuum (5×10^{-9} torr) is a condition suggested to induce partial reduction of species such as U^{6+} (Baer and Thomas, 1981; Wersin *et al.*, 1994). Thus, the estimate of proportions of U^{4+} and U^{6+} originally present on the surface of the medium may be biased somewhat toward an overestimation of the reduced uranium component.

The identification of chemical states depends primarily on the accurate determination of line energies. XPS spectra were curve-fitted by fixing the peak positions for the U^{4+} and U^{6+} components using reported literature values for these species. Data were fitted to U 4*f* photoelecton peaks by adjusting intensities and full width at half maxima (FWHM) until a reasonable agreement with the as-recorded spectrum was achieved.

IV. RESULTS

A. Oak Ridge, Tennessee, Y-12 Pathway 2 Permeable Reactive Barrier

1. Field Sampling

Fan core samples were located in the trench so they would include the iron-filled zone where it intercepted the contaminated groundwater, near sampling well DP-22 (see Figs. 1 and 2). Core MSE 1-1 intercepted only the fill above the ZVI trench and was discarded. No elevated radiation was associated with samples from this core. The second core series in fan 1 (core 1-1A) intercepted the top of the ZVI zone of the trench with approximately 50% recovery of the iron material and 100% recovery of the backfill interfaces. Adjacent area core samples were collected for mineralogical examination at the Oak Ridge National Laboratory ("ORNL fans," identified in Fig. 1) (see Phillips *et al.*, 2000).

2. Distribution of Radioactivity

Samples of core material were dried and crushed, and α and β particle activities were monitored with use of a Tennelec Series 5-E gas proportional counter. The section of core 1-1A at 17 to 19.6 ft had associated radioactivity that was above background level (see Table I). Due to particle depth in the counter, the activity measurements in Table I are subject to self-shielding; by comparison to an in-house reference material [U^{6+} deposited on magnetite (Fe_3O_4) particles], the maximum measured radioactivity in core 1-1A (found at a distance of 15 cm) is estimated to be equivalent to ~ 0.14 wt% as U. Relatively lower levels of activity were also found in the 20 to 24-ft section of fan 1 core 1-2. The third core from fan 1 intercepted both soil interfaces and ZVI but had only background levels of radioactivity. Fan 2 core samples were taken "downstream" from fan 1 (see Fig.1), but contained minimal radioactivity, with slightly elevated values detected within the 17- to 21-ft section (data not shown).

3. X-Ray Diffraction Data

A sample of "spent" (contaminant solution exposed) ZVI was collected from the 15-cm location of MSE core 1-1A (*cf.* Table I) and was examined by X-ray diffraction to identify crystalline phases that were present. Dominant phases identified were synthetic iron, quartz, calcite, and minor amounts of

Table I

Material Description and Radioactivity Estimates for Oak Ridge, Tennessee, Pathway 2 Site Material Sampled from Core 1-1A, Fan 1, at 20 to 24 Feet

Subsample[a] (cm from top)	Material	α (pCi/g)	β (pCi/g)	Total (pCi/g)
5	ZVI + clay	3.67	49.7	53.4
5 (D)[b]	ZVI + clay	2.95	45.8	48.7
10	ZVI	9.02	107	117
15	ZVI	37.7	382	420
15 (D)	ZVI	39.4	381	420
20	ZVI	9.09	127	136
20 (D)	ZVI	9.13	98.5	108
25	ZVI	0.98	3.93	4.91
30	ZVI	0.31	1.52	1.83
35	ZVI	0.01	1.49	1.50
40	ZVI	0.19	0.75	0.94

[a]Distance (cm) from top end of core.
[b]Duplicate sample.

magnetite. Several minor XRD lines were not assigned to a specific phase. Some ferric mineral salts, such as carbonate (siderite) and the sulfides (mackanawite and troilite), often form amorphous or poorly crystalline phases that are not detected readily by the XRD technique. Phillips *et al.* (2000) examined companion core samples from the Oak Ridge, Tennessee, Y-12 installation and reported finding extensive deposition of poorly crystalline akaganeite (β-FeOOH) and goethite (α-FeOOH) throughout the barrier. Goethite sorbs (and desorbs) U^{6+} at near-neutral pH values (Giammer and Hering, 2001). Several minor U oxide phases are considered possible (judged by the criterion of having one or more major reference spectral lines detected, within measurement uncertainty), but positive identification is not possible, as corroborating spectral features are obscured by the major constituents also present in the sample.

We also examined a sample of material said to represent the as-received ZVI used to fill the treatment trench. Perhaps surprisingly, the as-received material appeared to have at least as much magnetite phase as the "spent" media. Odziemkowski *et al.* (1998) indicated that surface coatings of magnetite form on ZVI over a period of time during anaerobic corrosion by the conversion of surface-bound $Fe(OH)_2$ deposits. It is speculated that magnetite may have formed on the as-received material due to the use of water quench of the recycled metal after first heating it to burn off surface deposits of oil. The "as-received" material also contained small amounts of hematite (α-Fe_2O_3).

4. SEM–EDS Data

We examined a sample of material said to represent the as-received ZVI used to fill the treatment trench. The material was noted to consist of highly irregular sizes and shapes, i.e., shavings and filings typical of recycled metal "scrap." Although predominantly ZVI (Fe with trace levels of Si), some particles were ferrous metal alloys (containing Cr, Mn), and there was even some small amount of nonferrous metal (Al) present.

"Spent" ZVI material from the 15-cm location of MSE core 1-1A was examined by SEM–EDS to determine the major elemental composition. In addition to Fe, there are phases that contain Zn, S, Cl, and Cu, as well as elements (Si, Al, K, Mg, etc.) associated with mica-quartz mineral assemblages.

Uranium-containing deposits in the 15-cm core 1-1A material were located using the Hitachi S-4500 SEM and backscattered electron imaging (BEI). Most of the U-containing deposits were very small (e.g., $\sim$ 1 to 30 μm in greatest dimension) and were distributed infrequently on the relatively large

ZVI substrate. This element of scale is apparent in comparing BSE images for a particle at a progression of magnifications, illustrated in Fig. 3. U-containing deposits are usually found in close proximity to surface deposits containing one or more of the elements Ca, S, C, O, Mg, Al, and Si. The latter groups

(*continues*)

Figure 3 (*continued*)

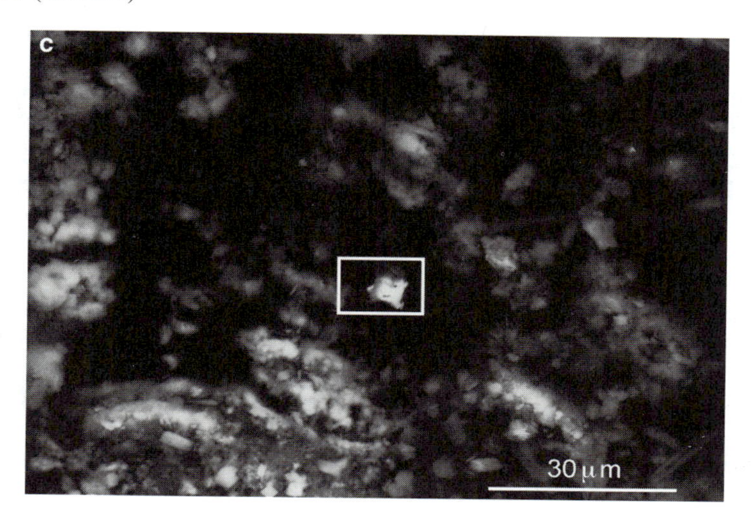

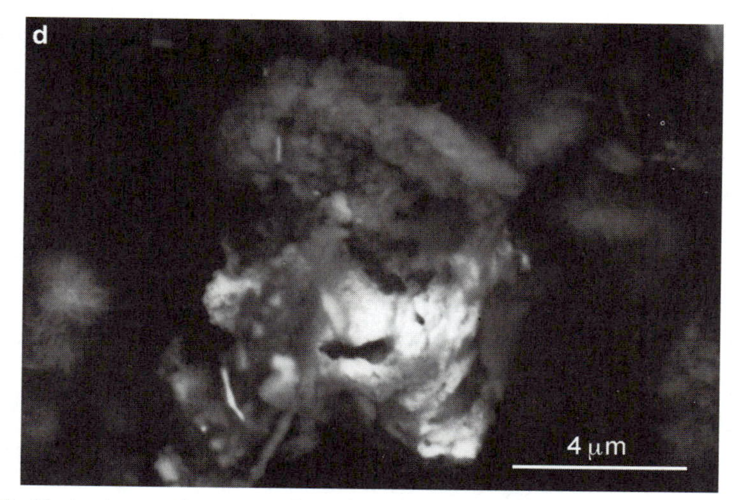

Figure 3 Photomicrograph montage of BSE images at increasing magnification (bright image at 7000X is an uranium-containing deposit). Material is from core 1-1A from the Oak Ridge, Tennessee, pathway 2 permeable reactive barrier.

of associated elements (with the possible exception of S) are not generally subject to reductive precipitation.

5. XPS Data

Material from fan 1, core 1-1A was examined by XPS. Only the samples at 15 and 20 cm from the top of the core had U that was detectable by this technique; this finding is consistent with the radioactivity estimates given in Table I. However, the relatively low U concentration in these samples challenged detection by the XPS technique, requiring long exposures (typically 3-4 h) of the samples to the X-ray excitation beam at high vacuum in order to accumulate an interpretable spectrum. These prolonged beam exposures appear to artifactually induce a partial reduction of U^{6+}, as evidenced by the XPS photoelectron peak shift observed for in-house reference material (low-level uranyl nitrate sorbed on magnetite) exposed under similar conditions. Wersin *et al.* (1994) similarly reported a slight shift (0.2–0.4 eV) to lower energy (i.e., toward U^{4+}) and a line broadening for the U 4*f* XPS photoelectron peak of U^{6+} compounds upon continued exposure to ultrahigh vacuum and the X-ray beam.

The curve-fitted XPS spectrum for the anoxic 15-cm sample from fan 1, core 1-1A is illustrated in Fig. 4. Note that K in the sample contributes to the composite photoelectron peak envelope, with the K 2*s* peak (at 377.9 eV) partially overlapping the U $4f_{7/2}$ photoelectron peak for U^{4+} (at 379.8 eV). After curve-fitting and peak deconvolution, U in the anoxic sample is

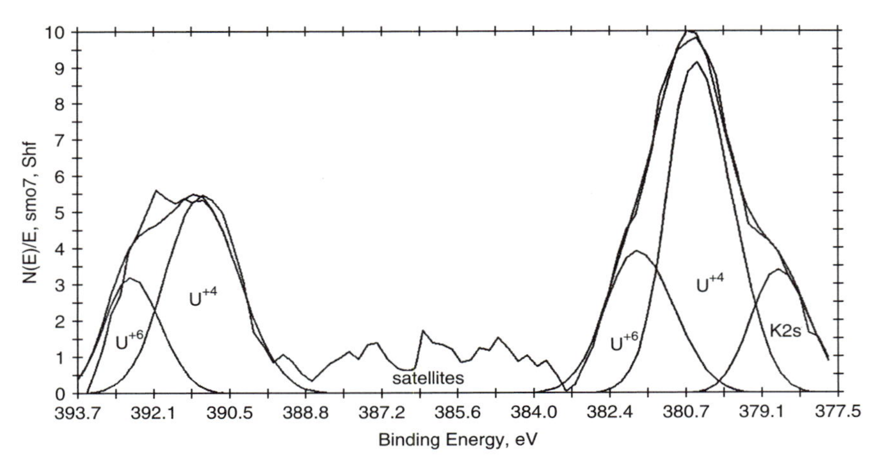

Figure 4 Curve-fitted XPS spectrum for the anoxic 15-cm subsample from the core 1-1A, fan 1, 20–24 ft, Oak Ridge, Tennessee, pathway 2 site.

determined to contain approximately equal amounts of U^{4+} (~55%) and U^{6+} (~45%). A duplicate sample of material from the same location in the core had been removed previously from the protective environment of the glovebox and allowed to air dry. This oxic sample was found by XPS to contain predominantly U^{6+} (~80%). As indicated previously, long exposure of the sample to the X-ray beam is believed to induce a partial reduction of U in both oxic and anoxic samples of material. The observation that U in the oxic-exposed sample was significantly more oxidized than the anoxic sample suggests that U present in the "as-found" (anoxic) material was at least partially reduced when it was collected (i.e., it is a mixed valent deposit).

Other elements present on the media surface, as identified by XPS methodology, include Si, Al, Ca, and K, suggesting the presence of a clay-like material interspersed with the scrap iron in the trench. Much of the Ca appears to be present as the minerals calcite and/or aragonite ($CaCO_3$) based on the C peak at 289 eV. Iron photoelectron peaks in XPS spectra fit an oxidized form such as Fe_2O_3, Fe_3O_4, or FeO(OH), but no Fe^0 is observed. Sulfur (S) is present at the near-surface in all of the core 1-1A samples examined. The photoelectron peak-binding energies observed are consistent with the presence of sulfide deposits, possibly iron sulfides.

B. Durango, Colorado, Bodo Canyon Permeable Reactive Barrier

1. Field Sampling

Chapter 15 describes the construction of PRBs installed at the uranium mill tailings repository near Durango, Colorado. PRB C was opened in July 1999, and samples were obtained, as described in Chapter 15, and split samples of the core material (ZVI foam plates, manufactured by Cercona of America, Dayton, OH) were sealed in plastic tubes under an Ar environment for shipment and subsequent examination at the Materials and Chemistry Laboratory in Oak Ridge, Tennessee.

2. Select Data for Core Samples

Table II summarizes select data for crushed samples of PRB C core material (black in color). The XRF estimate for U in the core samples is comparable to the mean values reported in Chapter 15 for similar material from this PRB. A comparison of gross activity values reported in Tables I and II indicates that the average loading of U in the Durango, Colorado, PRB C cores is significantly greater than that observed for the Oak Ridge, Tennessee, pathway 2 PRB. Note that the gross activity measurements for the PRB C material reported in Table II are for gas proportional counting of emitted α and β radiation; measurement of this weak radiation is subject to self-shielding by the irregular geometry of the material counted. By comparison to the corresponding activity measured for an in-house reference material, median gross activities for the samples reported in Table II range from $\sim$ 0.03 wt% U (sample 99–0159) to $\sim$ 0.14 wt% U (sample 99-0160). Chapter 15 also reports activity measurements for similar material as determined by γ spectroscopy (penetrating radiation, thus less susceptible to self-shielding); their reported γ activity data for PRB C material yield computed estimates for wt% total uranium within the medium ranging from $\sim$ 0.01 to 0.21 wt%.

Subsamples were taken of core material, dried, and ground. A semiquantitative survey by SEM–EDS indicated that the samples contained Fe, Al, Si, S, Ca, Cu, and V. Quantitative estimates for constituents in two of the core material samples, expressed as oxide equivalents, are presented in Table III. These elemental compositions for PRB C material are similar to those reported in Chapter 15.

Samples of crushed core material were examined by XRD. Synthetic iron was by far the predominant phase, with other crystalline phases as minor constituents. Iron oxide phases included hematite, magnetite, and possibly wustite. A sample taken from the middle portion of the inlet end of core 2 (MCL sample ID 99-0160; see Table II) contained the phases quartz (SiO_2), gypsum ($CaSO_4 \bullet 2H_2O$), calcite ($CaCO_3$), hematite, and magnetite. Some very minor lines are also present; these are not assigned to a specific phase, but are possibly due to clay-like (aluminosilicate) material used as a binder in the manufacture of ZVI foam.

3. XPS Data

The relatively higher U content and the more uniform particle size for the crushed ZVI foam specimens allowed XPS spectra to be accumulated more

Table II

Select Data for Core Samples Taken from Durango, Colorado, PRB C

MCL sample ID	Description	Gross activity (pCi/g)	Total U by XRF (wt%)
99–0159	1st core, bottom, inlet side	350	ND[a]
99–0161	1st core, outlet	191	ND
99–0160	2nd core, inlet	1030	0.174
99–0162	2nd core, outlet, bottom	654	0.073

[a]Value not determined.

Table III

XRF Data (Calculated as wt% Oxides) for Durango, Colorado, PRB C Cores

Species	Sample 99–0160	Sample 99–0160 D[a]	Sample 99–0162
Na_2O	1.48	1.66	1.30
MgO	0.05	<[b]	<
Al_2O_3	6.68	6.36	6.15
SiO_2	10.71	11.01	11.78
SO_3	7.84	6.61	5.77
Cl	0.014	<	0.017
K_2O	0.02	0.02	0.06
CaO	3.14	2.69	9.02
TiO_2	<	<	0.07
V_2O_5	2.77	2.25	0.95
Cr_2O_3	0.28	0.29	0.23
MnO	0.97	1.00	0.78
Fe_2O_3	64.32	66.74	66.62
Co_3O_4	0.03	0.05	0.03
NiO	0.07	0.07	0.07
CuO	0.08	0.07	0.12
ZnO	0.41	0.29	0.07
As_2O_3	0.03	0.02	<
SeO_2	0.04	0.03	<
ZrO_2	0.03	0.02	<
MoO_3	0.09	0.05	0.08
SnO_2	<	<	0.01
BaO	0.03	<	<
U_3O_8	0.23	0.18	0.09

[a]Duplicate sample.
[b]Not detected.

rapidly than was the case for Oak Ridge, Tennessee, pathway 2 PRB material. The shorter exposure to the conditions of the X-ray beam and high vacuum is expected to minimize the artifactual reduction of U within the instrument. Figure 5 presents a composite overlay for XPS spectra obtained on Durango, Colorado, PRB C core samples. All specimens examined (including sample 99-0160, both air-excluded and air-exposed material) are consistent with U being present nearly exclusively as the oxidized species, U^{6+}.

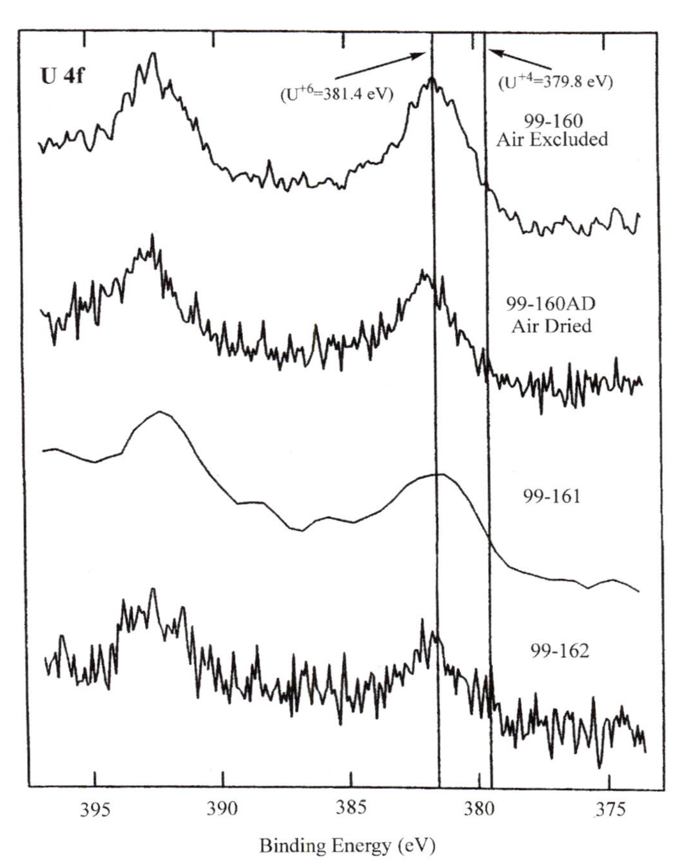

Figure 5 Composite overlay for XPS spectra obtained on Durango, Colorado, PRB C core samples. All specimens examined (including sample 99-0160, both air-excluded and air-exposed material) are consistent with U being present as U^{6+}.

Other surface species identified by XPS (for anoxic sample 99–0161) included oxidized Fe at ~711.5 eV, and S peaks centered at ~ 161.8 eV (probably sulfide species) and at ~ 169.0 eV (probably sulfate species), with the "sulfide" (reduced S) species predominant by a factor of nearly fourfold. Other surface features identified included Ca and weak intensity peaks for Si and Al.

V. DISCUSSION

We have examined core samples of PRBs, composed of ZVI, obtained from U.S. DOE sites located at the Y-12 Plant in Oak Ridge, Tennessee, and at the uranium mill tails repository near Durango, Colorado. ZVI media from these barrier installations were sampled after prolonged subterranean exposure to U-contaminated groundwaters.

Material from both installations had some similar properties. First, regardless of the probable precipitation mechanism(s) (e.g., reduction, sorption, or coprecipitation), the average deposition of U (the contaminant of principle concern) within the media is relatively modest (*e.g.*, < 0.2 wt% U for the more heavily laden Durango core material; see Table II). This value for U within the ZVI medium itself can be compared to the reported value of up to 2.1 wt% U in the oxidized corrosion product "slime" collected at the inlet side to one of the Durango, Colorado, PRBs (see Chapter 15). Giammer and Hering (2001) indicated that oxidized iron phases such as goethite may sorb >1 wt% U at near-neutral pH values. Deposition of U within the PRB at Oak Ridge, Tennessee, was more modest, with a maximum value estimated at ~ 0.14 wt% U, confined to a small portion of one core, near the contaminated water inlet (see data, Table I).

A second noteworthy feature is the extensive deposition of mineral phases within the ZVI, including amorphous iron sulfide and iron oxyhydroxide, and crystalline calcite and gypsum. Gypsum was much more prevalent in the ZVI from the Durango site, where the groundwater contained high levels of sulfate ion. These and similar mineral phases within ZVI PRB materials have also been observed by Phillips *et al.* (2000) for the Oak Ridge, Tennessee, installation (see Fig.1) and by Morrison *et al.* (2001; Chapter 15), for the Durango, Colorado, installation.

A third observation is that U deposited within the ZVI remains at least partially oxidized. This latter observation may be somewhat controversial in view of the predicted thermodynamic equilibrium outcome presented in Eq. (8) and is discussed further.

Fiedor *et al.* (1998) conducted a carefully controlled series of experiments in which ZVI coupons were exposed for various time intervals to synthetic groundwater. For the anoxic condition, the solution was purged with nitrogen containing 20% CO_2. The CO_2 gas maintained relatively constant values for solution phase pH (~ 6.5) and the bicarbonate ion concentration during the experiment, favoring the formation of soluble uranyl carbonate complexes (Meinrath *et al.*, 1996). Carbonate complexation renders uranyl ion soluble even under fairly strong reducing conditions (e.g., Eh $= -0.1$ V; see Bodek *et al.*, 1988). Soluble U was removed relatively rapidly under oxic conditions (i.e., when air was added to the purge gas mixture), with U^{6+} being bound to detached corrosion products ("rust"). Under anoxic conditions, soluble uranium was removed comparatively slowly, but U remained present on the metal coupon surface in the form of a mixed-valent U deposit (as identified by XPS). When the coupon with mixed valence U oxide was removed from the instrument and exposed to ambient atmosphere for several hours, the U phase became more oxidized. These observations are similar to those reported by Grambow *et al.* (1996), who suggested that U^{6+} in anaerobic brine is initially sorbed to magnetite and is then slowly and incompletely reduced at the iron surface. Analogously, Wersin *et al.* (1994) reported on the kinetically slow partial reduction of U^{6+} to U^{4+} at the surface of sulfide minerals under strongly reducing anoxic conditions. The latter authors concluded that the reduction of U is controlled kinetically and not thermodynamically.

A more recent investigation by Qiu *et al.* (2000) on the removal of contaminants from anoxic aqueous solution by reaction with iron surfaces under controlled conditions concluded that partially reduced Se^{4+} and Cr^{3+} are absorbed on the metal surface, whereas U is deposited as U^{6+}, i.e., without reduction. However, it may be noted that the authors maintained anoxic conditions by purging the solution with N_2 only (no added CO_2 to buffer the solution pH or to produce carbonate ion to keep uranyl ion in solution phase). Thus it is possible that the "unusually thick" deposit of U^{6+} observed on the Fe foil could be hydrolyzed uranyl ion, formed as the solution pH increases due to anaerobic corrosion of the Fe substrate [see Eq. (5)].

In contrast to the studies cited above, Gu *et al.* (1998) examined ZVI that had been batch contacted with a high concentration of uranyl ion and concluded that reductive precipitation caused by the ZVI was the predominant mechanism for the removal of soluble U. The supporting evidence cited for this conclusion was that excitation of the solid surface with ultraviolet radiation did not stimulate the visible fluorescence emission characteristic of many hydrous U^{6+} minerals. However, Morris *et al.* (1996) reported that sorption of U^{6+} onto iron oxyhydroxide surfaces deactivates (quenches) the usual uranyl ion fluorescence. Additionally, the redox condition at near-neutral pH values favors the oxidation of U^{4+} by Fe^{3+} (Rao and Sagi, 1962).

Amadelli *et al.* (1991) reported that photoreduction of uranyl solutions on TiO_2 yields a mixed valent uranium oxide, with a stoichiometry close to U_3O_8 (the latter would be a nonfluorescent compound). However, Adeloulas *et al.* (1999b) concluded that the reduction of U^{6+} by ZVI yields a poorly crystallized hydrated uraninite, $UO_2 \bullet nH_2O$. [Note that amorphous UO_2 is reported to be as much as five orders of magnitude more soluble than the crystalline form (Bodek *et al.*, 1988).] An amorphous uraninite may be susceptible to oxidation at the solid–liquid interface, and thus may appear to be partially oxidized by a surface-sensitive analytical technique, such as XPS. Indeed, natural and synthetic UO_2 typically has a partially oxidized surface coating (Duff *et al.*, 1999).

Given the aforementioned discussion, it is therefore not unexpected that U deposited on ZVI media under actual field service conditions would appear to be at least partially oxidized. A perhaps greater concern for the use of ZVI in a PRB to remove soluble U may be whether the surface deposition of various mineral phases during prolonged exposure to the aqueous system will ultimately limit its effectiveness by blocking access to reactive sites.

REFERENCES

Abdelouas, A., Lutze, W., and Nuttall, H. E. (1999a). Uranium concentration in the subsurface: Characterization and remediation. *In* "Reviews in Mineralogy" P. C. Burns, and R. Finch, (eds.), Vol. 38, Mineralogy Soc. of America.

Abdelouas, A., Lutze, W., Nuttall, H. E., and Gong, W. L. (1999b). Remediation of U(VI)-contaminated water using zero-valent iron. *Co. R. Acad. Sci. Paris Ser. II* **328**, 315–319.

Amadelli, R., Maldotti, A., Sostero, S., and Carassiti, V. (1991). Photodeposition of uranium-oxides onto TiO2 from aqueous uranyl solutions. *J. Chem. Soc. Faraday Trans.* **87**, 3267–3273.

Baer, D. R., and Thomas, M. T. (1981). "Use of Surface Analytical Techniques to Examine Metal Corrosion Problems", presented at the 181st National Meeting, American Chemical Society (ACS Symposium Series 189, Industrial Applications of Surface Analysis).

Bodek, I., Lyman, W. J., Reehl, W. F., and Rosenblatt, D. H. (eds.) (1988). "Environmental Inorganic Chemistry", p. 9.2.31. SETAC/Pergamon Press.

Cantrell, K. J., Kaplan, D. I., and Weitsma, T. W. (1995), Zero-valent iron for the in-situ remediation of selected metals in groundwater. *J. Hazard. Mater.* **42**, 201–212.

Charlet, L., Liger, E., and Gerasimo, P. (1998). Decontamination of TCE- and U-rich waters by granular iron: Role of sorbed Fe(II). *J. Environ. Engr.* **124**, 25–30.

Duff, M. C., Hunter, D. B., Bertsch, P. M., and Amrhein, C. (1999). Factors influencing uranium reduction and solubility in evaporation pond sediments. *Biogeochem.*, **45**, 95–114.

Farrell, J., Bostick, W. D., Jarabek, R. J., and Fiedor, J. N. (1999). Uranium removal from ground water using zero valent iron media. *Ground Water* **37**, 618–624.

Farrell, J., Wang, J., O'Day, P., and Conklin, M. (2001). Electrochemical and spectroscopic study of arsenate removal from water using zero-valent iron media. *Environ. Sci. Technol.* **35**, 2026–2032.

Fiedor, J. N., Bostick, W. D., Jarabek, R. J., and Farrell, J. (1998). Understanding the mechanism of uranium removal from groundwater by zero-valent iron using X-ray photoelectron spectroscopy. *Environ. Sci. Technol.* **32**, 1466–1473.

Giammer, D. E., and Hering, J. G. (2001). Time scales for sorption-desorption and surface precipitation of uranyl on goethite. *Environ. Sci. Technol.* **35**, 3332–3337.

Grambow, B., Smailos, E., Geckis, H., Mueller, R., and Hentschel, H. (1996). Sorption and reduction of uranium(VI) on iron corrosion products under reducing saline conditions. *Radiochim. Acta* **74**, 149–154.

Gu, B., Liang, L., Dickey, M. J., Yin, X., and Dai, S. (1998). Reductive precipitation of uranium (VI) by zero-valent iron. *Environ. Sci. Technol.* **32**, 3366–3373.

Halada, G. P., Eng, C., Francis, A. J., Dodge, C. J., and Gillow, J. B. (2001). "Mechanisms of interaction of uranium with corroded steel surfaces", presented at the 222nd National Meeting, American Chemical Society (ACS Symposium Series 116. Environmental Management Science Program).

Ho, C. H., and Miller, N. H. (1986). Sorption of uranyl species from bicarbonate solution onto hematite particles. *J. Colloid Interface Sci.* **110**, 165–171.

Katz, J. J., and Rabinowitch, E. (1951). "The Chemistry of Uranium", *Part I.* McGraw-Hill, New York.

Kertes, A. S., and Guillaumont, R. (1985). Solubility of UO_2: A comparative review. *Nucl. Chem. Waste Mgt.* **5**, 215.

Lackovic, J. A., Nikolaidis, N. P., and Dobbs, G. M. (2000). Inorganic Arsenic removal by zero-valent iron. *Environ. Eng. Sci.* **17**, 29–39.

Langmuir, D. (1978). Uranium solution-mineral equilibria at low temperatures with applications to sedimentary ore deposits. *Geochim. Cosmochim. Acta* **42**, 547.

Matheson, L. J., and Tratnyek, P. G. (1994). Reductive dechlorination of halogenated methanes by iron metal. *Environ. Sci. Technol.* **28**, 2045–2052.

Meinrath, G., Kato, K., Kimura, T., and Yoshida, Z. (1996). Solid-aqueous phase equilibria of uranium(VI) under ambient conditions. *Radiochim. Acta* **75**, 159–167.

Miyahara, K. (1992). "Sensitivity of uranium solubility to variation of ligand concentrations in groundwater. *J. Nuclear Sci. Technol.* **30**, 314–322.

Morris, D. E., Allen, P. G., Berg, J. M., Chisholm-Brause, C. J., Conradson, S. D., Donohoe, S. D., Hess, N. J., Musgrave, J. A., and Tait, C. D. (1996). Speciation of uranium in Fernald soils by molecular spectroscopic methods: Characterization of untreated soils. *Environ. Sci. Technol.* **30**, 2311–2331.

Morrison, S. J., Metzler, D. R., and Carpenter, C. E. (2001). Uranium precipitation in a permeable reactive barrier by progressive irreversible dissolution of zero valent iron. *Environ. Sci. Technol.* **35**, 385–390.

Moyes, L. N., Parkman, R. H., Charnock, J. M., Vaughan, D. J., Livens, F. R., Hughes, C. R., and Braithwaite, A. (2000). Uranium uptake from aqueous solution by interaction with goethite, lepidocrocite, muscovite, and mackinawite: An X-ray absorption spectroscopy study. *Environ. Sci. Technol.* **34**, 1062–1068.

Murphy, W. M., and Shock, E. L. (1999). Environmental aqueous geochemistry of actinides. *In* "Reviews in Mineralogy", (P. C. Burns, and R. Finch, eds.), Vol. 38. Mineralogy Soc. of America.

Odziemkowski, M. S., Schuhmacher, T. T., Gillham, R. W., and Reardon, E. J. (1998). Mechanism of oxide film formation on iron in simulating groundwater solutions: Raman spectroscopic studies, *Corr. Sci.* **40**, 371–389.

Phillips, D. H., Gu, B., Watson, D. B., Roh, Y., Liang, L., and Lee, S. Y. (2000). Performance evaluation of a zero valent iron reactive barrier: Mineralogical characteristics. *Environ. Sci. Technol.* **34**, 4169.

Qui, S. R., Lai, H.-F., Roberson, M. J., Hunt, M. L., Amrhein, C., Giancarlo, L. C., Flynn, G. W., and Yarmoff, J. A. (2001). Removal of contaminants from aqueous solution by reaction with iron surfaces. *Langmuir* **16**, 2230–2236.

Rao, G. G., and Sagi, S. R. (1962). A new reductimetric reagent: Iron(II) in a strong phosphoric acid medium: Titration of uranium(VI) with iron(II) at room temperature. *Talanta* **9**, 715.

Scanlan, J. P. (1977). Equilibria in uranyl carbonate systems – II. *J. Inorg. Nucl. Chem.* **39**, 635–639.

Scherer, M. M., Balko, B. A., and Tratnyek, P. G. (1999). The role of oxides in reduction reactions at the metal-water interface. *In* "Mineral-Water Interfacial Reactions, Kinetics and Mechanisms" (D. L. Sparks and T. J. Grundl, eds.), pp. 301–322. ACS Symposium Series 715, American Chemical Society, Washington, DC.

Science Applications International Corp. (SAIC) (1997). Phase I report on the Bear Creek Valley Treatability Study, Oak Ridge Y-12 Plant, Oak Ridge, Tennessee, Report Y-ER-285.

Tratnyek, P. G., Scherer, M. M., Johnson, T. L., Matheson, L. J. (2001). Permeable reactive barriers. *In* "Chemical Degradation Methods for Wastes and Pollutants: Environmental and Industrial Applications" (M. A. Tarr, ed.). Dekker, New York.

Tricknor, R. V. (1994). Uranium sorption on geological materials. *Radiochim. Acta* **64**, 229–236.

U.S. EPA (1998). Permeable Reactive Barrier Technologies For Contaminant Remediation, EPA/600/R-98/125.

Von Gunten, H. R., and Benes, P. (1995), Speciation of radionuclides in the environment. *Radiochim. Acta.* **69**, 1–29.

Wagner, C. D., Riggs, W. M., Davis, L. E., Moulder, J. F., and Muilenberg, G. E. (1979). "Handbook of X-Ray Photoelectron Spectroscopy." Perkin-Elmer Corp., Eden Prairie, MN.

Wersin, P., Hochella, M. F., Jr., Persson, P., Redden, G., Leckie, J.O., and Harris, D. W. (1994). Interaction between aqueous uranium (VI) and sulfide minerals: Spectroscopic evidence for sorption and reduction. *Geochim. Cosmochim. Acta* **58**, 2829–2843.

Part IV

Case Studies of Permeable Reactive Barrier Installations

Chapter 13

Design and Performance of a Permeable Reactive Barrier for Containment of Uranium, Arsenic, Selenium, Vanadium, Molybdenum, and Nitrate at Monticello, Utah

Stan J. Morrison,* Clay E. Carpenter,* Donald R. Metzler,† Timothy R. Bartlett,* and Sarah A. Morris*

*Environmental Sciences Laboratory,[1] Grand Junction, Colorado 81503
†U.S. Department of Energy, Grand Junction, Colorado 81503

A permeable reactive barrier (PRB) was installed at a site near Monticello, Utah, in June 1999 to remediate contaminated groundwater at a former uranium-processing mill. Laboratory and field column tests were used to evaluate reactive materials. Zero valent iron (ZVI) was selected for the installation because it effectively removes contaminants (arsenic, molybdenum, nitrate, selenium, uranium, and vanadium) present in the groundwater at this site and is available in large quantities at low cost. The PRB is 31 m long and contains 251 metric tons of ZVI. The installation includes slurry walls that direct contaminated groundwater in an alluvial aquifer through the PRB. The PRB was keyed at least 0.3 m into impermeable bedrock. After 1 year of operation, effluent from the PRB continued to meet regulatory concentration goals for all contaminants. Low iron (Fe) concentrations in

[1]Operated by MACTEC Environmental Restoration Services for the U.S. Department of Energy Grand Junction Office.

Handbook of Groundwater Remediation Using Permeable Reactive Barriers

effluent from the PRB contrasted with high Fe concentrations (30 mg per liter) in effluent from column tests. The difference is explained by the longer residence times in the PRB that led to higher pH values and precipitation of $Fe(OH)_2$. Groundwater mounded immediately upgradient of the PRB. The mounding resulted partly from a reduction in the width of the alluvial aquifer because of the impermeable slurry walls and possibly because of a zone of decreased permeability caused by steel sheet pile installation during PRB construction. The mounding appears to have stabilized and has not reached the ground surface. Results of tracer tests and downhole flow measurements suggest that groundwater is flowing though the PRB at nearly the average linear design velocity of 5.7 m per day.

I. INTRODUCTION

A uranium (U)-contaminant plume resulting from the processing of uranium- and vanadium-bearing ore at the Monticello Mill Tailings site, near Monticello in southeastern Utah, has migrated about 2400 m through an alluvial aquifer, contaminating about $3.7 \times 10^5 m^3$ of groundwater (Fig. 1). Arsenic (As), molybdenum (Mo), selenium (Se), and vanadium (V) derived from the ores and nitrate (NO_3) that was used in the milling operations have also contaminated the groundwater. Unacceptable risks to humans and animals could occur from chronic exposure to this contaminated groundwater. Because of the potential adverse health effects, the Monticello Mill Tailings site was placed on the national priorities list in 1989 and is being remediated in accordance with the U.S. Environmental Protection Agency (EPA) Comprehensive Environmental Response, Compensation, and Liability Act. The U.S. Department of Energy (DOE), EPA, and the state of Utah entered into a federal facilities agreement that specifies DOE as the lead agency and gives oversight authority to the EPA and the state of Utah.

In September 1998, a record of decision for an interim remedial action was signed by DOE, EPA, and the state of Utah for the Monticello Mill Tailings site. The interim remedial action, in part, specified installation of a permeable reactive barrier (PRB) for *in situ* treatment of contaminated groundwater. The PRB was installed in June 1999 and will be evaluated during a 5-year interim period as part of the final site remedy.

At the time of installation, the only other widely used groundwater cleanup technology involved extraction of the groundwater, *ex situ* treatment, and discharge or reinjection, a strategy often referred to as pump and treat. Pump-and-treat systems are costly and have generally failed to achieve regulated concentration limits (National Academy of Sciences, 1994). Therefore, a lower cost, more effective solution was sought.

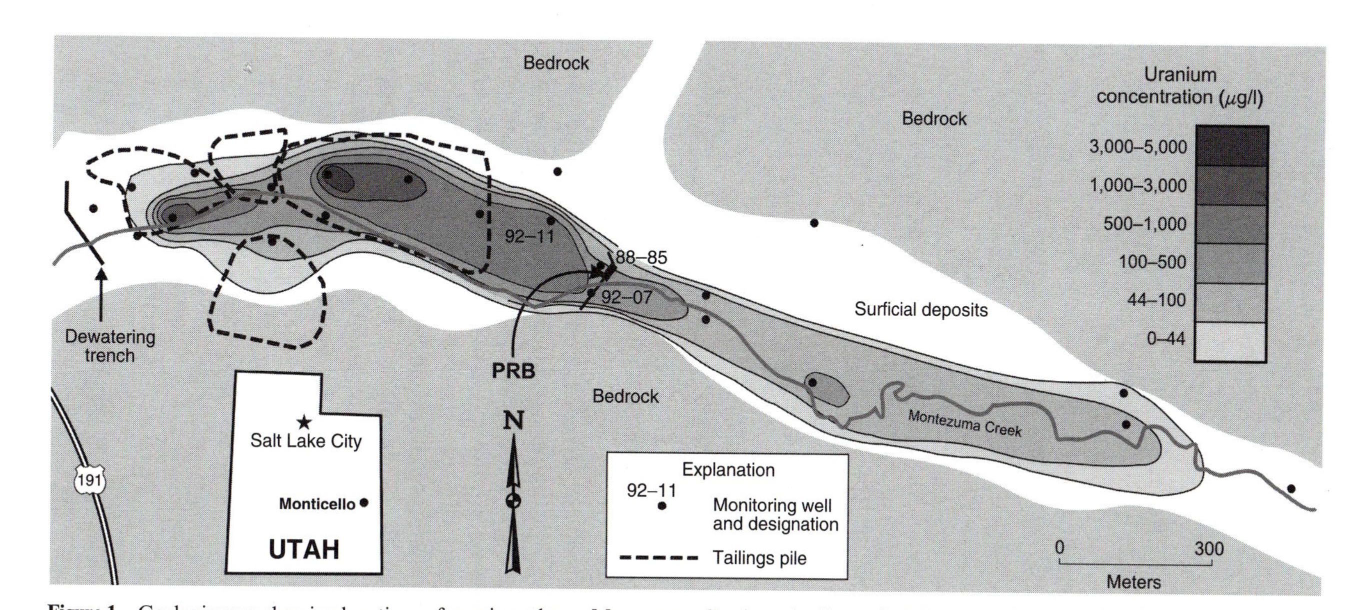

Figure 1 Geologic map showing locations of uranium plume, Montezuma Creek, and tailings pile before remediation of the Monticello Mill Tailings site and installation of a permeable reactive barrier in southeastern Utah.

By 1999, more than 40 PRBs had been installed at other sites, primarily for the remediation of groundwater contaminated by chlorinated organic solvents (Scherer *et al.*, 2000), and PRBs to treat U-contaminated groundwater had been installed at three sites. The first PRB to treat U-contaminated groundwater was installed at Durango, Colorado, in October 1995 (see Chapter 15); three PRBs, each with a different reactive material, were installed at Fry Canyon, Utah, in September 1997 (see Chapter 14); and two PRBs were installed at the Oak Ridge National Laboratory, Tennessee, in November and December 1997 (see Chapter 12). Analyses of groundwater samples exiting these PRBs indicate that zero valent iron (ZVI) removed U and associated contaminants from groundwater, but the monitoring periods were too short to evaluate long-term performance.

This chapter presents results of tests that preceded the decision to install a PRB at Monticello and an assessment of PRB performance after about 1 year of operation. Analytical results obtained during operation of the PRB at Monticello can help determine the feasibility of PRB installations for the remediation of groundwater at numerous industrial sites, weapon production facilities, and base metal mine and mill sites worldwide.

II. SITE DESCRIPTION

A. Former Uranium-Processing Mill Site

The Monticello mill processed U and V ores from the mid-1940s until 1960. Approximately 1 million metric tons of U ore was processed, and the resultant tailings were impounded at four locations on the 31-ha (Fig. 1). Much of the waste material was slurried to impoundments, and the wastes were in direct contact with the shallow aquifer in some areas, resulting in significant contamination of the groundwater.

From July 1997 to September 1999, 2.5 million m^3 of tailings and contaminated soil was relocated from the site to an engineered repository constructed 0.8 km south of the mill site. During tailings removal, large areas of the site and the alluvial aquifer were excavated to bedrock to ensure that the site was cleaned up to the mandated regulatory levels. Much of the alluvial aquifer was rebuilt using clean gravel by December 2000.

B. Site Geology and Geohydrology

The Monticello Mill Tailings site is located within the valley of Montezuma Creek. Perennial flow in the creek is about 28 liters/s and is regulated at the Monticello Reservoir 2.4 km west of the mill site. The watershed of the creek includes part of the Abajo Mountains that rise to 3350 m several kilometers farther west. Remedial action significantly modified the alignment and elevation of the creek from its original position on the mill site and in the area of the PRB.

Permeable sand and gravel within the buried paleo-channel of the creek constitute a shallow, unconfined aquifer (alluvial aquifer) beneath the site. Abundant cobbles and up to 15% silt and clay are also present in the alluvium. The paleo-channel is about 135 m wide near the PRB and controls most or all of the groundwater flow from the Monticello mill site (Fig. 2). Depth to bedrock ranges from about 3 to 10 m. Groundwater flows in the alluvial aquifer east and southeast down the valley from the mill site. The alluvial aquifer pinches out against hill-slope colluvium and bedrock flanking the valley. Thin, intermittent lenses of groundwater occur in colluvial deposits north of the alluvial aquifer and at the PRB site. This groundwater, which originates from the irrigation of farm land and other sources not related to the mill site, flows into the alluvial aquifer.

Low-permeability mudstone and siltstone beds of the Dakota Formation underlie the alluvial aquifer and isolate it from a deeper aquifer. The bedrock erosional surface at the base of the alluvial aquifer near the PRB slopes gently east to southeast and is relatively flat with mild undulations to 1 m. Saturation in the alluvium ranged from about 1 to 2 m before remediation of the area; depth to groundwater beneath the valley floor ranged from about 2 to 3 m.

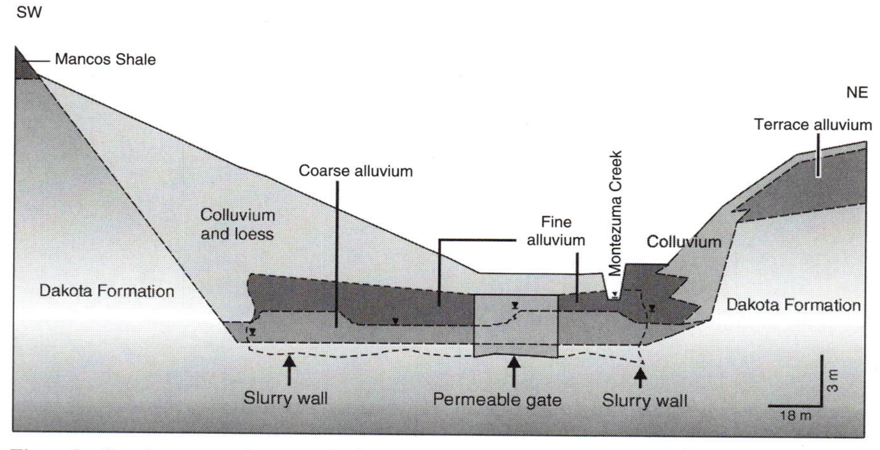

Figure 2 Southwest–northeast geologic cross section along the line of permeable reactive barrier orientation, view to the northwest, near Monticello, Utah. Montezuma Creek has been diverted from the locations shown in Fig. 1.

Large-scale aquifer dewatering operations and creek diversions between spring 1999 and fall 2000 affected groundwater flow throughout the PRB area. Reconstruction of the alluvial aquifer and creek has been completed and all water diversion has ceased. The amount of groundwater flowing into the mill site area (about 380 liters/min) has probably not been altered by construction activities. However, the creek bed has been lowered in elevation, which may reduce the volume of groundwater that enters the PRB area. Prior to remedial actions, groundwater flow to the PRB area was estimated to be about 189 liters/min.

C. Groundwater Geochemistry and Site-Related Contaminants

Mill-related contaminant concentrations in groundwater samples collected hydraulically upgradient of the PRB and preliminary remediation goals (PRGs) for contaminants at the Monticello Mill Tailings Site are presented in Table I. PRGs were proposed to EPA and the state of Utah and will be finalized in the record of decision. The concentrations of four contaminants (NO_3, Se, U, and V) exceeded the PRGs in the groundwater.

Calcium (Ca) is the dominant cation in the alluvial groundwater upgradient of the mill site. Anions are generally dominated by sulfate (SO_4), but

Table I

Monticello Mill-Related Contaminant Concentrations in Groundwater Upgradient of the Permeable Reactive Barrier (PRB) and after Exiting the PRB and Preliminary Remediation Goals (PRG) Established for the Site

Contaminant	Upgradient concentration[a] (μg/liter)	Concentration exiting the PRB[a] (μg/liter)	PRG (μg/liter)
Arsenic	10.3	<0.2	50
Manganese	308	177	730
Molybdenum	62.8	17.5	100
Nitrate	60,720	< 65.1	44,000
Selenium	18.2	0.1	10
Uranium	396	< 0.24	44
Vanadium	395	1.2	260

[a]Concentrations are from the October 2000 sampling. Upgradient concentrations are mean values for samples from five wells located 1 m upgradient of the PRB; concentrations exiting the PRB are mean values for samples from five wells located on the downgradient portion of the PRB.

bicarbonate is abundant. Alkalinity, SO_4, Ca, and sodium (Na) average about 250 mg/liter as calcium carbonate ($CaCO_3$), 500, 250, and 40 mg/liter, respectively. The total dissolved solids concentration is about 1000 mg/liter, pH is about 6.8, Eh is about 370 mV, and dissolved oxygen is about 6 mg/liter. Some of these values have been altered substantially by mill site operations and subsequent remediation activities. In particular, dissolved oxygen and oxidation–reduction potential for groundwater influent to the PRB are about 0.4 mg/liter and 80 mV, respectively, having been altered during operation of the mill.

III. LABORATORY AND FIELD TESTS

Laboratory batch tests were conducted on 14 sorbents containing amorphous ferric oxyhydroxide (AFO), phosphate, activated carbon, humate, or magnetite. Distribution ratios for the sorbents were all less than 1000 milliliters ml/g, suggesting that they may not be cost effective for the PRB. Flow-through experiments were conducted in small (150-mm-long by 15-mm-diameter) glass columns with five types of ZVI. In some experiments, ZVI was combined with limestone, cement, or magnesium hydroxide ($Mg[OH]_2$) to lower the iron (Fe) concentrations in column effluents by increasing the pH values. Groundwater collected from well 92–11 (Fig. 1) was used in the laboratory experiments. Uranium concentrations were typically less than 1 μg/liter in effluents from columns of ZVI operated for as many as 219 pore volumes. The results of these experiments led to the decision to use ZVI in the PRB; in addition to the effectiveness of ZVI in removing the contaminants, it is inexpensive and readily available. To determine the effectiveness of several types of ZVI with conditions as close to *in situ* as possible, larger scale column experiments were conducted at the field site with groundwater collected from well 88–85 located near the PRB site (Fig. 1).

Columns 1.2 m in height and 10 cm in diameter were used for the field tests. Groundwater was piped from well 88–85 (Fig. 1) through columns set up in a trailer 3 m away. Groundwater from well 88–85 was similar in major ion chemistry to that in well 92–11 used for the laboratory study (total dissolved solids of 2290 mg/liter in 88–85 compared to 2520 mg/liter in 92–11). Field tests were used to confirm laboratory results and to determine the design parameters more accurately. Five different types of ZVI supplied by four vendors were used in the tests. The five types of ZVI performed similar to one another, thus results of only one column containing one type of ZVI are presented. This column contained 16.7 kg of ZVI (mesh −8/+18) and had 52% porosity. Groundwater flux ranged from 20 to 80 ml/min, resulting in residence times ranging from 1 to 4 h; the residence time expected in the PRB was about 20 h. The pH value increased from about 6.8 in the influent to 7 in the effluent, Eh decreased from about 100 mV to −170 mV, and alkalinity decreased from about 350 to 250 mg/liter as $CaCO_3$.

Reactions with the ZVI were relatively fast. Analytical results of samples collected from ports placed along the column indicate that concentrations of As, Se, U, and V decreased to their effluent values within 15 cm of the inlet, corresponding to a residence time of 21 min (Fig. 3). Molybdenum and NO_3 reactions were slower; their concentrations decreased to the effluent values within 33 cm from the inlet, corresponding to a residence time of 45 min. These samples were collected after about 54 pore volumes and after some of the ZVI in the lower portion of the column was no longer effective. In other experiments that used fresh ZVI, residence times required for nearly complete removal of U and V were often less than 6 min.

Nearly 100% of the U and V; 70 to 80% of the Se, Mo, and As; and 60 to 100% of the NO_3 were removed from the groundwater throughout most of the experiment (Fig. 4). Some of the variation was due to changes in the flow velocity; slower flow rates resulted in increased removal efficiency. The results were encouraging and led to a decision to proceed with installation of the PRB.

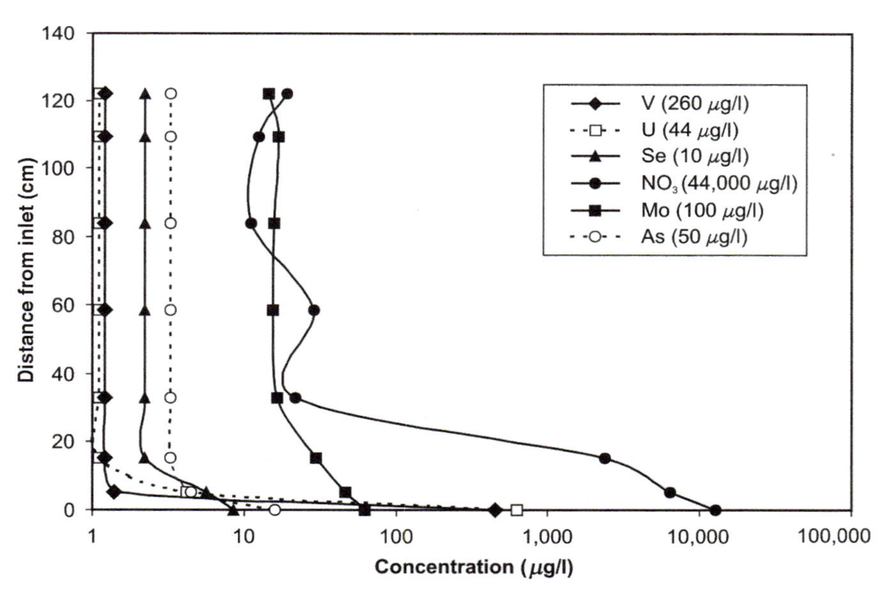

Figure 3 Contaminant concentrations along the flow path in the ZVI column; samples collected after 54 pore volumes at a groundwater flux of 31 ml/min. Preliminary remediation goals are shown in parentheses.

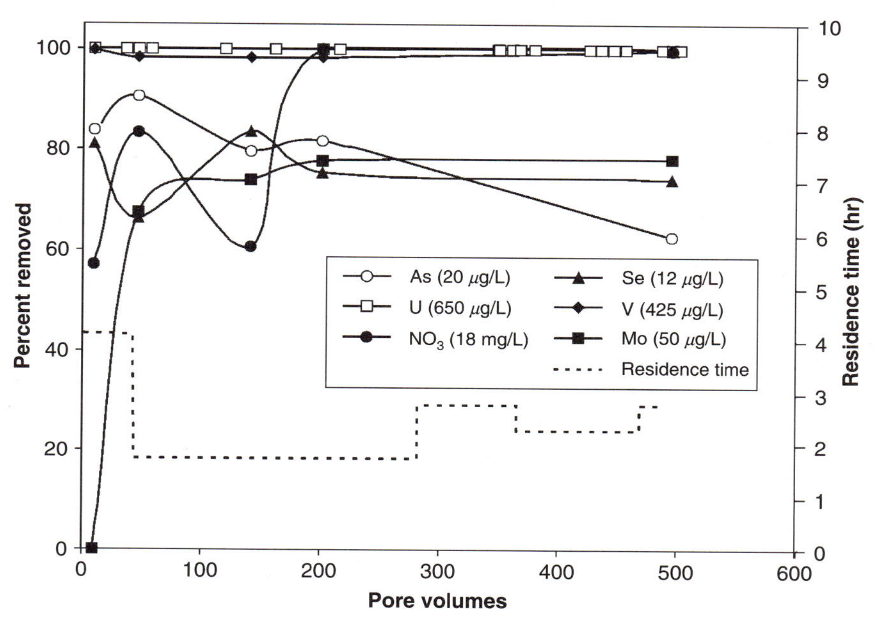

Figure 4 Contaminant concentrations in effluents from field column tests (influent concentrations are in parentheses) near Monticello, Utah.

Reaction of the groundwater with ZVI also caused some undesirable effects. Iron was observed in the effluent in concentrations that averaged 20 mg/liter and were as high as 32 mg/liter (Fig. 5). These concentrations exceeded a risk-based goal of 11 mg/liter and exceeded the concentration (about 1 mg/liter) that causes red coloration in the water. Manganese (Mn) was also released from the ZVI with the maximum Mn concentration (3 mg/liter) occurring in the first effluent sample; however, its concentration decreased rapidly (Fig. 5). Manganese is a trace component of ZVI and is likely released from dust particles in the early stage of column operation.

Changes in major ion concentrations between influent and effluent samples were also measured to assess the amount of mineral matter that precipitated in pore spaces. As minerals accumulate in pore spaces of the ZVI, performance may decrease because groundwater will increasingly flow under or around the PRB. More pore space is occluded by the deposition of minerals containing major elements than from minerals containing the contaminants because of higher concentrations of major elements in the groundwater. Average Ca concentrations decreased from about 325 mg/liter (8.1 mmol) to 250 mg/liter (6.3 mmol), and dissolved inorganic C (calculated from

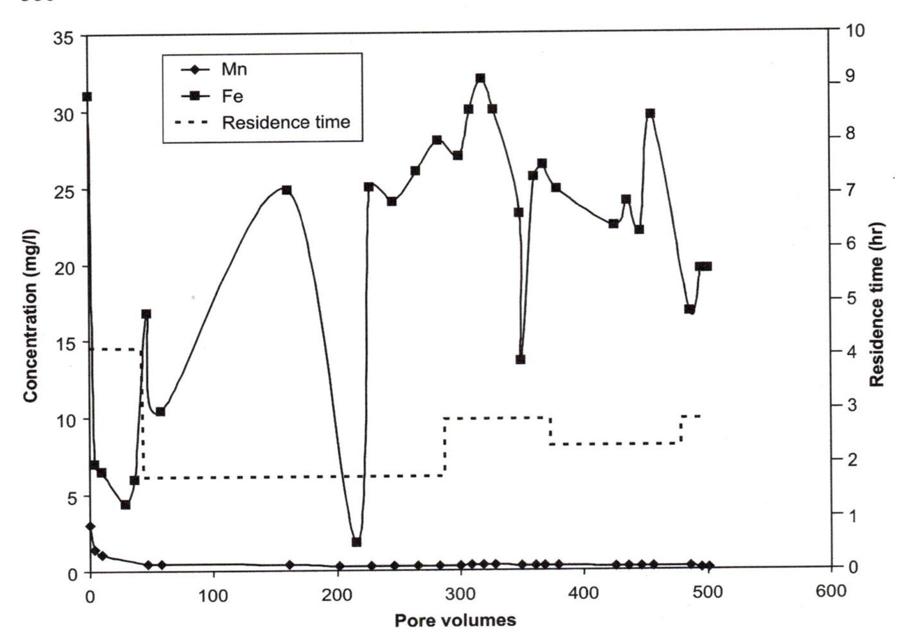

Figure 5 Iron and Mn concentrations in field column test effluents near Monticello, Utah, June through August 1998. Samples were acidified and unfiltered.

alkalinity) decreased from about 84 mg/liter (7.0 mmol) to 60 mg/liter (5.0 mmol) as the groundwater passed through the field columns. The decreases in Ca and alkalinity indicate that $CaCO_3$ was accumulating in the column. Sulfate showed no decrease in concentration, indicating little or no precipitation of sulfur-bearing minerals.

IV. PERMEABLE REACTIVE BARRIER DESIGN

A. Hydraulics

Flow modeling was conducted with MODFLOW (McDonald and Harbaugh, 1988) and MODPATH (Pollock, 1989) to predict (1) the extent of groundwater mounding upgradient of the PRB, (2) capture efficiency of the PRB relative to its orientation in the plume, and (3) flow velocity through the PRB. The PRB and the slurry wall system were designed to capture nearly the entire flux of groundwater leaving the mill site in the alluvial aquifer. The position of the creek in the model was based on preliminary design specifications for restoration of the mill site.

The groundwater flow model (without the PRB or the slurry walls) was calibrated to water level data collected throughout the plume area before remedial action at the mill site. The resulting hydraulic conductivity of the alluvial aquifer was about 0.01 cm/s, which is consistent with pump test data. This model was used to predict steady-state water table configurations for various PRB and slurry wall designs. Modeled groundwater flux at the proposed location of the PRB was 189 liter/min. Variables tested in the predictive models were length, orientation, and position of the slurry walls and the hydraulic conductivity, thickness, and position of the PRB. The model that best represented the installed system predicted a 0.6- to 0.9-m water table rise in an area extending not more than about 135 m hydraulically upgradient of the PRB. Mounding under this scenario was not expected to reach the ground surface. The length and width of the modeled PRB were 30 and 3 m, respectively; the model used a hydraulic conductivity of 0.03 cm/s for ZVI, which was similar to permeameter test results. Slurry walls were represented as impermeable vertical barriers that extended north and south from the PRB to the boundaries of the alluvial aquifer.

B. Width Of Permeable Reactive Barrier

The design width (dimension in the direction of groundwater flow) of a PRB is often based on the amount of groundwater contact time required for ZVI to reduce concentrations of contaminants to acceptable levels. This contact time is referred to as residence time. The residence time of many hours is commonly required for halogenated organic compounds. In contrast, U and other site- related contaminants (As, Mo, NO_3, Se, and V) in the Monticello groundwater reacted rapidly with ZVI in column tests; concentrations of most contaminants decreased to acceptable levels in less than 6 min.

The flux of groundwater through the PRB was estimated at 189 liter/min based on groundwater flow model results. The flow model also predicted a saturated thickness of 3 m, resulting in a cross-sectional area of 93 m^2. Using these values and a porosity for ZVI of 0.5 (an average value from multiple measurements), the 6-min residence time for U would require the PRB to be only 2.5 cm wide. Although some of the other contaminants required slightly longer residence times, it was apparent that a thin PRB would be acceptable if residence time was the only factor that needed to be considered.

Because residence time in the PRB decreases with ZVI dissolution, a thicker PRB is required than is predicted by residence time only. The effect of ZVI loss was estimated based on the average effluent Fe concentration (30 mg/liter) in the field column tests. Some of the Fe (perhaps most) would reprecipitate within the PRB; however, it was assumed that these secondary Fe minerals would not further reduce contaminant concentrations in the groundwater. On the basis of the groundwater flux rate (189 liter/min), the 1.2-m width of ZVI would be depleted in 117 years. Permeability reduction caused by mineral precipitation and loss of reactivity from passivation of ZVI surfaces will also decrease longevity, but no reliable methods were available to evaluate these processes. These factors need to be better understood to accurately predict long-term PRB performance.

The PRB was constructed with 1.2 m of 100% ZVI to accommodate ZVI loss for 117 years and because this width was amenable to the construction practices used. Gravel zones (0.5 m wide) were designed upgradient and downgradient of the 100% ZVI zone to help distribute groundwater flow through the system (Fig. 6). Upgradient zones included 13% (by volume) ZVI to initiate oxygen and mineral removal. The gravel zone was intended to extend the life of the PRB by providing additional pore space to accommodate mineral precipitates. The downgradient gravel zone included an air-sparging system. Data from field column tests indicate that Fe and Mn may be released from the PRB and may become mobile in the groundwater, but

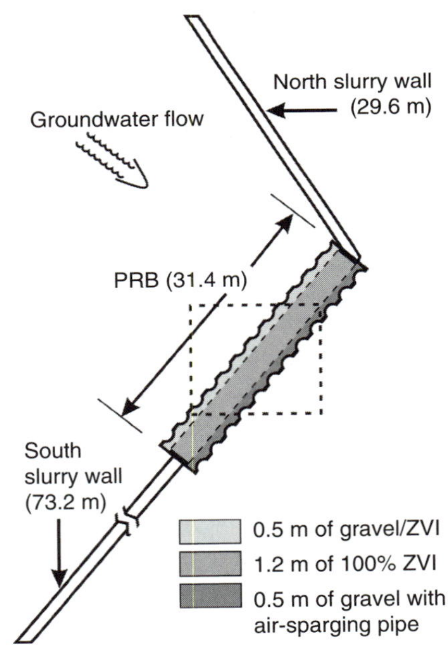

Figure 6 Diagram of the permeable reactive barrier and slurry wall design near Monticello, Utah. The corrugated perimeter of the PRB represents the outline of sheet pilings used for construction. The dashed line box represents the area depicted in Figs. 11 and 12.

concentrations can be decreased significantly by air sparging. The air-sparging system was installed as an option to precipitate Fe and Mn by increasing the concentration of dissolved oxygen but has not been used because Fe and Mn concentrations in samples of groundwater exiting the PRB have remained low.

V. CONSTRUCTION

A. Permeable Reactive Barrier

The PRB was constructed by driving steel sheet piling through alluvium and into impermeable bedrock, forming a rectangular box 31.4 m long by 2.2 m wide. The sheet pile box was constructed of Z-shaped steel sheet piling with an interlocking single jaw and was driven with a 127-ton crane and hydraulic vibratory pile-driving equipment until it would no longer penetrate. Once in place, the sheet piles were cut near the ground surface, and structural beams and cross bracing were installed (Fig. 7a). Native soils inside the box were excavated to about 0.3 m into the impermeable bedrock and were replaced with ZVI and the two gravel zones (Fig. 6). The depth from the top of the sheet pile to the bottom of the excavation was 3.4 to 4.0 m.

Sheet steel boxes were placed on the bedrock to keep the gravel zones and 100% ZVI zone separate during placement. The 1.2-m-high boxes were constructed from a 0.6-cm-thick steel plate cut in various lengths and welded together to fit the length of the excavated sheet pile box. As the ZVI and gravel were added, the steel sheet piling boxes were incrementally hoisted and were removed when the boxes were filled. After the reactive materials and gravel were placed in the boxes, the Z-shaped sheet pilings perpendicular to the groundwater flow (two 31.4-m sections) were removed to allow groundwater to flow through the PRB.

Materials placed in the excavation consist of an upgradient gravel/ZVI mix, a middle section of 100% ZVI, and a downgradient gravel zone that included the air-sparging system (Fig. 6). The upgradient gravel zone is 0.5 m wide and is composed of 13% by volume $-4/+20$ mesh ZVI mixed uniformly with 1.3 cm of gravel. The 1.2-m-wide middle section of the PRB contains 100% $-8/+20$ mesh ZVI; approximately 127 m^3 of ZVI with a loose-filled weight density of 1840 kg/m^3 was used. ZVI was placed in the excavation from super sacks (900-kg canvas bags with a drawstring release at the bottom) suspended from a 127-ton crane.

The downgradient gravel zone is 0.5 m wide, composed of 1.3-cm sieved gravel and includes an air-sparging system constructed of perforated polyvinyl chloride (PVC) pipe. The horizontal perforated PVC pipe was placed 0.3 m

(*continues*)

Figure 7 (*continued*)

Figure 7 Permeable reactive barrier and slurry wall construction near Monticello, Utah, (June 1999), (a) Open trench before reactive materials were emplaced, (b) Bentonite slurry in the trench for the south slurry wall.

from bedrock and was connected to an air-sparging feed pipe that can be used to force air from the ground surface through the perforated pipe. Three vertical air-sparging vents were attached to a perforated horizontal pipe placed 0.5 m below the top of the gravel to provide a path for the air to escape.

More than 70 monitoring wells were installed in or near the PRB to measure water table elevations and to collect samples for chemical analysis.

B. Slurry Walls

Slurry walls direct contaminated groundwater to the PRB (Figs. 6 and 7b). The south slurry wall is 73.2 m long and the north slurry wall is 29.6 m long (Fig. 6). To construct the slurry walls, soil/bentonite backfill was mixed in an excavated depression adjacent to the PRB.

As an excavator removed soil from the trench for the slurry walls, the excavated soil was placed in the mixing area. A water/bentonite slurry was pumped to the mixing area, and a bulldozer mixed the soil and slurry until homogeneous. The bentonite content of the soil/bentonite mix was 4% by volume. The water/bentonite slurry (6% by volume bentonite) was mixed in a separate tank. The water/bentonite slurry also served to maintain trench stability, to create a seal or skin on the trench walls, and to fill voids between the trench walls and the backfill mix. Slurry walls were constructed by placing the soil/bentonite mix in the trenches filled with the water/bentonite slurry. The displaced water/bentonite slurry was then added to the soil in the mixing area to generate additional soil/bentonite mix. Slurry walls are keyed 0.2 to 1.4 m into bedrock and range from 3.0 to 4.9 m deep. Laboratory permeability tests (ASTM, 2001) conducted on slurry wall materials indicate that the hydraulic conductivity was less than 10^{-7} cm/s.

The PRB and slurry walls were constructed on private land located downgradient from the Monticello Mill Tailings site. To avoid destroying a grove of established trees, the landowner requested that the north slurry wall end approximately 15 m before the north boundary of the alluvial aquifer was reached. Construction on the south slurry wall was stopped when no more alluvial materials were observed in the excavation.

VI. PERFORMANCE

A. Hydraulics

Groundwater levels in monitoring wells in the PRB and the surrounding area have been monitored periodically since installation of the system. Water table measurements in August 2000 indicated a losing stream condition and a groundwater divide south of the creek and west of the PRB (Fig. 8). Even before the PRB was built, this divide caused water to flow east toward the PRB area and to the southeast. Because Montezuma Creek now flows through a 30-m-long culvert under the north slurry wall, water level contours in that area may have changed.

1. Groundwater Mounding

A steep hydraulic gradient is present between the upgradient alluvial aquifer and the PRB (Fig. 9). The hydraulic head decreases by about 0.75 m

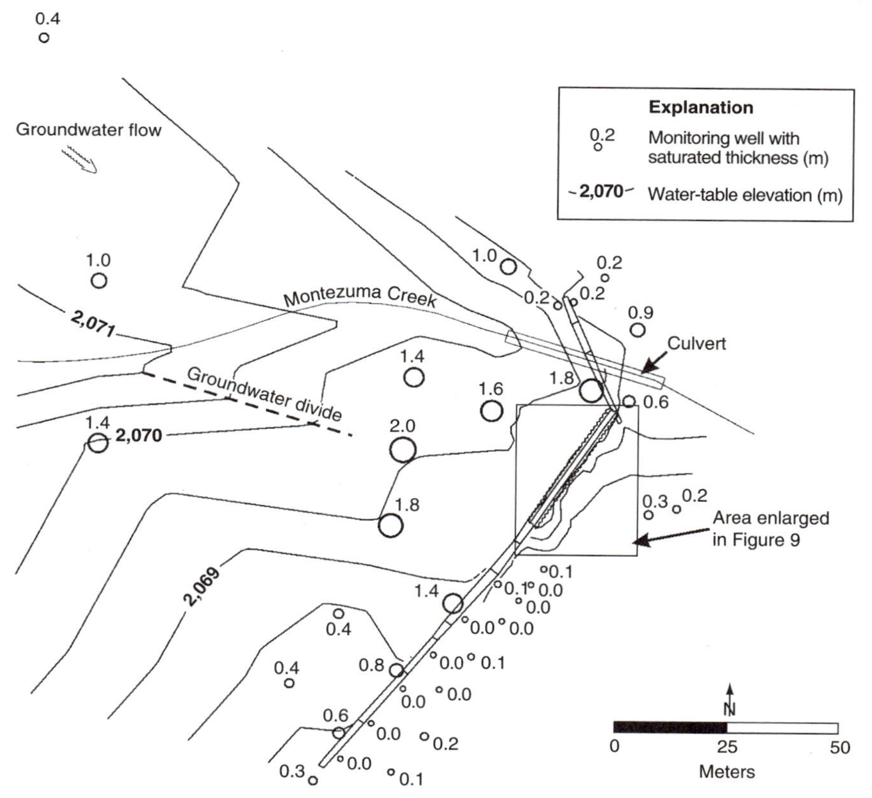

Figure 8 Water table elevations (contoured) and saturated thicknesses of alluvium (size of circle indicates relative amount of saturation) in and near the permeable reactive barrier, near Monticello, Utah, (August 2000).

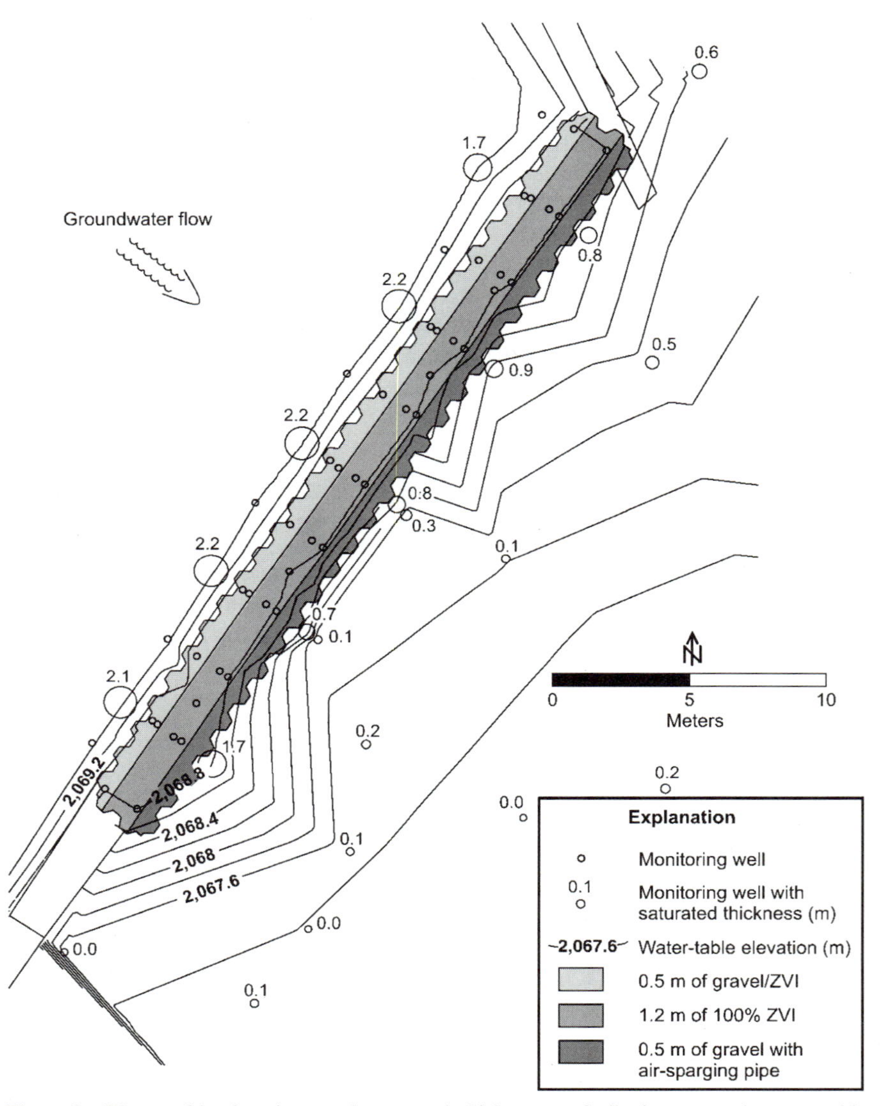

Figure 9 Water table elevations and saturated thicknesses of alluvium near the permeable reactive barrier, near Monticello, Utah, (August 2000). Size of circle indicates relative amount of saturation.

across the alluvial aquifer–PRB interface within 1 m hydraulically upgradient of the PRB. The water table within the PRB is essentially flat. Areas of the downgradient interface between the PRB and the alluvial aquifer also have steep hydraulic gradients. Up to 1 m of groundwater saturation is present 0.6 m downgradient of the PRB, but less than 0.3 m of groundwater saturation is present 6 or 7 m farther downgradient (Fig. 9). The hydraulic gradients at the PRB have remained nearly unchanged since monitoring began.

2. Groundwater Bypassing the PRB

Hydraulic gradients suggest that some groundwater flows around the ends of the slurry walls. This was anticipated for the north wall, where the slurry wall stopped 15 m short of the alluvial aquifer boundary. Water levels in the upgradient area near the south end of the south slurry wall are similar to pre-PRB levels. Water levels downgradient of the south slurry wall decreased immediately after installation and have remained low (Fig. 8). Apparently the south end of the south slurry wall was not fully keyed into low-permeability material and some groundwater flows around the wall.

Dewatering of the Monticello Mill Tailings site caused depression of the water table in the project area from April 1998 to April 1999 (Fig. 10). By July 1999, water levels immediately upgradient of the PRB (in wells 88–85 and 92–07) had rebounded to their historical values. The exact cause of the rebound is

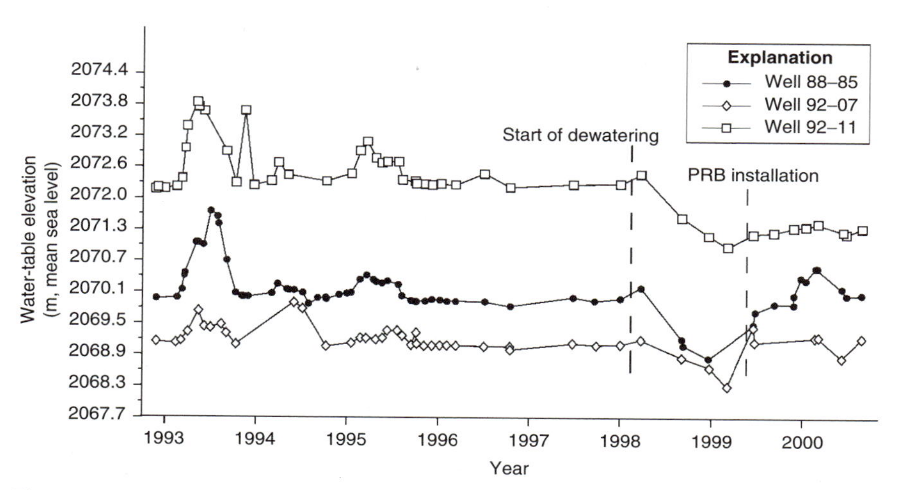

Figure 10 Hydrograph of three wells hydraulically upgradient of the permeable reactive barrier that have been monitored since 1992; well locations are in Fig. 1.

uncertain because of the perturbations made to the groundwater system by dewatering, repositioning of the Monticello Creek, and installation of the PRB and slurry walls.

3. Flow–Velocity Measurements

A colloidal borescope was used to help determine groundwater velocity through the PRB. This instrument measures the rate and direction of groundwater by recording the movement of natural colloids through a wellbore (Kearl, 1997). Six alluvial monitoring wells and four monitoring wells in the 100% ZVI portion of the PRB were tested in July 2000 to provide a broad perspective of the flow field and to identify possible flow anomalies. Tracer testing was also conducted during July 2000 to evaluate flow paths and travel times through the PRB. A tracer (either bromide or iodide) was injected into the groundwater through 4 of the upgradient alluvial monitoring wells, and groundwater samples were collected in 51 monitoring wells along nine transects through the PRB. Results of the tracer study and colloidal borescope measurements suggest that groundwater flow is heterogeneous and migrates through the PRB at velocities ranging from 0.6 to 5.5 m/day (N. Korte, personal communication), which compares favorably with the design flow rate of 5.7 m/day that was based on a hydrologic flow model.

B. Chemistry

1. Chemical Distributions

Analyses of groundwater samples from more than 70 monitoring wells (2 and 5 cm in diameter) were used to determine the spatial distribution of groundwater chemistry. Wells had 1.7-m screens in the lower portion of the alluvial aquifer; samples were collected monthly for 4 months and then quarterly for two periods before October 2000. Concentrations were nearly the same for all monitoring periods. The latest sampling data (October 2000) shown in Fig. 11 and Fig. 12 portray the spatial distributions of pH, alkalinity, and concentrations of eight constituents in and near the PRB.

In October 2000 (1 year and 4 months after installation of the PRB), concentrations of U in groundwater samples collected upgradient of the PRB ranged from 312 to 356 μg/liter (Fig.11). Concentrations decreased to a range of 65 to 283 μg/liter in samples from within the gravel/ZVI zone and were less than 0.4 μg/liter in samples from the 100% ZVI zone. Vanadium concentrations ranged from 354 to 413 μg/liter in samples collected upgradient of the PRB, decreased to a range of 1 to 78 μg/liter in the gravel/ZVI zone,

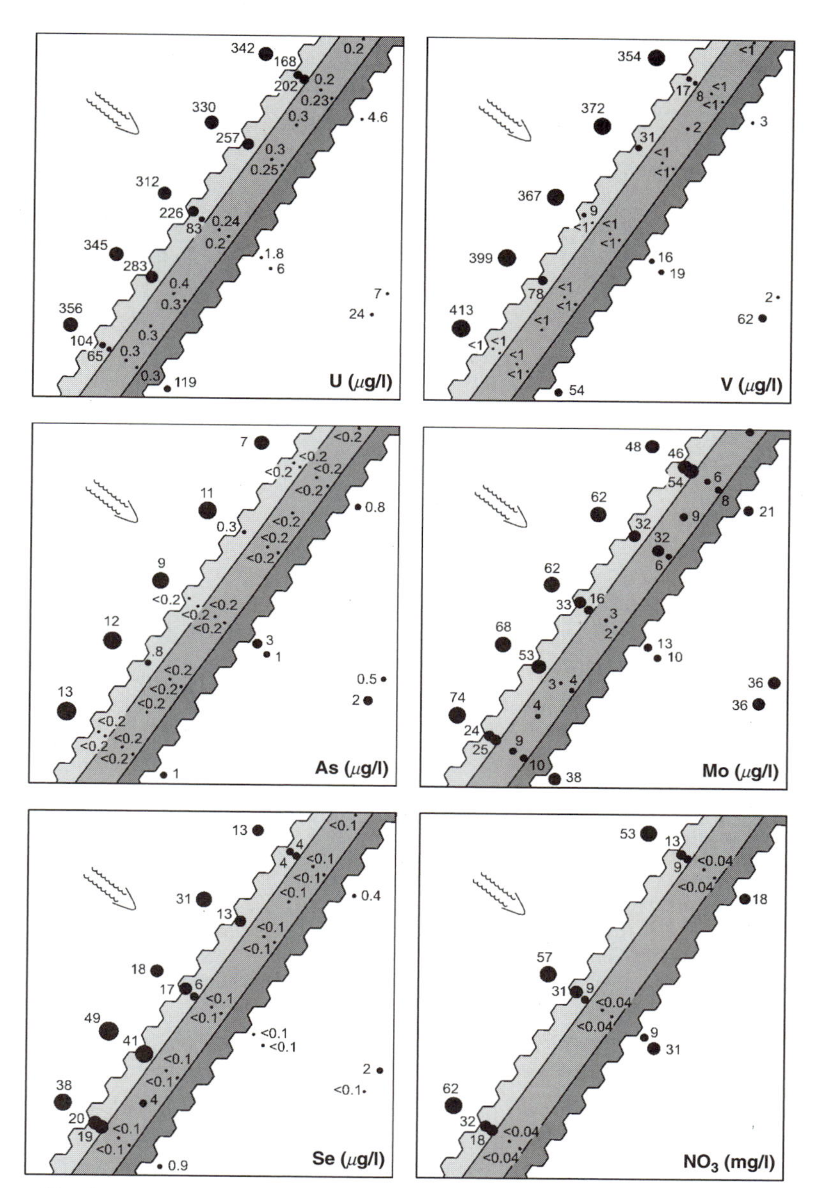

Figure 11 Concentrations of uranium (U), vanadium (V), arsenic (As), molybdenum (Mo), selenium (Se), and nitrate (NO_3) in groundwater samples from monitoring wells in and near the permeable reactive barrier (PRB) near Monticello, Utah (October 2000). Dot diameter is proportional to concentration. Arrow indicates groundwater flow direction. See Fig. 6 for location represented in Fig. 11 and explanation of shaded PRB components.

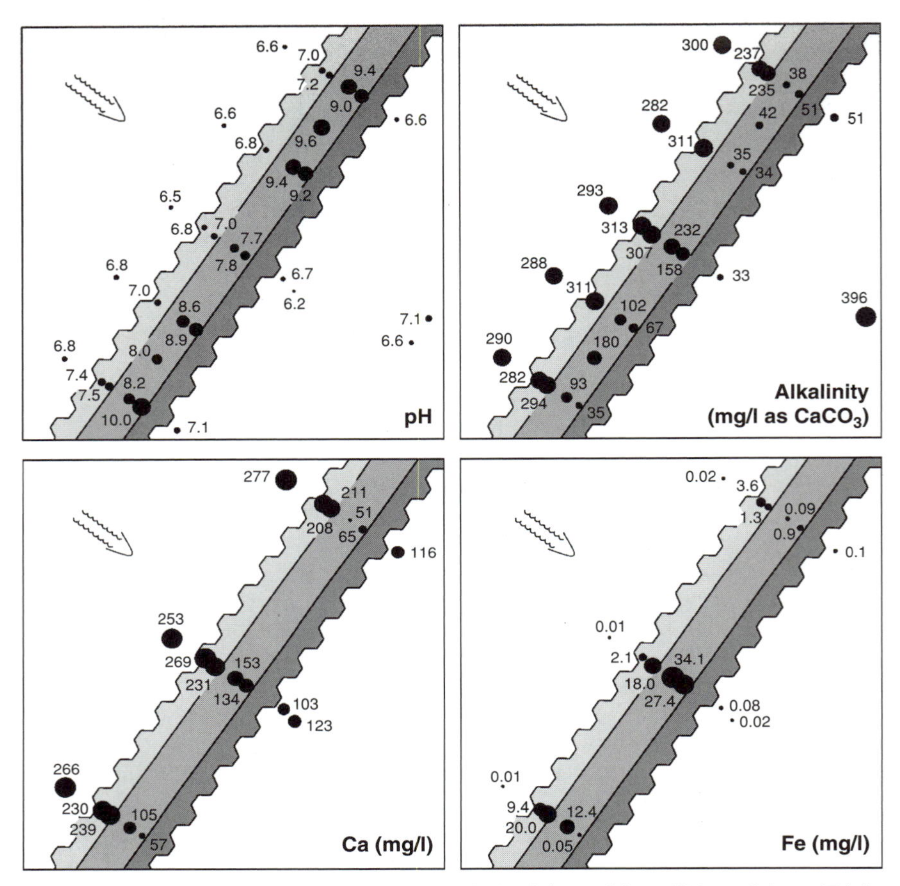

Figure 12 Values of pH and concentrations of alkalinity, calcium (Ca), and iron (Fe) in groundwater samples from monitoring wells in and near the permeable reactive barrier, near Monticello, Utah (October 2000). Dot diameter is directly proportional to concentration. Arrow indicates groundwater flow direction. See Fig. 6 for location represented in Fig. 12 and explanation of shaded PRB components.

and were less than 2 μg/liter throughout the 100% ZVI zone. Arsenic, Se, and NO_3 all had low concentrations in groundwater samples collected in the PRB (Fig. 11). The PRB was slightly less effective in decreasing Mo concentrations, as evidenced by concentrations as high as 32 μg/liter in samples from the 100% ZVI zone (Fig. 11). This pattern for Mo is consistent with the slower rate of reaction observed in column experiments (Fig. 3). Overall, the PRB met or exceeded expectations in reducing contaminant concentrations.

The pH values of upgradient groundwater samples ranged from 6.5 to 6.8 (Fig. 12). Values increased slightly to a range of 6.8 to 7.5 in groundwater samples from the gravel/ZVI zone and up to 10.0 in samples from the 100% ZVI zone. The increase in pH values is caused by the transfer of electrons from the oxidation of ZVI to aqueous hydrogen ion (H^+), reducing it to elemental hydrogen. The increase in pH values caused calcite and Fe-rich calcite to precipitate (Morrison *et al.*, 2001), which accounts for decreasing concentrations of alkalinity and Ca (Fig. 12). Dissolved Fe concentrations were lower than observed in the field column experiments. In the 100% ZVI zone, the groundwater sample with the highest pH value (10.0) had the lowest Fe concentration (0.05 mg/liter). Higher pH values probably caused precipitation of $Fe(OH)_2$, resulting in lower Fe concentrations. Shorter residence times (faster flow rates) are indicated in the middle portion of the 100% ZVI zone by higher Fe concentrations (34.1 and 27.4 mg/liter), lower pH values (7.7 and 7.8), higher Ca concentrations (153 and 134 mg/liter), and higher alkalinity (232 and 158 mg/liter as $CaCO_3$) (Fig. 12).

2. Downgradient Contamination

The U concentrations in groundwater samples from most wells 0.3 to 0.6 m downgradient of the PRB have been less than 30 μg/liter (Fig. 13). Concentrations are higher in wells downgradient of the PRB than in the PRB because U sorbed to the sediment is slowly being released. Samples from one monitoring well (open squares on Fig. 13), located 10.6 m from the south end of the PRB, have had U concentrations approaching those detected in the untreated groundwater at this site. This well has been low yielding and the water chemistry has been inconsistent with other downgradient wells. However, the U concentrations have decreased in groundwater samples from this well since installation of the PRB.

3. PRB Longevity

One of the most difficult issues related to PRBs is accurate predictions of how long they will remain effective. Three mechanisms have been identified that could cause a PRB containing ZVI to become ineffective over time: (1) dissolution of ZVI, (2) clogging by mineral precipitation, and (3) loss of ZVI reactivity (passivation). Mass balance calculations are presented in this section for a partial evaluation of longevity. While these estimates provide a basis for discussion, they should be used cautiously until the chemical mechanisms are better understood and long-term empirical data have been collected.

Longevity of the PRB based on dissolution of ZVI was calculated by assuming a dissolved concentration of 30 mg/liter based on results of column

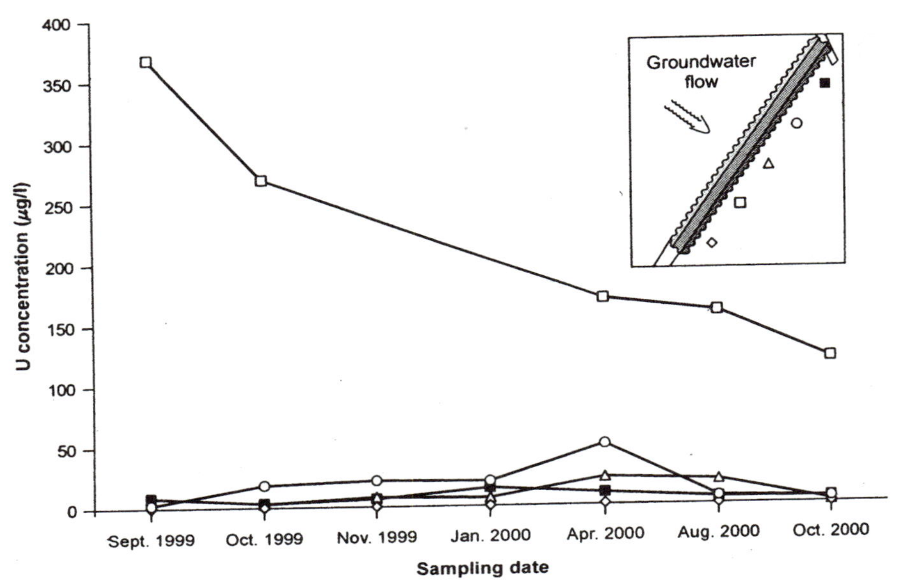

Figure 13 Temporal changes in uranium concentrations in groundwater samples from monitoring wells immediately downgradient of the permeable reactive barrier near Monticello, Utah. Sample locations on inset map correspond to data points.

tests. At this concentration, all the ZVI would dissolve in 117 years (Table II). Most of this Fe is likely redeposited in pores as Fe minerals. The change in porosity because of ZVI dissolution and secondary Fe mineral precipitation was not calculated, although porosity reduction would likely result.

Porosity reduction because of carbonate mineral precipitation was calculated by assuming that Ca lost from the groundwater as it flowed through the PRB was precipitated as $CaCO_3$. Using reasonable assumptions for groundwater flux (189 liter/min), porosity (0.5), and saturated thickness (1.2 m), the PRB could fail in as few as 10 years because of calcite deposition in pore spaces (Table II). This calculation assumes that all the porosity is available for calcite deposition. Although the calculation is sensitive to the groundwater flow rate and the average Ca loss, which are highly variable, it suggests that calcite precipitation is an important factor in the longevity of PRBs.

Passivation of ZVI could not be determined explicitly, but loss of about 35% of the reactivity efficiency was observed in the gravel/ZVI zone during the first 7 months (Fig. 14). If this loss of efficiency were to continue, the PRB could be expected to remain effective for about 36 years (Table II). The rate of

Table II

Calculations of Longevity for a Permeable Reactive Barrier Constructed near Monticello, Utah, June 1999

Mechanism affecting longevity	Rate	Longevity (years)
Dissolution of ZVI	8 kg/day	117
Precipitation of calcite	50 kg/day	10
Loss of efficiency	2.8% per year	36

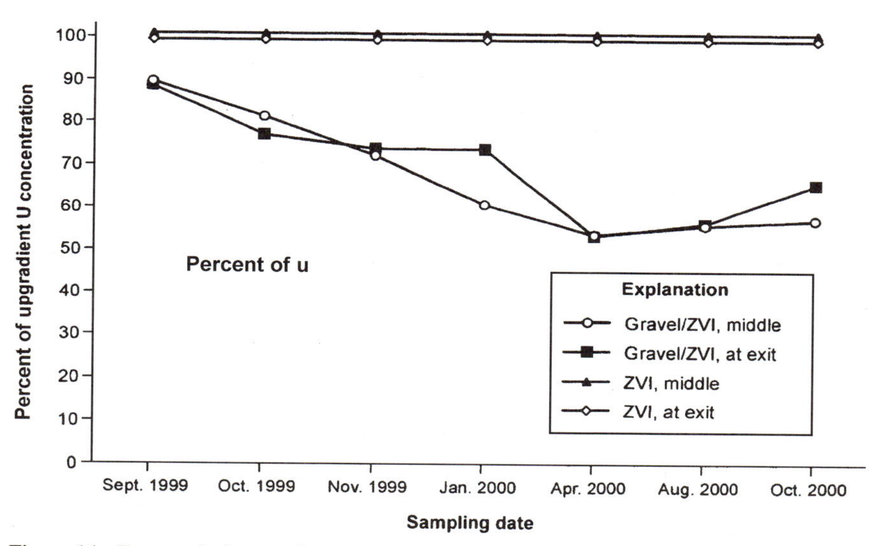

Figure 14 Temporal changes in uranium concentrations in groundwater samples from the permeable reactive barrier as a percentage of concentrations in samples from upgradient wells (untreated groundwater). Each point is the average of six wells.

reactivity loss appears to have stabilized after 7 months (Fig. 14); thus, this estimate will need to be revised after additional data are collected.

These calculations suggest that porosity reduction could significantly reduce the effectiveness of the Monticello PRB. Other metal oxides, hydroxides, and carbonates, as well as contaminant minerals, are also forming and will cause an additional reduction in porosity. A PRB installed at the Canadian Forces Base, Borden, Ontario, Canada, operated for 4 years with only trace amounts of calcite precipitation, suggesting that calcite precipitation is not an important factor in PRB longevity (O'Hannesin and Gillham, 1998). The Borden PRB had reductions in dissolved Ca concentrations similar to those of the Monticello PRB; however, it had a much lower groundwater flow rate. Understanding permeability reduction in PRBs is essential to the acceptance of this technology.

4. Chemical Mechanisms

Two different chemical mechanisms have been proposed for the uptake of U by ZVI. One proposed mechanism is the reduction of U(VI) to U(IV) and precipitation as U(IV) minerals (Cantrell *et al.*, 1995; Gu *et al.*, 1998; Morrison *et al.*, 2001). The other proposed mechanism is U(VI) adsorption to ferric oxyhydroxide that forms during the oxidation of ZVI (Fiedor *et al.*, 1998; Matheson and Goldberg, 1999). This mechanism is also discussed in Chapter 12. These two mechanisms are not mutually exclusive and both may occur in PRBs. Other redox-sensitive contaminants (As, Se, V, and Mo) present in the groundwater at Monticello are also capable of forming low-solubility reduced minerals and are known to adsorb to ferric oxyhydroxide; thus, uptake of these contaminants could result from either process. Nitrate can be reduced abiotically to ammonium by ZVI (Huang *et al.*, 1998).

pH values did not increase by more than a value of about 0.3 by reaction with ZVI in the field column experiments, but increased significantly in the PRB. In addition, Fe concentrations were high (about 27 mg/liter) in samples from the column experiments but were lower (about 0.2 mg/liter) in many samples of groundwater exiting the PRB. As noted previously, higher Fe concentrations correlate to lower pH values. A reaction progress model explained these apparent contrasts in chemistry between the column experiments and the PRB as a result of different residence times (Morrison *et al.*, 2001). The longer interaction between groundwater and ZVI in the PRB caused more reduction of aqueous H^+, which resulted in lower dissolved Fe concentrations because of precipitation of $Fe(OH)_2$. The variation in Ca concentrations and alkalinity because of the precipitation of carbonate minerals was also explained by this model.

VII. IMPLICATIONS AND CONCLUSIONS

Overall, the PRB appears to be performing well. After more than 1 year of operation, groundwater exiting the PRB installed at Monticello, Utah, met contaminant-concentration goals based on the monitoring results. The gravel/ZVI zone lost some reactivity during the first year.

About 0.75 m of groundwater mounding occurs immediately upgradient of the PRB, with a sharp gradient between groundwater in the alluvium and the PRB. Part of the mounding is due to a reduction in the cross-sectional area caused by the presence of the slurry walls and part may be due to a low-permeability zone caused by sheet pile installation. Some contaminated groundwater flowed around the slurry walls; the volume of groundwater that bypassed the wall is believed to be relatively small but data are insui;cient to quantify the amount.

A substantial effort was made to determine the flux of groundwater through the PRB using chemical concentrations in groundwater samples from more than 70 monitoring wells, tracer experiments, and flow direction measurements made with a downhole tool. Despite these efforts, some details of the flow system remain unclear. The uncertainty is attributable in part to the heavy construction that altered groundwater flow hydraulically upgradient of the PRB and in part because of the difficulty in making reliable interpretations from subsurface flow measurements.

Instrumentation was installed for the continuous monitoring of water levels over the period while the aquifer was reestablished. Groundwater bypassing both slurry walls will continue to be monitored. Mass balance calculations suggest that calcite precipitation within the PRB could cause significant groundwater diversion in as few as 10 years after installation. Additional measurements should be made for better determination of the groundwater flux through the PRB and to estimate longevity.

The PRB will not function indefinitely, and at some time it may release some of the contamination back to the groundwater. The performance will continue to be monitored and predictions of future performance improved as the chemical and physical processes become better understood. Should the PRB start to release contamination, two options exist: (1) removal of the PRB and disposal of the reactive material at an approved facility or (2) injection of a permanent sealant (such as grout or cement) into the pore spaces. The second option would require breeching the system to permit groundwater flow.

ACKNOWLEDGMENTS

This project was funded by the Accelerated Site Technology Deployment Program sponsored by the DOE Office of Science and Technology. Many people from several organizations were important to the successful completion of this project. A tri-party agreement and strong support from personnel in the EPA region 8 office (Paul Mushovic), the state of Utah (David Bird), and the DOE Grand Junction (Colorado) Office were essential to the successful completion of the PRB. Vern Cromwell was the DOE Grand Junction Office Project Manager in the early stages of the project. Kristen McClellen (MACTEC Environmental Restoration Services) was instrumental in integrating the PRB project with Operable Unit III of the Monticello Mill Tailings site. Personnel from the DOE Oak Ridge National Laboratory (Nic Korte and Gerry Moline) designed and directed the tracer study; Korte also provided colloidal borescope data. David Blowes (University of Waterloo, Canada), Brian Dwyer (Sandia National Laboratories/New Mexico), Will Goldberg (MSE Technologies, Inc.), and Dawn Kaback (Concurrent Technologies, Inc.) provided valuable insight during the design phase. This project would not have been possible without the cooperation of the landowner, Kedric Somerville.

REFERENCES

ASTM (2001). Standard test methods for measurement of hydraulic conductivity of saturated porous materials using a flexible well permeater, Test Method D5084–00. Am. Soc. Testing and Materials.

Cantrell, K. J., Kaplan, D. I., and Wietsma, T. W. (1995). Zero-valent iron for the in situ remediation of selected metals in groundwater. *J. Hazard. Mater.*, **42**, 201–212.

Fiedor, J. N., Bostick, W. D., Jarabek, R. J., and Farrell, J. (1998). Understanding the mechanism of uranium removal from groundwater by zero-valent iron using x-ray photoelectron spectroscopy. *Environ. Sci. Technol.* **32**, 1466–1473.

Gu, B., Liang, L., Dickey, M. J., Yin, X., and Dai, S. (1998). Reductive precipitation of uranium (VI) by zero-valent iron. *Environ. Sci. Technol.* **32**, 3366–3373.

Huang, C., Wang, H., and Chiu, P. (1998). Nitrate reduction by metallic iron. *Water. Res.* **32**, 2257–2264.

Kearl, P. M. (1997). Observations of particle movement in a monitoring well using the colloidal borescope. *J. Hydrol.* **200**, 323–344.

Matheson, L. J., and Goldberg, W. C. (1999). Spectroscopic studies to determine uranium speciation in ZVI permeable reactive barrier materials from the Oak Ridge Reservation, Y-12 plant site and Durango, CO, PeRT wall C. *In* "Supplement to EOS, Transactions 1999 Fall Meeting," p. F366. American Geophysical Union, Washington, DC.

McDonald, M. G., and Harbaugh, A. W. (1988). A modular three-dimensional finite-difference ground-water flow model. U.S. Geological Survey Techniques of Water-Resources Investigations Book Chapter 01–A1.

Morrison, S. J., Metzler, D. R., and Carpenter, C. E. (2001). Uranium precipitation in a permeable reactive barrier by progressive irreversible dissolution of zero-valent iron. *Environ. Sci. Technol.* **35**, 385–390.

National Academy of Sciences (1994). "Alternatives for ground water cleanup," Report of the National Academy of Sciences Committee on Ground Water Cleanup Alternatives, National Academy Press, Washington, DC.

O'Hannesin, S. F., and Gillham, R. W. (1998). Long-term performance of an in situ "iron wall" for remediation of VOCs. *Ground Water* **36**, 164–170.

Pollock, D. W. (1989). Documentation of computer programs to complete and display pathlines using results from the U.S. Geological Survey modular three-dimensional finite-difference ground-water model. U.S. Geological Survey, Open-File Report 89–381.

Scherer, M. M., Richter, S., Valentine, R. L., and Alvarez, P. J. (2000). Chemistry and microbiology of permeable reactive barriers for *in-situ* groundwater cleanup. *Crit. Rev. Environ. Sci. Technol.* **30**, 363–411.

Chapter 14

Field Demonstration of Three Permeable Reactive Barriers to Control Uranium Contamination in Groundwater, Fry Canyon, Utah

David L. Naftz,* Christopher C. Fuller,† James A. Davis,† Stan J. Morrison,‡ Edward M. Feltcorn,§ Geoff W. Freethey,* Ryan C. Rowland,* Christopher Wilkowske,* and Michael Piana†

*U.S. Geological Survey, Salt Lake City, Utah 84119
†U.S. Geological Survey, Menlo Park, California 94025
‡Environmental Sciences Laboratory,[1] Grand Junction, Colorado 81503
§U.S. Environmental Protection Agency, Radiation and Indoor Air, Washington, DC 20460

Historical uranium exploration, mining, and milling operations have resulted in elevated uranium (U) concentrations in many aquifers throughout the western United States. Permeable reactive barriers (PRBs) may represent a viable technology for the passive remediation of U-contaminated groundwater. In 1997, a funnel-and-gate design was used to construct three PRBs consisting of (1) bone char pellets (PO4), (2) zero valent iron (ZVI) pellets, and (3) amorphous ferric oxyhydroxide (AFO) to demonstrate the field-scale removal of dissolved U with PRBs. The demonstration site was located near

[1]Operated by MACTEC Environmental Restoration Services for the U.S. Department of Energy Grand Junction Office.

Fry Canyon, an abandoned U-upgrading site in southeastern Utah. Groundwater from the shallow, colluvial aquifer at this site contains U concentrations that exceed 20,000 μg/liter. Results from the first 3 years of PRB operation at Fry Canyon can be used in the design, operation, and monitoring of future, full-scale PRB installations in similar hydrologic and geochemical settings. Important hydrologic information gained during the demonstration includes (1) groundwater velocities measured with a heat pulse flow meter in the colluvial aquifer and ZVI PRB compare favorably with ionic tracer results; (2) small increases in water level elevations in PRBs result in increased groundwater velocities that could result in significantly different residence times; and (3) PRB thickness should be designed to account for the large seasonal and transient variations in groundwater velocities that were observed. Important geochemical information gained during the demonstration includes (1) chemical data in water from monitoring wells in the upgradient parts of PRBs can be used as "early warning systems" to detect future chemical breakthrough; (2) PRB design should account for seasonal and transient changes in water levels that may cause significant variations in influent U concentrations; (3) contaminant removal may be less efficient directly above the PRB/confining-unit interface, therefore, monitoring points in these areas are needed to document contaminant removal efficiencies; (4) during the 1130-day demonstration period, the ZVI PRB demonstrated the highest U removal efficiencies (mean $C/C_o = 0.007$) relative to the AFO (mean $C/C_o = 0.271$) and PO4 (mean $C/C_o = 0.412$) PRBs and treated about 1650 pore volumes of contaminated groundwater; (5) carbonate and noncarbonate mineral precipitation resulting from iron (Fe) corrosion is occurring in the ZVI PRB and could decrease future U removal efficiencies; and (6) degradation of groundwater quality exiting the PRBs should be expected immediately after installation; however, most constituents will return to baseline values after the first 1 to 1.5 years of PRB operation.

I. INTRODUCTION

Permeable reactive barriers (PRBs) represent a viable tool for *in situ* remediation of metal-contaminated groundwater associated with abandoned mine lands (AML). There are more than 25,000 inactive mine sites in the western United States and numerous exploration prospects, many of which are on federal lands (U.S. Geological Survey Abandoned Mine Lands Science Team, 1999). Because many AML sites are in remote areas, low maintenance remediation technologies, such as PRBs, are especially useful. A PRB can be installed in shallow aquifers (<5 m below land surface) with commonly available construction equipment that is transported easily to remote sites.

Because PRBs have no energy or short-term maintenance requirements, the costs typically associated with installing power lines and maintaining equipment for above ground treatment technologies are not incurred. PRBs treat contaminated groundwater below the land surface, which eliminates visual impacts from the use of equipment on the land surface. Damage from vandalism, livestock, and wildlife to above ground remediation infrastructures does not occur with PRBs.

As an example of mining-related waste, historical uranium (U) exploration, mining, and milling operations have resulted in elevated U concentrations in many aquifers (Landa and Gray, 1995). The Fry Canyon site (Fig. 1) was selected in 1996 as a field-scale demonstration site to assess the feasibility of treating U-contaminated groundwater with PRBs (Naftz *et al.*, 1999, 2000). This chapter presents PRB performance results during the first 3 years of operation, including (1) groundwater flow velocities; (2) U distributions and removal efficiencies; (3) hydrologic impacts on U influent and removal in PRBs; (4) total mass of U removed during PRB operation; and (5) suggestions for future PRB installations. The U.S. Environmental Protection Agency (USEPA)/Superfund and Office of Air and Radiation provided funding for this ongoing project.

II. SITE DESCRIPTION

A. Former Uranium Upgrading Site

The Fry Canyon site was constructed and operated by COG Minerals Corporation, a subsidiary of Colorado Oil and Gas Corporation. The purpose of this facility was to upgrade U from ore obtained primarily from three U mines in the White Canyon Mining District of southeastern Utah. The upgraded material was then transported about 115 km to the Texas-Zinc Minerals Corporation mill at Mexican Hat, Utah (Fig. 1).

The upgrader operated during 1957–1960 and processed about 45,400 metric tons of ore containing between 0.10 and 0.15% U_3O_8 (Utah Department of Health, 1987). Water for the operation was supplied from a large-diameter well completed in the colluvial material within and adjacent to Fry Creek (Fig. 1). About 36,000 metric tons of sand tailings, containing about 0.02% U_3O_8, were impounded on the site by the time the upgrader was closed (Utah Department of Health, 1987).

In 1962, the Fry Canyon site and associated water rights were acquired by the Basin Company for a copper (Cu) heap leaching operation. This operation used sulfuric acid to leach Cu into solution. The Cu was subsequently

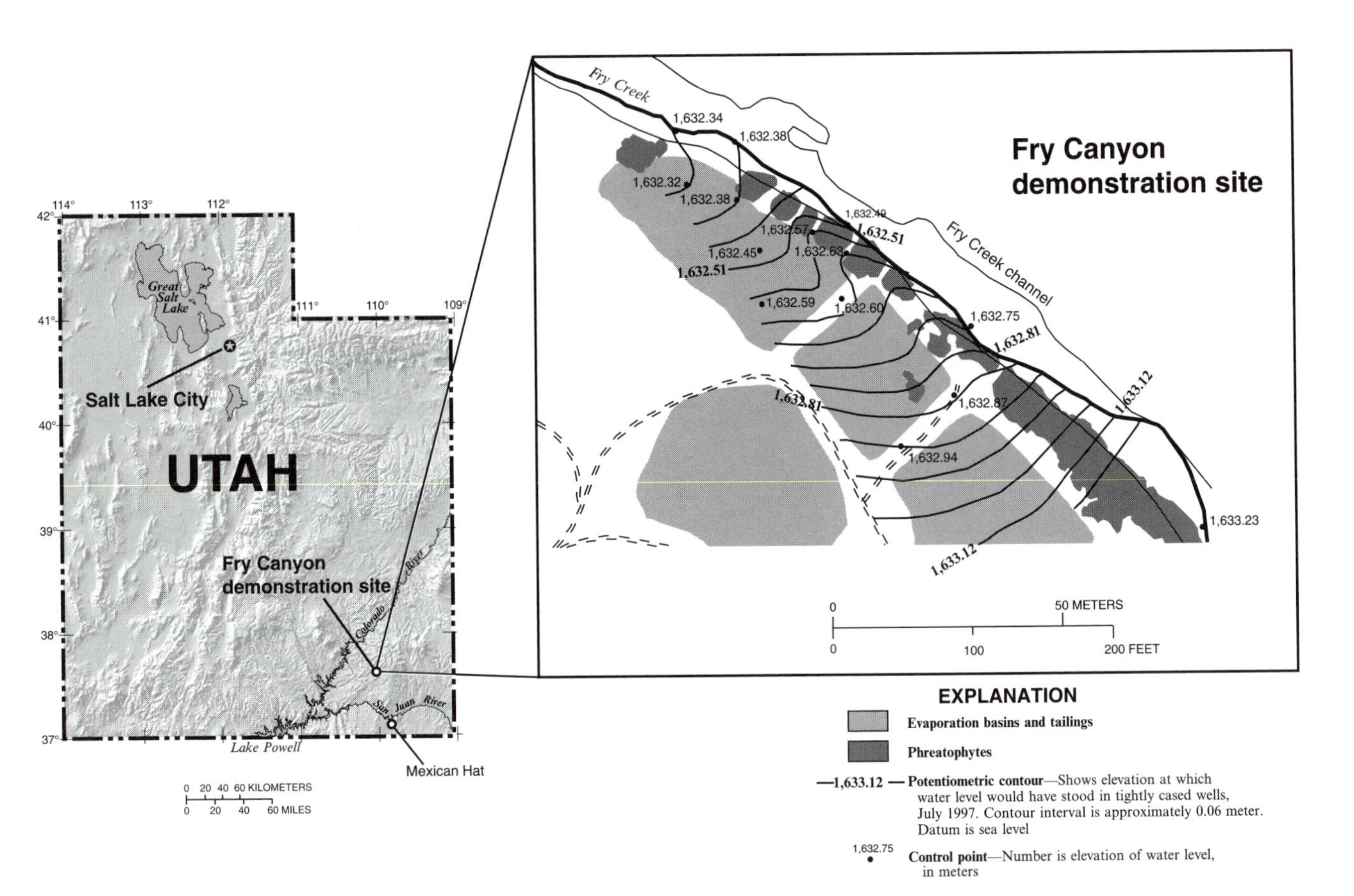

Figure 1 Location of the Fry Canyon demonstration site and elevation of the potentiometric surface in the colluvial aquifer before installation of permeable reactive barriers.

precipitated with hydrogen sulfide and collected on pieces of scrap iron (Fe), some of which is still at the site. Cu extraction operations ceased in 1968. In 1990 the site was given a no further remedial action planned (NFRAP) rating by the USEPA. The site remained inactive until 1996, when the site was considered for the field demonstration of PRBs for removal of U from contaminated groundwater (Naftz *et al.*, 1999, 2000; see Chapter 10).

B. Geohydrology and Uranium Concentration

The Fry Canyon demonstration site is located adjacent to Fry Creek (Fig. 1). Fry Creek is subject to large flash floods resulting from localized thunderstorms that generally occur during July through September. There is a perennial flow in Fry Creek that is fed by a series of springs about 250 m upstream from the ore upgrading facility. Measured discharge ranged from 0 to 0.0016 m^3/s; however, stream discharge can exceed 8.5 m^3/s during flood events.

Silt to gravel-size particles derived from nearby sandstone and shale formations make up the shallow, unconfined colluvial aquifer at the site. Maximum thickness of the colluvial deposits is 5.5 m, and saturated thickness of the aquifer ranges from about 0.6 to 1.2 m. Hydraulic conductivity (K) values range from 17 to 26 m/day, and transmissivity values range from 10 to 31.2 m^2/day. The *in situ* effective porosity for the aquifer is estimated to be 20 to 25% (Freethey *et al.*, 1994). The average linear velocity of water in the colluvial aquifer was estimated to be approximately 0.23 m/day (Naftz *et al.*, 2000). Underlying the colluvial aquifer is the Permian-age Cedar Mesa Sandstone, which is virtually impermeable compared to the colluvial aquifer.

The potentiometric contour map shows that the colluvial aquifer is recharged by subsurface inflow from Fry Creek upstream of the site, by precipitation directly on the site, and by runoff from the sandstone upslope from the site (Fig. 1). Groundwater discharges from the aquifer by seepage back into Fry Creek, by evaporation where the saturated sediments are near land surface, by riparian vegetation (primarily *Tamarisk*) transpiration, and by possibly limited downward leakage into the sandstone. Discharge measurements in Fry Creek in November 1997 indicate that about 12,335 to 18,500 m^3/year of groundwater seep into the stream along the 90-m reach adjacent to the demonstration site.

Concentrations of U in the contaminated part of the colluvial aquifer have exceeded 20,000 μg/liter. The high alkalinity (mean = 400 mg/liter as $CaCO_3$) coupled with measurable amounts of dissolved oxygen in the contaminated groundwater indicate that U is likely to be in the +6 oxidation state as a carbonate species. These concentrations were substantially higher than background U concentrations, which range from 60 to 80 μg/liter. Desorption experiments indicate that sediments from the contaminated part of the colluvial aquifer at the Fry Canyon site contain a large amount of U that can be desorbed readily (Naftz *et al.*, 2000). Additional descriptions of preinstallation site hydrology and geochemistry can be found in Naftz *et al.* (2000). Adsorption and formation of U–apatite surface complexes dominated U removal by apatite at dissolved U concentrations present at Fry Canyon (see Chapter 9).

III. METHODOLOGY

A. Laboratory Testing of Reactive Materials

Three groups of reactive materials were selected for U removal at Fry Canyon: (1) hydroxyapatite (bone meal, bone charcoal, and phosphate rock) (PO4), (2) zero valent iron (ZVI), and (3) amorphous ferric oxyhydroxide (AFO). These materials were selected because U removal was believed to occur by the following mechanisms: (1) PO4—precipitation of an insoluble uranyl phosphate phase or U-phosphate surface complexation, (2) ZVI—reduction of U(VI) to U(IV) and subsequent precipitation (Gu *et al.*, 1999), and (3) AFO—by adsorption to the ferric oxyhydroxide surface. Detailed results of these laboratory tests can be found in Naftz *et al.* (2000) and Chapter 9.

One reactive material was selected from each of the three material groups for field demonstration. Selection factors considered included (1) availability; (2) cost; (3) hydraulic conductivity; (4) structural strength; (5) extent, rate, and duration of U removal; (6) off-site migration potential of reactive material; (7) potential for rerelease of uranium; and (8) possible detrimental effects on groundwater quality. The material selected from the phosphate group was bonechar phosphate pellets manufactured by Cercona of America (Dayton, OH). The bone char pellets used in the PRB had a high K value (475 m/day) and treated 250 pore volumes (PVs) of water containing 2400 μg/liter dissolved U before 50% breakthrough (see Chapter 9).

Material selected from the ZVI group was Cercona iron foam pellets. The foam pellets had high K values (122 to 250 m/day) and high removal efficiencies from groundwater containing 2100 μg/liter dissolved U (Naftz *et al.*, 2000). The material selected for the AFO group was pea gravel (0.95 cm) mixed with AFO in the form of a slurry containing 13% AFO [as $Fe(OH)_3$]. The gravel and AFO slurry were mixed in the field at a ratio of 2 parts gravel to 1 part slurry, by volume. Column tests using gravel coated with 0.2% Fe indicated low U removal capacities relative to ZVI material (Naftz *et al.*, 2000); however, increasing the Fe concentration to 2% resulted in higher U removal capacities and excellent *K* values, which exceeded 915 m/day (Naftz *et al.*, 2000).

B. Design and Installation of PRBs

In order to accommodate field testing of all three reactive materials, three PRBs were designed for installation and concurrent "side-by-side" operation at the site. A funnel-and-gate design was chosen, consisting of three "permeable windows" where each of the PRBs would be placed, separated by "no-flow walls" and "no-flow wing walls" on each end to channel the groundwater flow into the PRBs (Fig. 2). Each PRB and no-flow boundary was placed onto the bedrock underlying the colluvial aquifer. The irregular surface of the bedrock did not present a consistent boundary between the reactive material and bedrock surface, and in places, as much as 5 cm above the bedrock may be devoid of reactive material. Construction equipment consisting of a trac-mounted backhoe and a bulldozer was chosen to install the PRBs. This design-and-installation technique was chosen for the following reasons: (1) amenable for multiple PRBs placed side by side; (2) low construction cost; (3) appropriate for shallow groundwater systems; and (4) transferability to other remote, abandoned mine sites with contaminated groundwater. Additional details of PRB construction at Fry Canyon can be found in Naftz *et al.* (2000).

Dimensions of each PRB were 2.1 m long by 0.9 m wide. The depth of each PRB was variable (PO4 and AFO = 0.97 m; ZVI = 1.1 m). A 0.46-m-wide layer of pea gravel was placed on the upgradient side of the PRBs to facilitate uniform flow of contaminated groundwater into each PRB. The 0.9-m thickness of each PRB at the Fry Canyon site was based on the following criteria: (1) large enough thickness to sample and map gradients in dissolved solutes across the PRB; (2) sufficient residence time for groundwater to interact chemically with the PRB material during changing hydrologic conditions; and (3) physical limitations of the trac hoe and trench box used during PRB construction.

Additional factors to consider when designing the thickness of a PRB include groundwater velocity, rate of contaminant removal by the reactive material, contaminant removal capacity of the barrier material, estimated mass of the contaminant in the groundwater plume, decreasing residence time after aging of the reactive material, and physical constraints of the trenching equipment (Gavaskar *et al.*, 1998; U.S. Environmental Protection Agency,

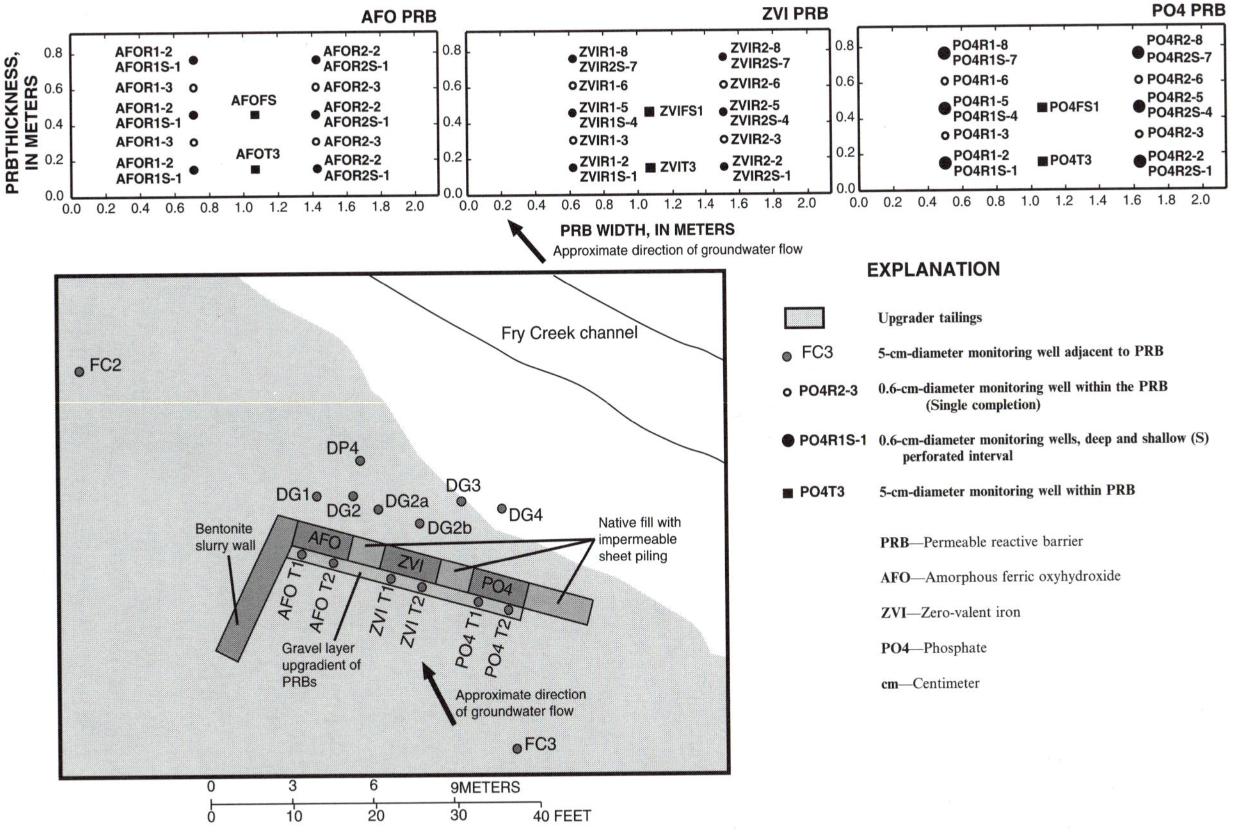

Figure 2 Location of the three permeable reactive barriers and monitoring wells, Fry Canyon, Utah.

1999; Naftz *et al.*, 2000). In an actual remediation project, these factors, and possibly others, would be used to determine the thickness of the PRB. Addressing all of these design criteria was not practical for the field demonstration of PRBs conducted during this study.

During trench excavation, a large bedrock nose that could not be excavated was encountered. This caused the trench orientation to be rotated, resulting in the construction of gate structures that intercepted groundwater flow at about 35° from perpendicular (Fig. 2). The nonperpendicular trench orientation caused noticeable differences in the direction and velocity of groundwater flow into the gate structures of each PRB and is discussed in detail later.

C. PRB Monitoring Network and Chemical Analysis

An extensive monitoring network was installed in each PRB consisting of 16, 0.6-cm-diameter polyvinyl chloride (PVC) wells located along two parallel flow paths and 4, 5-cm-diameter PVC wells (Fig. 2) for sample collection and monitoring of water levels and selected water quality parameters. Three sites along each flow path contained a shallow and deep 0.6-cm-diameter monitoring well (Fig. 2). Because of the proximity of wells to one another, limited purge volumes were extracted prior to sample collection. A 3.8-liter purge volume was removed from the 5-cm-diameter monitoring wells and 1 liter of water was removed from the 0.6-cm-diameter monitoring wells.

After purging, water samples were filtered on site with a 0.45-μm capsule filter and collected in acid- and field-rinsed polyethylene bottles. Samples for U analysis were acidified on site with ultrapure concentrated nitric acid. Each monitoring point contained a dedicated sampling tube to minimize cross contamination during sampling.

Dissolved U concentrations were measured by kinetic phosphorescence analysis (KPA) at the USGS Research Laboratories in Menlo Park, California (method detection limit 0.06 μg/liter; precision ±2.5%). Twelve pressure transducers were used to provide hourly measurements of water level elevations in selected wells within and adjacent to each PRB. Water level elevations in all 5-cm-diameter wells were monitored periodically using an electronic tape.

IV. PERFORMANCE

A. Hydraulics

1. Potentiometric Surface

The flow system was altered because of PRB construction. Hydraulic gradients were increased because the wing walls (Fig. 2) concentrate more groundwater from a wider area of the aquifer than was common in the natural state. As groundwater moves near the PRB, the hydraulic gradient becomes nearly flat because the hydraulic conductivity of the pea-gravel buffer zone and the wall materials is nearly 10 times larger than that of the aquifer.

The estimated capture zone during December 1998 for the installed PRBs is shown in Fig. 3. Even though this zone includes only about 1350 m^2, it represents the terminus of a drainage area nearly five times that size. The amount of water that enters and exits this capture zone was estimated to be no greater than 2.6 m^3/day (about 10% of the precipitation falling on the drainage area annually) or about 950 m^3 on an annual basis.

2. Groundwater Flow Velocities

Conservative ionic tracers, including bromide (Br) and lithium (Li), were used to measure groundwater velocities in the colluvial aquifer and within each of the PRBs during April through June 1999. A 340-liter slug of native groundwater containing a Br concentration of 1500 mg/liter was injected into well FC 3 on May 19, 1999. Autosamplers were used to collect samples from downgradient wells (PO4T2, PO4T1, ZVIT2, ZVIT1, and AFOT2). On the basis of Br peak arrival times in the downgradient wells, groundwater velocities in the colluvial aquifer ranged from 0.26 to 0.43 m/day (Fig. 4). This is within the range of groundwater velocities calculated from measured gradients in the colluvial aquifer (0.06 to 0.76 m/day) at Fry Canyon (Naftz *et al.*, 2000). Injected Br was not detected in water from wells PO4T2 and PO4T1, indicating a flow vector from well FC3 toward the ZVI PRB, which also agrees with the direction of groundwater flow indicated by the potentiometric surface in the colluvial aquifer (Fig. 3).

Four separate injections of conservative ionic tracers were conducted within the PRBs during April 1999 (three Br and one Li tracer tests). Because of the smaller travel distances, less tracer mass was injected. Br tracer volumes ranged from 26.5 to 77.6 liters, with Br concentrations between 232 and 300 mg/liter. Injections were separated by about 2 days to avoid interference of tracers between each PRB. Based on peak arrival times in downgradient

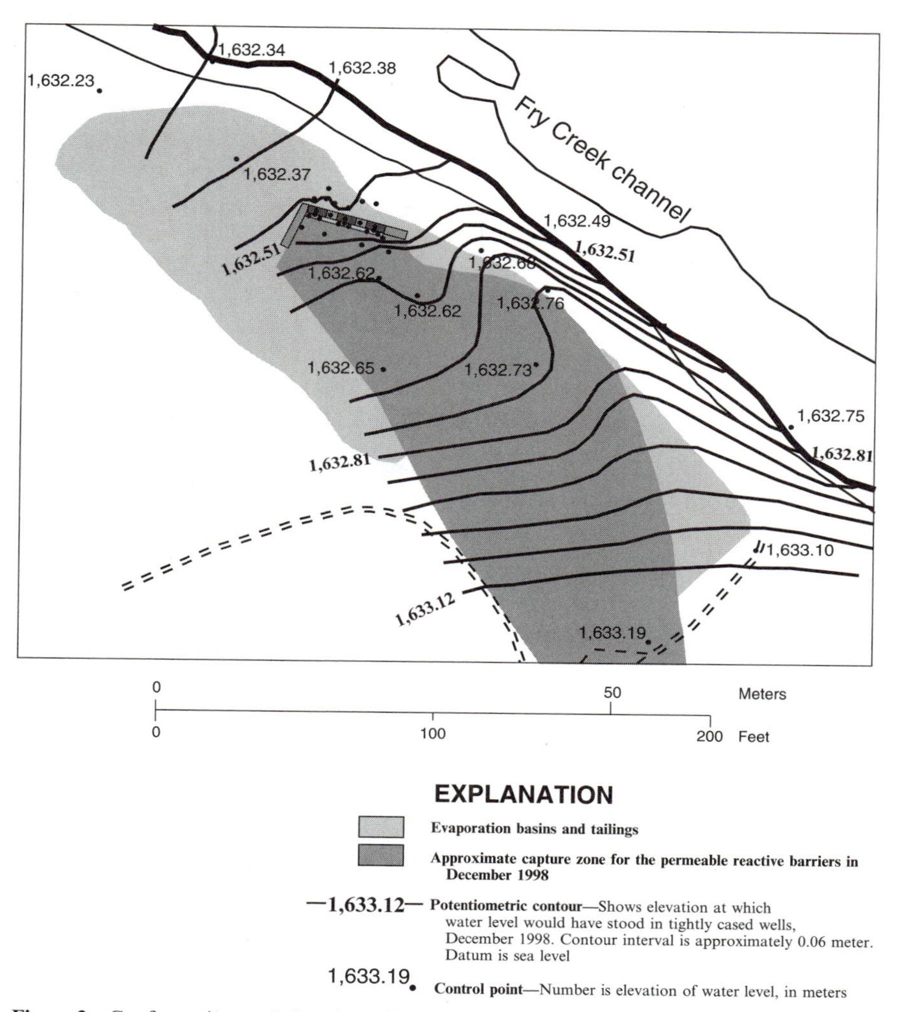

Figure 3 Configuration and elevation of potentiometric surface in the colluvial aquifer after installation of permeable reactive barriers at Fry Canyon, Utah, December 1998.

wells, groundwater velocities calculated from tracer data in PRBs ranged from 0.14 m/day in the PO4 PRB to 2.7 m/day in the AFO PRB (Fig. 4). The low groundwater velocity in the PO4 PRB is the result of water intersecting the PRB at less than a 90° angle. On the basis of tracer test results, water entering the left half of the PO4 PRB is flowing back into the gravel pack and

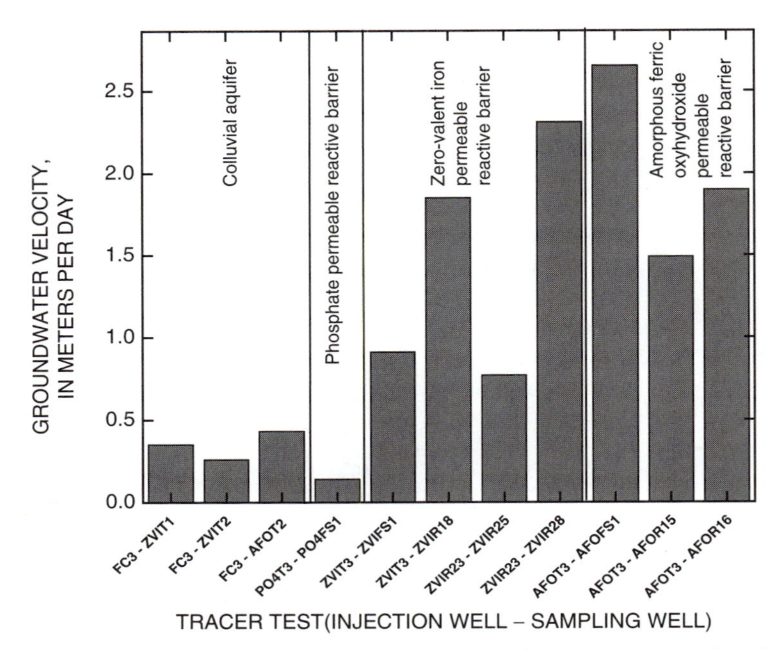

Figure 4 Summary of groundwater flow velocities calculated from tracer tests conducted at the Fry Canyon field demonstration site from April through June 1999.

entering the right side of the ZVI PRB. Water entering the ZVI and AFO PRBs does not migrate back into the gravel pack.

A K-V Associates Model 40 GeoFlo heat-pulse groundwater flowmeter (K-V Associates, Inc., 2001) was used to periodically monitor groundwater velocity and direction through each of the PRBs and the colluvial aquifer. The flowmeter measures the velocity and direction of water movement by emitting heat pulses and measuring subsequent temperature increases in the groundwater using a series of thermistors (Guthrie, 1986). Monitoring wells with at least a 5-cm inside diameter are needed for flowmeter measurements. Comparison of groundwater velocities measured with the flowmeter and calculated from ionic tracer tests were comparable for measurements conducted in the colluvial aquifer and the ZVI PRB (Fig. 5); however, flowmeter measurements in the PO4 and AFO PRBs were problematic and not presented.

Velocity trends measured by the flowmeter in the ZVI PRB indicate a significant variability ranging from less than 1 to more than 5 m/day (Fig. 6). Comparison of groundwater velocities to mean daily water level elevations within the ZVI PRB (well ZVIT3) indicates that increased velocity is closely

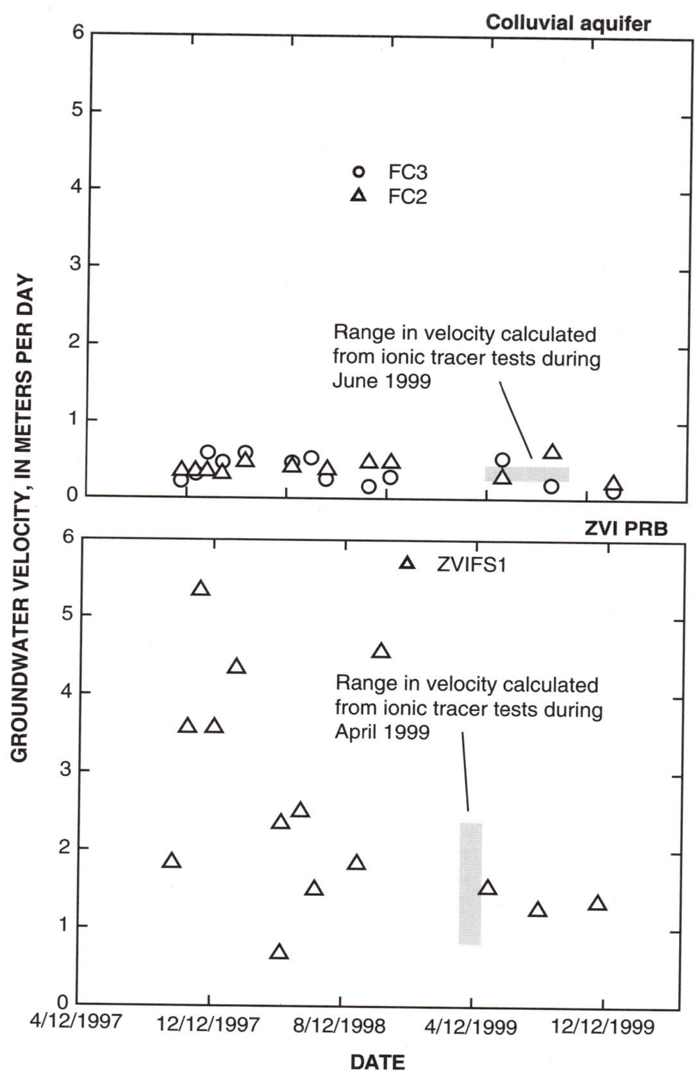

Figure 5 Groundwater velocities measured with a heat-pulse flowmeter and velocities calculated from ionic tracer tests, Fry Canyon, Utah.

correlated to increased water level elevations within the PRB (Fig. 6). For example, during early October 1997, a relatively low groundwater velocity (1.8 m/day) was measured in the ZVI PRB, corresponding to a significant

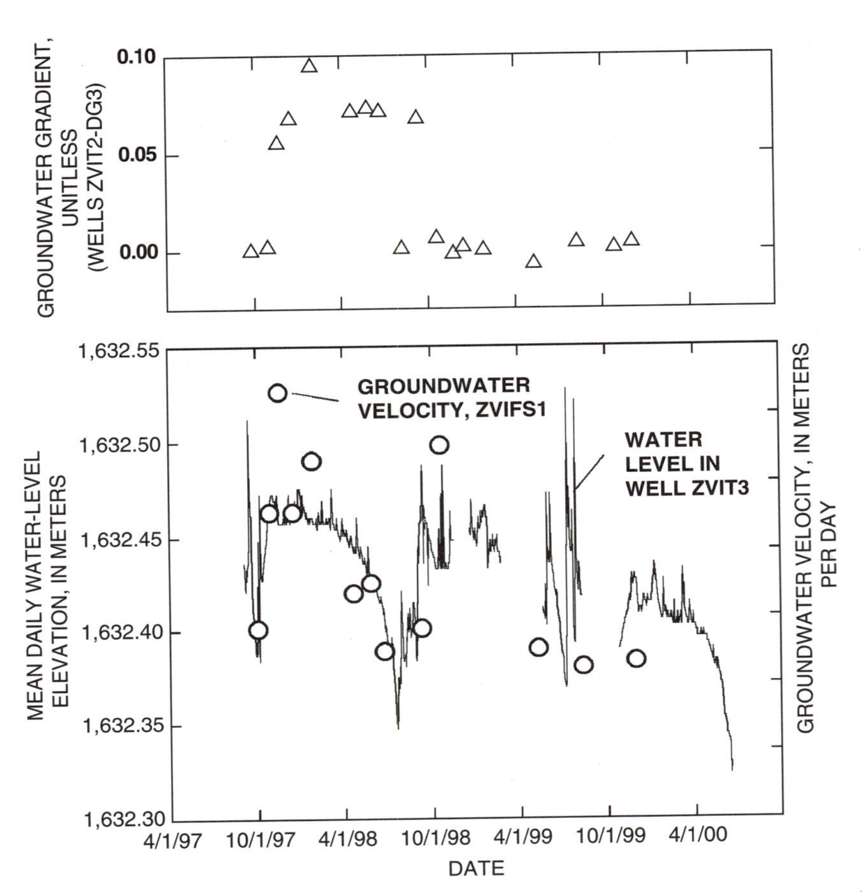

Figure 6 Continuous water level elevations in well ZVIT3, groundwater velocities measured with a heat-pulse flowmeter, and groundwater gradients, Fry Canyon, Utah.

decrease in water level elevation in the PRB. In late October 1997, an increased groundwater velocity (3.8 m/day) was measured and corresponded to an increase in the water level elevation of 5.79 cm.

Based on Darcy's law, groundwater velocity is directly proportional to the gradient. General trends in the groundwater gradient between wells ZVIT2 (upgradient of PRB) and DG3 (downgradient of PRB) agree with the measured changes in groundwater velocities (Fig. 6). For example, near-zero gradients are observed in early October 1997, corresponding to the low groundwater velocity. Larger groundwater velocities measured from late October 1997 through late January 1998 correspond to larger groundwater gradients. The large variations in groundwater velocities measured in the ZVI PRB indicate that the residence time of contaminated water in the PRBs could decrease significantly during storms and seasonal increases in water level elevations that may increase the gradients across the PRBs.

B. Chemistry

1. Contaminant Removal Efficiencies

Selected monitoring wells in each PRB were sampled 13 times from September 1997 through November 2000 to determine spatial distributions and removal efficiencies of U by each PRB. Contour maps of U concentration in each PRB during three selected monitoring periods (November 1997, May 1999, and December 1999) were used to evaluate the spatial differences of contaminant removal. The influent U concentration to the PO4 PRB ranged from 1400 to 7050 μg/liter (Fig. 7) and decreased by more than 4030 μg/liter after traveling through the first 0.15 m of reactive material. The right side of the PO4 PRB appears to provide more efficient U removal during May 1999 and December 1999 relative to the center and left parts of the PRB (Fig. 7). This is probably the result of variable groundwater velocities and different flow paths in the right and left sides of the PO4 PRB.

The influent U concentration to the ZVI PRB was larger than the PO4 PRB and ranged from 2650 to 12860 μg/liter (Fig. 8) and decreased significantly after traveling through the first 0.15 m of reactive material. For example, during the May 1999 monitoring period, the influent U concentration of more than 10000 μg/liter was reduced to < 0.06 μg/liter in a penetration distance of 0.15 m on the left side of the PRB. The December 1999 U distribution map indicates an increased path length (0.46 m) before complete U removal on the left side of the PRB (Fig. 8). This increased treatment length is probably the result of increasing U influent and the effects of aging on the ZVI reactive material that could include passivation of the reactive surface by mineral precipitation.

The influent U concentration to the AFO PRB was the largest of all three PRBs, ranging from 12890 to 19850 μg/liter during the three monitoring periods shown in Fig. 9. The performance of AFO PRB decreased significantly during the field demonstration. During the initial part of field demonstration (November 1997), very high U concentrations ($>$ 19000 μg/liter) were reduced to less than 1000 μg/liter over a distance of about 0.60 m into the PRB (Fig. 9). About 2 years later, similar U influent concentrations ($>$17000 μg/liter) were exiting the AFO PRB with U concentrations exceeding 5000 μg/liter.

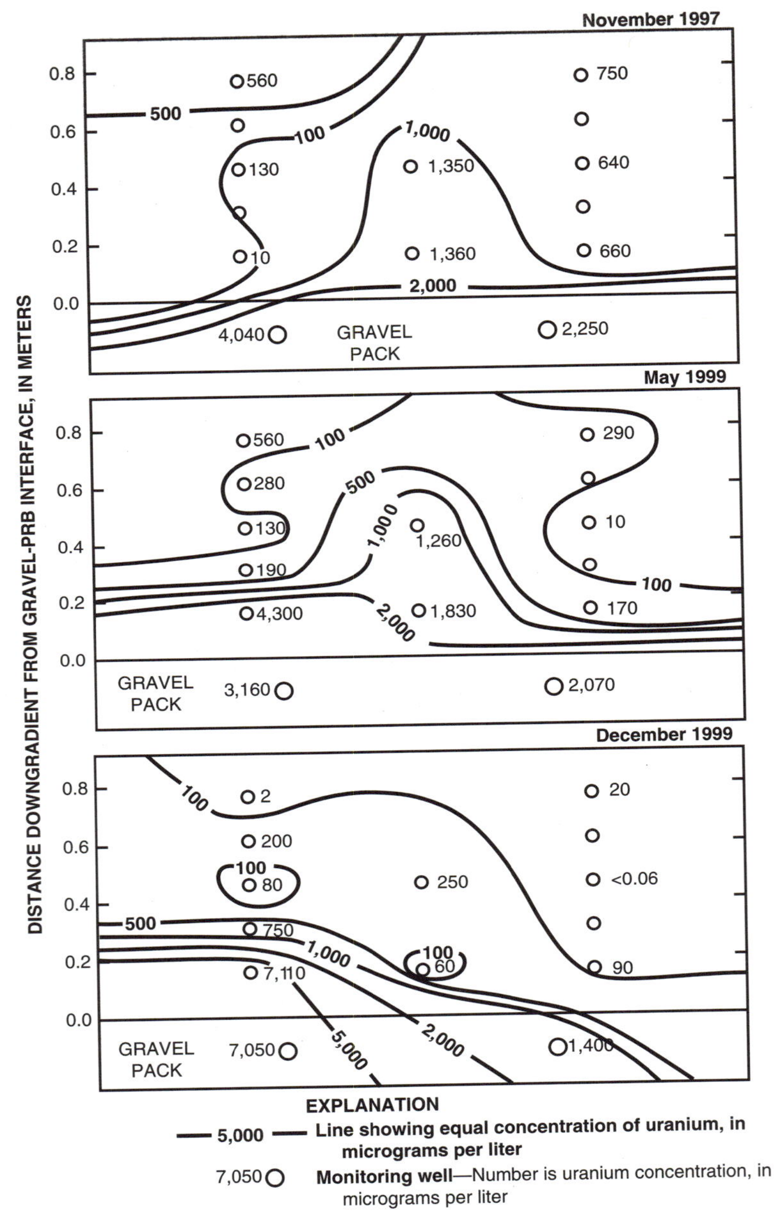

Figure 7 Uranium concentration in groundwater samples collected from the phosphate permeable reactive barrier during three different time periods, Fry Canyon, Utah.

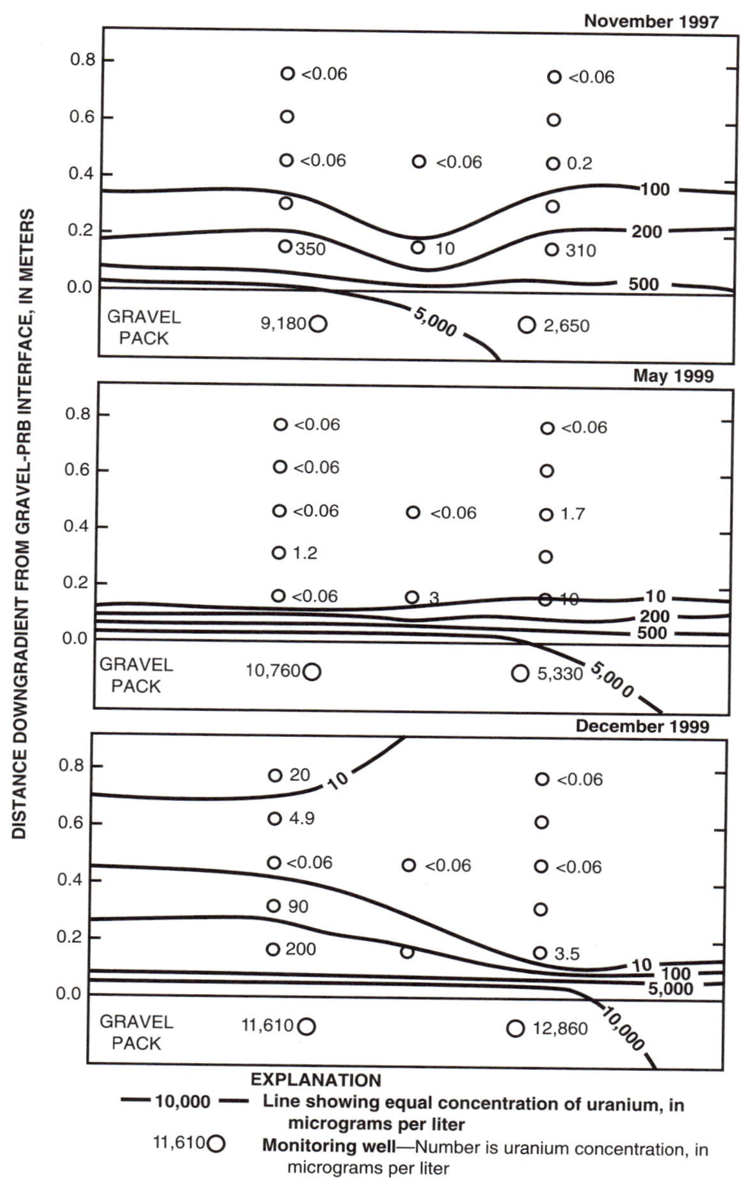

Figure 8 Uranium concentration in groundwater samples collected from the zero valent iron permeable reactive barrier during three different time periods, Fry Canyon, Utah.

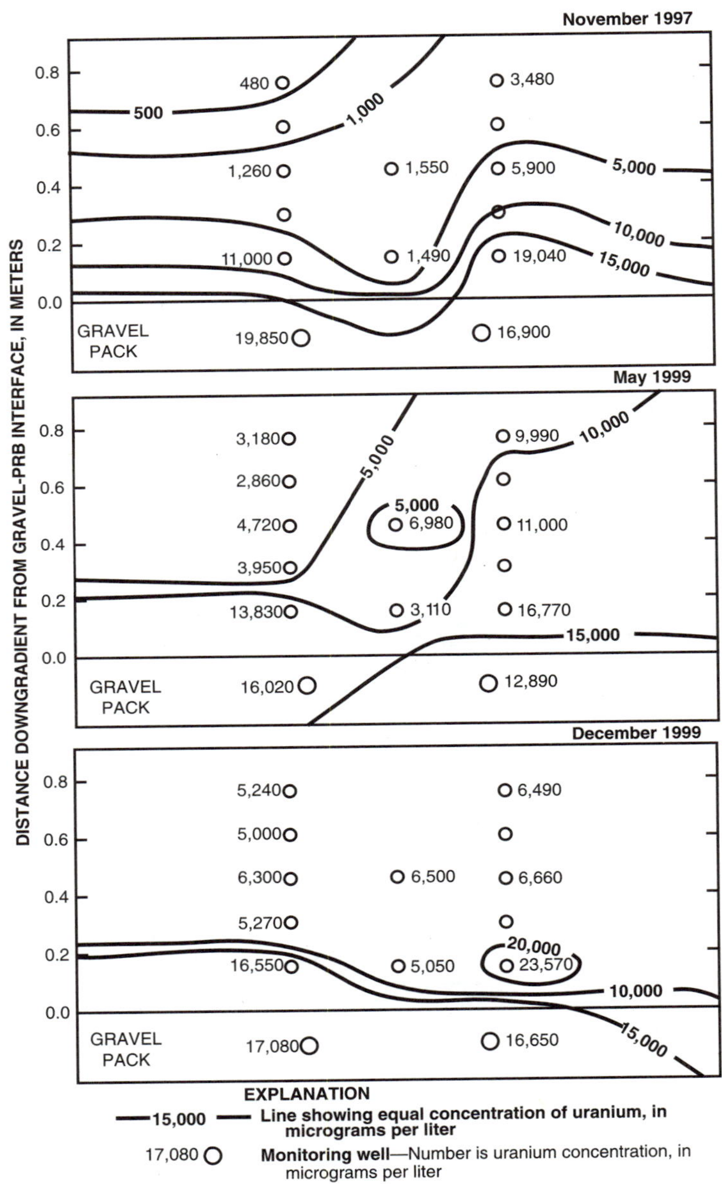

Figure 9 Uranium concentration in groundwater samples collected from the amorphous ferric oxyhydroxide permeable reactive barrier during three different time periods, Fry Canyon, Utah.

Uranium removal ratios (URR), unitless, were used for comparison of U removal efficiencies to detect early signs of PRB failure during the initial 39-month operating period (September 1997 through November 2000) (Fig. 10). The URR for flow paths 1 and 2 in each PRB was calculated by

$$URR = 1 - (U_{15\,cm}/U_{influent}), \qquad (1)$$

where $U_{15\,cm}$ is the U concentration in treated groundwater after traveling about 15 cm through the reactive material and $U_{influent}$ is the U influent concentration into each PRB. The shortest possible flow path was used because the upgradient section of each PRB will likely be the most sensitive to changes in contaminant removal efficiencies. A URR of 1.0 indicates complete (100%) U removal, whereas a URR less than 0.0 indicates that the U is being mobilized from the reactive material.

Uranium removal ratios for both flow paths in the ZVI PRB were consistently near 1.0 (Fig. 10), indicating high U removal efficiencies and no indication of onset of PRB failure. In contrast, URRs in both flow paths in the AFO PRB and flow path 1 in the PO4 PRB have decreased consistently since PRB installation. URRs <0.0 in both flow paths of the AFO PRB appear to indicate that U sorbed previously by the AFO material is now being desorbed

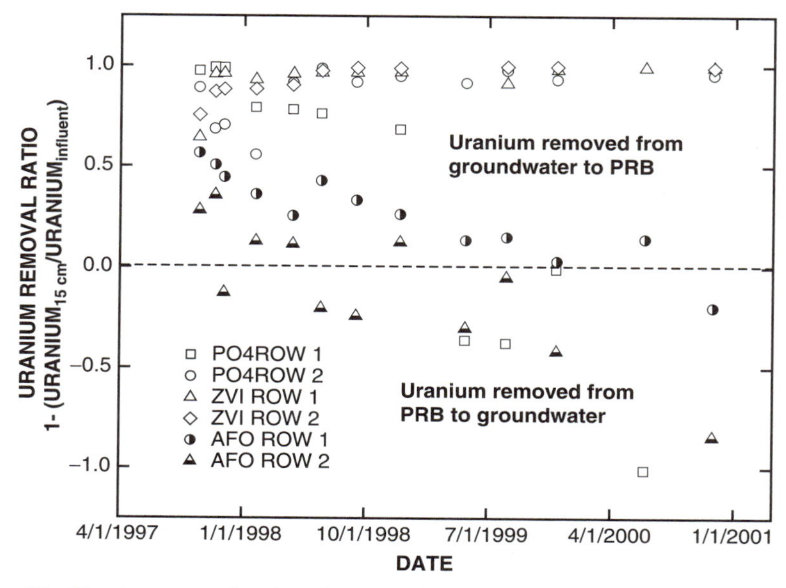

Figure 10 Uranium removal ratios after groundwater has traveled about 15 cm into each permeable reactive barrier at Fry Canyon, Utah.

by incoming groundwater with different chemistry (e.g., lower U concentrations). It is likely that the desorbed U is then sorbed further downgradient in the AFO PRB and that this process will continue until the U sorption equilibrium throughout the entire thickness of the AFO PRB has been attained. Given that desorption likely occurs in response to a decrease in U input concentration, an AFO PRB would need to be removed after sorption equilibrium is reached and (or) the contaminant plume passes to prevent rerelease of U back to the groundwater. The complex flow paths indicated by Br tracer tests make interpretation of the PO4 PRB more difficult. Two hydrologic processes causing variable U influent and removal in PRBs are discussed in the following section.

2. Hydrologic Impacts on U Input and Removal

Variable hydrologic conditions can have important impacts on PRB design and operation (Gavaskar *et al.*, 1998). Changes in water level elevations during the operating period have directly influenced the U influent to each PRB. This trend is illustrated by the comparison of continuous water level elevations measured in well ZVIT3 to the average $U_{influent}$ concentration measured in the gravel pack upgradient of the AFO PRB (Fig. 11). Higher

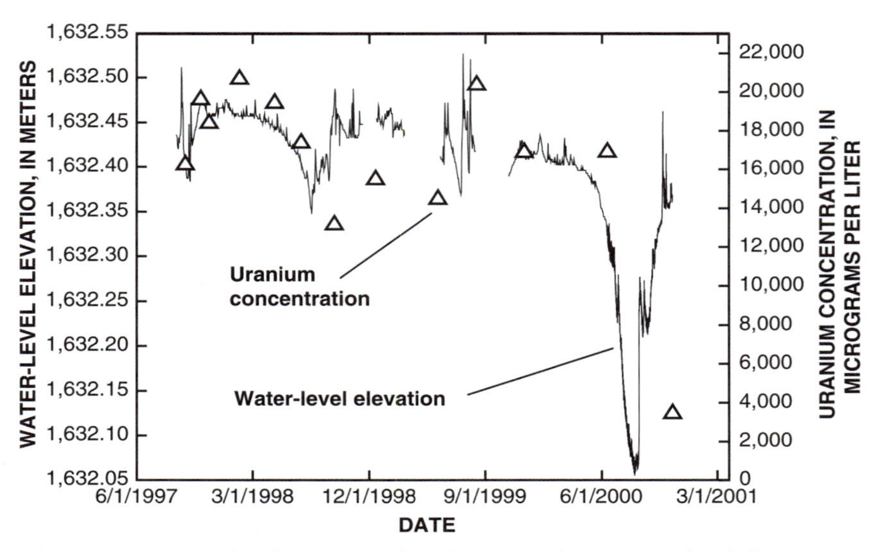

Figure 11 Water level elevation measured in well ZVIT3 and average uranium-influent concentration measured in wells AFOT1 and AFOT2, Fry Canyon, Utah.

groundwater levels are associated with increased $U_{influent}$. For example, one of the lowest $U_{influent}$ concentrations was measured in September 1998 and was preceded by the minimum seasonal water level elevation measured during July 1998 (Fig. 11). The highest $U_{influent}$ concentration was measured in early August 1999, about 14 days after a sharp increase in the water level elevation, presumably the result of a summer thunderstorm and resulting flash flood in Fry Creek.

Transient recharge events in the colluvial aquifer associated with either a higher stage in Fry Creek or infiltration of localized precipitation through the tailings material causes increased water levels. As the water level increases, previously unsaturated areas of the colluvial aquifer and tailings material are saturated and may have higher amounts of easily desorbed U relative to colluvial material that is saturated consistently. Laboratory experiments conducted with sediments collected from Fry Canyon indicated that large amounts of U could be desorbed easily (Naftz *et al.*, 2000).

A change in U concentrations with depth was present consistently in each of the PRBs during operation (Fig. 12). Water samples along flow paths from the deepest part of the ZVI and AFO PRBs have consistently higher U concentrations compared to shallow flow paths. For example, water samples from the deep monitoring points in the AFO PRB at 15, 46, and 76 cm along the flow path contain 660 to 4890 μg/liter more U than water samples from the shallow monitoring points.

One possible explanation for the elevated U concentrations in the deeper flow paths in AFO and ZVI PRBs is the result of water moving along the PRB/Cedar Mesa Sandstone interface. Because of the irregular surface of the Cedar Mesa Sandstone, it is possible that the deepest few centimeters of the PRB material were diluted with native material during placement of the reactive material. This process may have created a slightly less efficient treatment area, resulting in slightly lower efficiencies of U removal compared to overlying sections of the PRB. This same trend is not evident in the PO4 PRB with depth (Fig. 12) and may be a result of the slower groundwater velocities that are allowing longer contact times with the diluted reactive material at the PRB/Cedar Mesa Sandstone interface.

3. Number of Pore Volumes Treated

The ratio of U concentrations in treated groundwater (C) to influent (C_o) groundwater was compared to the estimated number of PVs treated from September 1997 through November 2000 in each PRB (Fig. 13). Measured U concentrations in water from wells PO4FS1, ZVIFS1, and AFOFS1 were used for C values, and average U concentrations from the monitoring points in the gravel pack, upgradient of each PRB, were used for C_o values. Changes

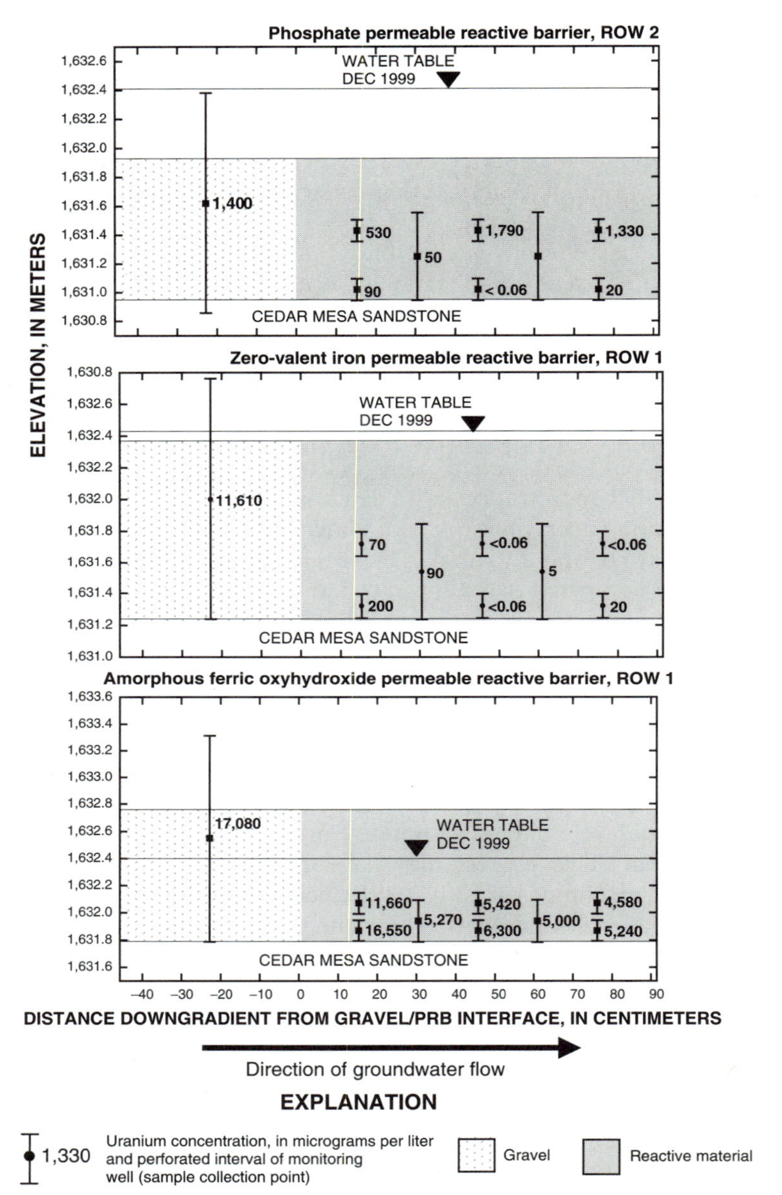

Figure 12 Cross section of uranium concentrations in permeable reactive barriers during December 1999, Fry Canyon, Utah.

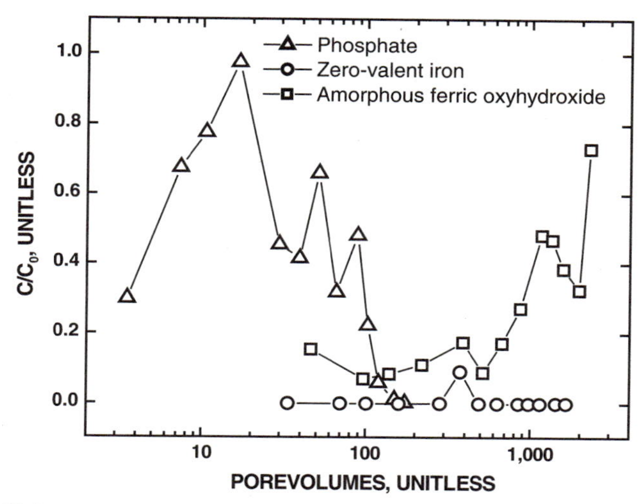

Figure 13 Efficiency of permeable reactive barriers determined by the ratio of uranium concentration in treated (C) groundwater to influent (C_o) groundwater in relation to the estimated number of pore volumes treated from September 1997 through November 2000.

in water level elevations were determined to control groundwater velocities in the ZVI PRB (Fig. 6); however, mean velocity estimates from the ionic tracer tests conducted during April 1999 (Fig. 4) were used as a representative velocity to calculate the number of PVs passing through each PRB.

Because of the low groundwater velocity measured in the PO4 PRB, a small number of PVs were treated (173) relative to the PVs treated by both ZVI (1650) and AFO (2270) PRBs (Fig. 13). After treating about 20 PVs, C/C_o values measured in the PO4 PRB have decreased substantially and have approached 0 during the latter stages of operation (Fig. 13). Without additional data, it is unclear if this trend to more efficient U removal is real or a result of stagnation or a change in the direction of groundwater flow in the PO4 PRB.

Trends in C/C_o values with increasing PVs in the AFO and ZVI PRBs are significantly different. Treatment efficiencies in the AFO PRB were excellent (C/C_o less than 0.2) during treatment of the first 670 PVs (Fig. 13). Decreasing treatment efficiencies were observed during the treatment of PVs 670 through 2270. A 50% breakthrough ($C/C_o = 0.50$) was attained in the AFO PRB after treatment of about 1160 PVs (Fig. 13). The ZVI PRB has treated about 1650 PVs without any noticeable reduction in treatment efficiencies (mean $C/C_o = 0.007$).

The total mass of U removed during PRB operation is another important factor that should be assessed when comparing the performance of individual

PRBs. Total mass removal is especially important at the Fry Canyon site because the groundwater velocities and influent U concentrations are different for each PRB. For example, early breakthrough of the AFO PRB could be the result of a higher U mass influent caused by increased concentrations and or groundwater flow velocities. The same assumptions used for PV calculations were also applied to the calculations of U mass removal. Additional assumptions included (1) PRB porosities did not change over time; (2) porosities were 70% (PO4 PRB), 56% (ZVI PRB), and 30% (AFO PRB) as measured in the laboratory by tritium breakthrough (PO4 and ZVI) and water displacement (AFO); (3) water level elevations measured during each monitoring period were constant and represented the saturated thickness of each PRB between monitoring periods; and (4) only the right half of the PO4 PRB had flow-through conditions.

U mass removal was normalized to the saturated thickness so that each PRB could be compared directly (Fig. 14). Although the C/C_o values were lower in the ZVI PRB, the AFO PRB removed more U mass per meter of depth until May 1999. This is a direct result of the higher U influent concentrations and greater estimated volumetric flow through the AFO PRB relative to the ZVI PRB. Decreasing trends in U mass removed in the AFO PRB began between June and September 1998 and coincided with the increasing trends in C/C_o values in the AFO PRB between PVs 520 and 670 (Fig. 13). The trend in mass of U removed per day by the ZVI PRB varied directly with the average U influent (Fig. 14), indicating that the high removal efficiency of ZVI was not affected by changes in influent U concentration.

Relative to AFO and ZVI PRBs, the U mass removed in the PO4 PRB was low (Fig. 14). This is caused by the combination of low groundwater velocities and U influent concentrations. During the 1130-day operating period, the total mass of U removed in each PRB was calculated to be PO4 (270 g), ZVI (13,400 g), and AFO (8320 g). The average saturated thickness in the AFO PRB was 0.51 m less than the average saturated thickness in the ZVI PRB, which resulted in a smaller total U mass removed during the 1130-day operating period.

4. Evidence for Mineral Precipitation in the ZVI PRB

As shown in the previous section, data indicate that the ZVI PRB is removing the highest percentage of influent U; however, aerobic and anaerobic corrosion of ZVI can result in large increases in the pH of the groundwater after entering the PRB (see Chapter 1). Depending on the groundwater chemistry, the pH increase can result in mineral precipitation. Cores collected from the ZVI PRB during May 1999 indicate that mineral precipitation is occurring. Minerals qualitatively identified in cores collected from the ZVI

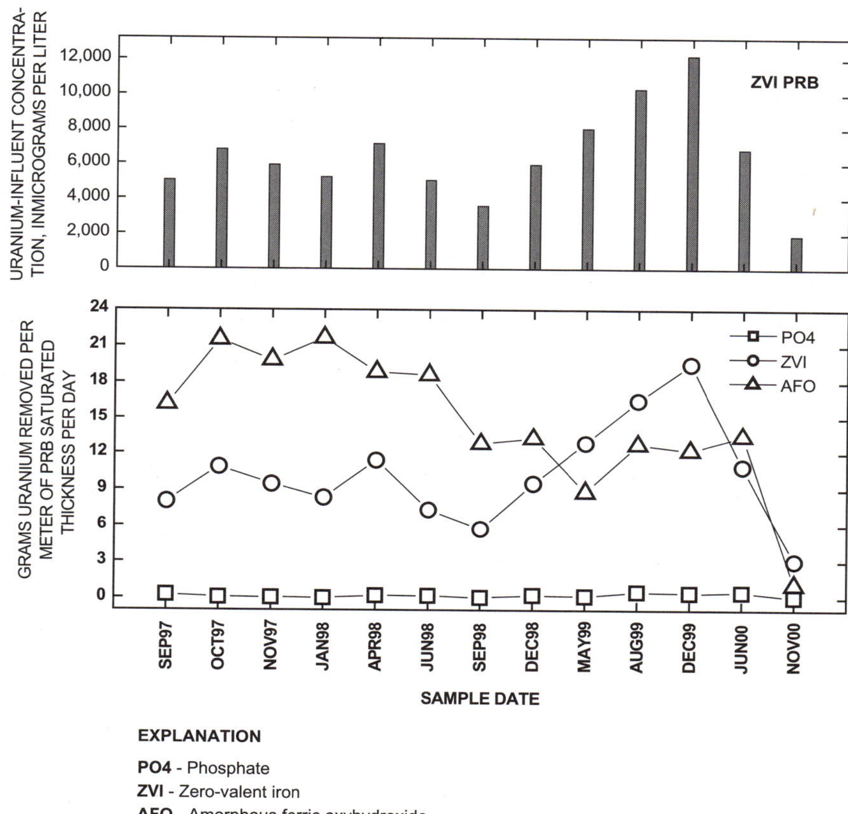

Figure 14 Mass of uranium removed by each permeable reactive barrier in relation to average U influent concentrations to the zero valent iron PRB from September 1997 through November 2000.

PRB during May 1999 included iron sulfide, calcite, and metallic iron. Mineral precipitation can decrease the permeability and reactivity of the ZVI PRB (Fig. 15) and result in a decrease in contaminant removal efficiencies (Gu *et al.*, 1999; see Chapters 11–13). In a series of laboratory column experiments, Gu *et al.* (1999) identified a number of mineral precipitates forming in ZVI material. Mineral precipitates included iron hydroxides, carbonates, and sulfides.

Changes in pH, calcium, and alkalinity along row 1 monitoring wells in the ZVI PRB were used to gain insight into the extent and type of mineral

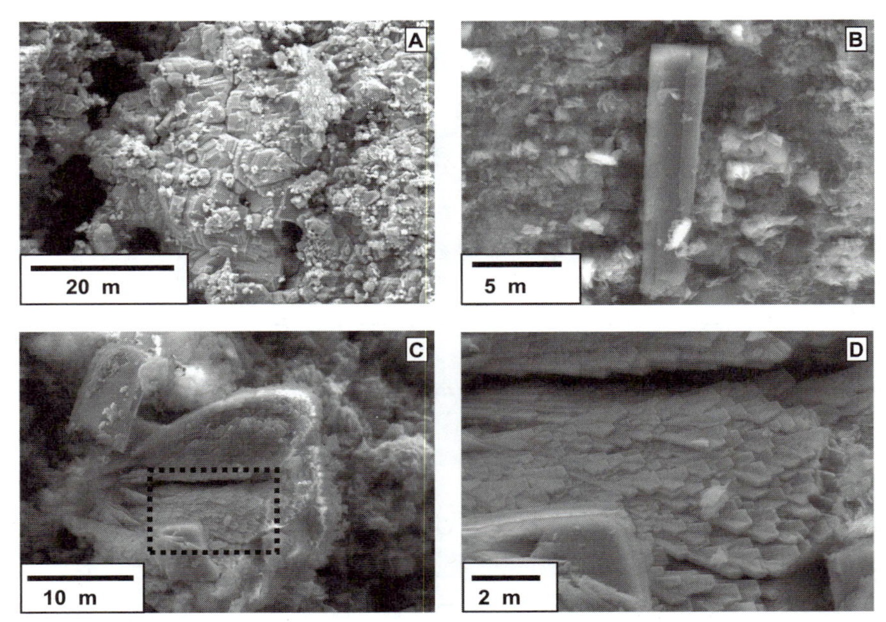

Figure 15 Scanning electron microscope photograph of solid-phase sample taken before installation (A) and after the zero valent iron permeable reactive barrier (ZVI PRB) had operated for about 21 months (B, C, and D), Fry Canyon, Utah. (A) Unreacted ZVI pellet surface showing metallic iron crystals and high porosity morphology; (B) calcite needle on surface of ZVI pellet; (C) calcite "fan" structure precipitated on surface of ZVI pellet; and (D) enlargement of outlined area shown in C.

precipitation. pH values increased significantly in the ZVI PRB during the three monitoring periods that were evaluated (Fig. 16). Large decreases in calcium (Ca) concentrations, as much as 72% after the incoming groundwater has penetrated 15 cm into the ZVI PRB, were associated with areas in the PRB displaying large pH increases (Fig. 16). Water samples from deeper monitoring wells completed directly above the ZVI PRB/Cedar Mesa Sandstone interface contain lower pH values and higher Ca concentrations compared to the shallow wells (Fig. 16). This probably results from a less efficient treatment zone along the PRB/Cedar Mesa Sandstone interface. Because of the irregular surface of the Cedar Mesa Sandstone, it is possible that the deepest few centimeters of the PRB material were diluted with native material during placement of the reactive material. This process may have mixed ZVI with native material, thereby decreasing the amount of Fe corrosion in this section of the PRB.

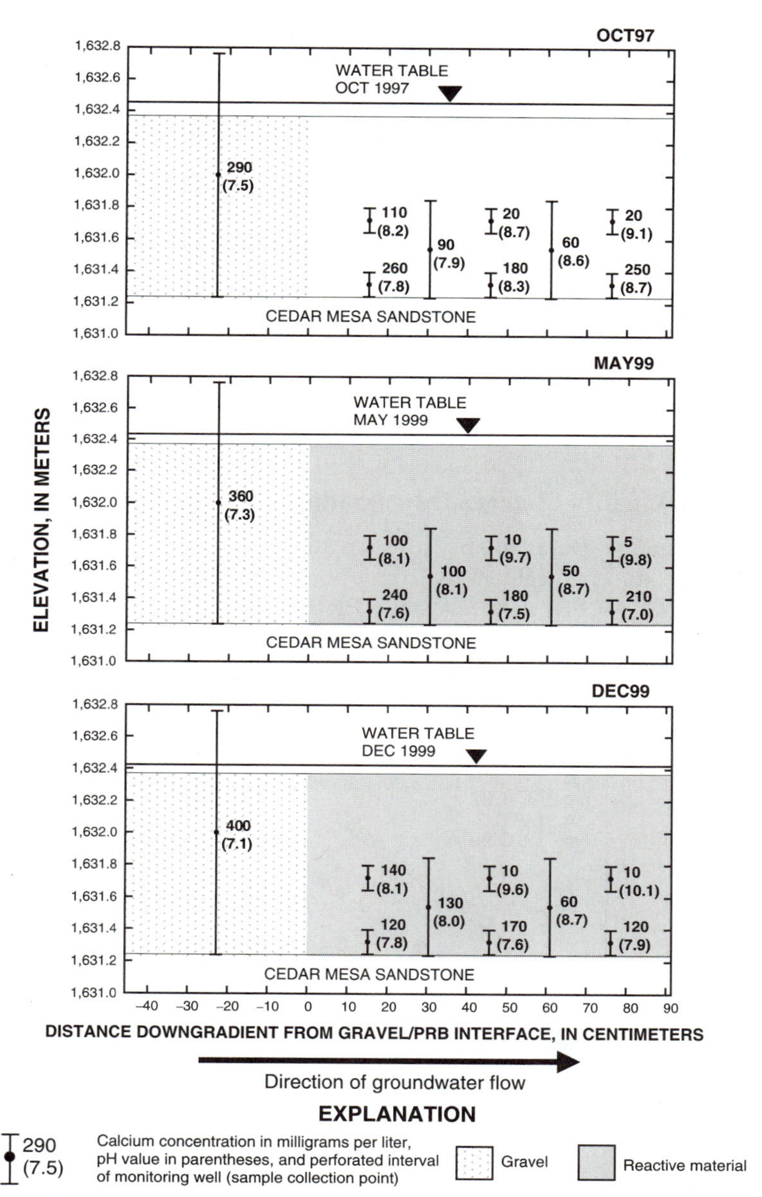

Figure 16 Cross section of pH values and calcium concentrations along row 1 monitoring wells in the zero valent iron permeable reactive barrier during three different time periods, Fry Canyon, Utah.

If calcite precipitation is assumed to be the dominant mineral sink for the observed losses of Ca in groundwater treated by the ZVI PRB, then decreases in alkalinity should correspond to losses in alkalinity on a 1:1 molar ratio. Alkalinity and Ca concentrations measured in row 1 monitoring wells were converted to millimoles (Fig. 17). Most water samples within the ZVI PRB plot below the reaction progress line, indicating that for the observed loss in Ca there is insufficient loss in alkalinity for the formation of calcite (Fig. 17). A number of plausible scenarios may be used to explain the trends observed in Fig. 17. For example, additional Ca-containing, noncarbonate mineral phases may be forming in the ZVI PRB. Geochemical modeling techniques were applied to water samples from the ZVI PRB to determine the plausible mineral phases that may be responsible for the observed Ca and alkalinity losses (see Chapter 10). Additional data are needed to provide a definitive answer.

5. Water Quality Changes Downgradient of PRBs

Degradation of water quality after treatment by PRBs is a concern. To address this concern, eight wells from about 1 to 3 m downgradient of the PRBs (Fig. 2) were monitored since PRB installation. Results from the downgradient monitoring wells were compared to mean water quality data from well FC3 located upgradient of the PRB installation (Fig. 18). With the

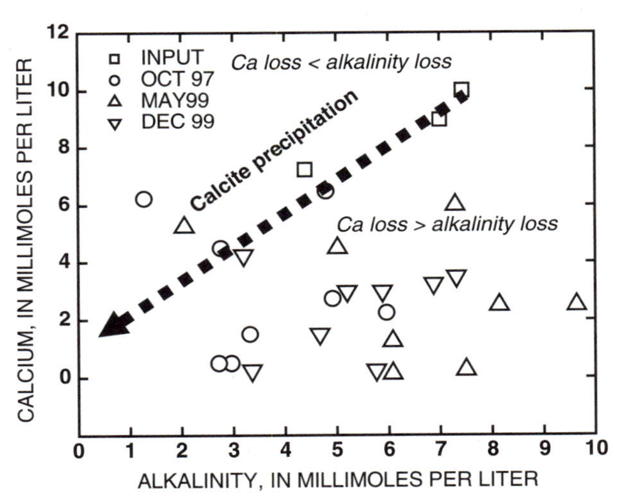

Figure 17 Calcium and alkalinity concentrations in water samples located upgradient (well ZVIT1) and within (row 1 monitoring wells) the zero valent iron permeable reactive barrier during October 1997, May 1999, and December 1999, Fry Canyon, Utah.

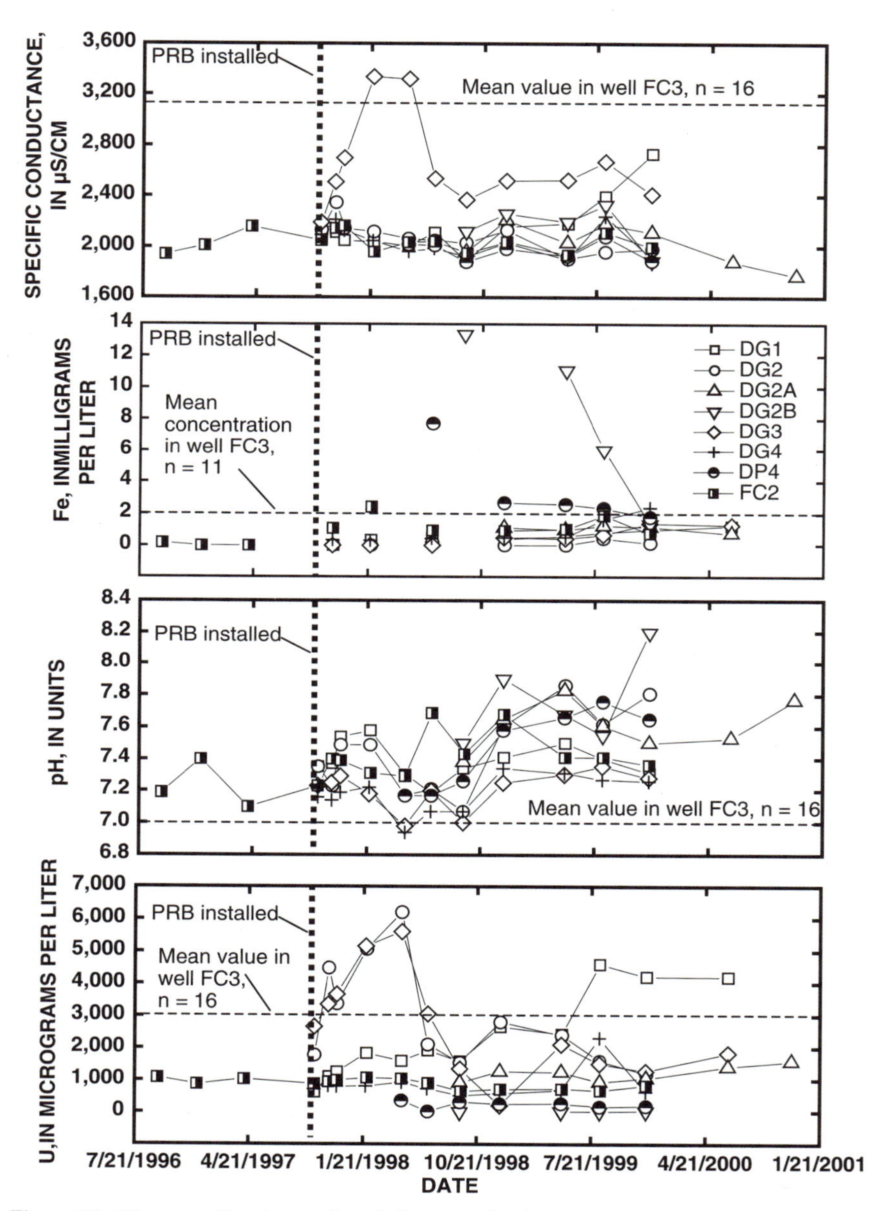

Figure 18 Water quality data collected from monitoring wells downgradient of permeable reactive barriers compared to data collected from well FC3, upgradient of the PRB installation, Fry Canyon, Utah.

exception of well DG3, the specific conductance (SC) value of groundwater exiting the PRBs is lower than the mean SC value in well FC3 (Fig. 18). An increase in SC was observed in well DG3 immediately after PRB installation and peaked about 5 months after PRB installation. After peaking, SC values in well DG3 returned to similar values measured in the other downgradient monitoring wells. It is likely that at least part of the observed decrease in SC is the result of mineral precipitation reactions that occur in the ZVI PRB.

Elevated Fe concentrations exiting from the ZVI PRB are a potential concern for downgradient water quality (see Chapter 13). With the exception of three samples from well DG2B and one sample from well DP4, the concentration of dissolved Fe in the downgradient monitoring wells is similar to the mean upgradient Fe concentrations (2 mg/liter) measured in well FC3 (Fig. 18). It is likely that the elevated Fe concentrations in samples from wells DG2B and DP4 are from the ZVI PRB; however, Fe concentrations in all downgradient wells decreased to background concentrations during the December 1999 monitoring period.

pH and U concentrations in water exiting the PRBs are important parameters to monitor because the PRBs at Fry Canyon were installed upgradient of previously contaminated sediments. In this location, treated water containing lower U concentrations has the potential to desorb U after exiting the PRBs and contacting the contaminated sediments. Increases in pH values in the treated groundwater could also cause U desorption from the contaminated sediments. The pH of groundwater exiting the PRBs is elevated compared to the mean pH of 7.0 units measured in water samples from well FC3 (Fig. 18). At least part of this pH increase may result from the aerobic and anaerobic Fe corrosion reactions occurring in the ZVI PRB. In general, the U concentration in the groundwater downgradient of the PRBs is at or below the mean U concentration measured in well FC3 (Fig. 18). Exceptions to this trend are elevated U concentrations in downgradient wells DG2 and DG3 for about 8 months after PRB installation and well DG1 beginning in August 1999 and continuing through the most recent sampling date (June 2000).

The transport of PO_4 from the PO4 PRB is also of concern to downgradient water quality. Phosphate concentrations measured in water samples from within the PO4 PRB have exceeded 80 mg/liter (Naftz *et al.*, 2000; Wilkowske *et al.*, 2001); however, the median PO_4 concentration in wells DG3 and DG4 (downgradient from the PO4 PRB) is < 0.2 mg/liter ($n = 22$). The low amount of PO_4 observed in the downgradient wells is probably a result of the large amounts of naturally occurring iron oxyhydroxides in the colluvial sediments downgradient from the PO4 PRB. Iron and manganese oxyhydroxides usually limit PO_4 concentrations in water to significantly less than 1 mg/liter (Hem, 1989).

V. APPLICATION OF FIELD DEMONSTRATION RESULTS

Results from the first 3 years of PRB operation at Fry Canyon can be applied to the design and operation of future, full-scale PRB installations in similar hydrologic and geochemical settings and are summarized according to hydraulics and chemistry.

A. Hydraulics

1. Groundwater velocities measured with a heat-pulse flowmeter compare favorably with velocities determined with ionic-tracer techniques in wells completed in the colluvial aquifer and ZVI PRB. Field demonstration results indicate that heat-pulse flowmeters may provide reliable results and should be considered as a method for the routine monitoring of groundwater velocities in selected PRBs. Velocity measurements conducted with heat-pulse flowmeters should be calibrated initially with ionic tracer methods and verified periodically.
2. Increases in groundwater velocities are related to increases in water level elevations and corresponding groundwater gradients within the ZVI PRB. The positive correlation between these parameters can be used to reconstruct detailed records of groundwater velocity by monitoring water table elevations continuously within a PRB. In addition, changes in the slope between water level measurements and groundwater velocity (measured by heat-pulse flowmeter or ionic tracers) could be used to assess changes in PRB performance as a result of plugging by mineral precipitation.
3. Groundwater velocities measured in the ZVI PRB ranged from less than 1 to more than 5 m/day. The treatment thickness of PRBs should be designed to accommodate the significantly shorter treatment times resulting from the increased groundwater velocities that may occur, especially in shallow aquifers with strong surface- and groundwater interactions.

B. Chemistry

1. The small-diameter (0.6-cm) monitoring wells located 15 cm downgradient from the gravel/AFO PRB interface were useful in providing "early warning signs" of PRB breakthrough. Monitoring wells should be installed within the upgradient areas of PRBs. Routine monitoring of these upgradient

wells can function as "early warning systems" for potential PRB breakthrough.

2. Increased water level elevations were related to increases in U influent concentrations to PRBs. Increased water level elevations also resulted in increased groundwater velocities in PRBs. The combined impacts of increasing U influent concentrations and increasing groundwater velocities resulting from increasing water level elevations should be considered during PRB design to allow for sufficient treatment thickness and residence time under changing hydrologic conditions.

3. Elevated U concentrations were measured in water samples from wells completed immediately above the PRB/bedrock interface relative to U concentrations measured in water samples from wells completed well above the PRB/bedrock interface. The increased U concentrations at the interface probably result from the irregular surface of the bedrock and the creation of a diluted and less efficient treatment zone during placement of the reactive material. During future installations, PRBs should be excavated into the confining layer interface to ensure capture of the entire vertical extent of the plume. In addition, PRB-monitoring networks should include monitoring points directly above the PRB/confining layer interface to document similar contaminant treatment efficiencies.

4. During the 1130-day demonstration period, the AFO PRB passed the largest number of PVs (2270), followed by the ZVI PRB (1650) and the PO4 PRB (173). The U removal efficiencies were excellent in the ZVI PRB during the entire demonstration period (mean $C/C_o = 0.007$); however, variable U removal efficiencies were observed in the AFO (mean $C/C_o = 0.271$) and PO4 (mean $C/C_o = 0.412$) PRBs. A 50% breakthrough ($C/C_o = 0.50$) was reached in the AFO PRB after treatment of about 1160 PVs. The total mass of U removed in each PRB was calculated to be PO4 (270 g), ZVI (13,400 g), and AFO (8320 g).

5. Evidence of mineral precipitation that could decrease future U removal efficiencies was found in solid-phase and groundwater samples from the ZVI PRB. Mineral phases identified in cores from the ZVI PRB included iron sulfide, calcite, and metallic iron. Aerobic and anaerobic corrosion of the ZVI has increased pH values and decreased Ca and total alkalinity concentrations in treated groundwater, resulting in mineral precipitation. Comparisons of Ca and alkalinity molar ratios in the influent and treated water indicate that a substantial amount of a Ca-containing, noncarbonate mineral phase is forming in the ZVI PRB.

6. Monitoring wells downgradient of the PRBs did not indicate significant, long-term degradation in the treated groundwater quality with respect

to baseline SC values, as well as baseline concentrations of Fe, U, and PO_4. Short-term increases in SC values and Fe and U concentrations were observed in selected wells following PRB installation. An overall increasing trend in pH values is observed in groundwater exiting the PRBs; however, it does not result in an increase in U desorption from the colluvial sediments downgradient of the PRBs.

REFERENCES

Freethey, G. W., Spangler, L. E., and Monheiser, W. J. (1994). Determination of hydrologic properties needed to calculate average linear velocity and travel time of ground water in the principal aquifer underlying the southeastern part of Salt Lake Valley, Utah. U.S. Geological Survey Water-Resources Investigations Report 92–4085.

Gavaskar, A. R., Gupta, N., Sass, B. M., Janosy, R. J., and O'Sullivan, D. (1998). "Permeable Barriers for Groundwater Remediation." Battelle Press, Columbus, OH.

Gu, B., Phelps, T. J., Liang, L., Dickey, M. J., Roh, Y., Kinsall, B. L., Palumbo, A.V., and Jacobs, G. K. (1999). Biogeochemical dynamics in zero-valent iron columns: Implications for permeable reactive barriers. *Environ. Sci. Tech.* **33**, 2170–2177.

Guthrie, M. (1986). Use of a Geo Flowmeter for the determination of ground water flow direction. *Ground Water Monitor. Rev.* **6**, 81–86.

Hem, J. D. (1989). Study and interpretation of the chemical characteristics of natural water. U.S. Geological Survey Water-Supply Paper 2254.

K-V Associates, Inc. (2001). Model 40 GEOFLO Heat-Pulse ground water flowmeter. Accessed March 9, 2001, at URL http://www.kva-equipment.com/model40.html.

Landa, E. R., and Gray, J. R. (1995). U.S. Geological Survey research on the environmental fate of uranium mining and milling wastes. *Environ. Geol.* **26**, 19–31.

Naftz, D. L., Davis, J. A., Fuller, C. C., Morrison, S. J., Freethey, G. W., Feltcorn, E. M., Wilhelm, R. G., Piana, M. J., Joye, J. L., and Rowland, R. C. (1999). Field demonstration of permeable reactive barriers to control radionuclide and trace-element contamination in ground water from abandoned mine lands. *In* "U.S. Geological Survey Toxic Substances Hydrology Program—Proceedings of the Technical Meeting, Charleston, South Carolina, March 8–12, 1999. Contamination from Hardrock Mining: U.S. Geological Survey Water-Resources Investigations Report 99–4018A" (D. W. Morganwalp, and H. T. Buxton, eds.), Vol. 1, pp. 281–288.

Naftz, D. L., Fuller, C. C., Morrison, S. J., Davis, J. A., Freethey, G. W., Rowland, R. C., Piana, M. J., Feltcorn, E. M., Wilhelm, R. G., and Blue, J. E. (2000). Field demonstration of permeable reactive barriers to remove dissolved uranium from groundwater, Fry Canyon, Utah, September 1997 through September 1998, Interim report. U.S. Environmental Protection Agency Report EPA 402-C-00–001.

U.S. Environmental Protection Agency (1999). In situ permeable reactive barriers: Application and deployment training manual. U.S. Environmental Protection Agency Report EPA 542-B-99–001.

U.S. Geological Survey Abandoned Mine Lands Science Team (1999). Abandoned Mine Lands Initiative—Providing Science for Watershed Issues. *In* "U.S. Geological Survey Open-File Report 99–321" (P. J. Modreski, Compiler), pp. 29–30.

Chapter 15

Collection Drain and Permeable Reactive Barrier for Treating Uranium and Metals from Mill Tailings near Durango, Colorado

Stan J. Morrison,* Donald R. Metzler,† and Brian P. Dwyer‡
*Environmental Sciences Laboratory,[1] Grand Junction, Colorado 81503
†U.S. Department of Energy, Grand Junction, Colorado 81503
‡Sandia National Laboratories, Albuquerque, New Mexico 87185

Contaminated water flowing from a uranium mill tailings repository near Durango, Colorado, is collected in a drain and piped to a system of permeable reactive barriers (PRBs). PRBs, which have been used to treat the water with zero valent iron (ZVI) since 1996, were constructed in two configurations: leach fields and baffled tanks. ZVI was used in three forms: foam plates, granular, and steel wool. All three forms of ZVI produced chemically reducing conditions and removed most contaminants [arsenic (As), cadmium, copper (Cu), molybdenum (Mo), nitrate, radium-226 (^{226}Ra), selenium (Se), uranium (U), vanadium (V), and zinc (Zn)], but equal volumes of foam plates or granular ZVI were more efficient than steel wool. A baffled tank containing ZVI removed contaminants more effectively than the leach field system and allowed easy replacement of the ZVI. Chemical distributions were determined in one PRB after 3 years of operation. The amount of contaminated

[1]Operated by MACTEC Environmental Restoration Services for the U.S. Department of Energy Grand Junction Office.

groundwater treated by this PRB could not be determined accurately because the flowmeter records were incomplete. However, we used the average concentrations of As, calcium (Ca), Se, U, V, and Zn in the solids to determine that between 238 and 489 m^3 of groundwater was treated. Chemical analysis indicated that significant quantities of Ca, inorganic carbon, and sulfur were deposited in the PRB in addition to the contaminants. Calcite, iron (Fe)–calcite, goethite, magnetite, ferrihydrite, and a U-V-silicon-rich Fe-oxide phase (or mixture of phases) were identified using X-ray diffraction and electron microprobe methods. Magnesium, manganese (Mn), ^{226}Ra, and strontium were deposited in the PRB, possibly in solid solution with carbonate minerals. Some constituents, including aluminum, chromium, Cu, Fe, Mn, and sodium, were partially leached from the ZVI foam plates. Concentrations of Fe in effluents from PRBs ranged from 7.69 to 115 mg/liter. Arsenic and Se concentrations in the ZVI foam plates were highest near the inlet, indicating rapid deposition in the PRB. Concentrations of other constituents, including Ca, Mo, U, V, and Zn, were also high near the inlet, but high concentrations extended some distance into the PRB, suggesting slightly longer residence times were required to remove them.

I. INTRODUCTION

Uranium (U) and associated constituents have contaminated groundwater at many U-ore milling sites and nuclear weapons facilities worldwide. Some concentrations of these contaminants are harmful to human health and the environment (see Chapter 13). Pump-and-treat systems, currently the most widely used groundwater remediation method, are costly and at many sites have not been effective in meeting groundwater quality standards (National Academy of Sciences, 1994). At the time this project was initiated in 1995 at the Bodo Canyon disposal site near Durango, Colorado, permeable reactive barrier (PRB) technology had been developed for the passive remediation of groundwater contaminated by chlorinated solvents (Gillham *et al.*, 1994) and acid mine drainage (see Chapter 2), but PRBs had not been installed to treat U and associated contaminants. Most PRB installations had been placed *in situ* without easy access for replacing the reactive material. In 1995, four PRBs featuring a unique design that allowed easy replacement of reactive material were installed at the disposal site for uranium mill tailings near Durango.

Laboratory research has shown that concentrations of redox-sensitive contaminants such as U and chromium (Cr) decrease when in contact with zero valent iron (ZVI) (Blowes *et al.*, 1997a; Cantrell *et al.*, 1995; Dwyer and Marozas, 1997). Chemical mechanisms responsible for the uptake of U are currently being debated in the scientific community (Fiedor *et al.*, 1998; Gu *et al.*, 1998; Matheson and Goldberg, 1999; Morrison *et al.*, 2001). Suggested mechanisms include (1) precipitation of reduced minerals such as uraninite (UO_2) and (2) formation of ferric oxyhydroxide [$Fe(OH)_3$] from oxidation of ZVI followed by adsorption on the newly formed $Fe(OH)_3$.

The Durango site provided an opportunity to study PRB designs for passive remediation of U-contaminated water. During construction of this uranium mill tailings repository in 1991 by the Uranium Mill Tailings Remedial Action (UMTRA) project, a contaminated seep developed that required treatment. A gravel collection drain was constructed in the disposal cell to collect contaminated seep water and pipe it to a lined retention pond where it was treated with lime and discharged to a drainage. In 1995, four PRBs were constructed to treat the contaminated water before it entered the retention pond (Fig.1). Inflow and outflow water chemistry has been monitored since 1996. Reactive material in one PRB was replaced with new material when flow through that PRB ceased because of pore plugging after 3 years of operation. This PRB was unearthed and the treatment material was sampled and analyzed extensively.

Data obtained from the analyses were used to evaluate the efficiency of PRBs in reducing aqueous concentrations of contaminants, to estimate relative uptake rates, and to determine the chemistry and mineralogy of material deposited in the PRBs. This chapter presents details of the Durango study, including PRB designs, construction methods, aqueous chemistry results for 4 years of sampling, and results of a detailed elemental and mineralogical analyses of the solids in one PRB after 3 years of operation. The collection drain and the PRB appear to be an effective system that could be used to treat contaminated groundwater at sites with favorable hydraulic gradients.

II. SITE DESCRIPTION

A. Uranium Mill Tailings Disposal Cell

A total of 1.9 million m^3 of uranium mill tailings was relocated in May 1991 from a former mill site adjacent to the Animas River in the city of Durango to the Bodo Canyon disposal site about 2.5 km away. Because a seep developed on the downgradient face of the repository during construction, a collection drain system was installed in the repository. An impermeable cap was placed on the repository to prevent surface water infiltration; however, water used during the construction process and still trapped in the pores of the tailings flowed into the drain system at about 2.6 liters/min.

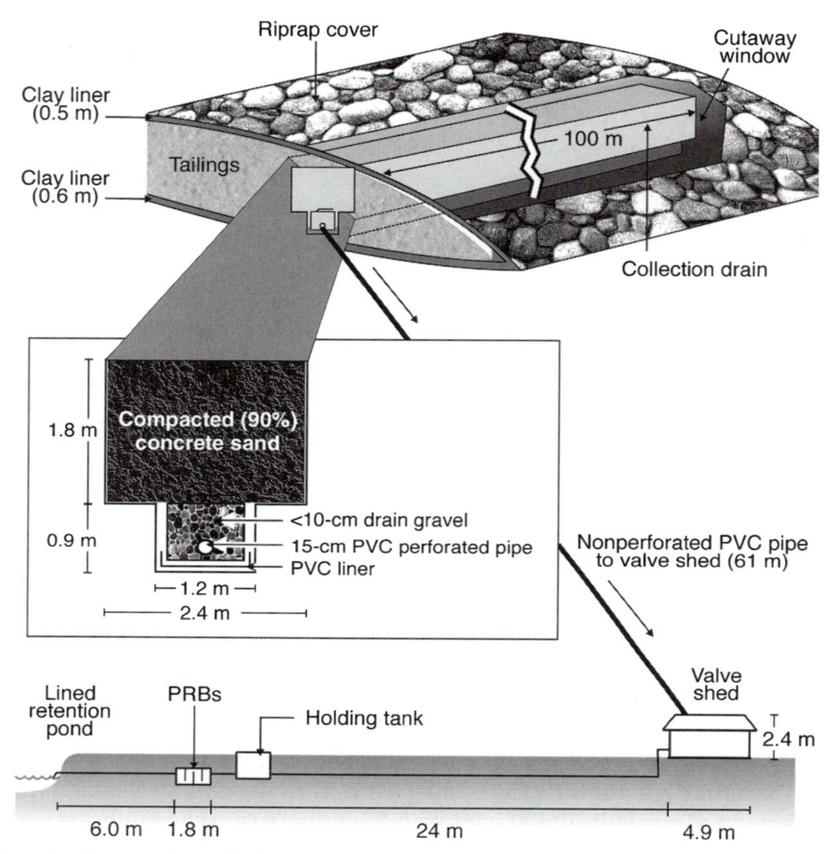

Figure 1 Cross sections of collection drain and permeable reactive barriers constructed at the Bodo Canyon disposal site near Durango, Colorado.

Samples of the contaminated seep water contained concentrations of contaminants [e.g., U, molybdenum (Mo), and selenium (Se)] that exceeded maximum contaminant levels for the UMTRA Project (*Federal Register*, 1995).

B. Design and Construction of the Collection Drain

The collection drain is 100 m long and consists of a 1.2-m-wide by 0.9-m-high gravel drain overlain by the 2.4-m-wide by 1.8-m-high collection portion that consists of compacted concrete sand (Fig. 1). The collection drain draws contaminated water from the surrounding U mill tailings. The contaminated water flows through a 15-cm-diameter perforated polyvinyl chloride (PVC) drainpipe that extends the length of the collection drain. The perforated PVC pipe connects to a nonperforated PVC pipe that transmits the water to the valve shed and then to a 2.4-m-diameter by 2.4-m-deep, capped, fiberglass holding tank. From the holding tank, water enters a valved manifold that directs the water into the desired PRB. Because heavy snowfall makes winter access to the site difficult, the system was shut off each year from about November through April. During the winter, contaminated groundwater collects in the collection drain.

C. PRB Design and Construction

In 1995, four PRBs (PRB A, PRB B, PRB C, and PRB D) were installed to treat the collected water with steel wool, copper mesh, and/or ZVI foam plates as the reactive material (Fig. 2; Table I). All four PRBs were placed

Figure 2 Four permeable reactive barriers were installed in 1995 near Durango, Colorado, and are shown just before backfilling of the system.

Table I

Description of Permeable Reactive Barriers at the Durango, Colorado, Site

Identifier	Design type	Reactant type	Notes
PRB A	Leach field	Steel wool and copper mesh	Never operated
PRB B	Leach field	Steel wool	
PRB C	Baffled tank	Cast ZVI foam plates	No flow through quadrants 2 and 3
PRB D	Baffled tank	Steel wool	No flow through quadrants 2 and 3
PRB E	Baffled tank	Granular ZVI	Still in operation; flow through all four quadrants

in a shallow excavation lined with plastic to prevent the release of contaminated water.

ZVI foam plates were manufactured by Cercona of America (Dayton, OH) by binding fine-grained ZVI with aluminosilicate (Mitchell, 1997). The surface area of this product is $1.7\,m^2/g$ with a bulk density of about $1.5\,g/cm^3$, the composition is more than 90% iron (Fe) as powdered ZVI, the porosity is about 0.5, and the hydraulic conductivity is about 0.05 cm/s. Steel wool was supplied in 1.3-cm-thick rolls by International Steel Wool; the copper mesh, also supplied by International Steel Wool, was 18 gauge and was composed of more than 97% copper.

PRB A and PRB B were constructed similar to septic leach fields; PRB A contained steel wool and copper mesh, and PRB B contained only steel wool (Fig. 3). Contaminated water was piped from the collection drain into the upper portion of the PRBs and flowed by gravity through reactive media to a PVC pipe that directed the outflow to the retention pond.

PRB C and PRB D were constructed in steel tanks with baffles that were intended to force the water to flow up and down through the PRB (Fig. 4). Each tank is 1.8 m long, 0.9 m wide, and 1.3 m deep. PRB C and PRB D are divided into four quadrants separated by flow baffles. Quadrant 1 is at the influent end of the PRB. A drainpipe connecting the bottom of quadrant 2 to the bottom of quadrant 4 was installed in both PRB C and PRB D during construction so that PRBs could be drained to prevent freezing during the winter season. Unfortunately, the drainpipes provided a route from quadrant 1 to quadrant 4, and much of the water bypassed quadrants 2 and 3. PRB C contained ZVI foam plates and PRB D contained steel wool. Covers for the steel boxes were sealed with sealant to the tops of the PRB walls and to two of the baffles and bolted down (Fig. 4). PRB C and PRB D were buried under about 0.9 m of soil. Residence time in PRB C filled with ZVI foam plates is about 3 h with a flow rate of 4 liters/min.

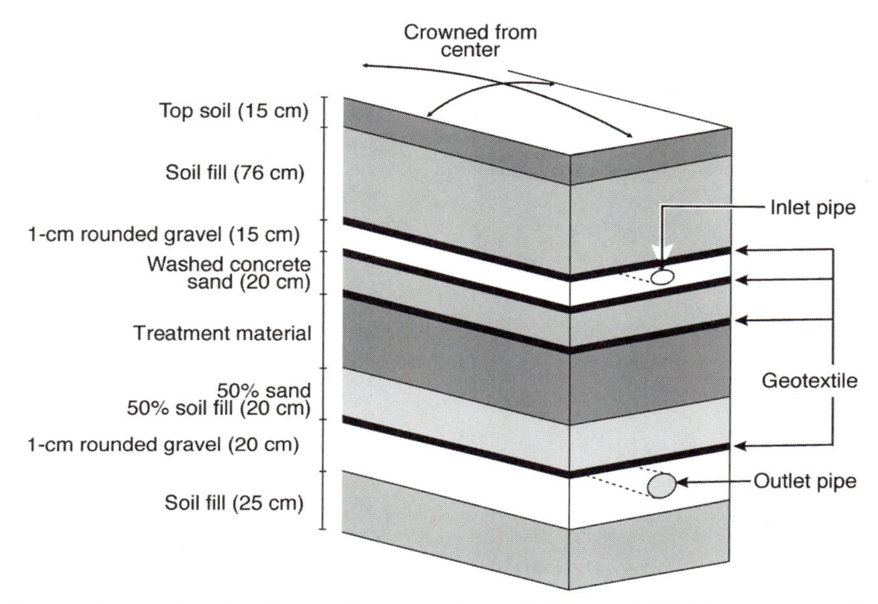

Figure 3 Design schematic of permeable reactive barrier (PRB) A and PRB B. Steel wool and copper mesh were the treatment materials for PRB A, and steel wool was the treatment material for PRB B.

ZVI foam plates in PRB C were oriented on their sides (vertical placement) in quadrant 1, but lay flat in quadrants 2, 3, and 4 (Fig. 4). Each plate is approximately 20 by 15 by 2.5 cm. Grooves were cut in the plates in quadrant 1 to increase permeability near the inlet. In July 1999, ZVI foam plates in PRB C were replaced with granular ZVI (Peerless Metals Products, mesh size $-8/+20$), and the PRB was renamed PRB E. At the same time, the drainpipe connecting quadrants 2 and 4 was removed to initiate up-and-down movement of the water, and a sampling port was added to collect samples from the bottom of quadrant 2.

Because of the low flow conditions (usually less than 2.6 liters/min), only one PRB was operated at a time; PRB A was never operated. PRB C was operated the longest (intermittently from May 1996 through May 1999) and treated about 490 m^3 of water as calculated from the vanadium (V) concentration in the solids (Fig. 5). PRB B treated 448 m^3 between April 1998 and May 1999; PRB D treated 112 m^3 between May and July 1999; and PRB E treated 192 m^3 between July 1999 and June 2000 (PRB E was still in operation after June 2000).

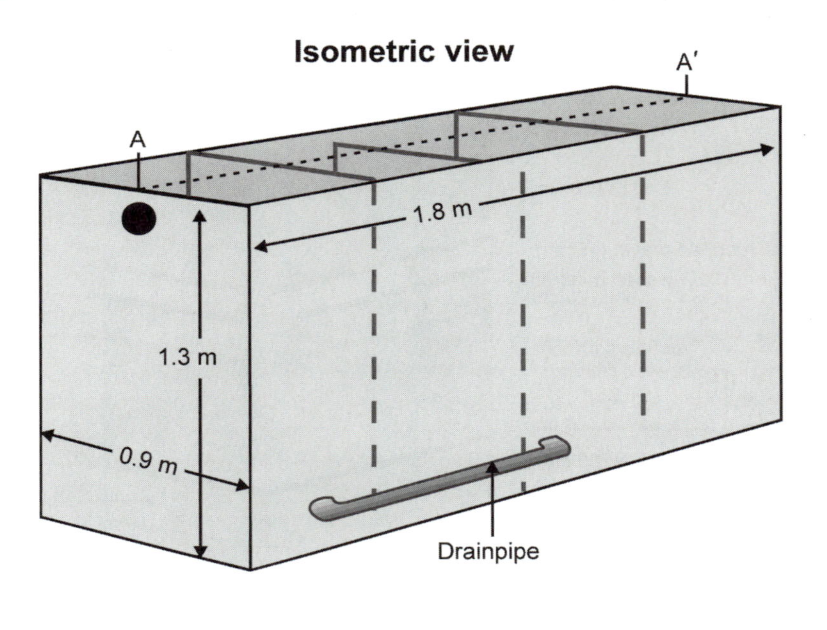

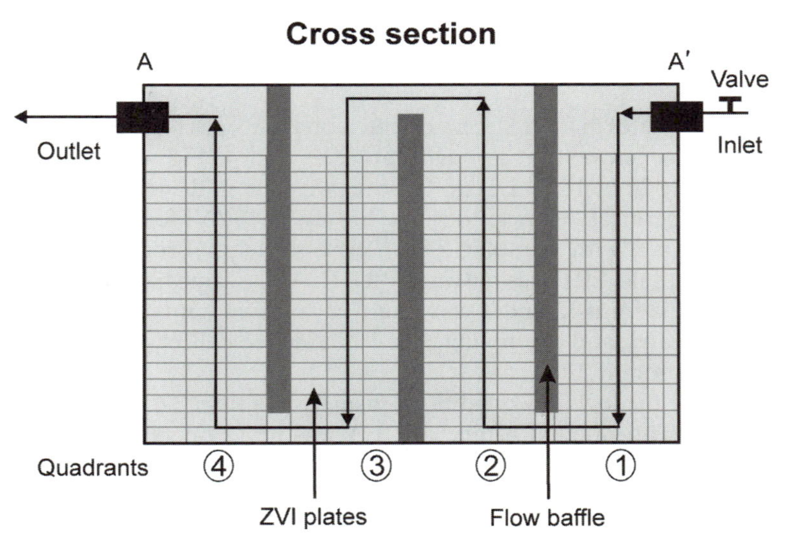

Figure 4 Isometric and cross section views of permeable reactive barrier (PRB) C and PRB D showing quadrant designations.

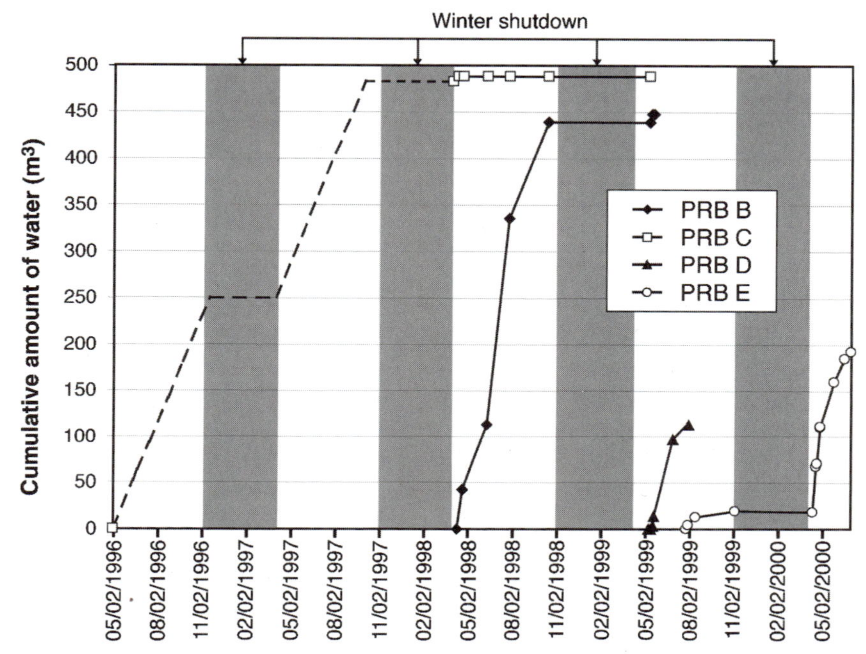

Figure 5 Record of permeable reactive barrier (PRB) operation. PRB A was never operated. Volume of treated groundwater was measured with a flowmeter for PRB B, PRB D, and PRB E. Dashed portion for PRB C is based on incomplete flow records; total volume is based on the concentration of vanadium in the solid fraction (see text). Shaded areas are winter shutdowns.

III. SAMPLING AND ANALYSIS METHODS

A. Water Sampling and Analysis

Flow rates were adjusted with a valve located in the valve shed (Fig. 1) and measured continuously with an in-line flow meter (records of flow rates were incomplete prior to March 1998). A peristaltic pump was used to collect samples from the outflow ends of the PRBs. About 10 liters of water was purged before collecting a sample. Samples of inflow water to all PRBs were collected in the holding tank by removing the cover of the tank and dipping a 1-liter Nalgene bottle in the water. Samples were filtered through a 0.45 μm filter in the field. Values of pH, Eh, electrical conductivity, and temperature were measured using probes in a flow-through cell or by swirling the probes in the holding tank. Alkalinity was measured in the field by sulfuric acid tritration on filtered samples. Dissolved oxygen was measured using the Hach Company AccuVac method (Hach, 1989).

Concentrations of aluminum (Al), barium (Ba), calcium (Ca), Cr, copper (Cu), Fe, magnesium (Mg), manganese (Mn), potassium (K), silicon (Si), sodium (Na), strontium (Sr), V, and zinc (Zn) in the water samples were analyzed by inductively coupled plasma atomic emission spectroscopy; concentrations of cadmium (Cd), Mo, and U were analyzed by inductively coupled plasma mass spectrometry; arsenic (As) and Se concentrations were determined by hydride-generation atomic absorption; concentrations of chloride (Cl), nitrate (NO_3), and sulfate (SO_4) were determined by ion chromatography; ammonium (NH_4) concentration was measured colorimetrically; total dissolved solids (TDS) concentrations were determined gravimetrically; concentrations of total organic carbon (TOC) were analyzed by coulometry; and concentrations of radium-226 (^{226}Ra) were measured by α spectrometry. Serial dilutions, spikes, and duplicate analyses were performed for quality control.

B. Solids Sampling and Analyses

PRB C was opened in August 1998, and several solids samples were collected from the top of the ZVI foam plates. PRB C was opened again in July 1999, after the end of its operating life (May 1999), and sampled extensively by coring and during removal of the ZVI foam plates. Three cores (two from quadrant 1 and one from quadrant 4) were collected by driving a core barrel with a handheld impact hammer or a sledge hammer. Splits of the cores were sealed in plastic tubes under an argon environment for electron microprobe examination. The same cores were examined with X-ray photoelectron spectroscopy (see Chapter 12).

Samples for chemical analyses were collected with a chisel and hammer during removal of the ZVI foam plates. Samples were collected at three depths: 1, 40, and 74 cm below the uppermost surface of the ZVI foam plates. Samples ranging in size from about 30 to 800 cm^3 were placed in plastic bags for transport to the laboratory where they were oven dried at 90 °C and then ground to powder. A split of the powder was digested in a mixture of concentrated nitric, perchloric, and hydrofluoric acids (ASTM, 1992) for analyses of Al, As, Ca, Cr, Cu, Fe, Mg, Mn, Mo, Na, Se, U, V, and Zn. A peroxide fusion was used to digest S so that all S was oxidized to SO_4. The same analytical methods used to analyze water samples were used to analyze digestates. Radiometric analyses were performed by γ ray counting using a germanium detector calibrated with commercial, cast-resin radiologic standards. Total inorganic carbon (TIC) concentrations were determined by analyzing carbon dioxide emission during acid digestion of the sample.

Polished, carbon-coated electron microprobe mounts were examined with a Cameca electron microprobe operated at 15 keV, 25 nanoamperes, with a 1 to 10-μm beam size, and 10-s counting on each element. Random powder mounts were used for X-ray diffraction examination. The X-ray beam was operated at 40 keV, 30 mA, and diffraction patterns were collected at a scan rate of 10° per minute.

IV. RESULTS

A. PRB B—LEACH-FIELD SYSTEM WITH STEEL WOOL

PRB B treated 448 m^3 of contaminated water (Fig. 5). Influent and effluent were sampled once in June and once in November 1998; Table II presents November results. Flow through the steel wool caused the water to be chemically reduced, as indicated by a decrease in Eh from 113 to 15 mV. The pH value increased from 6.77 to 7.22, alkalinity decreased by 20.7%, and the concentration of TDS decreased by 12.4%. Other significant changes in water chemistry include (1) a more than 90% decrease in concentrations of dissolved oxygen (DO), As, Cd, NH_4, NO_3, and V; (2) a more than 70% decrease in the concentrations of Cr, Se, and Zn; (3) a 65% decrease in the U concentration; and (4) an increase in the Fe concentration from 0.023 to 28.8 mg/liter.

B. PRB C—BAFFLED TANK WITH ZVI FOAM PLATES

1. Water and Gas Phase Chemistry

PRB C operated intermittently from May 1996 through May 1999; however, only 15 m^3 of contaminated groundwater was treated after October 1997 (Fig. 5). Results of quarterly samplings indicate that concentrations of U, NO_3, and Mo remained low throughout this period (Dwyer and Marozas, 1997; Morrison *et al.*, 2002). Table II presents results of the last sampling of PRB C in May 1999. At that time, the ZVI foam plates in PRB C were removing much of the contamination from the water. Most contaminants (As, Cd, Cu, NO_3, Se, U, V, and Zn) decreased to low concentrations; Mn and Mo concentrations also decreased, but to a lesser extent than the other contaminants. Electrical conductivity and concentrations of alkalinity, Ca, TDS, and SO_4 also decreased significantly. The dissolved Fe concentration

Table II

Analytical Results for Influent and Effluent Samples from PRB B and PRB C near Durango, Colorado[a]

	PRB B (November 1998) Flow rate = 2.3 liter/min			PRB C(May 1999) Flow rate = 0.3 liter/min		
Constituent	Influent	Effluent	% Removed	Influent	Effluent	% Removed
pH	6.77	7.22	na	6.72	7.36	na
Eh (mV)	113	15	na	74	−278	na
T (°C)	12.5	10.7	na	nd	nd	nd
EC (μS/cm)	3770	3620	na	3420	2,740	na
DO (mg/liter)	3.1	<0.2	>93.5	2.3	0.95	58.7
Alk (mg/liter)	590	468	20.7	540	104	80.7
As (μg/liter)	232	7.2	96.9	209	2.3	98.9
Ba (μg/liter)	16.6	25.6	na	13.8	165	na
Ca (mg/liter)	622	518	16.7	631	236	62.6
Cd (μg/liter)	10.8	<1	>90.7	9.0	<1.0	>88.9
Cl (mg/liter)	112	105	6.3	130	128	1.5
Cr (μg/liter)	13.4	3	77.6	10.6	16.4	na
Cu (μg/liter)	<5	<5	na	9.1	<5.0	>45.1
Fe (mg/liter)	0.023	28.8	na	0.029	7.69	na
K (mg/liter)	18.2	16.8	7.7	14.4	18.6	na
Mg (mg/liter)	86.7	79.9	7.8	71.7	57.4	19.9
Mn (μg/liter)	9450	5290	44.0	4700	726	84.6
Mo (μg/liter)	1160	613	47.2	1090	359	67.1
Na (mg/liter)	318	306	3.8	288	338	na
NH_4 (μg/liter)	55.5	<1	>98.2	137	655	na
NO_2 (μg/liter)	<1	<1	na	29	8	72.4
NO_3 (μg/liter)	1830	123	93.3	415	72.7	82.5
^{226}Ra (pCi/liter)	nd	nd	nd	2.57	1.53	40.5
Se (μg/liter)	370	48.4	86.9	202	5.9	97.1
Si (mg/liter)	27.3	18.2	33.3	24.9	0.74	97.0
SO_4 (mg/liter)	1730	1570	9.2	1700	1260	25.9
Sr (mg/liter)	3.28	2.52	23.2	3.06	2.18	28.8
TDS (mg/liter)	3640	3190	12.4	3500	2170	38.0
TOC (mg/liter)	nd	nd	nd	9.4	18.4	na
U (μg/liter)	8040	2810	65.0	6600	1.2	>99.9
V (μg/liter)	9090	630	93.1	8100	<6.0	>99.9
Zn (μg/liter)	834	101	87.9	1050	<4.0	>99.6

[a]EC, electrical conductivity; DO, dissolved oxygen; Alk, alkalinity in milligrams per liter of $CaCO_3$; TDS, total dissolved solids; TOC, total organic carbon; na, not applicable; nd, no data; mV, millivolts; μs/cm, microsiemens per centimeter; mg/liter, milligrams per liter; μg/liter, micrograms per liter; pCi/liter, picocuries per liter.

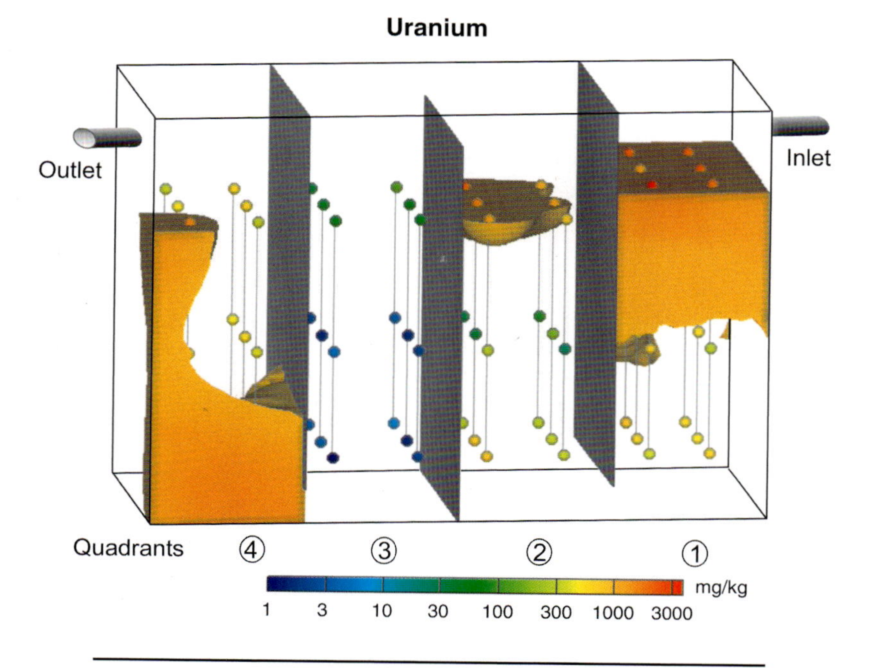

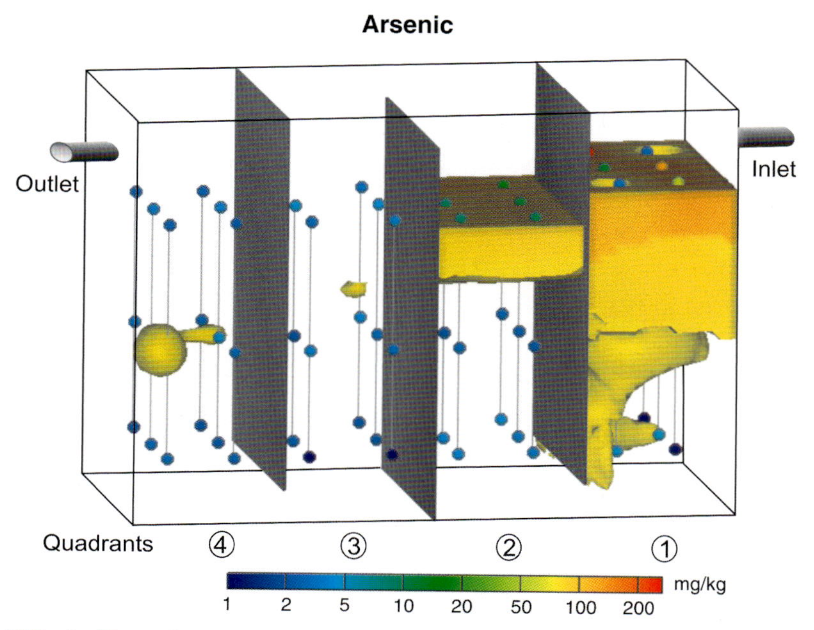

FIG. 6. Three dimensional views of uranium (U) and arsenic (As) concentrations in solids sampled from PRB C, July 1999. Regions where As and U concentrations are more than 75 mg/kg and 1,000 mg/kg, respectively, are colored. Thin vertical shaded areas represent flow baffles.

increased from 0.029 mg/liter in the influent to 7.69 mg/liter in the effluent. A black, odiferous precipitate that could be a sulfide mineral phase was observed in the effluent sampling pipe.

When the ZVI foam plates were jostled with a rock hammer, gas bubbles erupted through the standing water. The presence of this free gas phase may have caused a vapor lock, contributing to the plugging of the PRB. Readings on an MSA Model 360 combustible gas detector placed in the downgradient sampling access port indicated the presence of explosive gases in the PRB; qualitative measurements with colorimetric tubes (Sensidyne Gastec detectors) indicated the presence of hydrogen and methane.

2. Constituent Concentrations in Solid Phases

Relatively high U concentrations were present in solid samples of ZVI from quadrants 1 and 4, but solid samples from quadrant 3 and the lower portion of quadrant 2 had low concentrations of U in July 1999 (Fig. 6). The highest concentration of U (21,300 mg/kg) was detected in a sample scraped from the surface of the ZVI foam plates during the August 1998 sampling (1998 samples are not included in Fig. 6). High concentrations of U in samples from the top portion of quadrants 2 and 4 were probably caused by leakage across the tops of the baffles; the sealant used to prevent flow across the tops of the baffles may not have been effective. The highest average U concentration (1310 mg/kg) was detected in quadrant 1, and the second highest average concentration (812 mg/kg) was detected in quadrant 4 (Table III). Thus, the U in this PRB is most concentrated near the inlet but is also dispersed throughout the PRB. Low concentrations of U in quadrant 3 but high concentrations in quadrant 4 are due to the presence of the drainpipe connecting quadrants 2 and 4 (Fig. 4).

A thin window Geiger–Müller detector, optimized to minimize background radiation response, was used to detect γ and β radiation during the excavation of PRB C. This instrument detected β-emitting material in the ZVI foam plates. Subsequent spectroscopy analysis of excavated material indicated that most of the β radiation was from protactinium-234 (^{234}Pa) (Table IV). Protactinium-234 is a short-lived isotope in the ^{238}U decay chain and is a strong β emitter. Because of potential health hazards, a radiation protection program was implemented. Control measures included use of Tyvek protective clothing, an exclusion zone, radiation monitoring, and decontamination procedures.

The V distribution in PRB C was similar to the U distribution, with the highest average concentrations occurring in the upper portion of quadrant 1. Samples from quadrant 1 had an average V concentration of 4160 mg/kg (Table III). Samples from quadrants 2 and 4 were also enriched but at a lower

Table III

Average Constituent Concentration by Quadrant for Permeable Reactive Barrier C near Durango, Colorado[a]

Quadrant	Al (mg/kg)	As (mg/kg)	Ca (mg/kg)	Cr (mg/kg)	Cu (mg/kg)	Fe (mg/kg)
Quadrant 1	12,800	105	23,400	1530	1,230	695,000
Quadrant 2	12,500	82	12,700	1440	1,150	801,000
Quadrant 3	13,400	51	3,750	1590	1,520	807,000
Quadrant 4	12,100	55	18,100	1620	1,150	766,000
Original material	17,600	53	580	1990	935	833,000
	Mg (mg/kg)	Mn (mg/kg)	Mo (mg/kg)	Na (mg/kg)	^{226}Ra (pCi/g)	S (mg/kg)
Quadrant 1	401	5,300	539	1870	1.24	2,254
Quadrant 2	379	6,160	538	1590	0.73	1,455
Quadrant 3	335	4,770	381	2450	0.53	604
Quadrant 4	406	5,600	471	1820	0.64	1,230
Original material	434	8,150	326	2810	0.53	52
	Se (mg/kg)	TIC (mg/kg)[b]	U (mg/kg)	V (mg/kg)	Zn (mg/kg)	
Quadrant 1	103	10,160	1,310	4160	672	
Quadrant 2	40	5,320	765	2030	298	
Quadrant 3	9	1,290	51	194	122	
Quadrant 4	8	6,980	812	1060	247	
Original material	9	134	19	100	176	

[a]Each quadrant contains approximately 525 kg of material.
[b]Total inorganic carbon.

Table IV

Spectroscopy Analysis of Samples from Quadrant 1 of Permeable Reactive Barrier C near Durango, Colorado in picocuries per gram (pCi/g)[a]

Radionuclide	Sample 1	Sample 2	Sample 3	Sample 4	Sample 5
Protactinium-234	103.37	61.57	86.76	442.74	65.83
Thorium-234	527.38	630.77	20.81	107.42	298.29
Uranium-235	67.18	73.95	3.04	13.37	41.65

[a]All samples were collected in the upper 6 cm.

concentration; however, the average V concentration in samples from quadrant 3 was only slightly higher than the concentration in the original material (unused foam plates), which is consistent with the water bypassing this quadrant. The highest V concentration (104,000 mg/kg) was identified in a sample scraped from the surface of the ZVI foam plates in quadrant 1 during the August 1998 sampling.

Arsenic was deposited primarily in quadrant 1 with a lesser amount in quadrant 2 (Fig. 6, See color plate). Like As, Se was deposited mostly in quadrant 1. Molybdenum and Zn concentrations were highest in samples from quadrant 1 but exceeded concentrations of the original material in quadrants 2 and 4; concentrations of Mo and Zn in samples from quadrant 3 were similar to the original material. Chromium and Mn concentrations were higher in the original material than in samples from PRB C; the highest Cu concentration was in quadrant 3 (Table III). Radium-226 precipitated near the inlet, but concentrations were low and reflect low concentrations in the influent (Table III). The highest Ca concentration (54,800 mg/kg) in an individual sample was detected in a sample from the middle portion of quadrant 1; samples from quadrant 1 also had the highest average Ca concentrations (Table III). Aluminum and Na concentrations were lower in samples from all four quadrants than the original material but were highest in quadrant 3.

3. Estimate of Amount of Groundwater Treated

Although the flow through PRB C was monitored using an in-line flowmeter, the records are incomplete and only a rough estimate of the amount of contaminated groundwater treated could be made based on information contributed by the field staff (Fig. 5). The concentrations of contaminants measured in the solids provided an independent means of determining the amount of water that was treated. This calculation requires the following data: (1) concentration of the constituent in the solids, (2) total mass of the solids (2250 kg), and (3) aqueous concentrations of the constituent in the influent and effluent. The aqueous concentrations were reasonably constant so the average concentration was used. The volumes of treated water calculated from concentrations of As, Ca, Se, U, V, and Zn were relatively consistent, ranging from 238 to 489 m^3 (Table V). This method of calculating treated volume from solid-phase concentrations is particularly useful for traditional *in situ* PRBs because the groundwater flow velocity is rarely known.

4. Mineralogy

The appearance of the ZVI foam plates changed markedly from its original fresh silver color. A black and orange, fine-grained, disaggregated material

Table V

Volume of Water Treated by Permeable Reactive Barrier C near Durango, Colorado, as Determined from Concentration of Constituents in Solids

Constituent	Average in-out concentration (mg/liter)[a]	Average ZVI concentration (mg/kg)[b]	Mass in ZVI (kg)[c]	Volume (m^3)[d]
Arsenic	0.21	20	0.05	238
Calcium	66	13,900	31.3	473
Selenium	0.20	31	0.07	350
Uranium	5.3	716	1.61	304
Vanadium	8.1	1,760	3.96	489
Zinc	1.1	159	0.36	327

[a] Average difference between influent and effluent concentrations for all sampling periods.
[b] Average concentration in solids minus concentration in original ZVI foam plates.
[c] Assumes 2250 kg for the mass of ZVI foam plates.
[d] Calculated volume of groundwater treated.

covered much of the upper surface of the ZVI foam plates and was about 0.5 cm thick in quadrant 1. Below this disaggregated material, the ZVI foam plates were black but were intact and structurally similar to unused foam plates. X-ray diffraction, electron microprobe, and petrographic methods were used to examine the mineralogy and elemental composition of the ZVI foam sampled from PRB C.

X-ray diffraction analysis of the ZVI foam plates identified ZVI, calcite ($CaCO_3$), magnetite (Fe_3O_4), goethite (FeOOH), and ferrihydrite [approximately $Fe(OH)_3$] (Table VI). Three unidentified peaks were present that are believed to be due to mineral phases in the aluminosilicate binder. All samples contained magnetite and calcite.

ZVI particles occur as spheres and shards, and some composite particles exhibit intergrowths of ZVI and aluminosilicate binder. Six distinct mineral phases were identified by electron microprobe analysis based on element stoichiometry and comparison with X-ray diffraction results: (1) ZVI; (2) magnetite; (3) goethite; (4) Ca–Fe carbonate [$(Ca, Fe)CO_3$]; (5) a phase (or mixture of phases) containing variable proportions of Fe, Si, Cu, and S; and (6) iron oxide with U. ZVI is typically composed of more than 98 weight % Fe; often contains several tenths of a percent each of Mn, Si, Zn, and Cu; and has less than 0.05 weight % U and V. The aluminosilicate binder contains large concentrations of Si, Fe, and Al and low concentrations of Ca and Na. Magnetite contains up to 3.02 weight % Mn, U concentrations that are less than the detection limit, and low concentrations of most of the other

Table VI

Summary of X-ray Diffraction Analysis of Samples from Permeable Reactive Barrier C near Durango, Colorado[a]

Sample location	Calcite	Goethite	Magnetite	Ferrihydrite
Top of quadrant 1	+		+	X
Middle of quadrant 1	+	X	X	
Bottom of quadrant 1	o	?	?	
Top of quadrant 4	X	X	X	
Middle of quadrant 4	X		+	
Bottom of quadrant 4	+	+	+	

[a]Major (X), moderate (+), minor (o), uncertain (?) because of line interference but likely of minor concentration.

constituents. Most samples contain some goethite (Table VI). Electron microprobe results indicate that goethite contains 37.03 to 66.89 weight % Fe, to 1.64 weight % S, to 2.75 weight % Cu, and U concentrations that are less than the detection limit in all analyses.

Electron microprobe analyses of 28 carbonate minerals indicate compositions ranging from nearly end-member $CaCO_3$ to about 50 atom % Fe (Fig. 7).

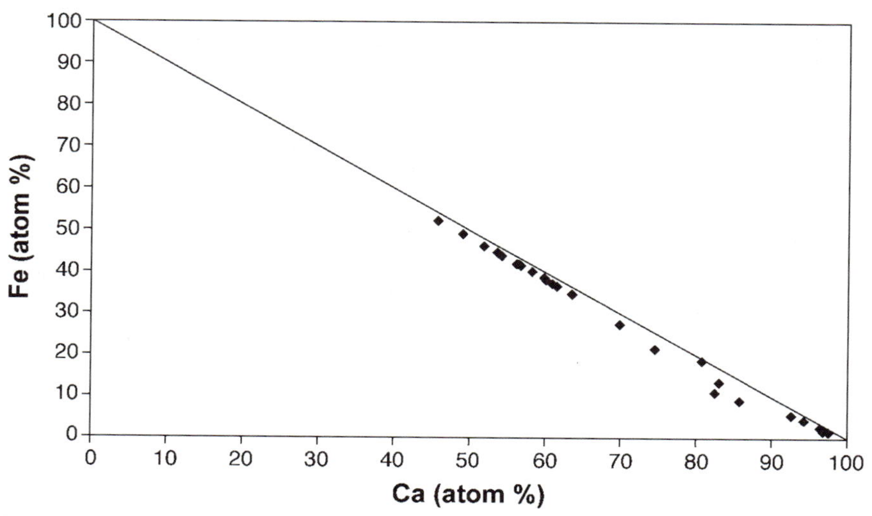

Figure 7 Fe–Ca composition (atom %) of carbonate minerals identified in PRB C with an electron microprobe.

The carbonates had V concentrations to 0.99 weight %, Mn to 3.01 weight %, Mg to 0.34 weight %, As to 1.25 weight %, and Cu to 2.05 weight %. Vanadium concentrations in the carbonate minerals correlated with Fe concentrations. Uranium concentrations are typically low in the carbonate phases, but four samples had U concentrations of more than 0.1 weight %. The Fe–Si–Cu–S-rich phase is possibly a carbonate mineral that contains variable proportions of Fe, Si, Cu, and S and is associated with a phase containing Fe, S, and Cu.

An iron oxide-like material with a porous texture was also identified that contained U concentrations ranging from 1.01 to 10.62 weight % and V concentrations to 7.83 weight %. This material also had high (excess of 1 weight %) and widely variable concentrations of Al, Ca, and Si and may be a mixture of submicron-sized phases (Fig. 8).

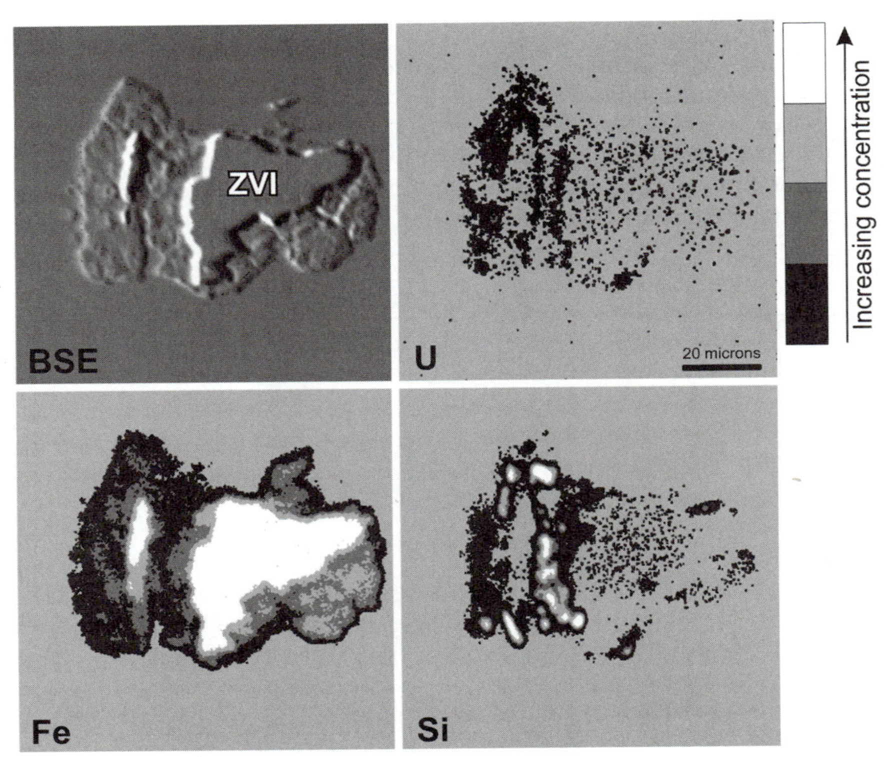

Figure 8 Backscatter electron image (BSE) and uranium (U), iron (Fe), and silicon (Si) maps analyzed by an electron microprobe of a corroded zero valent iron (ZVI) grain from the upper surface of PRB C. An Fe oxide-like mineral coats the ZVI and has a heterogeneous distribution of U and Si that may indicate discrete domains smaller than the beam diameter of 5 μm.

Complex intergrowths of carbonate, iron oxide, and ZVI were observed on the margins of ZVI particles where the material was corroded. Particles that had completely reacted to iron oxide or carbonate were rare. Many composite particles of ZVI foam plate were completely unreacted.

C. PRB D—Baffled Tank with Steel Wool

PRB D operated in 1999 and treated 112 m^3 of contaminated water. The design of PRB D was identical to PRB C except that it contained steel wool instead of ZVI foam plates (Table I). PRB D was sampled twice; analytical results for November 1998 samples are representative (Table VII). The steel wool in PRB D removed more than 90% of the As, U, and V and more than 75% of the Cd, Se, and Zn. The ^{226}Ra concentration decreased by 54.8%, and the Cu concentration decreased by more than 64.3%. Values of pH increased from 6.88 to 7.59, and Eh values decreased from 398 to 20 mV. Concentrations of some major ions (e.g., Ca, Mg, Sr, and Si) also decreased. The Fe concentration increased to 115 mg/liter and alkalinity increased slightly.

D. PRB E—Baffled Tank with Granular ZVI

In July 1999, the ZVI foam plates in PRB C were replaced with granular ZVI, the drainpipe was removed, and the PRB was renamed PRB E. PRB E was sampled only once, in April 2000, while it was flowing at 4.2 liter/min (residence time about 9 h). More than 97% of the As, Cd, U, V, and Zn was removed from the groundwater by granular ZVI (Table VII). More than 81.3% of the Cr, 88.1% of the Mo, 59.6% of the Se, and 44.3% of the Mn were also removed. The pH value increased from 6.51 to 7.28, and the Eh value decreased from 408 to 6 mV. Concentrations of some major constituents (e.g., Ca, Sr, and Si) also decreased. The Fe concentration increased from 0.034 to 110 mg/liter, and SO_4 and alkalinity concentrations increased slightly.

V. DISCUSSION

Data collected during this study were useful for comparing the efficiency of ZVI for contaminant removal, determining the nature of mineral precipitation, estimating relative rates of contaminant uptake, determining if constituents had been leached from the ZVI, and evaluating various PRB designs.

Table VII

Analytical Results for Water Samples from PRB D and PRB E near Durango, Colorado[a]

	PRB D (May 1999) Flow rate = 2.5 liter/min			PRB E (April 2000) Flow rate = 4.2 liter/min		
Constituent	Influent	Effluent	% Removed	Influent	Effluent	% Removed
pH	6.88	7.59	na	6.51	7.28	na
Eh (mV)	398	20	na	408	6	na
T (°C)	10.6	11.8	na	8.3	8.2	na
EC (μS/cm)	2700	2300	na	3780	3700	na
DO (mg/liter)	9.1	0.1	98.9	0.44	0.35	20.5
Alk (mg/liter)	590	644	na	561	620	na
As (μg/liter)	202	13.9	93.1	186	4.1	97.8
Ba (μg/liter)	14.6	<4	>72.6	18.9	20.7	na
Ca (mg/liter)	636	568	10.7	612	566	7.5
Cd (μg/liter)	8.7	<1	>88.5	9.9	<0.2	>98.0
Cl (mg/liter)	132	120	9.1	103	111	na
Cr (μg/liter)	9.3	5.3	43.0	14.4	<2.7	>81.3
Cu (μg/liter)	14	<5	>64.3	4.3	<3.1	>27.9
Fe (mg/liter)	0.047	115	na	0.034	110	na
K (mg/liter)	15.7	15.1	3.8	16.5	16.4	0.6
Mg (mg/liter)	81.9	74.8	8.7	75.8	74.7	1.5
Mn (μg/liter)	6140	4010	34.7	5850	3260	44.3
Mo (μg/liter)	1150	472	59.0	1180	140	88.1
Na (mg/liter)	314	292	7.0	302	298	1.3
NH_4 (μg/liter)	6.3	38.4	na	168	279	na
NO_3 (μg/liter)	232	822	na	1730	1370	20.8
^{226}Ra (pCi/liter)	5.29	2.39	54.8	nd	nd	nd
Se (μg/liter)	237	41.7	82.4	337	136	59.6
Si (mg/liter)	25.9	23.2	10.4	24.9	21.6	13.3
SO_4 (mg/liter)	1720	1670	2.9	1470	1530	na
Sr (mg/liter)	3.07	2.55	16.9	3.39	3.04	10.3
TDS (mg/liter)	3580	3320	7.3	3420	3280	4.1
TOC (mg/liter)	14.2	15.3	na	5.1	5.6	na
U (μg/liter)	7430	738	90.1	5540	0.4	>99.9
V (μg/liter)	9120	495	94.6	8800	1.4	>99.9
Zn (μg/liter)	941	223	76.3	1600	16.3	99.0

[a]EC, electrical conductivity; DO, dissolved oxygen; Alk, alkalinity in milligrams per liter of $CaCO_3$; TDS, total dissolved solids; TOC, total organic carbon; na, not applicable; nd, no data; mV, millivolts; liters/min, liters per minute; μS/cm, microsiemens per centimeter; mg/liter, milligrams per liter; μg/l, micrograms per liter; pCi/liter, picocuries per liter.

A. Efficiency of Contaminant Removal in PRBs

Significant contaminant removals were achieved in all PRBs (Tables II and VII). The changes in chemistry of the contaminated groundwater caused by interaction with the ZVI foam plates in PRB C are consistent (thermodynamically permissible) with precipitation of UO_2, V_2O_3, As_2S_3, ZnS, $FeSe_2$, $FeMoO_4$, $CaCO_3$, $Fe(OH)_2$, $MnCO_3$, and FeS (Morrison *et al.*, 2002). Other chemical mechanisms are also possible, including coprecipitation and adsorption. Some of the Mg, Mn, ^{226}Ra, and Sr was likely coprecipitated with calcite.

The NO_3 concentration decreased and the NH_4 concentration increased in PRBs C and E, suggesting that some of the nitrate was reduced to NH_4. In PRB C, NO_3 concentrations decreased slowly for the first 100 pore volumes and then remained low (Dwyer and Marozas, 1997; Morrison *et al.*, 2002). This behavior may be attributable to the development of nitrate-reducing bacteria in PRB C. In contrast, both NO_3 and NH_4 concentrations decreased in PRB B (Table II) and increased in PRB D (Table VII).

Concentrations of As, Mo, ^{226}Ra, Se, U, V, and Zn are highest in quadrant 1 of PRB C, and all these constituents but Se are lowest in quadrant 3 (Table III). Constituents were removed from the inflowing groundwater as it passed through quadrant 1; water flowed from quadrant 1 into the bottom of quadrant 2 and through the drainpipe into the bottom of quadrant 4, bypassing quadrant 3. Groundwater flowed up through quadrant 4 and out the exit pipe. Some water probably leaked across the seal at the top of the partition between quadrants 1 and 2 to cause high levels of some constituents (e.g., U and As; Fig. 6) at the top of quadrant 2.

B. Mineral Precipitation in PRB C

Precipitation of minerals will cause a reduction in permeability, leading to an increased head difference. A maximum of about 0.6 m of head was available to drive water through PRB C. As minerals and gases built up in the system, flow gradually decreased and ceased after about 3 years. Other than nitrogen, which can volatilize, constituents that decreased in concentration between influent and effluent water samples must continue to reside in the solid materials of the PRB.

Decreases in concentrations of TDS and electrical conductivity between influent and effluent samples indicate that salts had precipitated in all the PRBs (Tables II and VII). The composition of precipitates was estimated based on the changes in water chemistry (Fig. 9). Precipitation is dominated

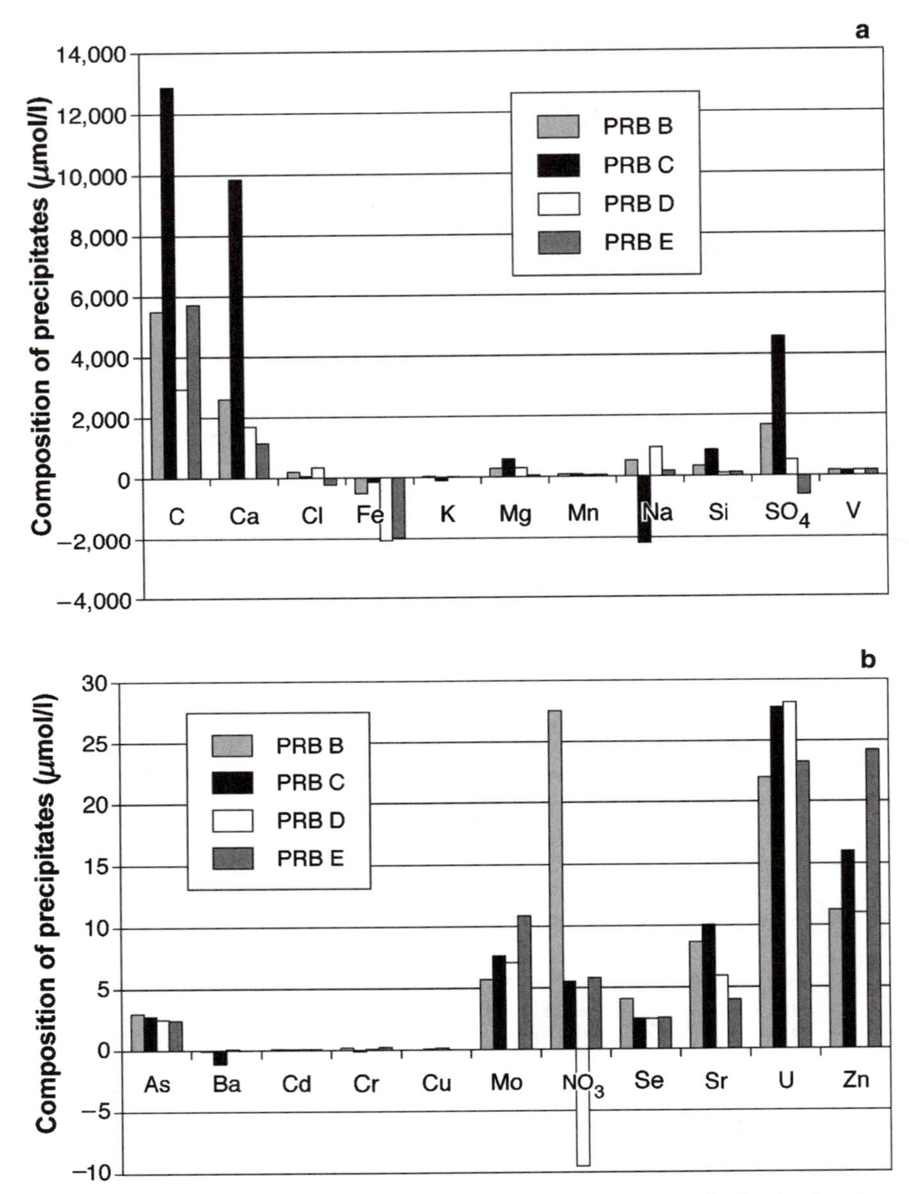

Figure 9 Composition of materials precipitated in the four permeable reactive barriers based on the difference between influent and effluent water compositions. Data are from Tables II and VII. Composition is expressed as micromoles of a constituent per liter of treated water. (a) Major constituents (b) minor constituents. Carbon was calculated from alkalinity.

by Ca and carbon (C), consistent with the formation of calcite ($CaCO_3$); calcite was identified in PRB C with X-ray diffraction. Carbonate mineral precipitation is caused by an increase in pH values as ZVI corrodes. Insufficient Ca is present in all four PRBs to support all the C as calcite (Fig. 9a). Calcium and C concentrations, which were determined by the chemical analysis of solids, show a linear relationship with a small excess of C (Fig. 10). Apparently, some of the carbonate combined with Fe released from ZVI to form Ca–Fe solid solution, as evidenced by a wide range of Ca and Fe compositions detected with the electron microprobe (Fig. 7). Some S also precipitated, probably caused by the reduction of SO_4 to sulfide (Fig. 9a; Table III). Vanadium precipitated in all four PRBs in nearly equivalent amounts ranging from 159 to 173 μmol/liter (Fig. 9a). Significant amounts of As, Mo, Se, Si, U, and Zn were deposited in the PRBs (Fig. 9b). Some of the Si is associated with U-rich corrosion products of ZVI (Fig. 8).

C. Relative Rates of Contaminant Uptake

The rate at which contaminants are removed from aqueous solutions is an important factor in evaluating the feasibility of PRBs. If uptake rates are fast,

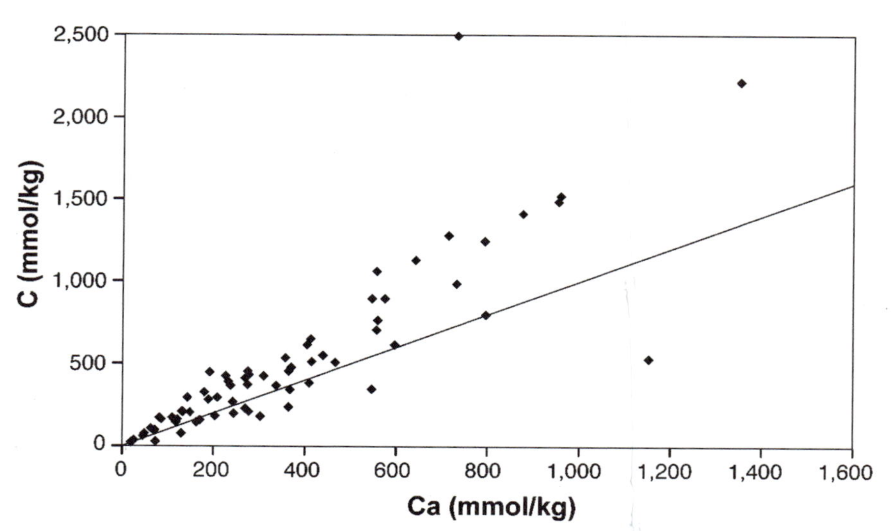

Figure 10 Relationship of calcium to carbon determined by the chemical analysis of solids in samples from permeable reactive barrier C. The line indicates equal molar concentrations.

less ZVI (at a lower cost) is needed. Laboratory studies have shown that U and V residence times are relatively short with nearly complete removal in less than 6 min; Mo removal rates are slower (see Chapter 13). The distribution of constituents in PRB C solids indicates the relative rate of uptake by ZVI. Higher concentrations of constituents in samples collected close to the inlet suggest relatively fast uptake rates, whereas slower rates are indicated for constituents that are distributed over longer distances.

The high concentrations of As (Fig. 6) and Se in the upper portion of quadrant 1 suggest that the uptake kinetics for these constituents are relatively fast (Table III). Uranium (Fig. 6) and V are concentrated near the inlet but are also distributed throughout the PRB, suggesting that their uptake reactions were slower than the reactions for As and Se. The distributions of Mo and Zn are similar to U and V and suggest similar rates of uptake.

Calcium was most concentrated in the middle portion of quadrant 1. This distribution may result from reaction rates for carbonate precipitation, which are slower than uptake reactions for U and V. This finding is consistent with observations from column tests performed for the Monticello, Utah, PRB that resulted in calcium carbonate deposited downgradient from the U deposition (see Chapter 13).

D. Leaching from ZVI Foam Plates

Some reactive materials contain compounds that can be harmful if they are dissolved in groundwater. In contrast, the dissolution of ZVI can be beneficial by providing additional pore space to maintain adequate flow. The distribution of chemicals in the solids of PRB C provides information on the extent of leaching.

The Cr concentration was higher in the original ZVI foam plates than in the ZVI foam plates sampled in PRB C, indicating that a small amount of Cr had been leached (Table III). The decrease in Cr concentrations in the ZVI foam plates is consistent with an effluent concentration that was higher than the influent concentration for the May 1999 water sampling (Table II). In the other three PRBs, however, effluent concentrations of Cr were lower than influent concentrations, indicating that Cr was precipitated by the ZVI (Tables II and VII). Laboratory studies elsewhere have shown that ZVI is effective in reducing Cr concentrations in groundwater (Blowes *et al.*, 1997a);

ZVI has also been used in full-scale PRBs to remediate Cr-contaminated groundwater (Blowes *et al.*, 1997b).

Manganese concentrations in the original ZVI foam plates were higher than Mn concentrations in the plates removed from PRB C (Table III). Mn concentrations in the effluent, however, were lower than the influent concentrations in May 1999 (Table II), indicating that the ZVI foam plates were concentrating Mn at that time. Perhaps a portion of the Mn was associated with highly reactive particles and was released from the ZVI early in the operation of the PRB.

Aluminum concentrations were lower in samples from all four quadrants than in the original material and were highest in samples from quadrant 3 (Table III). This distribution indicates leaching of some of the aluminosilicate binder. The aluminosilicate binder also contains Na. The Na concentration is also highest in the original ZVI foam plates and in quadrant 3, suggesting that Na also leached from the binder. The amount of leaching is minimal, and concentrations of Al and Na in the effluent samples were relatively low.

Dissolved Fe concentrations were always higher in effluent samples than in influent samples, indicating that ZVI was dissolving (Fig. 9a). Concentrations of Fe in effluent samples ranged from 7.69 mg/liter in PRB C to 115 mg/liter in PRB D (Tables II and VII). The lower Fe concentration in the effluent from PRB C may be a result of a slower flow rate that provided time for Fe to precipitate as hydroxide or sulfide compounds. The concentration of dissolved Fe in effluent samples indicates the minimum amount of Fe released from the ZVI; additional Fe may release but become fixed by minerals in the PRB. Dissolved Fe can be an aesthetic problem because it causes red coloration in the water and will prevent some uses of the water. Iron concentrations exceeding 11 mg/liter can result in adverse health effects if used as a drinking water supply.

E. Comparison of PRB Designs

Overall, ZVI foam plates and the granular ZVI appear to be more effective than steel wool on an equal volume basis. Although PRB C was flowing at a slower rate (0.3 liter/min) than PRB D (2.5 liter/min) during the samplings reported in Tables II and VII, data for U, Mo, and NO_3 for the previous 3 years indicate that these contaminants were removed more effectively by PRB

C even at higher flow rates (Dwyer and Marozas, 1997). Granular ZVI used in PRB E was effective for removing Mo, U, V, and Zn even at the higher flow rate of 4.2 liters/min.

The leach field design used in PRB B was effective in lowering concentrations of most contaminants but not as effective as baffled tanks. Replacement of reactive material is easier in baffled tanks than in the other PRBs and no advantages are apparent for the leach field design.

VI. LESSONS LEARNED AND CONCLUSIONS

PRBs containing ZVI are effective in removing U and associated contaminants from groundwater that has been in contact with U mill tailings. PRB C at Durango operated from May 1996 through May 1999 with average U concentrations decreasing from about 6,600 μg/liter in influent samples to less than 10 μg/liter in effluent samples. The high concentrations of As, Cr, Se, U, and V measured in the ZVI foam plates (up to 2.13 weight %) are consistent with the substantial decreases in contaminant concentrations between influent and effluent water.

The cost benefit of a PRB is directly related to the length of time that it will operate effectively. Mineral precipitation caused by the interaction of contaminated water with ZVI can reduce the permeability of the ZVI. As permeability is reduced, preferential flow paths develop, which can lead to early failure of the PRB. Mineral precipitation in the ZVI is expected because corrosion causes an increase in the pH value as a result of the reduction of aqueous protons (see Chapter 1). The rise in the pH value causes precipitation of carbonate minerals, such as calcite and siderite. The location of mineral precipitation within the PRB depends on the flow rate and the precipitation reaction rates. Carbonate minerals in PRB C were located throughout the PRB and not just at the fluid/ZVI interface. Hydrogen gas, which is a by-product of corrosion, and methane gas, which is produced by microbes, may also cause some reduction in permeability.

Concentrations of some constituents (Al, Cr, Fe, Mn, and Na) appear to have leached from the ZVI foam plates, but concentrations of these constituents in the effluent water are not likely to exceed regulatory standards.

Overall, the collection drain and PRB system effectively treated U-contaminated groundwater from a disposal cell containing mill tailings and could efficiently treat U-contaminated groundwater at other sites with a favorable hydraulic gradient. The baffled tank system used at the Durango site provides flexibility because the ZVI can be replaced easier than in permanent *in situ* PRBs. Because the reactive material was containerized, it took only about 8 h to replace the ZVI in PRB C and to reinitiate flow.

The U that accumulates in the PRB has been removed from its daughter isotopes and its low-energy γ emission is not a health threat. However, β emissions due to protactinium-234 are strong enough to require special handling and disposal. Spent ZVI from the Durango site is disposed of at a repository near Grand Junction, Colorado, operated by the U.S. Department of Energy (DOE) Long-Term Surveillance and Maintenance Program. Disposal options for the reactive material must be considered when installing new PRBs.

Most PRBs rely on natural flow gradients to move groundwater and do not have a metering system to gauge the flux. The amount of groundwater being treated is extremely difficult to quantify in these systems. As demonstrated in this study, the composition of some constituents in the ZVI can be compared with the decrease in concentrations in the groundwater resulting from reaction with ZVI in the PRB to calculate the amount of water treated. Use of calculations from solid-phase compositions may be the most reliable method for determining groundwater flux given the large uncertainty in making subsurface flux measurements with existing tools.

ACKNOWLEDGMENTS

This work was funded by the DOE UMTRA Ground Water Project under the auspices of the DOE Grand Junction Office. The work was conducted in cooperation with the DOE Long-Term Surveillance and Maintenance Program that generously delayed final construction on the site to support longevity testing of the PRBs. We thank Diane Marozas (Sandia National Laboratories/New Mexico) for the inspiration to install the PRBs at Durango. We appreciate many hours of field work undertaken by Sarah Morris and Greg Smith (MACTEC Environmental Restoration Services) and by Bruce Thomson (University of New Mexico) and his graduate students. Our thanks also to Oak Ridge National Laboratory personnel (Bob Schlosser, Doug Pickering, and Nic Korte) and MSE, Inc., personnel (Leah Matheson and Will Goldberg) for assistance with sampling of solid materials in PRB C. The project benefited from skilled mineralogical examinations by Erich Petersen (University of Utah).

REFERENCES

ASTM (1992). Standard practice for total digestion of sediment samples for chemical analysis of various metals. Designation D4698–92. *American Society for Testing and Materials*, 585–589.

Blowes, D. W., Ptacek, C. J., and Jambor, J. L. (1997a). In-situ remediation of Cr(VI)-contaminated groundwater using permeable reactive walls: Laboratory studies. *Environ. Sci. Technol.* **31**, 3348–3357.

Blowes, D. W., Puls, R. W., Bennett, T. A., Gillham, R. W., Hanton-Fong, C. J., and Ptacek, C. J. (1997b). In-situ porous reactive wall for treatment of Cr(VI) and trichloroethylene in groundwater. *In* "International Containment Technology Conference Proceedings," pp. 851–857, St. Petersburg, FL.

Cantrell, K. J., Kaplan, D. I., and Wietsma, T. W. (1995). Zero-valent iron for the in situ remediation of selected metals in groundwater. *J. Hazard. Mater.* **42**, 201–212.

Dwyer, B. P., and Marozas, D. C. (1997). In situ remediation of uranium contaminated groundwater. *In* "International Containment Technology Conference Proceedings," pp. 844–850 St. Petersburg, Florida.

Federal Register (1995). Title 40 *Code of Federal Regulations* Part 192, "Health and Environmental Protection Standards for Uranium and Thorium Mill Tailings." **53**, 2854–2871.

Fiedor, J. N., Bostick, W. D., Jarabek, R. J., and Farrell, J. (1998). Understanding the mechanism of uranium removal from groundwater by zero-valent iron using x-ray photoelectron spectroscopy. *Environ. Sci. Technol.* **32**, 1466–1473.

Gillham, R. W., Blowes, D. W., Ptacek, C. J., and O'Hannesin, S. F. (1994). Use of zero-valent metals in in-situ remediation of contaminated ground water. *In* "In-Situ Remediation: Scientific Basis for Current and Future Technologies," (G. W., Gee and N. R., Wing, eds.), pp. 913–930, Battelle Press, Columbus, OH.

Gu, B., Liang, L., Dickey, M. J., Yin, X., and Dai, S. (1998). Reductive precipitation of uranium (VI) by zero-valent iron. *Environ. Sci. Technol.* **32**, 3366–3373.

Hach (1989). Oxygen, dissolved, HR; proprietary method, *In* "*Water Analysis Handook*," pp. 442–444. Hach Company, Loveland, CO.

Matheson, L. J., and Goldberg, W. C. (1999). Spectroscopic studies to determine uranium speciation in ZVI permeable reactive barrier materials from the Oak Ridge Reservation, Y-12 plant site and Durango, CO, PeRT wall C. *In* "Supplement to EOS, Transactions 1999 Fall Meeting," p. F366. American Geophysical Union, Washington, DC.

Mitchell, P. (1997). Reactive walls. "*Industrial Wastewater*," July/August, pp. 33–37.

Morrison, S. J., Metzler, D. R., and Carpenter, C. E. (2001). Uranium precipitation in a permeable reactive barrier by progressive irreversible dissolution of zero-valent iron. *Environ. Sci. Technol.* **35**, 385–390.

Morrison, S. J., Metzler, D.R., and Dwyer, B. P. (2002). Removal of As, Mn, Mo, Se, U, V, and Zn from groundwater by zero-valent iron in a passive treatment cell; reaction progress modeling. *J. Contam. Hydrol.* **56**, 99–116.

National Academy of Sciences (1994). "Alternatives for ground water cleanup," Report of the National Academy of Sciences Committee on Ground Water Cleanup Alternatives, National Academy Press, Washington, DC.

Chapter 16

In Situ Reduction of Chromium-Contaminated Groundwater, Soils, and Sediments by Sodium Dithionite

Cynthia J. Paul, Faruque A. Khan, and Robert W. Puls
U.S. Environmental Protection Agency, Office of Research and Development, National Risk Management Research Laboratory, Ada, Oklahoma 74820

Laboratory studies were conducted to characterize the extent of chromium (Cr) contamination in the groundwater and underlying soils and sediments of a chrome-plating shop at the United States Coast Guard (USCG) Support Center near Elizabeth City, North Carolina. Most of the mobile hexavalent Cr [Cr(VI)] is present in the vadose zone just above the capillary fringe and in the upper zone of the aquifer under oxidizing conditions. The reduction of Cr(VI) to trivalent Cr [Cr(III)] through *in situ* redox manipulation (ISRM) in the presence of a reductant is an innovative technique for the remediation of Cr(VI)-contaminated sediments and groundwater. Three liquid reductants (sodium dithionite, *l*-ascorbic acid, and free hydroxylamine) were evaluated in laboratory studies to determine their ability to detoxify and immobilize Cr(VI) *in situ*. Sodium dithionite proved to be the most effective at reducing Cr(VI) to Cr(III) with the least adverse side effects. A field-scale pilot study was conducted in July 1999 using sodium dithionite to evaluate the effectiveness of reducing Cr(VI) *in situ*. Postinjection monitoring results show aqueous Cr(VI) concentrations decreased from up to 5 to <0.01 mg/liter in all wells within the 1-m cylindrical treatment zone. Continued Cr(VI) reduction was seen 48 weeks postinjection, indicating that long-term reduction is occurring. Analyses of postexperimental sediment cores indicate that nonmobile, reduced structural iron (Fe) within the silicate minerals present in the treated sediments may be contributing to the long-term remediation of Cr(VI).

I. INTRODUCTION AND BACKGROUND

A. Chromium Oxidation, Mobility, and Toxicity

While naturally occurring chromium (Cr) has been found at elevated levels in numerous locations (Palmer and Wittbrodt, 1991), Cr contamination of subsurface soils and groundwater has increased greatly in the past several years due to wastes produced by Cr electroplating, leather tanning, steelworks, and other industrial uses (DOE, 1992; USEPA, 1989). In subsurface systems, Cr exists primarily in trivalent [Cr(III)] and hexavelent [Cr(VI)] oxidation states (Richard and Bourg, 1991). Aqueous Cr(VI) is present in anionic forms and is generally soluble over a wide pH range (Chen and Hao, 1998). The oxidized species is much more mobile, soluble, and toxic than the reduced species, Cr(III), which is relatively insoluble (Palmer and Wittbrodt, 1991; Palmer and Puls, 1994; Pettine *et al.*, 1998). Cr mobility is controlled by several factors, including pH, dissolved inorganic carbon (DIC), sulfate (SO_4^{2-}), and phosphate (PO_4). Cr(III) adsorption increases as pH increases (Korte *et al.*, 1976; Rai *et al.*, 1987), whereas Cr(VI) adsorption increases with decreasing pH (Bartlett and James, 1979; Davis and Leckie, 1980). Studies have shown decreased CrO_4^{2-} adsorption and increased mobility with increasing dissolved inorganic carbon, sulfate, and phosphate (Stollenwerk and Grove, 1985; Zachara *et al.*, 1988, 1989) due to competition for binding sites.

Cr(VI) is considered teratogenic (Abbasi and Soni, 1984), mutagenic (Bonatti *et al.*, 1976; Paschin *et al.*, 1983), and carcinogenic (Waterhouse, 1975; Ono, 1988). According to the International Agency for research on Cancer (IARC), Cr(VI) is considered a powerful carcinogen and its presence in water is cause for major concern. The national primary drinking water standard set by the USEPA for total Cr is 100 ppb; however, the state of California has lowered their maximum contaminant level (MCL) to 50 ppb. No separate drinking water standard for Cr(VI) has been established.

B. Mechanisms of Chromium Reduction

Reduction of Cr(VI) to Cr(III) represents an effective mechanism for detoxification of Cr(VI) in contaminated environments (Chen and Hao,

1998). According to Puls *et al.* (1994), the reduction of Cr from Cr(VI) to Cr(III) can only occur in the presence of another redox couple that provides the required three electrons. The ferrous iron/ferric iron [Fe(II)/Fe(III)] redox couple exists in natural subsurface systems, as do others, which results in reduction (Richard and Bourg, 1991) and subsequent precipitation as (Fe, Cr) $(OH)_3$ (Puls *et al.*, 1994). Cr(VI) is a strong oxidant and is reduced in the presence of electron donors such as Fe(II) minerals, reduced sulfur, and soil organic matter (Palmer and Puls, 1994). The reduction of Cr(VI) by Fe(II) is illustrated as

$$HCrO_4^- + 3\,Fe^{2+} + 7\,H^+ \rightleftharpoons Cr^{3+} + 3\,Fe^{3+} + 4\,H_2O. \tag{1}$$

Reaction kinetics are very fast, going to completion within 5 min even in the presence of dissolved oxygen (DO) (Eary and Rai, 1988). At $pH > 4$, Cr(III) precipitates with Fe(III) in a solid solution with the general composition $Cr_xFe_{1-x}(OH)_3$ (Sass and Rai, 1987; Amonette and Rai, 1990). Numerous minerals containing Fe(II) exist for the potential reduction of Cr(VI) in the environment (Palmer and Puls, 1994) that increase the potential for natural attenuation of Cr(VI)-contaminated sites. Potential reductants of Cr(VI) include adsorbed ions, mineral constituents, aqueous species, and organic matter (Palmer and Puls, 1994). Research has shown that Fe oxides (i.e., hematite) and Fe-containing silicates (i.e., nontronite) can reduce Cr(VI) (White and Hochella, 1989; Eary and Rai, 1987). White and Yee (1985) proposed that Fe(III) is reduced to Fe(II) at the mineral surface. Cr(VI) in the solution is then reduced by Fe(II), resulting in the production of more Fe(III), thereby establishing a continuous cycle.

Other mechanisms exist for Cr(VI) reduction or removal from solution. Adsorption mechanisms include those discussed earlier, activated carbon (Han *et al.*, 2000), and various soil components, including Fe and aluminum (Al) oxides and hydroxides (Ainsworth *et al.*, 1989; Schultz *et al.*, 1987), kaolinite (Griffin *et al.*, 1977), and sphagnum peat moss (Sharma and Forster, 1993). Research has also been conducted to evaluate microbial reduction as a low-cost mechanism for the reduction of Cr(VI) in the subsurface (Martin *et al.*, 1994). Some bacterial species may uptake and accumulate a significant amount of metal ions, resulting in the transfer of metals from a contaminated matrix to biomass (Smith *et al.*, 1994). Another potential mechanism for the microbial removal of metal is a dissimilatory reduction of metal oxyanions such as CrO_4^{2-} (Lovely, 1993). Palmer and Puls (1994) state that both aerobic and anaerobic microbial reduction can occur; however, anaerobic reduction is more common. Chen and Hao (1998) showed that reduction of Cr(VI) to Cr(III) in a biological system generally occurs within a neutral pH system when compared to chemical reduction methods.

C. Chemical Reduction and Immobilization

Abiotic reduction by manipulation of the oxidation–reduction (redox) status of the vadose and saturated zones is an approach for *in situ* remediation of a redox-sensitive contaminant like Cr(VI). A spatially fixed zone is created in the aquifer via reduction of structural Fe in the aquifer where the redox environment is substantially more reducing than the rest of the aquifer (Amonette *et al.*, 1994). As groundwater enters the reducing zone or permeable reactive barrier (PRB), redox-sensitive constituents [e.g., Cr(VI)] are reduced (see Chapter 6). Reduction of mobile and toxic Cr(VI) to an insoluble and less toxic Cr(III) precipitate has been demonstrated to occur in the aqueous system (Palmer and Puls, 1994; Amonette *et al.*, 1994). Under alkaline to slightly acidic conditions, Cr(III) is immobilized in the soil because it precipitates as a fairly insoluble hydroxide (Palmer and Puls, 1994). This "natural attenuation" of Cr(VI) is of interest because it suggests that strict water quality standards do not have to be attained everywhere within a site. Where "natural attenuation" is insufficient to prevent environmental degradation, it may be possible to rejuvenate the "natural attenuation" capacity of an aquifer or the vadose zone by incorporating a reductant into these zones.

In previous studies, ISRM techniques were used for plume treatment, where a vertical reducing zone was created and the contaminant plume would pass through, resulting in treatment (see Chapter 6). In this study, a modification of this technique was used to deal with a source zone located in unsaturated soils and the capillary fringe area or upper part of an aquifer that was contributing to the Cr source.

Fruchter *et al.* (2000), Istok *et al.* (1999), and Amonette *et al.* (1994) described the development of an ISRM method. This method utilizes naturally occurring Fe in soils and sediments to create a spatially fixed reducing zone or permeable barrier within the aquifer (Amonette *et al.*, 1994) by injecting chemicals into the aquifer that can reduce Fe(III) to Fe(II). Redox-sensitive contaminants, such as Cr(VI), are immobilized or precipitated as they migrate through the reducing zone. This reaction is illustrated in Eq. (2) (Fruchter *et al.*, 2000), where CrO_4^{2-} is precipitated as a solid hydroxide [e.g., $Cr(OH)_3$].

$$HCrO_4^-(aq) + 3\,Fe^{2+}(s) + 8\,H_2O \rightleftharpoons Fe_3Cr(OH)_{12}(s) + 5\,H^+. \qquad (2)$$

Naturally occurring Fe in minerals and oxide coatings is the most abundant redox-sensitive element in soils and sediments (Fruchter *et al.*, 2000). According to Amonette *et al.* (1994) mineral phases with the highest specific surface areas, particularly Fe oxyhydroxides and Fe-bearing silicate clays, have the most accessible forms of Fe. Oxide phases dissolve when the Fe is reduced and may precipitate as siderite ($FeCO_3$) if enough CO_3 is present. If not, the Fe becomes soluble and will migrate within the aquifer until it is adsorbed or reprecipitated as Fe(III). In order to create a spatially fixed reducing zone in an aquifer, a reagent capable of reducing the structural Fe in clay layer silicates is required (Amonette *et al.*, 1994). The reagent must also be relatively nontoxic in both its original and reacted forms and create few side effects in the subsurface. Reduced sulfur species vary in their reactivity and toxicities, but generally yield safe oxidized products such as sulfate (Amonette *et al.*, 1994).

In preliminary laboratory studies conducted at the Robert S. Kerr Environmental Research Center (RSKERC) by Shiau and LaFever (1997), several reagents (sodium borohydride, *l*-ascorbic acid, hydroxylamine, and sodium dithionite) were evaluated for their ability to reduce Cr(VI) on contaminated soils and sediments. Although sodium borohydride reduced Cr(VI) to Cr(III), it decomposed rapidly and produced large amounts of H_2, even under basic conditions. Results showed *l*-ascorbic acid to be the most effective at reducing Cr(VI); however, only a small amount was reduced to Cr(III). O'Brien and Woodbridge (1997) and Goodgame and Joy (1987) showed that the reduction of Cr(VI) to Cr(III) using *l*-ascorbic acid is very slow, with intermediaries of Cr(V) and Cr(IV) being produced. Hydroxylamine and sodium dithionite reduced more than 95% of Cr(VI) present on contaminated soils to Cr(III) in batch studies (Shaiu and LaFever, 1997). Amonette *et al.* (1994) conducted a laboratory study to evaluate dithionite ion ($S_2O_4^{2-}$), as a potential reagent for use in the creation of an ISRM. The injected dithionite reacts with structural Fe in oxyhydroxide and Fe-bearing layer silicate mineral phases, thereby reducing the Fe(III) to Fe(II). The overall reaction using dithionite is described in Eq. (3) (Amonette *et al.*, 1994; Istok *et al.*, 1999):

$$S_2O_4^{2-}(aq) + 2\,Fe(III)(s) + 2\,H_2O \rightleftharpoons 2\,SO_3^{2-}(aq) + 2\,Fe(II)(s) + 4\,H^+. \quad (3)$$

In addition to the reduction reaction, dithionite undergoes a disproportionation reaction that yields thiosulfate ($S_2O_3^{2-}$) and bisulfate (HSO_4^{2-}) anions. Anions such as $S_2O_3^{2-}$ and HCO_3^- (where bicarbonate is used as a buffer) will be precipitated with calcium (Ca) present in the native aquifer materials. Also, some small portions of these anions may oxidize to yield sulfate. Results of both laboratory and field-based studies using dithionite have been published that show dithionite to be effective as a reductant without the adverse side effects of other Fe-reducing reagents (Fruchter *et al.*, 2000; Istok *et al.*, 1999; Amonette *et al.*, 1994).

D. Chromium Reoxidation

Very few mechanisms exist for the reoxidation of Cr(III) back to Cr(VI) (Palmer and Puls, 1994). Dissolved Oxygen (DO) and manganese dioxides (MnO_2) are the only two constituents in the environment known to oxidize Cr(III) to Cr(VI) (Palmer and Puls, 1994; Eary and Rai, 1987). Based on studies conducted by Schroeder and Lee (1975) and Eary and Rai (1987), the oxidation of Cr(III) by DO is not likely. While the oxidation of Cr(III) to Cr(VI) has been observed in soils and sediments, it is relatively slow (Palmer and Puls, 1994). Chromium oxidation appears to occur mainly at surface coatings, crack deposits, or in finely disseminated grains (Richard and Bourg, 1991; Bartlett and James, 1979). An increase in the rate and amount of Cr(III) oxidation has been observed as pH decreases and surface area to solution volume increases (Palmer and Puls, 1994). A correlation has been observed between the amount of Cr(III) oxidized by soils and the amount of hydroquinone-reduced manganese in soils (Bartlett and James, 1979). They suggested that oxidation of Cr(III) is the result of an interaction with manganese dioxides, which also has been verified by others (Fendorf and Zasoski, 1992; Riser and Bailey, 1992; Eary and Rai, 1987). Eary and Rai (1988) and Puls and Powell (1992) proposed that the reduction of Cr(VI) to Cr(III) consists of the hydroxide solid solution composition of $Cr_xFe_{x-1}(OH)_3$. The hydroxide surface inhibits oxidation in two ways: (1) it is a redox stable sink for oxidizable, soluble Cr(III) and (2) it forms a physical barrier between aqueous Cr(III) and Mn surfaces (Fendorf, 1995).

E. Objectives of Current Study

This chapter presents the results of laboratory experiments and a field-scale pilot study to evaluate the *in situ* immobilization and detoxification of Cr(VI)-contaminated soils and sediments, which were acting as a continual source for groundwater contamination at the United States Coast Guard (USCG) Support Center. Based on laboratory studies, sodium dithionite was selected for the field pilot test to evaluate its efficiency in immobilizing Cr(VI) near the source area beneath the old plating shop. The objectives of the pilot test were to evaluate the effectiveness of sodium dithionite as a reducing agent under field conditions and to examine the impact on Cr(VI) present in the soils in the unsaturated zone and capillary fringe zone sediments and groundwater. It was also important to evaluate any adverse impacts, including solubilization or mobilization of selected Resource Conservation and Recovery Act (RCRA) metals caused by the reductant treatment. Results of this study may provide an alternative method for source control and groundwater remediation at sites with Cr(VI) contamination.

II. SITE DESCRIPTION

A. Background and Previous Study

1. Chromium Source Area

The field site is located at the USCG Support Center, Elizabeth City, North Carolina, about 100 km south of Norfolk, Virginia, and 60 km inland from the outer banks of North Carolina. The center is located on the southern bank of the Pasquotank River, about 5 km southeast of Elizabeth City. A former electroplating shop located inside a hangar was in use for approximately 30 years until 1984. Wastes that had been managed inside the shop included sulfuric acid (H_2SO_4), cyanide (Cn), cadmium (Cd), Cr, and lead (Pb) (Kearney, 1986). In 1988, while the old plating shop was being remodeled for office space, a demolition contractor discovered a hole approximately 35 cm in diameter in the concrete floor slab below a tank that had contained chromic acid, indicating the release of chromic acid and sulfuric acid wastes and associated organic solvents into the subsurface beneath the shop. The rupture of a water main in the hangar in 1983, approximately 5 years prior to the demolition, had possibly spread contaminants into the soils and groundwater beneath the shop. Soil was excavated beneath the floor of the former plating shop and was found to contain metal concentrations above background concentrations. The amount of waste(s) released is unknown, however, the subsurface of the former plating shop was partially remediated in 1988 by removal of the top 2 ft of soil. It was believed that the majority of the contaminated soil at the site was removed at that time. Clean fill material topped with a layer of coarse gravel was used to replace the soil excavated from the plating shop.

2. Chromium Distribution in Soil and Groundwater

Initial site characterization was conducted beneath the plating shop by the National Risk Management Research Laboratory (NRMRL), USEPA, in 1990 and 1991 to determine the extent of soil and aquifer contamination. In general, the highest concentrations of total Cr in the soils beneath the shop were detected in the shallower soils 0.9–1.2 m above the water table, with a maximum concentration of 10,200 mg/kg. Comparison with background concentrations indicated that elevated Cr levels extended from inside the former shop to approximately 45 m north of the shop (Fig.1).

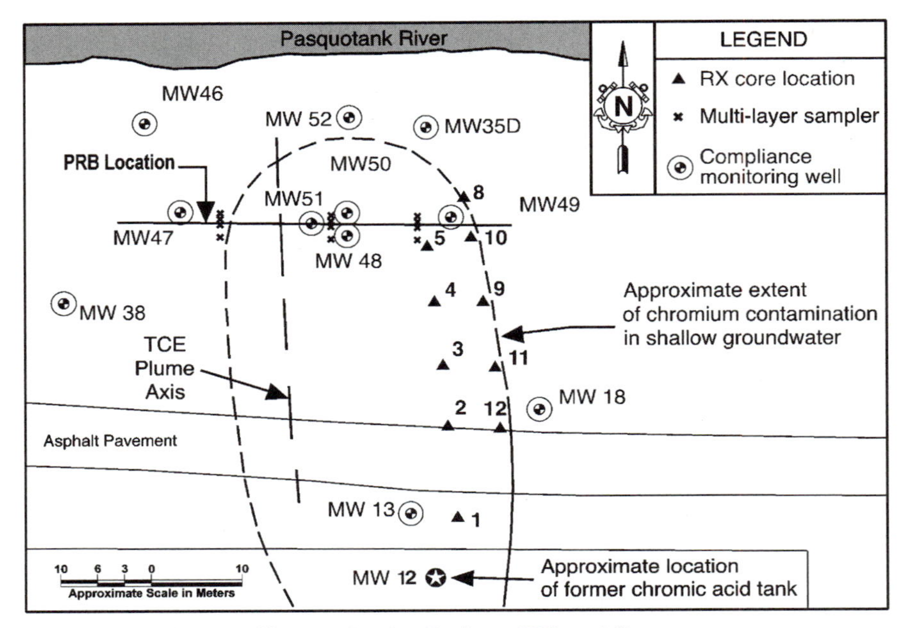

Figure 1 Site map showing Cr plume, PRB, and Cr source area.

Groundwater movement is generally toward the Pasquotank River, and groundwater levels in the vicinity of the plating shop range from 1.8 to 2.1 m below land surface (bls). Historical data from the electroplating shop suggest that elevated Cr(VI) levels in the capillary fringe area had contaminated the groundwater, creating a mobile Cr(VI) plume that overlapped an existing trichloroethylene (TCE) plume. Concentrations in excess of 10 mg/liter Cr(VI) and 19 mg/liter TCE had been detected in the groundwater since 1991 (Puls *et al.*, 1999b).

The rupture of a subsurface fire suppression water line adjacent to the shop in 1994 is suspected to have caused an artificial rise in the water table, solubilizing some of the Cr(VI) in the remaining soils beneath the shop. This likely mobilized additional contaminants in the soil and flooded the fill material and gravel floor inside the shop. A CrO_4^{2-} residue remained on the surface materials after the water receded. Historical Cr(VI) concentrations in groundwater from a monitoring well (MW12) that had been installed inside the shop in 1991 are presented in Table I.

Table I

Historical Cr(VI) and SO_4^{2-} Groundwater Concentrations (mg/liter) for MW12 inside Plating Shop, USCG, Elizabeth City, NC

Sample date	Cr(VI)	SO_4^{2-}
February 1991	1.60	82.0
April 1992	0.800	87.0
June 1993	1.41	86.0
June 1994	28.0	140
May 1995	27.0	121
June 1997	4.40	143
June 1998	3.00	128
June 1999	4.25	150

3. Corrective Actions: Installation of PRB in 1996 to Treat Mobile Chromium Plume

Preliminary laboratory studies using soil from the site and a subsequent field-scale pilot study determined zero valent iron (ZVI) to be effective in remediating Cr-contaminated groundwater. Details of the laboratory experiments and pilot study can be found in Puls *et al.* (1996, 1999b) and in Powell *et al.* (1994, 1995). In June 1996, a continuous PRB consisting of 100% ZVI was installed at the site to intercept and treat the Cr(VI)/TCE plume (Fig. 2).

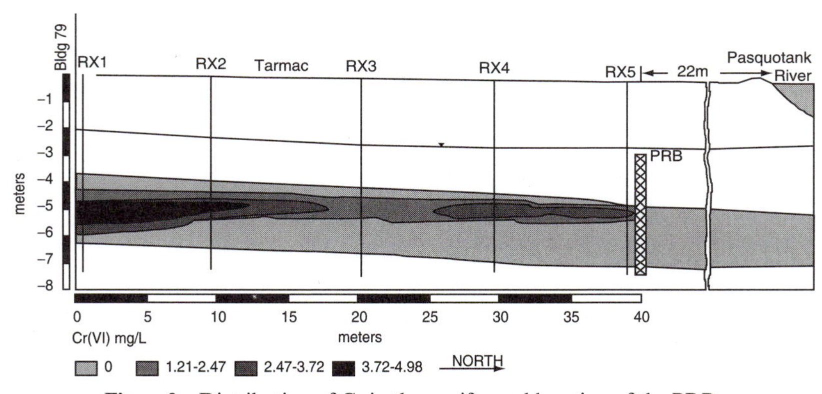

Figure 2 Distribution of Cr in the aquifer and location of the PRB.

A trencher was used to install granular Fe to a depth of 8 m. The PRB was 0.6 m thick and approximately 50 m long. A total of 350 tons of ZVI was used in the installation (Puls *et al.*, 1999a). Four years of postinstallation monitoring have shown the *in situ* PRB wall technique to be a very effective method for reducing and immobilizing contaminants such as Cr(VI) and the dehalogenation of TCE present in groundwater. Results of long term performance monitoring of the PRB at the USCG are reported in Puls *et al.* (1999a).

While the PRB wall has proven to be effective for reducing and immobilizing the Cr plume, NRMRL researchers, the state of North Carolina, and the USCG felt it was necessary to treat the residual Cr beneath the old plating shop, which was providing a continual contaminant source to the plume. This required additional site characterization and laboratory studies due to a water main break in 1994 and subsequent redistribution of the Cr beneath the old plating shop.

B. Site Characterization for Present Study

1. Installation of Monitoring Wells

Eight 1.3-cm inside diameter(i.d.) monitoring wells (S2, S3, S4, S5, S6. S7, S8, S9), one 5.1-cm i.d. monitoring well (S10), and one 5.1-cm i.d. injection well (S1) were installed in May 1998 (Fig. 3). Depths of the 1.3-cm wells are 3.0 m with a screened interval of 1.2 to 3.0 m below existing grade. Wells S1 and S10 were completed to a depth of 2.7 m with screened intervals from 1.5 to 2.7 m below existing grade. All well casing was schedule 40 polyvinyl chloride (PVC) with 0.25-mm slotted screens. The wells were installed using Geoprobe direct push methods with a natural sand pack around the well screen. Wells were fully developed after installation, with final turbidity values of < 5 nephelometric turbidity units (NTUs).

2. Soil and Sediment Sampling

The water main break in the immediate vicinity of the chrome plating shop in 1994 and the subsequent release of large amounts of water into the soils beneath the shop resulted in desorption of soluble Cr(VI) and its redistribution throughout the underlying soils. In an effort to recharacterize the distribution and extent of total Cr and Cr(VI), soil cores were collected from more than 32 locations beneath the plating shop. In September 1997, 18 cores were collected using Geoprobe sampling equipment. Two 1.2-m-long, 4.3-cm-diameter, stainless-steel core barrels, lined with plastic sleeves, were advanced per soil core location and soil collected to a depth of 2.4 m bls.

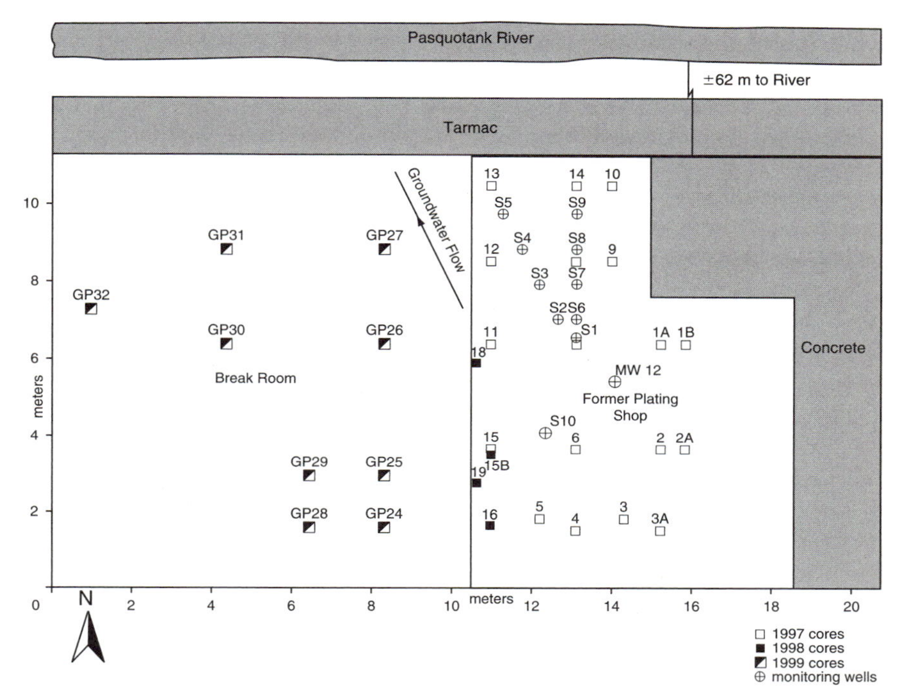

Figure 3 Location of monitoring wells and soil core inside plating shop.

Upon withdrawing each 1.2-m-long core barrel from the subsurface, the plastic sleeve filled with the soil core was extracted from each core barrel and cut into 15-cm subsections. Samples were capped and sealed with tape to prevent moisture loss. No special preservation techniques were used to store the soils after collection. Laboratory tests had been conducted previously to compare Cr(VI) values between fresh (preserved moist) and air-dried samples from the site. No significant differences in Cr(VI) concentrations were observed in the extraction results (Puls *et al.*, 1994). Therefore, it was felt that no special preservation techniques were necessary to store the soils after collection. In March 1998, four cores were obtained using a hand auger along the west wall of the shop to determine if there had been redistribution of Cr onto the soils beneath an adjacent break room that had previously been uncontaminated. Initial metals analysis showed that Cr was present in soils underlying the west wall of the shop and appeared to have spread beneath the reclaimed portion of the shop. In order to further delineate the extent of Cr, NRMRL personnel assisted a contractor for the USCG in collecting 13 soil cores from beneath the break room using Geoprobe technology in February 1999. All soil cores were transported to the NRMRL laboratory in Ada, Oklahoma, for processing and analysis.

III. METHODS

A. Laboratory Studies

1. Soil Processing and Analysis

All soil cores were cut into 15-cm intervals, air-dried, homogenized, and passed through a 2-mm sieve for initial screening of total Cr concentration using X-ray fluorescence (XRF). Based on results of the XRF screening, samples from the most contaminated depths (0.5–0.6 and 1.7–1.8 m) were analyzed for total metals concentrations by inductively coupled plasma emission spectroscopy (ICP-ES) . Analytical methods can be found in Puls and Paul (1997).

2. Selective Extractions

Laboratory batch extractions were conducted to selectively remove Cr(VI) from contaminated soils in order to delineate the extent of the soluble, easily mobilized, Cr(VI) beneath the shop. Laboratory extractions were performed on each 15-cm soil sample to determine Cr(VI) concentrations. Due to the simultaneous release of chromic and sulfuric acids and because SO_4^{2-} is a coadsorbing anion (Zachara *et al.*, 1988), sulfate concentrations were determined at 0.3-m intervals on soil cores containing the highest Cr(VI) concentrations. A modification of the extraction procedure of Bartlett and James (1988) was used for Cr(VI) and sulfate extractions. Details of the extraction method can be found in Puls *et al.* (1994).

3. Reduction Studies

Laboratory studies were conducted using sodium dithionite ($Na_2S_2O_4$), *l*-ascorbic acid ($C_6H_8O_6$), and free hydroxylamine (FH-50) to assess their effectiveness in reducing and immobilizing Cr(VI). Soil samples from cores 6, 7, and 15 (0.5–0.6 m depth) were used in the reduction studies. Studies were conducted using 5.0 g soil and 25 ml 0.05 *M* reductant. Samples were shaken, and Eh and pH were measured immediately. Following a 24-h equilibration, Eh and pH were remeasured. Samples were centrifuged and filtered through 0.2-μm, 25-mm filters. Cr(VI) concentrations were determined using the diphenylcarbazide method, and total metals were determined by ICP-ES. Following reductant studies, extractions were conducted on residual soils using NaH_2PO_4 (Bartlett and James, 1988) and Na_2CO_3–NaOH (James *et al.*, 1995) to determine if any exchangeable Cr(VI) or any insoluble forms of Cr(VI) remained on the soils.

Results of Cr(VI) reduction studies showed that all three reducing solutions evaluated reduced Cr(VI), effectively; however, *l*-ascorbic acid only reduced a small fraction of the Cr(VI) to Cr(III), with most being reduced to intermediary Cr(V) and/or Cr(IV). Eh values for *l*-ascorbic acid remained constant at about 80 mV, whereas values for hydroxylamine increased from an initial 80 to 120 mV . The ability of a reducing agent to reduce native Fe to Fe^{2+} was a major requirement for selecting a reducing agent to use in the field-scale pilot test. As Fe(II) becomes oxidized, Cr(VI) is reduced to Cr(III). Sodium dithionite reduced significantly more native Fe to Fe^{2+} than *l*-ascorbic acid. No Fe^{2+} was detected in the posthydroxylamine extracts. Based on the results of this study, sodium dithionite was selected for use in the pilot study.

B. Field-Scale Pilot Study

1. Chemical Injection Procedure

The pilot study was conducted to evaluate the effectiveness of sodium dithionite as a reducing agent under field conditions and to examine the impact on Cr(VI) present in unsaturated soils immediately above the groundwater table or upper groundwater zone. It was also important to evaluate any adverse side affects, including any solubilization or mobilization of selected RCRA metals caused by the sodium dithionite injection. While the main focus was on treating the upper groundwater zone, an area 0.6 m above the groundwater table was also targeted in the study. The goal was to create a treatment zone from 1.2 m bls (in the vadose zone) to 2.4 m bls (in the saturated zone) and approximately 1.0 m in radius.

a. Permitting Process

Two permits were required by the North Carolina Department of Environmental and Natural Resources (NCDENR) prior to conducting the pilot study. The sodium dithionite injection required a permit to construct and/or use a well for injection for class 5I wells. The bromide tracer injection required a similar permit for class -4S22,5R21, SS23,5T, SX15,5Z wells. NRMRL personnel assisted the USCG in obtaining these permits by supplying hydrogelogic site description, chemicals and amounts used, injection procedure, well locations and construction logs, monitoring procedures, and other required information.

b. Reductant and Concentration Used

A reducing agent consisting of 0.05 *M* sodium dithionite ($Na_2S_2O_4$) buffered with 0.1 *M* potassium bicarbonate ($KHCO_3$) and a tracer 0.025 *M* potassium bromide (KBr) was injected into an existing 5.1-cm i.d. injection well (S1), 2.7 m deep and screened from 1.4 to 2.7 m, inside the electroplating shop. Depth to water table ranges from 1.8 to 2.1 m in the area beneath the plating shop.

c. Volume Injected

The total volume of liquid reductant was approximately 1874 liters. Estimated totals of chemicals used in the field study were 1874 liters distilled and deionized water, 17 kg sodium dithionite, 19 kg potassium bicarbonate, and 5.63 kg bromide tracer.

d. Equipment Used

The sodium dithionite solution was injected using a Pump Drive low-flow centrifugal pump via Tygon tubing that was inserted into the 55-gallon polyethylene drum containing the injection solution. A Cole Parmer direct reading flowmeter was used to maintain the flow rate of 1 liter per minute, and a Cole Parmer utility gauge was used to maintain a steady injection pressure of $\approx 2.0 \times 10^4$ Pa. The configuration of the injection setup is shown in Fig. 4.

2. Postinjection Monitoring

a. Groundwater Sampling Methodology

Ten monitoring wells were used to monitor the site. Eight downgradient monitoring wells were divided into two transects. Transect 1 consisted of wells S2, S3, S4, and S5, whereas transect 2 consisted of wells S6, S7, S8, and S9. Upgradient wells, MW12 and S10, were also monitored as well as the injection well, S1. Groundwater samples were collected one week prior to the injection in order to determine baseline levels of contaminants and geochemical characteristics of the groundwater. Monitoring of these wells was continued during the injection and at 24 h, 2 weeks, 4 weeks, 8 weeks, 32 weeks, and 48 weeks following the injection experiment.

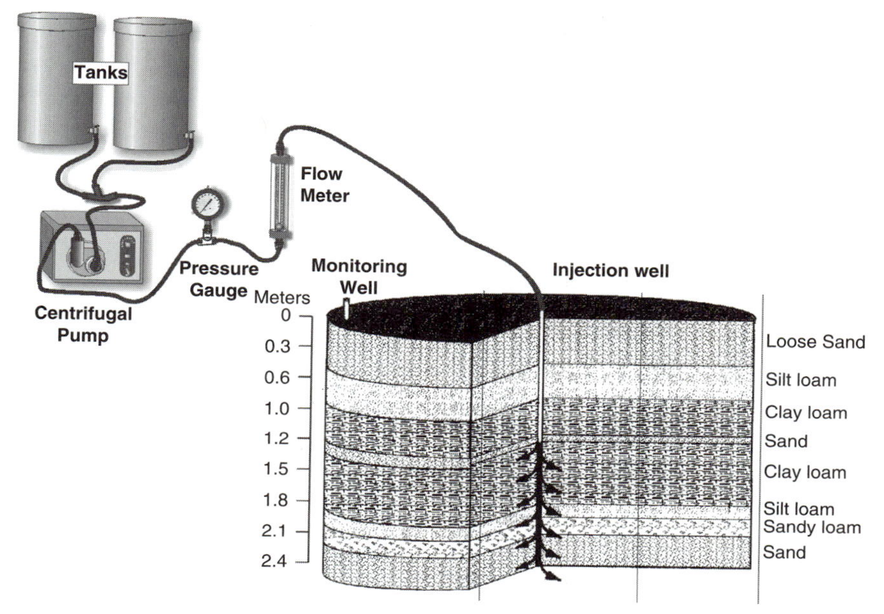

Figure 4 Injection setup during pilot study.

b. Groundwater Analysis and Parameter Measurements

All groundwater samples were collected using a GeoTech peristaltic pump and Teflon-lined 0.6-cm i.d. tubing dedicated to each well. Tubing intake was generally 7.5 cm above the bottom of the well. The flow rate during purging and sampling was approximately 100 ml/min. Water quality indicator parameters were monitored continuously and readings were recorded every 3 min during purging. Temperature, specific conductance, Eh, pH, and DO were monitored using a GeoTech multiprobe flow cell system fitted with Orion electrodes and meters (Model 720A). Turbidity was monitored using a portable Hach 2100P turbidimeter. Purging was considered complete when three successive readings taken at 3-min intervals were within ± 0.1 for pH, ± 5 mV for Eh, $\pm 3\%$ for specific conductance, $\pm 10\%$ for DO, and $\pm 10\%$ for turbidity. Alkalinity, Cr(VI), and Fe(II) were analyzed at the site using Hach and Chemetrics field test kits. Groundwater samples were collected and transported to the NRMRL laboratory for analysis of total metals, chloride (Cl), nitrate (NO_3), SO_4, and bromide (Br) concentrations.

c. *Postinjection Soil and Sediment Sampling*

Forty-eight hours following the injection, four soil cores, PDS1, PDS2, PDS3, and PDS4 (Fig. 5), were collected around the radial treatment zone using Geoprobe sampling equipment to evaluate Cr(VI) concentrations and the effectiveness of the reductant in the unsaturated soils and in the capillary fringe zone sediments. PDS1, PDS2, and PDS3 soil samples were collected from the treatment zone on the downgradient side, and PDS4 was collected from the upgradient side, at the edge of the treatment zone. Eight weeks after the injection, PDS5 was collected from inside the treatment zone and PDS6 was collected from outside the treatment zone (Fig. 5). All the cores were divided into 15-cm sections, air dried, and homogenized. Duplicate samples from each section were used to determine the Cr(VI) concentrations in sediments using the method described in Section III, A, 2.

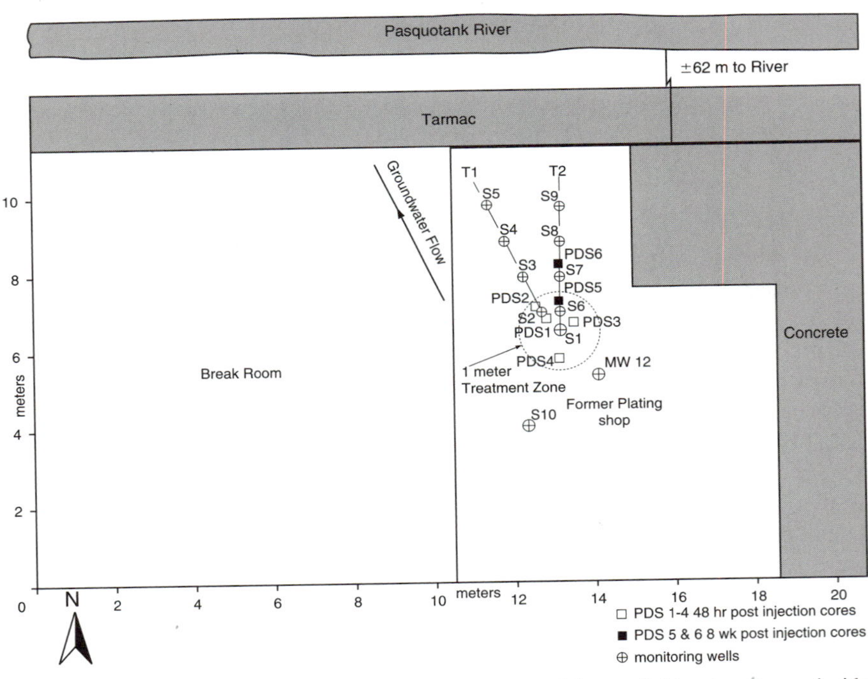

Figure 5 Postinjection core locations, monitoring wells, and 1-m radial treatment zone inside plating shop.

IV. RESULTS AND DISCUSSION

A. Groundwater

1. Geochemical Parameters

Following the injection, specific conductance values increased significantly in the monitoring wells inside the treatment zone from ≈950 μS/cm to greater than 2000 μS/cm in wells S2 and S6 at 24 h postinjection. Values remained near background in wells outside the treatment zone. Dissolved constituents such as Fe(II), manganese (Mn), and sulfate ions and other major constituents of the injected solution such as sodium (Na), potassium (K), and Br may have contributed to the increased specific conductivity of the groundwater. At 48 weeks postinjection, specific conductance values had returned to near preinjection levels.

Preinjection pH and Eh (SHE) ranged from 6.1 to 6.8 and 410 to 445 mV, respectively. Groundwater pH was stable after the injection period and remained in the range of near neutral. Potassium bicarbonate ($KHCO_3$) was used to buffer the reductant solution in order to maintain near-neutral pH conditions in the treatment zone. The injected reductant considerably affected the redox conditions in the treatment zone. A significant difference in redox environments is reflected in data for the various monitoring wells within the treatment zone. Eh (SHE) values were significantly lower than the preinjection values in the radial treatment zone (Fig. 6).

Preinjection DO concentrations measured with a DO electrode across the site were highly variable and ranged from 3.48 to 0.19 mg/liter. DO values at 24 h postinjection ranged from 0.09 to 0.32 mg/liter in all monitoring wells and remained low in the wells within the treatment zone at 48 weeks postinjection. No Fe(II) was detected in any of the wells prior to the injection; however, the presence of Fe(II) in the treatment zone following the injection indicated the presence of an anoxic environment in the treatment zone. Throughout the monitoring period, monitoring wells (S2 and S6) near the injection well maintained lower redox conditions compared to other monitoring wells inside the treatment zone. The reductant may have been progressively losing its reactivity as it traveled the radial distance. Eh (SHE) values outside the treatment zone remained close to background levels.

2. Bromide Tracer Test

Background Br^- concentrations were measured in order to establish a baseline to determine when breakthrough of the reductant occurred. Within 3 h, elevated concentrations of Br^- were observed in monitoring wells S2 and

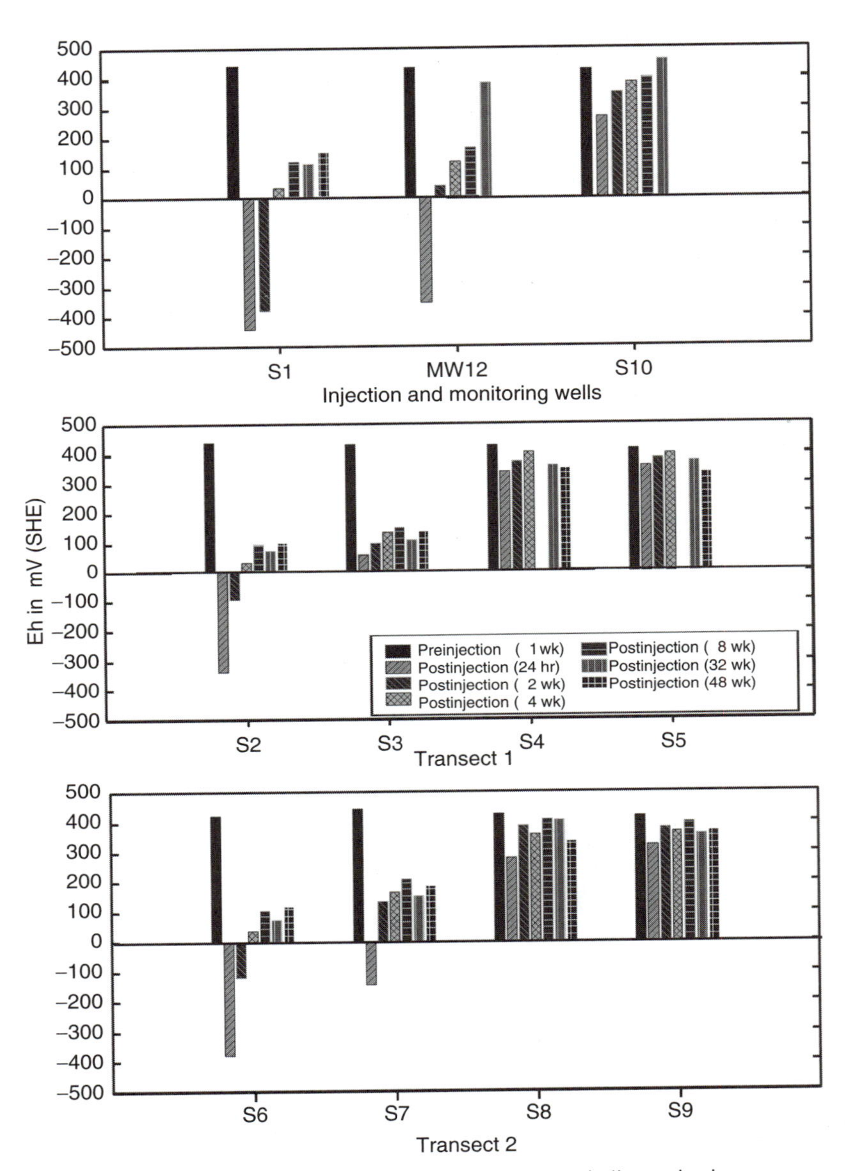

Figure 6 Eh values (mV) in groundwater during periodic monitoring.

S6. The radial distance of these wells is 0.5 m from the injection well. Br^- was detected in all wells in the treatment area in 10 h. Postinjection monitoring data show that the increased levels of Br^- arrived at variable concentrations in all the treatment zone monitoring wells, whereas the concentrations of Br^- remained at background levels outside the treatment zone in monitoring wells S4, S5, S8, S9, and S10 . These data suggest that the radius of influence of the injected solution was approximately 1 m.

3. Chromium, Iron, and Sulfate Concentrations

Initial Cr(VI) concentrations ranged from 1.25 to 7.25 mg/liter, whereas concentrations in the reduced zone were below the detection limit (0.01 mg/liter) following the injection and at 48 weeks postinjection (Fig. 7). A significant amount of dissolved Fe was detected in the postinjection groundwater samples, suggesting that secondary Fe was being reduced and liberated from solid phases into the ground water. Dissolved Fe, presumably the ferrous form liberated by the reductant, was responsible for the reduction and precipitation of Cr(VI) to Cr(III). In addition, structural Fe(III) within the layer of silicate minerals may have been reduced in the treatment zone to an insoluble Fe(II), creating a spatially fixed reducing condition for continued Cr(VI) reduction.

Lower Cr(VI) concentrations outside the treatment zone were observed at 32 and 48 weeks following injection, indicating an increase in the reducing zone downgradient from the injection point. Variations in concentrations of dissolved Fe (Fig. 8), manganese, and sulfate ions were observed in groundwater samples collected from the monitoring wells. The values decreased steadily; by 48 weeks postinjection, values were similar to preinjection concentrations.

Background sulfate concentrations ranged from 145 to 166 mg/liter. The only increase was seen in the wells inside the treatment zone, with the highest concentrations of 2500–5000 mg/liter observed 24 h postinjection. All sulfate concentrations had decreased to near background levels by 48 weeks postinjection (Fig. 9).

4. Other Metals

The potential for mobilization of certain metals that fall under RCRA was of concern to the NCDENR. Therefore, these metals [arsenic (As), barium (Ba), copper (Cu), cadmium (Cd), nickel (Ni), lead (Pb), and selenium (Se)] were monitored throughout the pilot study and during postinjection monitoring. Concentrations in groundwater were determined by ICP-ES except Hg. The levels of Ba and Cd in wells S2, S3, S6, and S7, inside the treatment zone, were slightly elevated above preinjection levels through the fourth week of monitoring. Copper values in wells S3 and S7 increased from preinjection

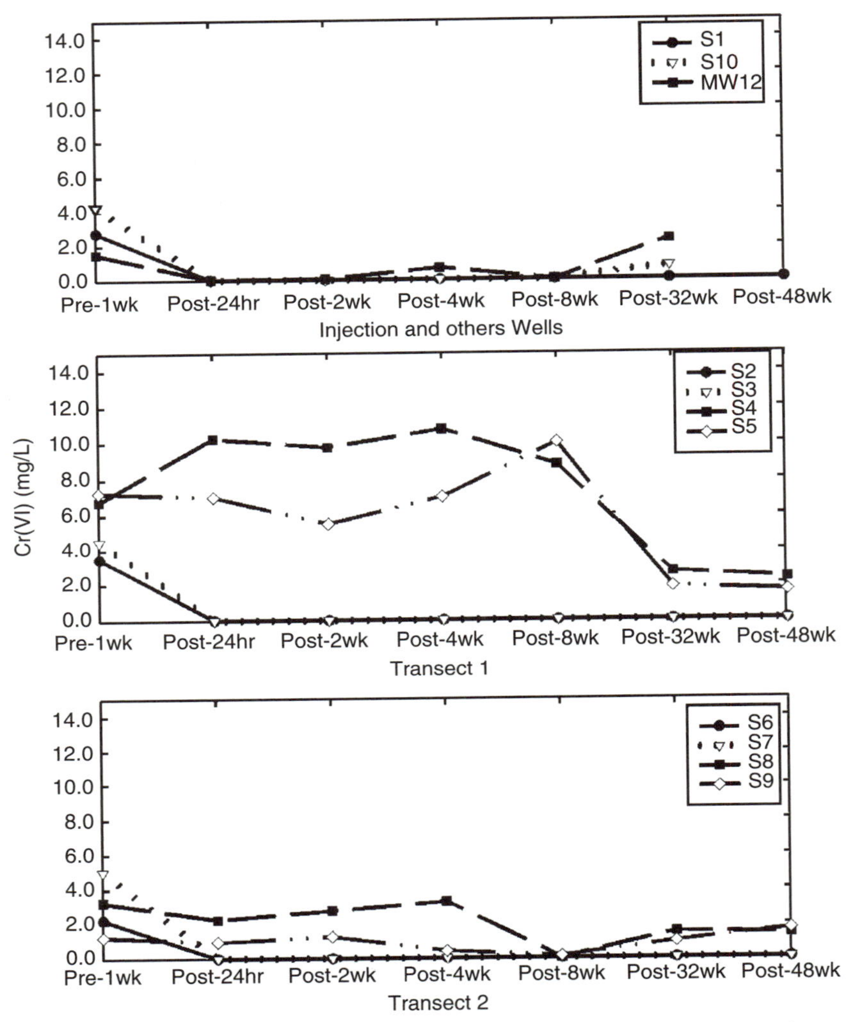

Figure 7 Cr(VI) concentrations (mg/liter) in groundwater during periodic monitoring.

levels below the detection limit (< 0.0054 mg/*kg*) to 0.533 and 0.765 mg/liter, respectively, at 24 h postinjection. However, Cu values decreased steadily and were back to preinjection levels by the eighth week of monitoring. The increased concentrations of these cations could be a function of cation-exchange reaction with the K and Na contents of the applied reductant. No increase was seen in concentrations of As, Ni, Pb, and Se throughout the

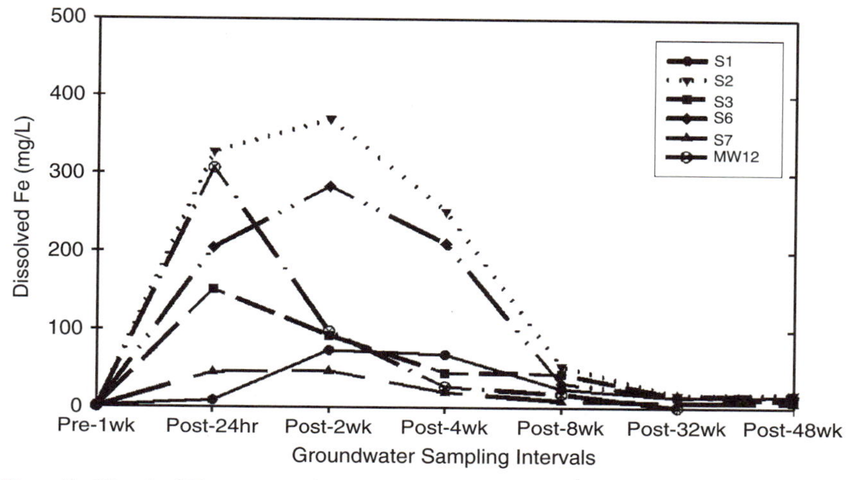

Figure 8 Dissolved Fe concentrations (mg/liter) in groundwater during periodic monitoring.

postinjection monitoring. Although Ba, Cd, and Cu levels were elevated early in postmonitoring periods, they remained localized within the treatment zone. Postinjection concentrations for Ba, Cd, and Cu outside the treatment zone were similar to preinjection values. Data indicate that the applied reductant did not solubilize or mobilize any of the evaluated RCRA metals outside the treatment zone and those elevated within the treatment zone were back to preinjection levels by 8 weeks following injection.

B. Chromium Concentration and Long-Term Fate of Cr(III) in Soils and Sediments

Figure 10 compares the distribution of Cr(VI) in postinjection soils within (PDS-5) and outside (PDS-6) the treatment zone. Below the 1.2-m depth, Cr(VI) values in the treatment zone ranged from 0 to 37 mg/kg. PDS1 at the 1.2- to 1.4-m depth was the only downgradient sample to exceed the Cr(VI) soil target concentration (78 mg/kg) of NCDENR (1999). Sediment samples from the upgradient core (PDS4) contained higher Cr(VI) at the same depth. Results suggest that sodium dithionite was more effective in reducing Cr(VI) in the capillary zone downgradient from the injection than upgradient.

Column and batch tests had been conducted on soils and sediments from several locations at the site from both saturated and unsaturated zones prior

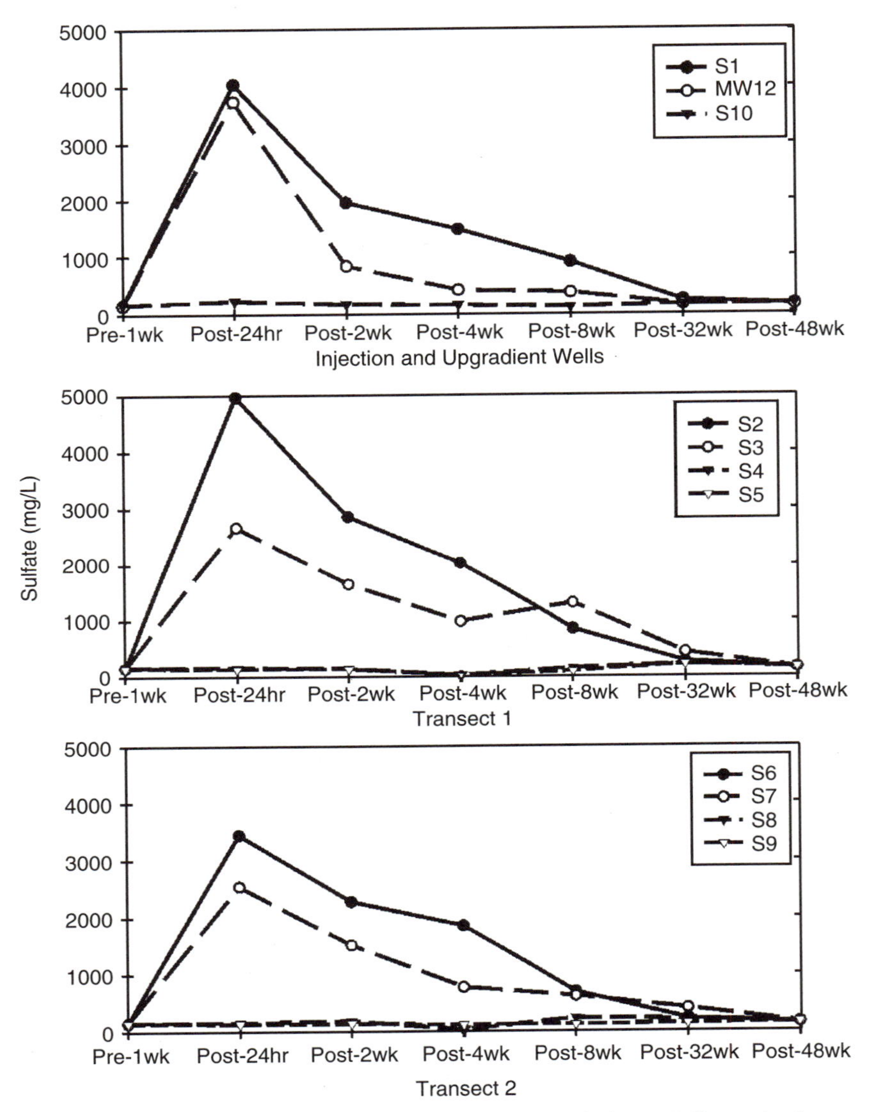

Figure 9 SO_4^{2-} concentrations (mg/liter) in groundwater during periodic monitoring.

to the pilot study in order to evaluate the potential for reoxidation of Cr. Results showed no oxidation of Cr(III) to Cr(VI), indicating that the potential for reoxidation of reduced Cr(III) at this site is unlikely. In a similar study conducted at the Department of Energy's Hanford site in Washington, Fruchter *et al.* (2000) used dithionite injection to create an ISRM for chromate reduction in the saturated zone. The treatment zone they created has remained anoxic and chromate values have remained below detection for 3.5 years following the injection.

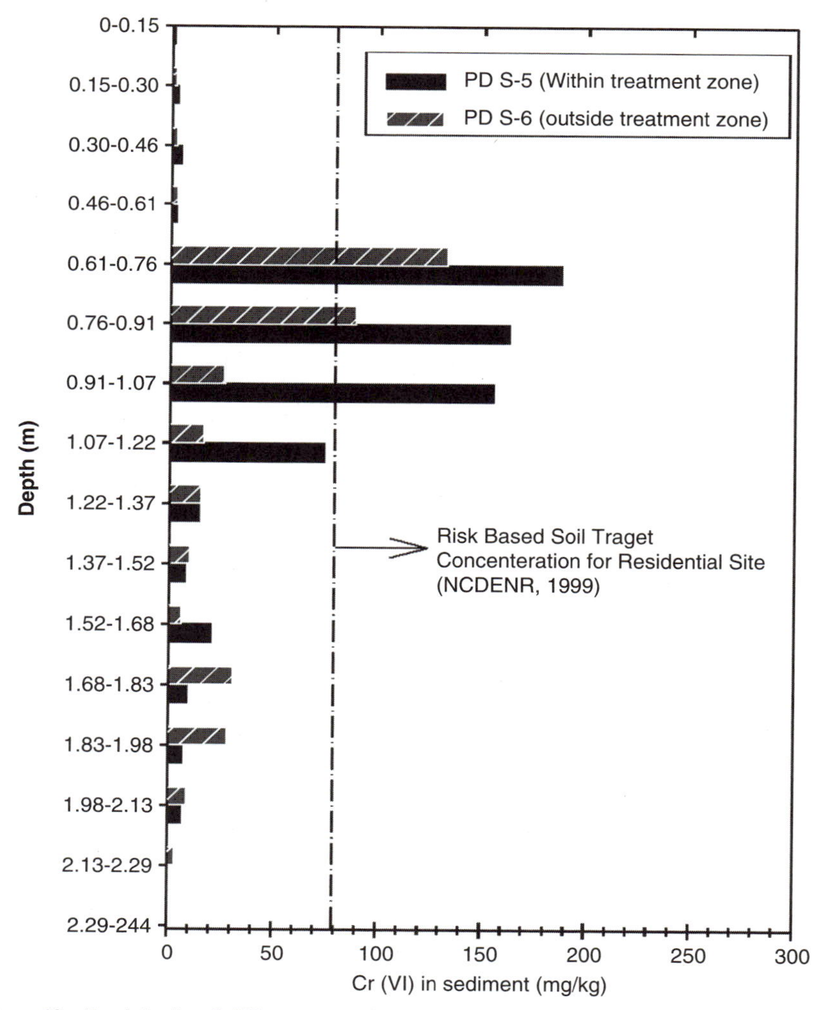

Figure 10 Postinjection Cr(VI) concentrations (mg/kg) in downgradient soils within treatment zone.

C. Cost Comparison with Other Treatment Methods

While installation costs for various treatments are somewhat comparable, the long-term operation and maintenance costs are significantly less with ISRM or chemical injection for treatment of a source area. Because this treatment technique requires no pumping or aboveground treatment following installation, operation and maintenance costs are very low. Estimated savings of using chemical injection to treat the source area combined with a PRB to treat the plume at the USCG site is $1,500,000 compared to using a PRB alone. Treating the source area essentially eliminates two replacement PRBs (assuming a 10-year PRB lifetime), reducing cleanup time from 30 to 10 years. Estimated savings compared with traditional pump and treat techniques with an estimated 30-year cleanup time is $7,500,000 (Table II). Fruchter *et al.* (2000) conducted a cost benefit analysis that estimated using ISRM instead of traditional pump and treat at the

Table II

Cost Comparisons Using Chemical Injection for Source Control in Conjunction with a PRB for Plume Treatment and Other Treatment Methods

	Estimated cost
Iron wall with source control (chemical injection)	
Iron wall (construction, design, characterization, reporting)	$1,000,000
Chemical injection (sodium dithionite)	$750,000
Two years monitoring	$250,000
Total	$2,000,000
2-year cleanup time	
Iron wall without source control	
First iron wall (construction, design, characterization, reporting)	$1,000,000
Two Replacement walls (10-year life span each)	$1,500,000
Monitoring (30 years)	$1,000,000
Total	$3,500,000
30-year cleanup time	
Pump and treat (precipitation, flocculation) without source control	
Capital costs	$2,000,000
Operating costs	$7,500,000
Total	$9,500,000
30-year cleanup time (conservative estimate)	

Hanford site could result in a savings of 60%. The estimated cost using chemical injection in conjunction with a PRB assumes that one application of sodium dithionite will clean up the source area and the groundwater. Within two years, downstream contamination will have flushed through the existing downgradient permeable reactive wall.

V. CONCLUSIONS

A. Residual Effects

Postinjection results indicate that sodium dithionite was effective in creating a treatment zone at the site. This was evident by the production and maintenance of low Eh conditions in the treatment zone. Continued Cr(VI) reduction was seen 48 weeks postinjection, indicating that long-term reduction occurred. Nonmobile, reduced, structural Fe within silicate minerals present in treated sediments may contribute to long-term remediation of Cr(VI). The applied reductant was also very effective in reducing the concentration of Cr(VI) from the capillary fringe zone sediment matrices. The effectiveness of the reducing zone is expected to be long term because the Cr(III) should remain immobilized with no potential for reoxidation at this site.

B. Full-Scale Implementation and Recommendations

Based on the results of this field study, the use of sodium dithionite proved to be a viable option for a full-scale remedial approach for the source area in the electroplating shop at the USCG Support Center. NRMRL personnel presented the results of this study to the USCG and officials from NCDENR. They have selected ISRM, using sodium dithionite, as the full-scale treatment method to remediate the source area in the electroplating shop at the USCG support center, Elizabeth City, NC. Fruchter *et al.* (2000) have also recommended using dithionite for full-scale treatment of a Cr(VI) plume at the Hanford site, Washington. This treatment method is also being considered at other locations, including a commercial chrome plating plant in Vancouver, Washington.

C. Potential Application to Other Contaminants

ISRM is being investigated for the treatment of chlorinated solvents such as trichloroethylene and perchloroethylene with field testing at Fort Lewis,

Washington, as well as proposed testing at Moffett Federal Airfield in California. Further studies will be necessary in order to determine the effi;cacy of using ISRM for other contaminants.

ACKNOWLEDGMENTS

The authors acknowledge the field support of Ms. Mary Sue McNeil, Mr. Frank Beck, and Mr. Patrick Clark of the National Risk Management Research Laboratory, USEPA; the laboratory assistance of Ms. Wendy Hightower, Ms. Stephanie Locke, and Ms. Angela Ott, Environmental Research Apprenticeship Program, East Central University; the analytical support provided by ManTech Environmental Research Services Corp; and the graphics support provided by Ms. Martha Williams, Orbital Astronomical Observatory (OAO). We also thank Mr. Murray Chappell, Mr. J. P. Messier, and Commander Michael Herring, USCG, for their continued field assistance and project support.

DISCLAIMER

Although the research described in this article has been funded wholly or in part by the U. S. EPA, it has not been subjected to the agency's peer and administrative review and therefore may not necessarily reflect the views of the agency; no official endorsement may be inferred.

REFERENCES

Abassi, S. A., and Soni, R. (1984). Teratogenic effects of chromium(VI) in the environment as evidenced by the impact of larvae of amphibian *Rand tigrina*: Implications in the environmental management of chromium. *Int. J. Environ. Stud.* **23**, 131–137.

Ainsworth, C. C., Girvin, D. C., Zachara, J. M., and Smith, S. C. (1989). Chromate adsorption on goethite: Effects of aluminum substitution. *Soil Sci. Soc. Am J.* **53**, 411– 418.

Amonette, J. E., and Rai, D. (1990). Identification of noncrystalline (Fe,Cr)(OH)3 by infrared specroscopy. *Clays Clay Miner.* **38**, 129–136.

Amonette, J. E., Szecsody, J. E., Schaef, H. T., Templeton, J. C., Gorby, Y. A., and Fruchter, J. S. (1994). Abiotic reduction of aquifer materials by dthionite: A promising in-situ remediation technology. *In* "Proceedings of the 33rd Hanford Symposium on Health and the Environment-In Situ Remediation: Scientific Basis for Current and Future Technologies" (G. W. Gee and N. R. Wing, eds.), pp 851–882. Battelle Press, Columbus, OH.

Bartlett, R. J., and James, B. R. (1979). Behaviour of chromium in Soils. III. Oxidation. *J. Environ. Qual.* **8**, 31–35.

Bartlett, R. J., and James, B. R. (1988). Mobility and bioavailability of Cr in soils. *In* "Chromium in the Natural and Human Environment" (J. Nriagu and Nieboer, eds.) pp. 267–304. Wiley, New York.

Bonatti, S., Meini, M., and Abbondandolo, A. (1976). Genetic effects of potassium chromate in *Schizosaccharomyces pombe. Mutat. Res.* **38**, 147–149.

Chen, J. M. and Hao, O. J. (1998). Microbial chromium(VI) reduction in critical reviews. *Environ. Sci. Technol.* **28**(3), 219–251.

Davis, J. A., and Leckie, J. O. (1980). Surface ionization and complexation at the oxide/water interface. 3. Adsorption of anions. *J. Colloid Interface Sci.* **74**, 32–43.

Department of Energy, DOE (1992). Chemical Contaminants on DOE Lands and Selection of Contaminant Mixtures for Subsurface Science Research. DOE/ER-0547T.

Eary, L. E., and Rai, D. (1988). Chromate removal from aqueous wastes by reduction with ferrous iron. *Environ. Sci. Technol.* **22**, 972–977.

Eary, L. E., and Rai, D. (1987). Kinetics of chromium(III) oxidation to chromium(VI) by reaction with manganese dioxides. *Environ. Sci. Technol.* **21**, 1187–1193.

Fendorf, S. E. (1995). Surface reactions of chromium in soils and waters: Environmental soil chemistry. *Geoderma.* **67**, 55–71.

Fendorf, S. E., and Zasoski, R. J. (1992). Chromium(III) oxidation by B-MnO2. 1. Characterization. *Environ. Sci. Technol.* **26**, 79–85.

Fruchter, J. S. (2000). In-Situ Redox Manipulation for Treatment of Chromate and Trichloroethylene in Groundwater. Abiotic In-Situ Technologies for Groundwater Remediation Conference Proceedings. Dallas, TX. EPA/625/R-99/012. U.S. Environmental Protection Agency, Cincinnati, OH.

Fruchter, J. S., Cole, C. R., Williams, M. D., Vermeul, V. R., Amonette, J. E., Sszecsody, Istok, J. D., and Humphrey, M. D. (2000). Creation of a subsurface permeable treatment zone for aqueous chromate contamination using in situ redox manipulation. *Ground Water Monitor. Remediat.* Spring, 66–77.

Goodgame, D. M. L., and Joy, A. M. (1987). EPR study of the Cr(V) and radical species produced in the reduction of Cr(VI) by ascorbate. *Inorgan. Chim. Acta* **135**, 115–118.

Griffin, R. A., Au, A. K., and Frost, R. R. (1977). Effect of pH on adsorption of chromium from landfill-leachate by clay minerals. *J. Environ. Sci. Health* **A12**, 431.

Han, I., Schlautman, M. A., and Batchelor, B. (2000). Removal of hexavalent chromium from groundwater by granular activated carbon. *Wat. Environ. Rsch.* **72**, 29–39.

Istok, J. D., Amonette, J. E., Cole, C. R., Fruchter, J. S., Humphrey, M. D., Szecsody, J. E., Teel, S. S., Vermeul, V. R, Williams, M. D., and Yabusaki, S. B. (1999). In situ redox manipulation by dithionite injection: Intermediate-scale laboratory experiments. *Ground Water* **37**, 884–889.

James, B. R., Petura, J. C., Vitale, R. J., and Mussoline, G. R. (1995). Hexavalent chromine extraction from soils: A comparison of five methods. *Environ. Sci. Technol.* **29**, 2377–2381.

Kearney, A. T., Inc. (1986). RCRA Facility Assessment Report, U.S. Coast Guard Support Center.

Korte, N. E., Skopp, J., Fuller, W. H., Hiebla, E. E., and Alesii, B. A. (1976). Trace element movement in soils: Influence of soil physical and chemical properties. *Soil Sci.* **122**, 350–359.

Lovely, D. R. (1993). Dissimilatory metal reduction. *Annu. Rev. Microbio.* **47**, 263.

Martin, C., Boone, D. R., and Palmer, C. D. (1994). Chromate-resistent microbes from contaminated soil and their potential for bioaugmented reduction of Cr(VI). *In* "Proceedings of the Eight National Outdoor Action Conference and Exposition, Minneapolis, MN, May 23–25, 1994. National Ground Water Association, 191–204.

North Carolina Department of Environment and Natural Resources (NCDENR), (1999). Guidelines for Determining Soil and Groundwater Clean-up Levels at RCRA Hazardous Waste Sites, December.

O'Brien, P., and Woodbridge, N. (1997). A study of the kinetics of the reduction of chromate by ascorbate under aerobic and anaerobic conditions. *Polyhederon* **16**, 2081–2086.

Ono, B. I. (1988). Genetic approaches in the study of chromium toxicity and resistance in yeast and bacteria. *In* "Chromium in the Natural and Human Environments" (J. Nriagu and E. Nieboer, eds.), Vol. 20, pp. 351–368. J Wiley, New York.

Palmer, C. D., and Puls, R. W. (1994). Natural Attenuation of Hexavalent Chromium in Ground Water and Soils. EPA/540/S-94/505.

Palmer, C. D., and Wittbrodt, P. R. (1991). Processes affecting the remediation of chromium-contaminated sites. *Environ. Health Perspect.* **92**, 25–39.

Paschin, Y. V., Kozachenko, V. I., and Sal'nikova, L. E. (1983). Differential mutagenic response at the HGPRT locus in V-79 and CHO cells after treatment with chromate. *Mutat. Res.* **122**, 361–365.

Pettine, M., Barra, I., Campanella, L., and Millero, F. J. (1998). Effect of metals on the reduction of chromium (VI) with hydrogen sulfide. *Wat. Res.* **32**, 2807–2813.

Powell, R. M., Puls, R. W., Hightower, S. K., and Sabatini, D. A. (1995). Coupled iron corrosion and chromate reduction: Mechanisms for subsurface remediation. *Environ. Sci. Technol.* **29**, 1913–1922.

Powell, R. M., Puls, R. W., and Paul, C. J. (1994). Chromate reduction and remediation utilizing the thermodynamic instability of zero-valence state iron. *In* "WEF Specialty Conference Series Proceedings: Innovative Solutions for Contaminated Site Management," pp. 485–495. Miami, FL.

Puls, R. W., and Paul, C. J. (1997). Multi-layer sampling in conventional monitoring wells for improved estimation of vertical contaminant distributions and mass. *J. Contamin. Hydrol.* **25**, 85–111.

Puls, R. W., Paul, C. J., Clark, D. A., and Vardy, J. (1994). Transport and transformation of hexavalent chrome through soil and into ground water. *J. Soil Contam.* **3**, 203–224.

Puls, R. W., and Powell, R. M. (1992). Acquistion of representative ground water quality samples for metals. *Ground Water Monit. Rem.* **12**, 167–176.

Puls, R. W., Blowes, D. W., and Gillham, R. W. (1999a). Long-term performance monitoring for a permeable reactive barrier at the U.S. Coast Guard Support Center, Elizabeth City, North Carolina. *J. Hazard. Mater.* **68**, 109–124.

Puls, R. W., Powell, R. M., Paul, C. J., and Blowes, D. (1999b). Groundwater remediation of chromium using zero-valent iron in a permeable reactive barrier. *In* "Innovative Subsurface Remediation: Field Testing of Physical, Chemical, and Characterization Technologies" (M. L. Brusseau, D. A. Sabatini, J. S. Gierke, and M. D. Annable, eds.), pp. 182–194. Oxford Univ. Press, American Chemical Society.

Puls, R. W., Paul, C. J., and Powell, R. M. (1996). In situ immobilization and detoxificaiton of chromate-contaminated ground water using zero-valent iron: Field experiments at the USCG Support Center, Elizabeth City, North Carolina. *In* "Proceedings of Great Lakes Geotechnical and Geoenvironmental Conference in-Situ Remediation of Contaminated Sites," pp. 23–37. University of Illinois at Chicago.

Rai, D., Sass, B. M., and Moore, D. A. (1987). Chromium (III) hydrolysis constants and solubility of chromium (III) hydroxide. *Inorgan. Chem.* **26**, 345–349.

Richard F. C., and Bourg, A. C. M. (1991). Aqueous geochemistry of chromium: A review. *Water Res.* **25**, 807–816.

Riser, J. A., and Bailey, G. W. (1992). Spectroscopic study of surface redox reactions with manganese oxides. *Soil Sci. Soc. Am. J.* **56**, 82–88.

Sass, B. M., and Rai, D. (1987). Solubility of amorphous chromium(III)-iron(III) hydroxide solid solutions. *Inorg. Chem.* **26**, 2228–2232.

Schroeder, D. C., and Lee, G. F. (1975). Potential transformations of chromium in natural waters. *Water Air Soil Pollut.* **4**, 355–365.

Schultz, M. F., Benjamin, M. M., and Ferguson, J. F. (1987). Adsorption and desorption of metals on ferrihydrite: Reversibility of the reaction and sorption properties of the regenerated solid. *Environ. Sci. Technol.* **21**, 863–869.

Sharma, D. C., and Forster, C. F. (1993). Removal of hexavalent chromium using sphagnum moss peat. *Water Res.* **27**, 1201.

Shiau, B. J., and LaFever, M. S. (1997). In Situ Remediation of Ground Water Using Permeable Reactive Barriers, final report, WA-03–SF-111, U.S. Environmental Protection Agency, Ada, OK.

Smith, L. A., Alleman, B. C., and Copley-Graves, L. (1994). Biological treatment options. *In* "Emerging Technology for Bioremediation of Metals" (J. L. Means and R. E. Hinchee, eds.), p. 1. CRC Press, Boca Raton, FL.

Stollenwerk, K. G., and Grove, D. B. (1985). Adsorption and desorption of hexavelent chromium in an alluvial aquifer near Telluride, Colorado. *J. Environ. Qual.* **14**, 150–155.

U.S. Environmental Protection Agency (EPA) (1989). "Evaluation of Groundwater Extraction Remedies." EPA/540/2–89/054. EPA, Washington, DC.

Waterhouse, J. A. H. (1975). Cancer among chromium platers. *Br. J. Cancer* **32**, 262.

White, A. F., and Hochella, M. F. (1989). Electron transfer mechanisms associated with the surface oxidation and dissolution of magnetite and ilmenite. *In* "Proceedings 6th International Symposium on Water–Rock Interaction," pp. 765–768.

White, A. F., and Yee, A. (1985). Aqueous oxidation-reduction kinetics associated with coupled electron-cation transfer from iron containing silicates at 25°C. *Geochim. Cosmochim. Acta* **49**, 1263–1275.

Zachara J. M., Ainsworth, C. C., Cowan, C. E., and Resch, C. T. (1989). Adsorption of chromate by subsurface soil horizons. *Soil Sci. Soc. Am. J.* **53**, 418–427.

Zachara, J. M., Cowan, C. E., Schmidt, R. L., and Ainsworth, C. C. (1988). *Clays Clay Miner.* **36**, 317–326.

Chapter 17

Sulfate Reduction Permeable Reactive Barriers to Treat Acidity, Cadmium, Copper, Nickel, and Zinc: Two Case Studies

Rick McGregor,[*] Shawn Benner,[†] Ralph Ludwig,[‡] David Blowes,[§] and Carol Ptacek[||]

[*]XCG Consultants Ltd., 50 Queen Street North, Suite 904, Kitchener, Ontario, Canada N2H 6P4
[†]Department of Geologic and Environmental Sciences, Stanford University, Stanford, California 94305
[‡]United States Environmental Protection Agency, Office of Research and Development, Natural Risk Management Research Laboratory, Subsurface Protection and Remediation Division, 919 Kerr Research Drive, PO.Box 1198, Ada, OK, 74820
[§]Department of Earth Sciences, University of Waterloo, Waterloo, Ontario, Canada N2L 3G1
[||]Environment Canada, NHRI, Burlington, Ontario, Canada L7R 4L7

Sulfate reduction and metal sulfide precipitation in permeable reactive barriers provide an alternative solution to the widespread problem of acid mine drainage and heavy metal contamination associated with mining and mine waste. This chapter describes two field-scale studies using municipal compost as the reactive mixture to promote sulfate reduction within permeable barriers. The Nickel Rim permeable barrier was designed to treat effluent containing potential acidity (as dissolved ferrous iron) from decommissioned mine wastes; the Vancouver barrier was designed to treat groundwater impacted by elevated concentrations of heavy metals, including cadmium, copper, nickel, and zinc. Both reactive barriers treated the impacted groundwater to acceptable concentration levels while decreasing the net acidity within the groundwater. Sulfate reduction rates within both barriers decreased with time; however, rates were sufficient to attenuate heavy metals below regulatory guidelines. These field installations indicate that reactive barrier design should address aquifer and barrier heterogeneities, effects of fluctuating water tables, groundwater temperature, and the potential for reductive dissolution reactions in the aquifer downgradient of the barrier. Cost analysis of the permeable reactive barriers indicates that barrier systems offer a cost-effective treatment option for acid drainage and heavy metals in both urban and remote environments.

I. INTRODUCTION

Activities associated with mining and metal processing can have both short- and long-term impacts on the environment. The storage of by-products, such as tailings and waste rock, from the mining of base metal ores can result in the long-term release of dissolved constituents to the surrounding surface and underlying groundwater. Ore processing can also result in similar impacts if proper storage and shipping procedures are not followed. Many of these impacts occur in remote areas or are of such a large scale that traditional approaches to containment (i.e., liners and covers) and treatment (i.e., pump and treat) become cost prohibitive. In addition, time frames for remediating these impacts can range from years to decades, resulting in potentially high operating costs for active treatment methods such as pump and treat.

Permeable reactive barriers (PRBs) have emerged over the past decade as an alternative to traditional methods of treating groundwater impacted by dissolved organic and inorganic constituents. In most applications, the PRB is placed in the path of a migrating plume. Reactive materials within the barriers are selected to promote geochemical reactions that result in the destruction or stabilization of the contaminants. Ideally, the materials should be sufficiently reactive to remove the contaminant to the desired concentration level while providing years to decades of effective treatment. In order to ensure that the groundwater plume flows through the reactive barrier, the permeability of the reactive material should be greater than the permeability of the surrounding aquifer during the life span of the barrier. Finally, the reactive materials should be cost competitive with other groundwater remediation treatment programs. A variety of PRBs have been developed and demonstrated for the effective treatment of dissolved metals. These barriers use a variety of chemical and biological mechanisms to remove contaminants of concern from groundwater. These mechanisms include sulfate reduction via biologically mediated reactions (Waybrant *et al.*, 1998), reductive precipitation (Blowes and Ptacek, 1992), alkaline precipitation (McMurtry and Elton, 1985), adsorption precipitation (see Chapter 14), and adsorption (Morrison and Spangler, 1993). This chapter provides two case studies of PRBs installed to intercept and treat groundwater impacted by acid drainage and heavy metals.

II. BACKGROUND

A. Acid Mine Drainage

Oxidation of sulfide minerals within mining waste creates an acidic effluent that often contains high concentrations of ferrous iron [Fe(II)], sulfate (SO_4), and toxic heavy metals (Fig. 1). For example, the oxidation of the sulfide mineral pyrite (FeS_2) may be expressed as

$$2\,FeS_2 + 7\,O_2 + 2\,H_2O \Rightarrow 4\,SO_4{}^{2-} + 2\,Fe^{2+} + 4\,H^+ \quad (1)$$

In addition to the potential for acid generation, mine wastes often contain high concentrations of trace and heavy metals and metalloids, including arsenic (As), cadmium (Cd), cobalt (Co), copper (Cu), nickel (Ni), lead (Pb), and zinc (Zn). Effluent generated by this type of reaction often enters surrounding aquifers where buffering reactions of the aquifer material can raise the effluent pH to the 4 to 7 range (Blowes and Ptacek, 1994). Upon discharge to surface water bodies, Fe(II) may be oxidized to ferric iron [Fe(III)] and hydrolyze to form ferric iron [Fe(III)] (oxy)hydroxide precipitates:

$$4\,Fe^{2+} + O_2 + 10\,H_2O \Rightarrow 4\,Fe(OH)_3 + 8\,H^+. \quad (2)$$

This reaction may regenerate acidic conditions and therefore represents potential acidity within the groundwater system. Once acidity is released, it may mobilize toxic trace metals that may impact the water ecosystem adversely. The discharge from mine wastes often continues for decades or even centuries (Dubrovsky *et al.*, 1984).

B. Sulfate Reduction

Organic carbon-based remediation systems, including engineered wetland systems and bioreactors, have been used to treat surface water runoff at mining sites that have been affected by acid mine drainage. These systems have achieved varying degrees of success (Wakao *et al.*, 1979; Hedin, 1989; Dvorak *et al.*, 1992; Machemer and Wildeman, 1992; Egar and Wagner, 1995; Christensen *et al.*, 1996). The primary mechanism for treatment in organic carbon-based systems is the microbially mediated conversion of sulfate to

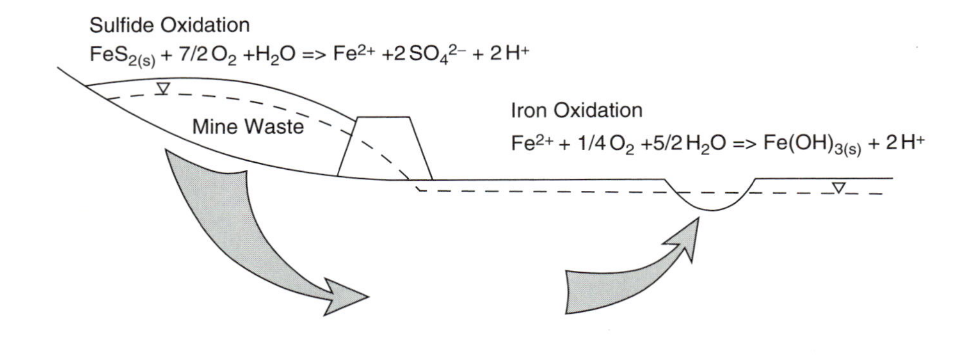

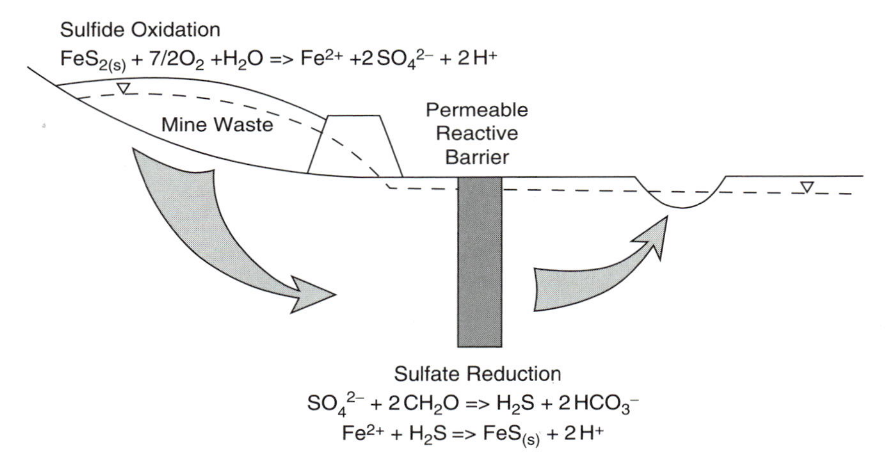

Figure 1 Schematic diagram of decoupled sulfide and iron oxidation associated with acid mine drainage and treatment using sulfate reduction in a permeable reactive barrier.

sulfide by sulfate-reducing bacteria (Machemer and Wildeman, 1992; Benner *et al.*, 1999). A simplified reaction involving the reduction of SO_4 and oxidation of a typical organic substrate (CH_2O) can be expressed as

$$SO_4{}^{2-} + 2\,CH_2O \Rightarrow H_2S + 2\,HCO_3{}^{-}. \tag{3}$$

The microbially mediated reaction between the sulfate and the organic substrate results in the production of hydrogen sulfide (H_2S) and bicarbonate alkalinity ($HCO_3{}^{-}$). The bicarbonate produced regulates the pH, promoting conditions supportive of sulfate-reducing bacteria. The sulfide produced reacts with dissolved metals, forming relatively insoluble metal sulfides following the general reaction:

$$H_2S + Me^{2+} \Rightarrow MeS_{(s)} + 2\,H^+, \quad (4)$$

where Me^{2+} denotes bivalent metals such as Cd, Co, Cu, Fe, Ni, Pb, and Zn.

The use of organic carbon-based sulfate-reducing systems in a subsurface PRB application is attractive because subsurface conditions are generally conducive to effective and controlled sulfate reduction (Fig. 1). These conditions include the absence of broad temperature swings and excessively low temperatures, which may adversely impact the activity of sulfate reducing bacteria in colder climates; reduced susceptibility to oxygen intrusion; and constant and predictable influent flow rates and residence times.

Although sulfate reduction coupled with sulfide precipitation is the predominant heavy metal removal process within an organic carbon-based, sulfate-reducing PRB, other mechanisms may also contribute to metal removal. These mechanisms include precipitation of metal carbonate phases (Benner *et al.*, 1999; Dvorak *et al.*, 1992) and direct adsorption onto the reactive material within the barrier.

This chapter reviews results from two reactive barrier installations for acid mine drainage. The primary objective of the Nickel Rim installation was to remove the acid-generating potential, predominantly in the form of Fe(II), from the groundwater. Results presented in this chapter indicate that the barrier is effective at removing Fe at concentrations as high as 1000 mg/liter. Long-term performance of the barrier is also summarized. The primary objective of the second barrier, the Vancouver installation, is the removal of trace metals, especially Cu, from the pore water. Monitoring indicates that the barrier removes trace metals to low microgram per liter concentrations.

III. SITE DESCRIPTIONS

A. Nickel Rim Site

1. Hydrogeology and Geochemistry

The Nickel Rim tailings impoundment is located 25 km northeast of Sudbury, Ontario, Canada. Deposition of silty-sand size tailings in a narrow bedrock valley resulted in an impoundment that covers approximately 9.4 ha and reaches a maximum depth of 10 m (Johnson *et al.*, 2000). Tailings contain an average of 3.0 wt % sulfur, principally as pyrrhotite ($Fe_{1-x}S$) in a predominantly silicate gangue (Bain *et al.*, 2000; Johnson *et al.*, 2000). Jambor and

Owens (1993) also identified minor amounts of other sulfides, including chalcopyrite ($CuFeS_2$), pentlandite (Ni_8S_9), pyrite (FeS_2), and marcasite (FeS). Sulfide oxidation has resulted in low pH conditions (<3.0) and high concentrations of Fe, SO_4, and other heavy metals in tailings pore water (Bain *et al.*, 2000; Johnson *et al.*, 2000).

Groundwater discharging from the tailings has impacted the aquifer downgradient of the tailings (Fig. 2). This alluvial aquifer is composed of sandy gravel and is situated in a bedrock valley. Groundwater velocity within the aquifer is estimated to be 15 m/year with the groundwater plume containing high concentrations of SO_4 (1000–5000 mg/liter) and Fe (200–2000 mg/liter) and having a pH range of between 4 and 6 (Bain *et al.*, 2000). The mobility of other heavy metals such as Cu and Ni is not as high as that of SO_4 and Fe due to attenuation reactions within the tailings and aquifer and therefore groundwater entering the PRB has not been impacted by other heavy metals as of yet.

2. Installation

The Nickel Rim PRB was installed in July 1995 using traditional cut-and-fill excavation techniques. The PRB contained a mixture of municipal compost (20 vol%), leaf mulch (20%), wood chips (9%), gravel (50%), and limestone (1%). The reactive materials were mixed using a conveyor belt and excavator. Following mixing, the mixture was placed in the trench using a excavator.

The barrier was keyed into the granite bedrock valley at the base and sides and had a total width of 20 m, a depth of 3.5 m, and a thickness of 4 m. Zones of sand, approximately 1 m thick, were installed at the upgradient and downgradient sides of the organic carbon mixture. A clay cap, approximately 0.4 m

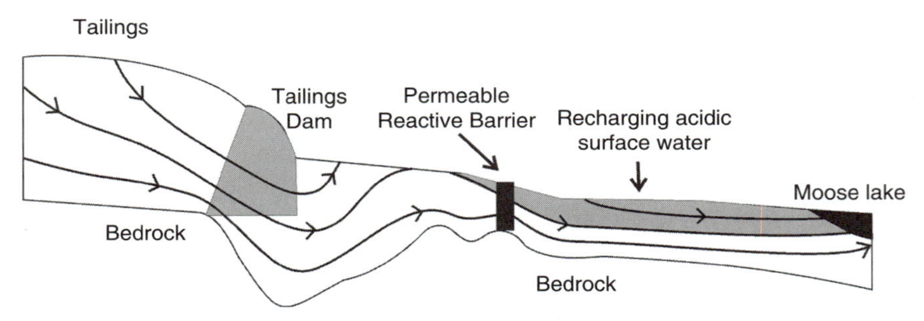

Figure 2 Cross-sectional profile along transect parallel to groundwater flow showing the Nickel Rim aquifer and the location of the barrier installation, Sudbury, Ontario, Canada.

thick, was installed over the barrier to prevent the direct infiltration of acidic surface runoff. However, acidic runoff does enter the aquifer on both upgradient and downgradient sides of the barrier, which results in a complex flow regime downgradient of the barrier. Treated water exiting the downgradient side of the barrier occupies the lower half of the aquifer, whereas the upper half of the aquifer is occupied by infiltrated acidic water, which has not undergone treatment. Additional details on construction and placement methods can be found in Benner *et al.* (1997).

B. Vancouver Site

1. Hydrogeology and Geochemistry

The site is situated on the shoreline of a marine inlet and is underlain by a thick (> 40 m) layer of nearshore and deltaic deposits consisting of sand and gravel with minor amounts of silt and clay. A surficial layer of sediments is composed of fill from unknown sources that was deposited with time to expand the shoreline and raise the land surface. This fill layer contains layers of ore concentrate and elemental sulfur.

Shallow sediments form an unconfined aquifer that extends to a depth of approximately 20 m (Ludwig *et al.*, in 2002). Metal contamination within the groundwater is confined to the upper 15 m of the aquifer, with the highest concentrations present in the upper 6 m of the aquifer. A freshwater–saltwater interface is present near the shoreline at a depth of approximately 15 m below the ground surface. The water table within the unconfined aquifer is influenced by tidal changes in the adjacent marine inlet with the range of tidal fluctuations dependent on the season. During the fall and spring, tidal fluctuations of more than 4.1 m have been measured in monitoring wells adjacent to the inlet. The fluctuations are dampened with increased distance from the shoreline with no measurable tidal effects in the aquifer at a distance of approximately 500 m inland. Water table fluctuations at the barrier site are approximately 1.45 m or approximately 35% of the fluctuation measured in monitoring wells adjacent to the inlet (Ludwig *et al.*, 2002). On the basis of bail tests and estimates based on the response of the aquifer to tidal changes, the hydraulic conductivity in the upper 15 m of the aquifer ranges from 10^{-2} to 10^{-3} cm/s (Hvorslev, 1951; Serfes, 1987). The hydraulic gradient across the site is toward the inlet during low tide and flow away from the inlet during high tide. The average hydraulic gradient (based on a 71-h water level average) was calculated at 0.001 m/m toward the inlet (Serfes, 1991).

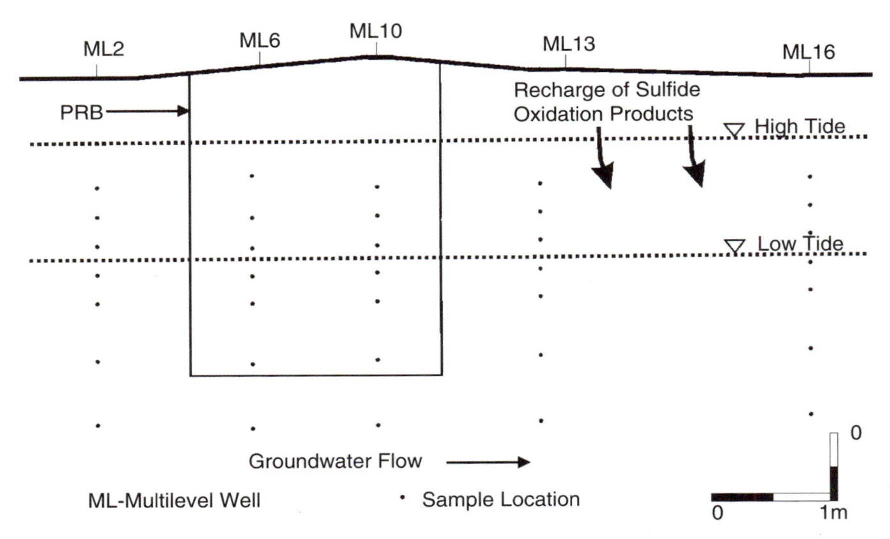

Figure 3 Cross-sectional profile along transect parallel to groundwater flow showing the site aquifer and the location of the barrier installation, Vancouver, British Columbia, Canada.

2. Installation

During March 1997, a demonstration barrier was installed approximately 50 m inland from the shoreline (Fig. 3). Due to the location of the barrier, groundwater and soil contamination were present on the upgradient and downgradient sides of the barrier, as well as underneath the barrier. The barrier was installed using traditional cut-and-fill techniques that utilized a backhoe. The barrier was approximately 6.5 m deep, 2.5 m wide, and 10 m long. Excavation initially involved benching down approximately 1 m to a depth just above the high-tide water table. A guar gum-based slurry was pumped into the trench to prevent trench collapse and to allow emplacement of reactive media. Guar gum, a natural food additive, will naturally biodegrade in most environments or can be broken down using an enzyme mixture. The reactive mixture utilized in the barrier consisted of 15% leaf compost, 84% pea gravel, and 1% limestone (by volume). This mixture was selected based on the results of previous laboratory and field studies (Benner *et al.*, 1997; Waybrant *et al.*, 1998). Pea gravel was added to achieve a minimum desired hydraulic conductivity of 10^{-1} cm/s within the barrier. The limestone was added to ensure suitable initial pH conditions for the establishment of a sulfate-reducing bacteria population within the barrier. The compost, pea gravel, and limestone were mixed using an excavator bucket. Following mixing, the reactive medium was placed into the trench using both a clamshell and an excavator bucket.

IV. METHODS OF ASSESSMENT

Samples of groundwater from both reactive barriers were collected from bundle-style multilevel monitoring wells that were installed upgradient, within, and downgradient of the barriers along transects roughly parallel to groundwater flow. Each bundle consisted of a series of discrete sampling depths distributed from the water table to the bottom of the aquifer at the Nickel Rim site and down to a depth of 8 m below ground level (mbgl) at the Vancouver site. Groundwater samples were collected using a peristaltic pump with Telfon tubing and were filtered in line using 0.45-μm cellulose acetate membranes. Detailed descriptions of well installation, water sampling, and field and laboratory analysis can be found in McGregor *et al.* (1999) for the Vancouver site and in Benner *et al.* (1997) for the Nickel Rim site.

Samples of secondary precipitates within the PRBs were collected by withdrawing polyvinyl chloride (PVC) tubes that were installed during installation of the PRB. The precipitates were immediately sampled from the surface of PVC tubes following the removal of the PVC tubes and frozen to minimize oxidation reactions. Continuous cores of the aquifer material and reactive mixtures were collected using a modified version of the Starr and Ingleton (1992) method. The cores were immediately sealed and frozen to limit exposure to atmospheric conditions and shipped to the laboratory for chemical and mineralogical analyses. Mineralogical tests conducted on solid samples collected from the tubing surface included X-ray diffraction (XRD), scanning electron microscopy, and reflected light microscopy. Detailed methodology for mineralogical and chemical sampling of the Nickel Rim and Vancouver PRBs can be found in Herbert *et al.* (1998) and McGregor *et al.* (1999), respectively. A detailed microbiological study was also completed on the Nickel Rim PRB (Benner *et al.*, 2000). Chemical nonsequential extractions were carried out on aquifer solids using methods described by Tessier *et al.* (1979) and McGregor *et al.* (1998) to determine the reducible iron fraction of the solids.

Groundwater chemistry was interpreted with the aid of the geochemical speciation/mass transfer computer code MINTEQA2 (Allison *et al.*, 1990). The database used was consistent with the WATEQ4F model (Ball and Nordstrom, 1991) with updated values for the jarosite phases (Alpers *et al.*, 1989). For reduction/oxidation (redox) couple calculations, field-measured Eh values (corrected to the standard hydrogen electrode, SHE) were used for groundwater outside of the reactive barrier. Within the barrier, field-measured dissolved sulfide concentrations were used to calculate sulfide mineral saturation indices.

V. RESULTS

A. Nickel Rim Site

Installation of the PRB at the Nickel Rim site resulted in immediate improvement in groundwater quality. Nine months after installation, concentrations of SO_4 and Fe downgradient of the PRB declined dramatically; SO_4 declined by >1000 mg/liter and Fe declined from as high as 1000 mg/liter to less than 50 mg/liter, with some points within the barrier <1 mg/liter. The barrier also reduced potential acidity (Fig. 4) while producing significant amounts of alkalinity, with alkalinity concentrations increasing from <100 mg/liter (as $CaCO_3$) to generally greater than 1000 mg/liter as groundwater entered the downgradient aquifer. Sulfate is removed by bacterially mediated sulfate reduction. Populations of sulfate-reducing bacteria within the barrier are approximately five orders of magnitude higher than within the adjacent aquifer (Benner *et al.*, 2000). The primary mechanism of Fe removal is the precipitation of mackinawite (FeS) (Benner *et al.*, 1999, Herbert *et al.*, 2000).

Long-term performance of the Nickel Rim PRB is shown in Fig. 5. There is a downward trend in the rate of SO_4 and Fe removal with time; after 3 years, SO_4 removal has declined by approximately 30% whereas Fe removal has declined by approximately 50%. However, at the 3-year point, the barrier continues to remove >250 mg/liter Fe from the pore water. Mass balance calculations indicate that only a portion of the total organic carbon within the barrier has been consumed (Benner *et al.*, 1997, 2002). The decline in treatment efficiency is attributed to consumption of the more readily oxidizable organic carbon within the barrier. This conclusion would suggest that the barrier will continue to promote sulfate reduction for tens of years into the future. However, performance of the barrier over time will likely be controlled by the declining rate of sulfate reduction.

In addition to an overall downward trend for barrier performance with time, the Nickel Rim barrier exhibits variations in the rate of SO_4 and Fe removal both seasonally and spatially within the barrier. A series of cross sections perpendicular to groundwater flow direction for SO_4 concentrations in May and October of 1998 are shown in Fig. 6, 3 years after installation. Within each transect there is a high degree of variability in SO_4 concentration.

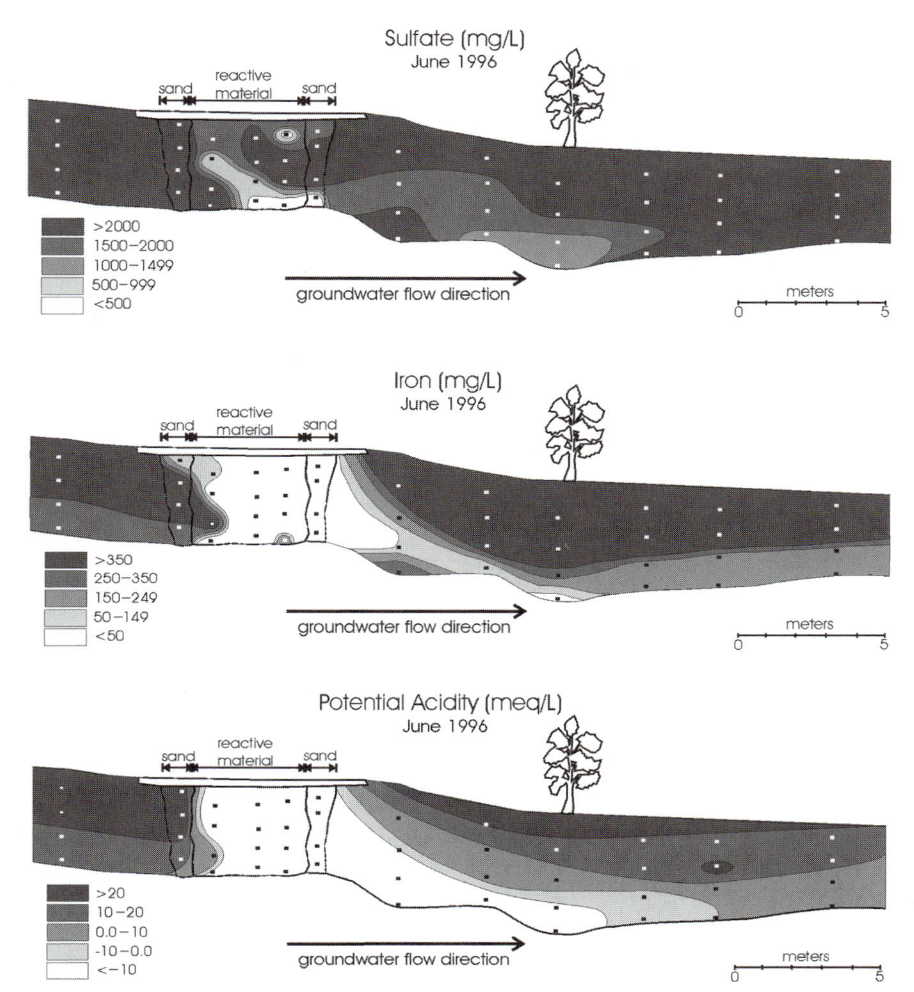

Figure 4 Cross-sectional view of the Nickel Rim permeable reactive barrier and adjoining aquifer showing concentrations of (a) sulfate and (b) iron in groundwater and (c) calculated "potential acidity" 9 months after installation. There is no vertical exaggeration. Reproduced with permission from Benner *et al.* (1997). Copyright © NGWA.

Comparing upgradient and downgradient transects, it is also apparent that the amount of SO_4 removal varies with location within the barrier; those zones of highest concentration on the upgradient side do not necessarily correlate with zones of highest concentration on the downgradient side of the barrier. Spatial variation in the removal of SO_4 within the barrier is

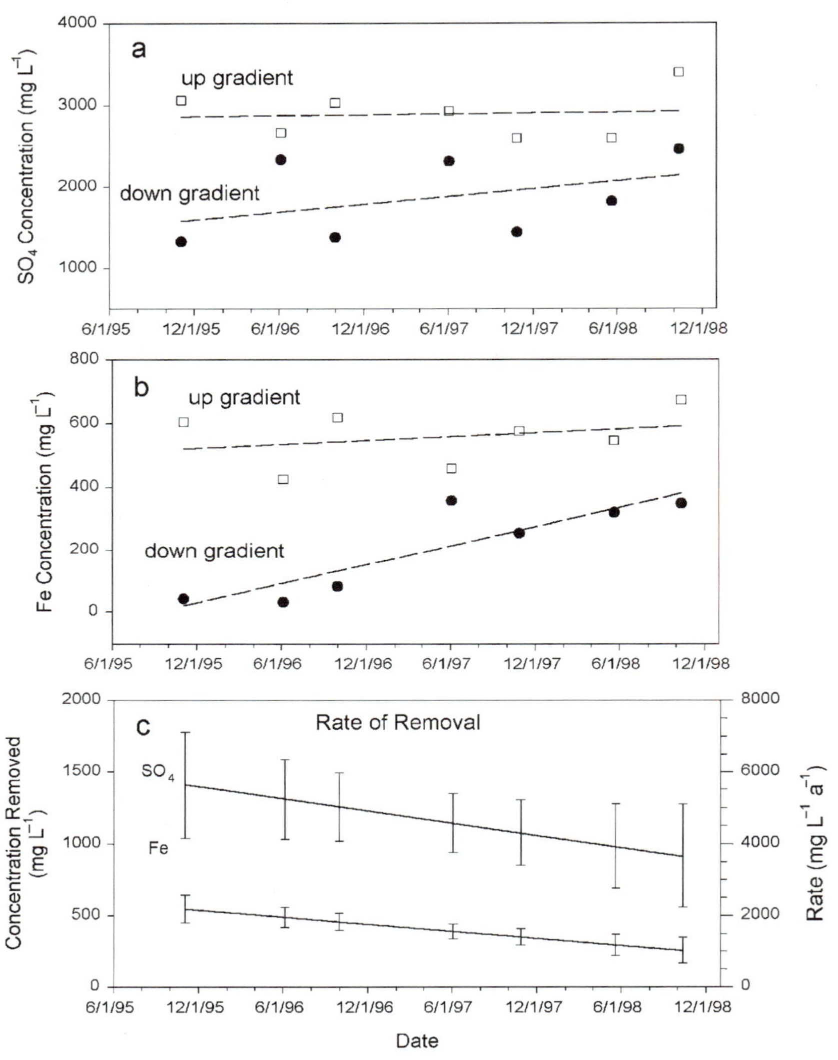

Figure 5 Vertically averaged concentrations of (a) sulfate and (b) iron in water from well nests upgradient and downgradient of the Nickel Rim permeable reactive barrier versus time. Removal of (c) sulfate and iron based on the difference (upgradient – downgradient) in concentrations versus time. Error bars reflect $\pm$ 1 SE. From Benner *et al.* (2002).

attributed to differences in groundwater velocity (Benner *et al.*, 2001). Where SO_4 removal is greatest, groundwater velocities are low; where SO_4 removal is lowest, groundwater velocities are high. Tracer test data corroborated by flow

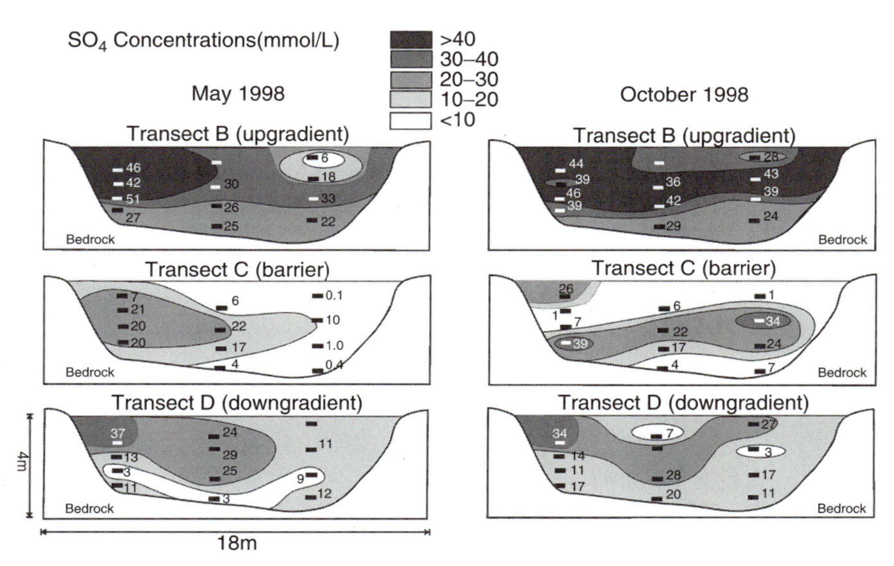

Figure 6 Cross-sectional profiles of sulfate concentrations along transects perpendicular to groundwater flow upgradient, within, and downgradient of the Nickel Rim permeable reactive barrier in May and October 1998 (flow direction is out of the page).

modeling indicate that the observed variations in treatment efficiency can be attributed to a factor of 3 difference in groundwater velocity (Benner *et al.*, 2001).

There is an observable variation in the rate of treatment between summer and winter months (Fig. 7). Rates in the fall are, on average, nearly twice the rates in the spring. This seasonal variation is attributed to fluctuating groundwater temperatures (Benner *et al.*, 2002). Groundwater temperatures vary within the barrier from an average low in the winter of 3°C to an average high in the summer of 15°C.

B. Vancouver Site

Initial results from the first samples collected 3 months after installation indicate that heavy metals such Cd, Cu, Ni, and Zn are being effectively removed from solution. Subsequent samples collected after 7, 9, 12, 15, and 21 months following installation indicate that the removal efficiency for heavy metals increased with time and that many of the heavy metals in the effluent water concentrations are below the analytical detection limit.

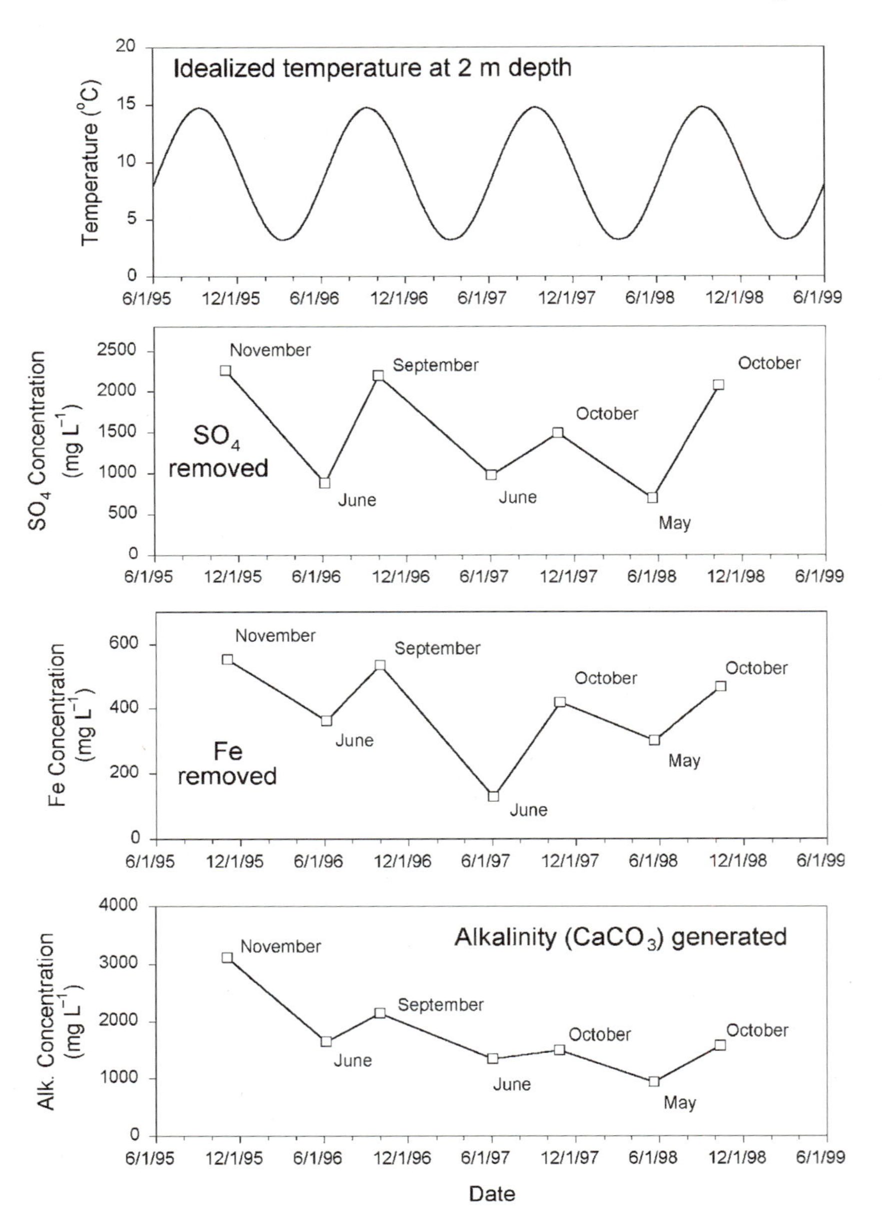

Figure 7 Temperature and average concentrations of sulfate and iron removed, and alkalinity added versus time. Based on well RW23 concentrations and concentrations of all points ($n = 12$) within the barrier. From Benner *et al.* (2002).

Table I

Summary of Heavy Metal Concentrations within the Influent (Well ML2) and the Reactive Barrier (Well ML10) for Upper and Lower Portions of the Vancouver PRB

Heavy metal	Average shallow concentration (μg/liter)		Average deep concentration (μg/liter)	
	ML2	ML10	ML2	ML10
Cadmium	21	0.2	10	<0.1
Copper	720	17	8570	1
Nickel	50	6	210	1
Zinc	2780	140	2050	<2

Results are based on data obtained from multilevel wells, ML-2, ML-6, ML-10, and ML-13, comprising the center transect through the barrier (Fig. 3). In general, metals removal is more effective in the lower half of the wall than in the upper half (Table I). This is attributed to the better maintenance of sulfate-reducing conditions in the lower half of the wall where the effects of tidal influences and oxygen intrusion are limited.

1. Cadmium

Average dissolved Cd concentrations in the groundwater entering the reactive barrier during the 21 months of the study ranged from 8 to 71 μg/liter. The average influent-dissolved Cd concentration varied spatially and temporally with the concentration of Cd increasing with depth and during the wet season. Geochemical calculations (Ludwig *et al.*, 2002) suggest that dissolved Cd entering the PRB would be predominantly in the form of free cadmium [Cd^{2+}] and cadmium sulfate complexes.

As the groundwater flows into the PRB, the average Cd concentration decreased to below the method detection limit (<0.1 μg/liter) for the majority of sampling periods, with the exception of the 3- and 21-month sampling periods when the average effluent Cd concentration was 0.2 μg/liter (Fig. 8A). Removal efficiency for dissolved Cd ranged from 96.4 to 99.8% during the study period (Fig. 9). Geochemical modeling of PRB pore water suggests that the pore water is supersaturated with respect to the cadmium sulfide greenockite [CdS] at the cadmium detection limit of 0.1 μg/liter (Ludwig *et al.*, 2002). A mineralogical analysis of sulfide precipitates found on PVC tube surfaces did not isolate any cadmium-bearing phases (Ludwig *et al.*, 2002).

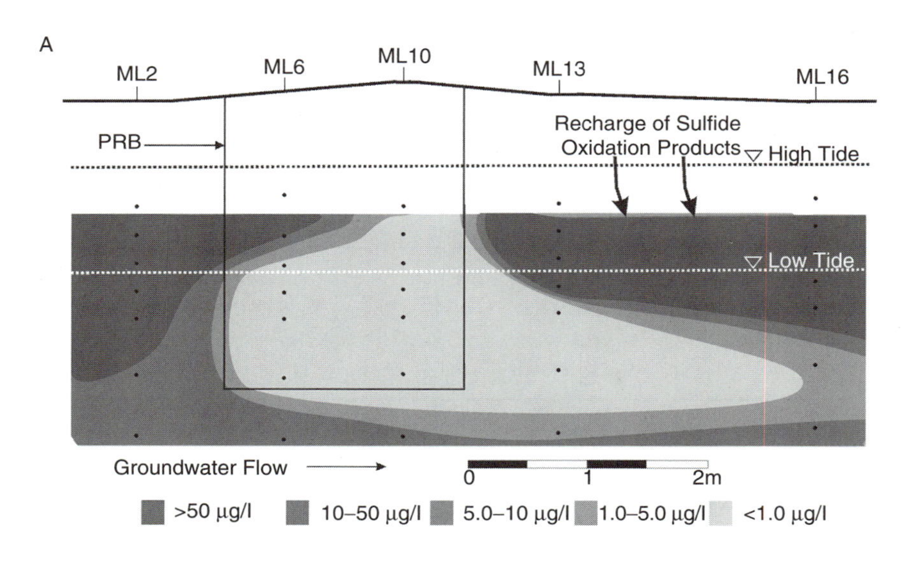
A
ML2
ML6
ML10
ML13
ML16
PRB
Recharge of Sulfide
Oxidation Products
High Tide
Low Tide
Groundwater Flow
0
1
2m
>50 μg/l
10–50 μg/l
5.0–10 μg/l
1.0–5.0 μg/l
<1.0 μg/l

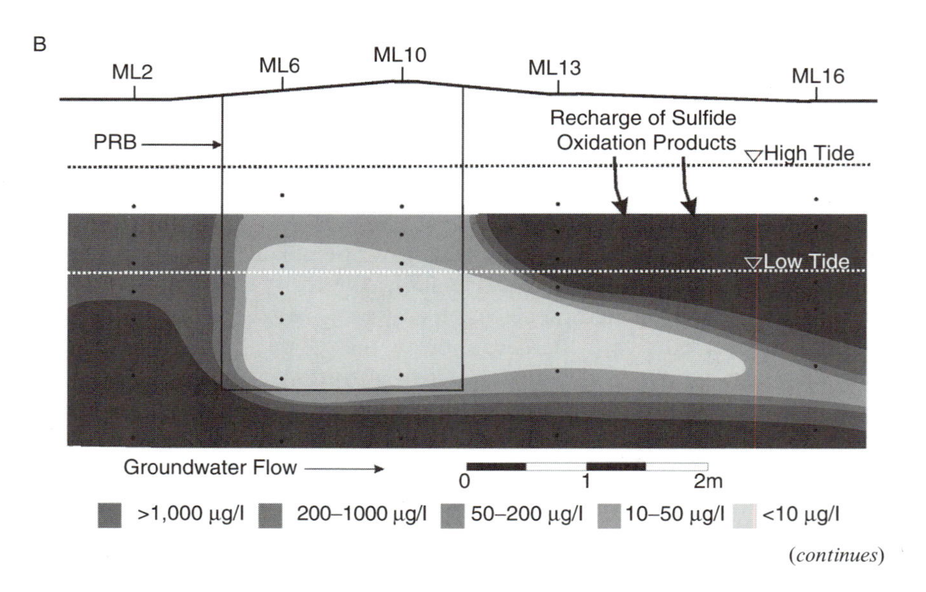
B
ML2
ML6
ML10
ML13
ML16
PRB
Recharge of Sulfide
Oxidation Products
High Tide
Low Tide
Groundwater Flow
0
1
2m
>1,000 μg/l
200–1000 μg/l
50–200 μg/l
10–50 μg/l
<10 μg/l

(*continues*)

Figure 8 *(continued)*

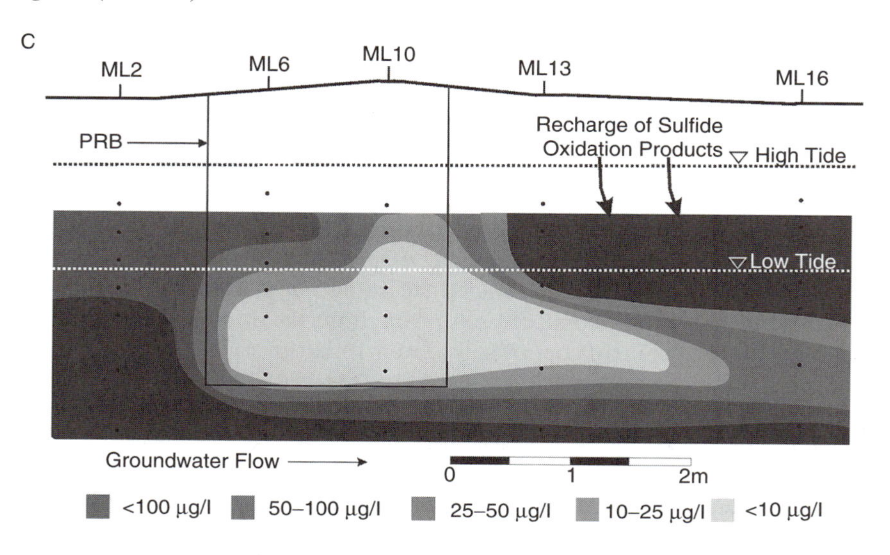

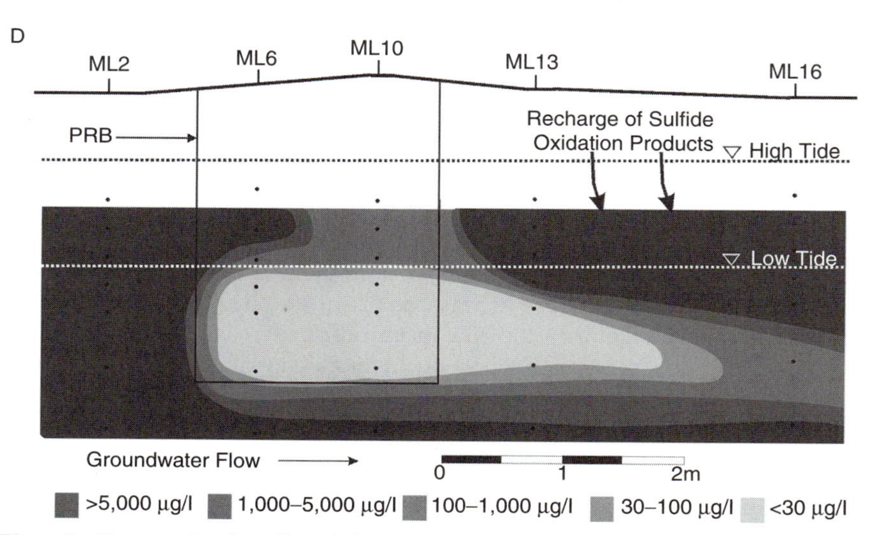

Figure 8 Cross-sectional profiles of the Vancouver permeable reactive barrier and adjoining aquifer showing concentrations of (A) cadmium, (B) copper, (C) nickel, and (D) zinc 21 months after installation.

2. Copper

Average dissolved Cu concentrations in groundwater entering the barrier during the 21 months of the study (based on data from well ML-2) ranged from 2250 to 4640 μg/liter. Influent dissolved Cu concentrations varied spatially and temporally, with Cu concentrations increasing with depth and during the wet season (i.e., winter). Calculation of mineral saturation indices for groundwater entering the PRB indicates that groundwater is at or near saturation with respect to the carbonate minerals malachite [$Cu_2(OH)_2(CO_3)$] and azurite [$Cu_3(OH)_2(CO_3)_2$] (Ludwig *et al.*, 2002). Mineralogical analysis of aquifer material at the site shows an accumulation of malachite and azurite on carbonate shell fragments present throughout the aquifer. These mineral phases appear to control copper solubility within the aquifer (Ludwig *et al.*, 2002).

As groundwater flows through the PRB, dissolved Cu removal efficiency ranged between 98.2 and 99.9% with effluent copper concentrations below 10 μg/liter with some concentrations being below the method detection limit of 1.0 μg/liter (Figs. 8B and 9). Copper removal is most pronounced in the lower half of the barrier. Geochemical modeling (Ludwig *et al.*, 2002) indicates groundwater within the barrier is at or near saturation with respect to the copper sulfide phases covellite [CuS] and chalcopyrite [$CuFeS_2$]. XRD analysis of precipitates collected from PVC tubing installed within the barrier confirmed the presence of covellite, suggesting that covellite or less crystalline precursors control copper concentrations within the barrier (Ludwig *et al.*, 2002).

3. Nickel

Average dissolved Ni concentrations within groundwater entering the PRB ranged from 130 to 290 μg/liter during the 21-month period. As was observed with the other heavy metals, the dissolved Ni concentration varied with location and time, with the Ni concentration increasing with depth and during heavy precipitation. Geochemical modeling of the aquifer system failed to predict any solid-phase controls for the mobility of Ni within the aquifer. Ludwig *et al.* (2002) suggested that dissolved Ni is controlled by adsorption/desorption and coprecipitation reactions associated with iron hydroxides present within the aquifer.

Following treatment by the reactive mixture, dissolved Ni concentrations decreased to near the detection limit (<1 μg/liter) in the core of the PRB in all sampling periods, as illustrated for 21-month data in Fig. 8C. Removal efficiencies ranged from 93.5 to 99.4% during the 21-month period (Fig. 9). Geochemical modeling conducted by Ludwig *et al.* (2002) suggests that pore

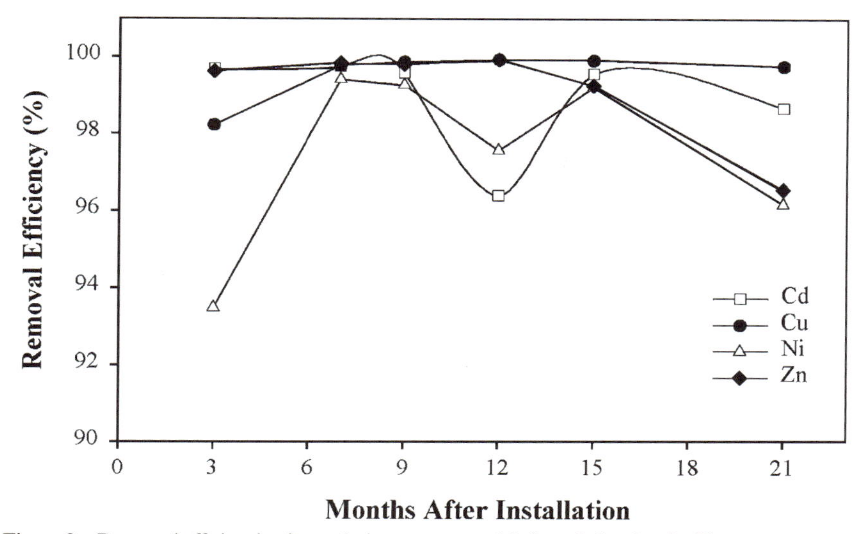

Figure 9 Removal efficiencies for cadmium, copper, nickel, and zinc for the Vancouver permeable reactive barrier versus time.

water within the PRB is supersaturated with respect to millerite [NiS], however, Ni-bearing phases were not isolated during mineralogical analysis of sulfide precipitates found on PVC tube surfaces. These precipitates were determined to be mackinawite [$Fe_{1-x}S$] and covellite. Poorly crystalline iron sulfide phases such as mackinawite have been shown to incorporate, either by adsorption or by coprecipitation, other heavy metals such as Ni into/onto their structure (Morse and Cornwell, 1987; Morse and Arakaki, 1993).

4. Zinc

Average dissolved Zn concentrations in groundwater entering the barrier ranged from 2,400 to 14,200 μg/liter, with the concentration varying spatially and temporally during the 21-month period. As with the other heavy metals, the most pronounced treatment occurred within the lower half of the barrier (4.6 to 6.0 mbgl) where concentrations were consistently below the method detection limit (<2.0 μg/liter). Data for the 21-month sampling period is provided in Fig. 8D. Removal efficiencies for dissolved Zn ranged from 96.6 to 99.9% during the 21-month period (Fig. 9).

Ludwig *et al.* (2002) suggested that dissolved Zn within the aquifer pore water is controlled by coprecipitation and adsorption reactions associated with secondary phases such as iron (oxy)hydroxides and jarosite that have been identified within the aquifer. Equilibrium calculations suggest that groundwater within the PRB is supersaturated with respect to the zinc sulfide sphalerite [ZnS] and that sphalerite or a less crystalline precursor would be expected to precipitate (Ludwig *et al.*, 2002). The presence of sphalerite or other zinc-bearing phases has not been confirmed by mineralogical analyses.

5. Sulfate Reduction and pH Buffering Reactions

In addition to removing heavy metals from solution, field determinations of groundwater pH and Eh indicate that the PRB is effective at maintaining Eh and pH conditions that support bacterially mediated sulfate reduction (Chapelle, 1993; Connell and Patrick, 1968). The average pH of groundwater entering the PRB (i.e., well ML2) ranged from 6.30 to 6.61, suggesting that the groundwater pH is buffered by carbonate phases such as dolomite and calcite (Blowes and Ptacek, 1994). The average pH of the groundwater within the PRB ranged between 6.45 and 6.70. The increase in pH is likely attributable to a combination of the effect of the limestone component in the barrier and the production of bicarbonate by sulfate-reducing bacteria. Alkalinity increases within the barrier, which is consistent with the production of bicarbonate during sulfate reduction and carbonate dissolution. However, the specific contribution of sulfate reduction to the increase in alkalinity relative to that of limestone dissolution is unknown. Alkalinity (as $CaCO_3$) within the groundwater upgradient of the PRB ranged from 80 to 150 mg/liter during the 21-month period. As groundwater flowed through the PRB, the alkalinity increased, ranging from 200 to 480 mg/liter. Ludwig *et al.* (2002) noted increases in Ca and Mg concentrations within the PRB and attributed the increases to the dissolution of the carbonate minerals calcite [$CaCO_3$], and dolomite [$CaMg(CO_3)_2$]. The dissolution of carbonate minerals such as calcite and dolomite within the PRB is predicted by geochemical modeling.

The average Eh of the groundwater flowing into the PRB ranged from between 370 and 500 mV during the 21 months of the study. The average Eh of groundwater within the PRB ranged from −28 to 217 mV. Field measurements of dissolved sulfide within the groundwater confirmed the presence of sulfide within and downgradient of the barrier. Dissolved sulfide concentrations, immediately upgradient of the PRB at well ML-2, were below the method detection level ($<1.0\ \mu g$/liter) during all sampling periods. Sulfide concentrations within the PRB were as high as 5800 μg/liter, indicating sulfate-reducing activity. Dissolved sulfide concentrations decreased to near or below the method detection limit ($<1\ \mu g$/liter) 5 m downgradient of the PRB.

VI. LESSONS LEARNED

A. Effects of Fluctuating Water Table

The Vancouver PRB is installed in an unconfined aquifer affected by tidal fluctuations of the adjacent marine inlet. McGregor *et al.* (1999) reported that the water table near the PRB has a maximum tidal fluctuation of 1.45 m between high and low tides. The fluctuating water table appears to have an impact on the treatment efficiency for heavy metals within the shallowest portions of the barrier. For example, the average dissolved concentration for Cu within the shallow portion of the barrier was 17 μg/liter, whereas in the deeper portion of the PRB, the average Cu concentration was 1.5 μg/liter, even though the influent concentration of Cu was significantly higher in the deeper portion of the barrier (8,570 μg/liter) than in the shallow portion of the barrier (720 μg/liter). The chemistry of groundwater entering (well ML2) and exiting (well ML10) the shallow and deeper portions of the PRB 21 months after installation is summarized in Table I. Shallow data represent the average of the uppermost three monitoring points, which are located within the upper 1.5 m of the barrier. Deep data represent the average of the two bottommost monitoring points within the PRB (4.25 and 6.0 mbgl).

The difference between shallow and deep removal efficiencies are attributed to the flushing of oxygen-rich pore gas through the shallow barrier related to the tidal-induced rising and lowering of the water table. Measurements of Eh and dissolved sulfide during the 21-month sampling period show a significant difference between lower and shallow portions of the barrier. The average Eh of the shallow groundwater is 194 mV within the barrier, whereas groundwater within the lower portions of the barrier has an average Eh of 61 mV. The Eh of groundwater entering the barrier is relatively constant throughout the depth profile; shallow groundwater has an Eh of 422 mV compared to 437 mV for deep groundwater. The average dissolved sulfide concentration of shallow groundwater within the barrier is 90 μg/liter compared with an average sulfide concentration of 1,400 μg/liter in the lower portions of the barrier. The sulfide concentration within groundwater entering the barrier is below the method detection limit (<1 μg/liter).

To overcome the lower level of treatment in the shallow portion of the aquifer, a full-scale design of the barrier should incorporate a thicker barrier for the shallow portion of the aquifer. Consideration should also be given to capping the barrier, which will limit the diffusion of oxygen into the PRB and the infiltration of surface runoff into the PRB.

B. Reductive Dissolution of Iron Oxides Downgradient of PRB

Dissolved Fe within groundwater entering the Vancouver PRB averaged 80 μg/liter during the six sampling periods. Dissolved Fe concentrations exiting the PRB averaged 31,100 μg/liter 3 months after installation (Fig. 10). The dissolved Fe concentration exiting the PRB 7 months after installation averaged 3200 μg/liter. This decreasing trend with time in the dissolved Fe concentration exiting the barrier continued for the remainder of the 21-month period, with a minimum average concentration of 960 μg/liter being measured 21 months after installation (Fig. 10). Speciation of the dissolved Fe indicated that Fe(II) is the predominant species within the groundwater.

Geochemical modeling of the PRB system suggests that reductive dissolution reactions are occurring within and downgradient of the barrier. Chemical and mineralogical analysis of aquifer solids confirmed the presence of ferric iron (oxy)hydroxides such as goethite within the shallow aquifer. The goethite and other ferric iron (oxy)hydroxides were present as coatings on silicates and carbonate minerals within the aquifer. Modeling results using Eh and pH values measured within the PRB indicate that ferric iron (oxy)hydroxides goethite and ferrihydrite are unstable under the geochemical conditions that exist within groundwater exiting the barrier and reductive dissolution will release ferrous iron to the pore water. The organic carbon within the PRB may act as an electron donor. However, mass balance calculations using the total amount of Fe available as determined by the chemical extraction methods of Tessier *et al.* (1979) and McGregor *et al.* (1998) and the amount of dissolved iron within groundwater downgradient of

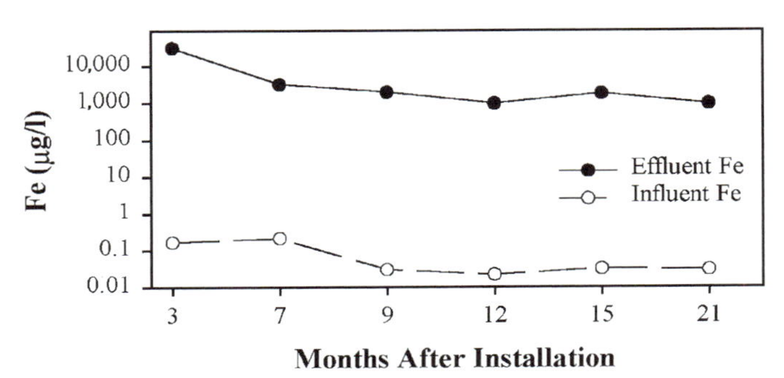

Figure 10 Iron concentrations immediately downgradient of the Vancouver permeable reactive barrier versus time.

the PRB indicate that less than 0.1% of the available reducible Fe on the aquifer solids was released over the 21-month period. The relatively small mass of Fe released to groundwater relative to the total mass of Fe available for release on the aquifer solids is likely a function of the crystal structure of the Fe-bearing minerals and kinetics related to Fe reduction. Extraction tests, which are designed to provide a "worst-case" scenario, use extractants that are highly acidic and strongly reducing. Therefore, the extractants dissolve crystalline and poorly crystalline Fe (oxy)hydroxides in a relatively short period of time (i.e., hours). Groundwater flowing out of the PRB has a neutral pH and is moderately reducing and therefore will not reduce as much Fe (oxy)hydroxides as the chemical extractants used in laboratory studies.

The reduction of Fe(III) (oxy)hydroxides releases dissolved Fe(II) to the groundwater. Dissolved Fe(II) represents potential acidity due to the potential for Fe(II) to oxidize and hydrolyze when exposed to oxygen. The oxidation and hydrolysis of every mole of Fe(II) will release up to 2 mol of acid [Eq. (2)]. However, the potential acidity of the Fe(II) derived from reduction dissolution reactions will be offset by the consumption of acidity and the production of alkalinity during the reduction of the Fe (oxy)hydroxides as illustrated by

$$4\,Fe(OH)_3 + CH_2O + 7\,H^+ \Rightarrow 4\,Fe^{2+} + HCO_3^- + 10\,H_2O. \qquad (6)$$

Therefore, reductive dissolution reactions within and downgradient of the PRB are a net neutral reaction. These reactions, combined with alkalinity produced within the PRB due to sulfate reduction reactions, result in effluent groundwater with a greater neutralizing potential than groundwater entering the barrier.

Associated with the reductive dissolution of Fe (oxy)hydroxides is the potential for release of trace metals that have been adsorbed onto or coprecipitated with the Fe (oxy)hydroxides. Based on data collected during the study, it was not possible to determine if trace metals within the groundwater exiting the PRB were due to incomplete treatment within the PRB or due to reductive dissolution reactions.

C. Flow Heterogeneities

A high degree of spatial heterogeneity was shown in the degree of treatment across the Nickel Rim PRB (Fig. 6). These heterogeneities are attributed to variations in hydraulic conductivity. The impact of these variations on treatment effectiveness has not been quantified. However, any variation in hydraulic conductivity will result in preferential flow, producing zones of higher flow where greater consumption of treatment capacity will occur.

Zones of slower flow will be correspondingly underutilized for treatment. Therefore, any preferential flow will compromise barrier performance. Flow modeling indicates that this problem will be much greater in relative thick barriers like the two described here (Benner *et al.*, 2001). To optimize the performance of larger barriers, every effort, such as strict quality assurance/quality control mixing and construction programs, should be made to achieve a homogeneous reactive mixture.

D. Declining Rates with Time

1. Nickel Rim Site

The efficiency of sulfate reduction and metal sulfide precipitation has declined over time in the Nickel Rim PRB with the average efficiency of sulfate removal declined by 30% and the efficiency of Fe removal declined by 50% over a 3-year period (Fig. 5). It is anticipated that this removal efficiency will continue to decline as the barrier ages. These declines are attributed to consumption of the more reactive fraction organic carbon material within the barrier. Despite this decline, the barrier removed an average of >200 mg/liter Fe from the water in the third year. In systems such as Nickel Rim, where the metal flux is very high, declining rates must be anticipated and the barrier designed accordingly.

2. Vancouver Site

Metal removal rates, as calculated by Ludwig *et al.* (2002), are shown in Fig. 11 for the 21-month evaluation period. Metal removal rates decreased during the 21-month period, leveling off after 1 year of installation. This trend is attributed primarily to decreases in metal concentrations entering the barrier. Sulfide ore-containing soils immediately upgradient of the barrier location were excavated prior to installation of the barrier, therefore the heavy metal concentrations within the shallow groundwater have decreased with time. Based on dissolved metal concentration differences during the 21-month period between well ML2, immediately upgradient of the barrier, and well ML10, within the barrier, sulfate reduction rates within the barrier were calculated at greater than 17 $mmol\,l^{-1}year^{-1}$ (Ludwig *et al.*, 2002). This rate is lower than rates measured by others (Eger and Wagner, 1995; Benner *et al.*, 2002) for similar systems. However, calculated sulfate reduction may be greater if excess sulfide is being transported downgradient of the PRB or hydrogen sulfide is volatilizing to the vadose zone and precipitating as elemental sulfur.

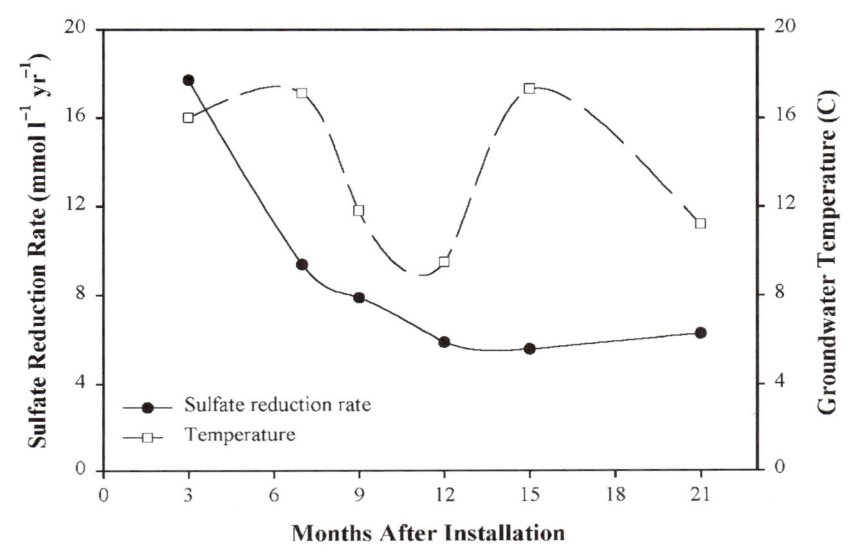

Figure 11 Inferred minimal sulfate reduction rates and groundwater temperature versus time based on metal concentration differences between ML2 and ML10 at the Vancouver PRB.

E. Installation Costs

1. Nickel Rim Site

Installation costs for the Nickel Rim PRB were approximately \$35,000 (U.S.) with a unit cost of approximately \$200/m^3. The cost breakdown for this installation was approximately half for materials and half for the actual installation, which included labor and equipment. Approximately half of the material costs were associated with shipping material to the site.

2. Vancouver Site

Design and installation costs for the Vancouver PRB were approximately \$31,000 (U.S.), which converts to a unit cost of approximately \$190/m^3. These costs include design, materials, and equipment costs but do not include site characterization or monitoring costs. Approximately two-thirds of the costs were associated with the installation of the barrier system and the remainder was associated with design and materials. The higher installation costs relative to the Nickel Rim site were associated with the guar gum, which required specialized equipment and staff to be on-site to prepare the gum.

VII. SUMMARY

Two compost-based PRBs were installed in aquifers impacted by acid mine drainage and heavy metals at sites near Sudbury, Ontario (Nickel Rim), and Vancouver, British Columbia. Monitoring of the PRBs indicates that heavy metals, including Cd, Cu, Ni, and Zn, are being removed effectively from solution by sulfide precipitation and coprecipitation reactions. Associated with sulfate reduction is a net decrease in the acidity of the groundwater. Rates of sulfate reduction decreased with time within both barrier systems; however, the barriers were still able to remove heavy metals to limits acceptable to regulatory agencies.

In order to design and install an effective PRB, the geology, hydrogeology, and geochemistry of the aquifer system must be well characterized and understood. Design specifications should account for heterogeneities within the aquifer and barrier as well as the effects of fluctuating water tables and the possibility of reductive dissolution of ferric iron (oxy)hydroxide phases within the aquifer downgradient of the reactive barrier. Permeable reactive barriers provide a cost-effective strategy for the long-term treatment of acid mine drainage and heavy metals in both urban and remote settings.

ACKNOWLEDGMENTS

The authors thank the numerous undergraduate and graduate students, as well as M. Choi, E. Pringle, and R. Booth, who aided in the collection and interpretation of field data. Funding for these studies was provided by Environment Canada, Falconbridge Ltd., and NSERC. Finally, the authors thank the reviews of J. West, D. Naftz, B. Bruce, and C. Allen for their comments, which enhanced this chapter greatly.

REFERENCES

Allison, J. D., Brown, D. S., and Novo-Gradac, K. J. (1990). MINTEQA2/PRODEFA2-A geochemical assessment model for environmental systems: Version 3.0 users manual. U.S. Environmental Protection Agency, Athens, GA.

Alpers, C. N., Nordstrom, D. K., and Ball, J. W. (1989). Solubility of jarosite solid solutions precipitated from acid mine waters, Iron Mountain, California, U.S.A. *Sci. Geol. Bul.* **42**(4), 281–298.

Bain, J. G., Blowes, D. W., Robertson, W. D., and Frind, E. O. (2000). Modelling of sulfide oxidation with reactive transport at a mine drainage site. *J. Contamin. Hydro.* **41**(1–2), 23–47.

Ball, J. W., and Nordstrom, D. K. (1991). User's manual for WATEQ4F with revised thermodynamic database. U.S. Geological Survey Open-File Report 91–183.

Benner, S. G., Blowes, D. W., Gould, W. D., Herbert, R. B, Jr., and Ptacek, C. J. (1999). Geochemistry of a permeable reactive barrier for metals and acid mine drainage. *Environ. Sci. Tech.* **33**, 2793–2799.

Benner, S. G., Blowes, D. W., and Molson, J. W. H. (2001). Modeling preferential flow in permeable reactive barriers: Implications for performance and design. *Ground Water* **39**(3), 371–379.

Benner, S. G., Blowes, D. W., and Ptacek, C. J. (1997). A full-scale porous reactive wall for prevention of acid mine drainage. *Ground Water Monitor. Remediat.* **17**, 99–107.

Benner, S. G., Blowes, D. W., Ptacek, C. J., and Mayer, K. U. (2002). Rates of sulfate reduction and metal sulfide precipitation in a permeable reactive barrier. *Appl. Geochem.* **17**(3), 301–320.

Benner, S. G., Gould, D., and Blowes, D. W. (2000). Microbial populations associated with the generation and treatment of acid mine drainage. *Chem. Geol.* **169**, 435–448.

Blowes, D. W., and Ptacek, C. J. (1992). Geochemical remediation of groundwater by permeable reactive walls: Removal of chromate by reduction with iron bearing solids. *In* "Proceedings of the Subsurface Restoration Conference, Third International Conference on Groundwater Quality Research," pp. 214–216, Dallas, TX.

Blowes, D. W., and Ptacek, C. J. (1994). Acid neutralization mechanisms in inactive tailings impoundments. *In* "Short Course Handbook on Environmental Geochemistry of Sulfide Mine- Wastes" (J. L. Jambor and D. W. Blowes, eds.), pp. 271–292. Waterloo, Ontario.

Chapelle, F. H. (1993). "Groundwater Microbiology and Geochemistry." Wiley, New York.

Christensen, B., Laake, M., and Lien, T. (1996). Treatment of acid mine water by sulfate-reducing bacteria; results from a bench-scale experiment. *Wat. Res.* **30**(7), 1617–1624.

Connell, W. E., and Patrick, W. H. (1968). Sulfate reduction in soil: Effects of redox potential and pH *Science* **159**, 86–87.

Dubrovsky, N. M., Morin, K. A., Cherry, J. A., and Smyth, D. J. A. (1984). Uranium tailings acidification and subsurface contaminant migration in a sand aquifer. *Water Poll. Res. J. Canada* **19**(2), 55–89.

Dvorak, D. H., Hedin, R. S., Edenborn, H. M., and McIntire, P. E. (1992). Treatment of metal-contaminated water using bacterial sulphate reduction: Results from pilot-scale reactors. *Biotech. Bioeng.* **40**(2), 609–616.

Eger, P., and Wagner, J. (1995). Sulphate reduction for the treatment of acid mine drainage: Long term solution or short term fix? *In* "Proceedings of Sudbury '95—Mining and the Environment" (T. P. Hynes and M. C. Blanchette, eds.), Vol. **2**, pp. 515–525, Sudbury, Ontario, Canada.

Hedin, R. S. (1989). Treatment of coal mine drainage with constructed wetlands. *In* "Wetlands Ecology and Conservation: Emphasis in Pennsylvania" (S. K. Majumdar, R. P. Brooks, F. J. Brenner, and R. W. Tiner, eds.), pp. 349–362. The Pennsylvania Academy of Science.

Herbert, R. B., Jr., Benner, S. G., and Blowes, D. W. (2000). Solid phase iron-sulfur geochemistry of a reactive barrier for treatment of mine drainage. *Appl. Geochem.* **15**, 1331–1343.

Herbert, R. B., Jr., Benner, S. G., Pratt, A. R., and Blowes, D. W. (1998). Surface chemistry and morphology of poorly crystalline iron sulfides precipitated in media containing sulfate-reducing bacteria. *Chem. Geol.* **144**, 87–97.

Hvorslev, M. J. (1951). "Time Lag and Soil Permeability in Groundwater Observations." U.S. Army Corps Engineers, Waterways Experimental Station Bull. 36, Vicksburg, MS.

Jambor, J. L., and Owens, D. R. (1993). "Mineralogy of the Tailings Impoundment at the Former Cu-Ni Deposit of Nickel Rim Mines Ltd., Eastern Edge of Sudbury Structure, Ontario." Division Report MSL 93–4 (CF), CANMET, EMR, Canada.

Johnson, R. H., Blowes, D. W., Robertson, W. D., and Jambor, J. L. (2000). The hydrogeochemistry of the Nickel Rim mine tailings impoundment, Sudbury, Ontario. *J. Contamin. Hydro.* **41**(1–2), 49–80.

Ludwig, R., McGregor, R. G., Blowes, D. W., Benner, S. G., and Mountjoy, K. (2002). A permeable reactive barrier for treatment of heavy metals, *Ground Water*. **40**(1), 59–66.

Machemer, S. D., and Wildemann, T. R. (1992). Adsorption compared with sulphide-precipitation as metal removal processes from acid mine drainage in a constructed wetland. *J. Contamin. Hydro*. **9**, 115–131.

McGregor, R. G., Blowes, D. W., Jambor, J. L., and Roberston, W. D. (1998). Mobilization and attenuation of heavy metals within a nickel mine tailings impoundment near Sudbury, Ontario, Canada. *Environ. Geol*. **36**(3–4), 305–319.

McGregor, R. G., Blowes, D. W., Ludwig, R., Pringle, E., and Pomeroy, M. (1999). Remediation of a heavy metal plume using a reactive wall. *In* "Proceedings from the Fifth International In Situ and On-Site Bioremediation Symposium" (A. Leeson and B. C. Alleman, eds.) Vol. **5**(4), pp. 19–24. San Diego, CA.

McMurty, D., and Elton, R. O. (1985). New approach to in-situ treatment of contaminated groundwaters. *Environ. Prog*. **4**, 168–170.

Morrison, S. J., and Spangler, R. R. (1993). Chemical barriers for controlling groundwater contamination. *Environ. Prog*. **12**, 175–181.

Morse, J. W., and Arakaki, T. (1993). Adsorption and coprecipitation of divalent metals with mackinawite (FeS). *Geochim. Cosmochim. Acta* **57**, 3635–3640.

Morse, J. W., and Cornwell, F. J. (1987). Analysis and distribution of iron sulfide minerals in recent anoxic marine sediments. *Mar. Chem*. **22**, 55–69.

Serfes, M. E. (1987). Interpretation of tidally affected ground- water flow systems in pollution studies. *In* "Proceedings NWWA/API Conference on Petroleum Hydrocarbons and Organic Chemicals in Ground Water—Prevention, Detection and Restoration," pp. 55–73. Houston, NWWA.

Serfes, M. E. (1991). Determining the mean hydraulic gradient of ground water affected by tidal fluctuations. *Ground Water* **29**(4), 549–555.

Starr, R. C., and Ingleton, R. A. (1992). A new method for collecting core samples without a drill rig. *Ground Water Monitor. Rev*. **41**, 91–95.

Tessier, A., Campbell, P. G. C., and Bisson, M. (1979). Sequential extraction procedure for the speciation of particulate trace metals. *Anal. Chem*. **51**(7), 844–851.

Wakao, N., Takahashi, T., Sakurai, Y., and Shiota, H. (1979). The treatment of acid mine water using sulphate-reducing bacteria. *J. Ferment. Tech*. **57**, 445–452.

Waybrant, K. R., Blowes, D. W., and Ptacek, C. J. (1998). Selection of reactive mixtures for use in permeable reactive walls for treatment of mine drainage. *Environ. Sci. Tech*. **32**, 1972–1979.

Subject Index

D

R

S